Electronic Properties of Organic Conductors

Takehiko Mori

# Electronic Properties of Organic Conductors

Springer

Takehiko Mori
Tokyo Institute of Technology
Tokyo
Japan

ISBN 978-4-431-55263-5 ISBN 978-4-431-55264-2 (eBook)
DOI 10.1007/978-4-431-55264-2

Library of Congress Control Number: 2016940345



Printed on acid-free paper

This Springer imprint is published by Springer Nature
The registered company is Springer Japan KK

# Preface

Since the discovery of electrical conduction in organic materials, more than half a century has passed. During this period, highly conducting stable materials have been realized, and even organic superconductors have been discovered. Organic semiconductor devices such as organic light-emitting diodes, organic transistors, and organic solar cells have been developed. Some of them are sold in the market. Chemical constituents of organic materials are complicated in comparison with inorganic materials, but the electronic structure is simple. Organic conductors afford an excellent playground of various intriguing physics. This book provides an overview of electrical conduction and the related phenomena in organic materials and elucidates the logic behind them.

This book is based on a Japanese book originally published in 2013. Organic conductors have been studied somewhat independently of organic semiconductors, but this book deals with both fields. The focus is placed on the physics and chemistry of organic materials that conduct electricity. This is a rapidly growing field, but the fundamental aspects are emphasized. This book is intended to be more than a textbook but less than a review.

Basic knowledge of solid-state physics is necessary to understand the properties of these materials. In a physics department, solid-state physics is treated in the latter part of the curriculum, because this field requires knowledge of many other fields such as quantum mechanics, electromagnetism, and statistical mechanics. In this book, however, knowledge of these fields is not assumed. Although there are many excellent textbooks for solid-state physics, the earlier part of this book is, in a sense, an introductory textbook in solid-state physics for chemists with no background in physics.

Chapter 1 is a brief description of quantum chemistry. The systematic description is not intended, but emphases are placed on the aspects related to materials design. The Hückel approximation is extensively described as a fundamental to understand $\pi$-conjugated molecules. In Chap. 2, the band theory is introduced as a natural extension of the Hückel approximation. In Chap. 3, the energy band theory is used to understand the conducting phenomena and the Fermi surface of organic materials. Chapter 4 outlines the theory of magnetism as a starting point of the next chapter. Chapter 5 describes various phenomena beyond the band theory and provides a plain explanation of electron correlation. Starting from the molecular orbital theory considering the electron–electron Coulomb interaction, electron correlation is examined both in molecules and solid states. Then magnetic interaction is derived from the Hubbard model. Finally, competition and coexistence of conduction electrons and localized electrons are investigated. Chapter 6 is an introductory treatment of superconductivity. Chapter 7 describes various aspects of organic charge-transfer salts. Chapter 8 deals with organic semiconductors and the devices involved. Accordingly, this book deals with several independent items. The fundamental aspects of energy bands are described in Chaps. 1–3. A concise description of strong correlation is afforded in Chaps. 4–6. Various aspects of organic conductors and semiconductors are discussed in Chaps. 7 and 8. Examples provide derivations of most equations and unit conversions.

The author is grateful to Profs. Kenji Yonemitsu (Chuo Universiy), Toshikazu Nakamura (Institute for Molecular Science), John Schlueter (Argonne National Laboratory and National Science Foundation), and Hatsumi Mori (University of Tokyo) for many variable suggestions. The author is, however, responsible for the entire contents. The author is grateful to Masaki Kamei, who is the editor of the Japanese version, particularly for preparing beautiful drawings.

Spring 2016 Takehiko Mori

# Contents

# Chapter 1
# Quantum Chemistry of Molecules

**Abstract** Quantum chemistry is a starting point of molecular design. This chapter outlines the fundamentals of the molecular orbital theory, with particular emphasis on the Hückel approximation of π-conjugated molecules, which will be mentioned in the later chapters.

**Keywords** Quantum chemistry · Schrödinger equation · Hamiltonian · Atomic orbital · Secular equation · Hückel approximation · Molecular orbital · Highest occupied molecular orbital (HOMO) · Lowest unoccupied molecular orbital (LUMO) · Hückel rule · Aromatic · Irreducible representation · Uncertainty principle

## 1.1 Principles of Quantum Mechanics

Quantum chemistry provides an important background to design functional materials showing intriguing electronic, magnetic, and optical properties. This chapter provides a brief survey of quantum chemistry from the viewpoint of materials design. Basic concepts of quantum mechanics are outlined at the end of this chapter. Expert readers are advised to skip these parts.

Quantum mechanics is derived from the following two "theorems." Although called "theorems," these relations are extracted from experiments.

**Waves with frequency ν are regarded as a particle with energy $E = h\nu = \hbar\omega$.**

Here, $h$ is **Planck's constant**, and $\hbar = h/2\pi$. This relation was proposed by Planck (1900) because electromagnetic waves with frequency $\nu$ have to be considered as particles (photons) with energy $E = h\nu$. For example, suppose one measures the temperature of a hot body like a blast furnace. Usual thermometers melt at such a high temperature, but the color of the radiation gradually changes from red to white as the temperature rises. Therefore, the higher the temperature of the furnace, the shorter is the wavelength of the radiation. Similarly, even a



T. Mori, *Electronic Properties of Organic Conductors*,
DOI 10.1007/978-4-431-55264-2_1

room-temperature material emits long wavelength far infrared light, namely thermal radiation in the order of several 100 $cm^{-1}$. When we suppose a "black body" which absorbs all incident light and thermal equilibrium is attained with the radiated light, we can thermodynamically obtain the wavelength distribution. However, classical thermodynamics results in short-wavelength divergence; the light intensity becomes infinity in the short-wavelength limit. When we assume the light energy is quantized to $E = h\nu$, we can correctly account for the experimentally obtained wavelength distribution. The light intensity follows the Planck distribution (Eq. 2.58), which goes to zero at short wavelengths.

Additional experimental evidence of the above relation comes from Einstein's model of the **photoelectric effect** (1905). When ultraviolet light with a certain wavelength is shone on a material, an electron incorporated in the material with the binding energy $\Phi$ is ejected, where the kinetic energy of the electron $E_k$ satisfies the relation, $h\nu = E_k + \Phi$, according to the energy conservation. This phenomenon, called photoelectron spectroscopy, is used to investigate electronic states of materials (Sect. 8.7).

The second theorem of quantum mechanics is

**A particle with momentum $p$ exhibits a wave-like property with a wavelength $\lambda = h/p$ (de Broglie wave).**

In general, a relation (wavelength) × (frequency) = (velocity), $\lambda\nu = c$, holds for light. If Planck's relation $E = h\nu$ is applied to a particle with mass $m$, the energy $E = mc^2$ of the special relativity leads to $mc^2 = h\nu = hc/\lambda$, which is reduced to $mc = h/\lambda$. In analogy with the definition of the momentum $p = mv$, the left-hand side $mc$ is substituted by $p$ to afford the de Broglie relation $\lambda = h/p$. Therefore, a particle like an electron has a wavelength. Evidence of this relation is provided by an electron microscope, where an electron beam behaves like a wave and is diffracted by materials.

*Example 1.1* Calculate wavelength of an electron accelerated by a voltage of 144 V.

Electron energy

$$eV = \frac{p^2}{2m} = \frac{1}{2m}\left(\frac{h}{\lambda}\right)^2$$

leads to

$$\lambda = \frac{h}{\sqrt{2\,\mathrm{meV}}} = \frac{1.2264\,\mathrm{nm}}{\sqrt{V}}.$$

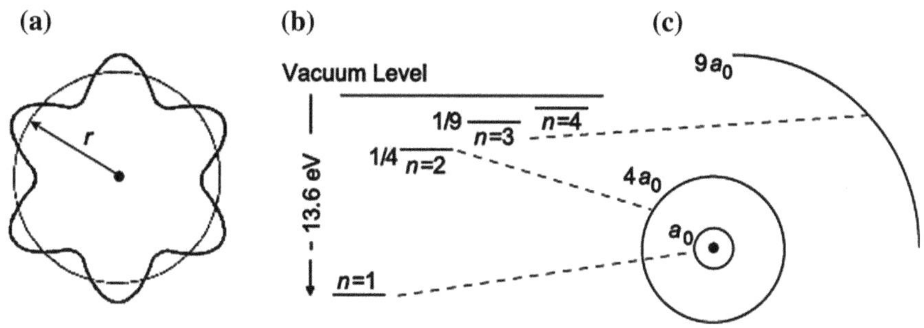


**Fig. 1.1** Atomic orbitals and energies of a hydrogen atom

Substitution of $V = 144$ V gives the wavelength $\lambda = 0.1022$ nm, which is comparable to the typical atomic spacing in a crystal.

Atomic spectrum of hydrogen consists of discrete lines, because the atomic orbitals are standing waves with integer times of the electron wavelength $2\pi r = n\lambda$ (Fig. 1.1a), where $r$ is the radius of the atomic orbital. The de Broglie wave was first introduced in order to account for this phenomenon.

*Example 1.2* Consider a hydrogen-like atom with the atomic number $Z$, in which an electron rotates around a nucleus with $+Z$ charge. Assume a circular orbit and the balance between the Coulomb attraction from the nucleus and the centrifugal force.

$$\frac{Ze^2}{4\pi\varepsilon_0 r^2} = \frac{mv^2}{r} = \frac{p^2}{mr} \tag{1.1}$$

Estimate the atomic radius $r$ and the energy $E$.

Inserting $\lambda = h/p$ in $2\pi r = n\lambda$, we obtain $2\pi r = nh/p$. From Eq. 1.1, $p$ is deleted using this relation.

$$\frac{Ze^2}{4\pi\varepsilon_0 r^2} = \frac{1}{mr}\left(\frac{nh}{2\pi r}\right)^2 = \frac{n^2h^2}{4\pi^2 mr^3}$$

This relation is reduced to

$$r = \frac{\varepsilon_0 h^2}{\pi m Z e^2} n^2 = \frac{a_0}{Z} n^2. \tag{1.2}$$

$a_0 = 0.529$ Å is called **Bohr radius**, which is the average radius of the hydrogen 1s orbital. The energy is obtained as a sum of the kinetic energy and the potential energy.

$$E = \frac{p^2}{2m} - \frac{Ze^2}{4\pi\varepsilon_0 r} = -\frac{Ze^2}{8\pi\varepsilon_0 r} = -\frac{mZ^2e^4}{8\varepsilon_0^2 h^2}\frac{1}{n^2} = -\frac{Z^2 R_0}{n^2}. \tag{1.3}$$

$R_0$ is the **Rydberg constant** (13.6 eV) corresponding to the energy of the hydrogen 1s orbital for $Z = 1$. The successive energies are 1, 1/4, 1/9, … of $R_0$ for $n = 1, 2, 3, \ldots$, respectively, (Fig. 1.1b), and the radius $r$ is 1, 4, 9, … times of $a_0$ (Fig. 1.1c). Since the energy is negative even for infinite $n$, there are an infinite number of bound states. Here, the radius increases in proportion to $n^2$. For $Z \neq 1$, the atomic orbital shrinks according to $r \propto 1/Z$, and the energy deepens following $E \propto -Z^2$.

These two "theorems" represent the following particle-wave duality.

| | | |
|---|---|---|
| Particle | (Energy, | Momentum) |
| | ↕ | ↕ |
| Wave | (Frequency, | Wavelength) |

From this, the wave-like properties are converted to the particle-like properties, and vice versa. Equations of quantum mechanics are ordinary equations of waves, but when transformed to the equations related to the energy and momentum using the above relations, we obtain unconventional relations. However, it is noteworthy that everything is derived from the above two conversion equations. The meaning will be investigated at the end of this chapter.

## 1.2 Conversion of Energy Units

Conversion of energy units is summarized in Fig. 1.2. It is convenient to represent electron energy in eV (electron volt). This is the energy when an electron is accelerated by 1 V voltage. From $\lambda\nu = c$ and $eV = h\nu$, we obtain $\lambda = hc/eV$, and

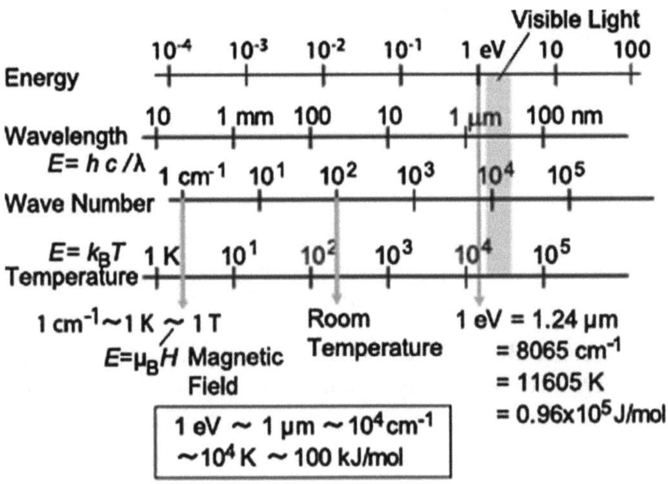


**Fig. 1.2** Conversion of energy units

1 eV corresponds to the energy of light with 1.24 μm wavelength. Alternatively, the energy of light with $\lambda$ nm wavelength is represented by

$$E = \frac{1240\,\text{nm}}{\lambda}\text{eV}.$$

For example, 400 nm light has the energy of 1240 nm/400 nm = 3.1 eV. Visible light has 400–800 nm wavelength, which approximately corresponds to 1.5–3.0 eV. Inverse of 1.24 μm affords 1 eV = 8065 $\text{cm}^{-1}$. The unit of thermal energy is given by 1 eV = 11,605 K using $eV = k_BT$. Then, the thermal energy at room temperature is approximately 0.03 eV. These relations are roughly approximated to

$$1\,\text{eV} \sim 10^4\,\text{cm}^{-1} \sim 10^4\,\text{K}.$$

This also corresponds to the light energy with about 1 μm wavelength, which is a little smaller than the energy of visible light (Fig. 1.2). Electron Zeeman energy $E = \mu_BH$ gives magnetic field 1 T = 0.7 K, so 1 T and 1 K approximately correspond to the same energy. When the Avogadro number is multiplied by $e$, we obtain 1 eV = $9.6 \times 10^4$ J/mol = 96 kJ/mol, so 1 eV is approximately equal to 100 kJ/mol. This is the typical energy of chemical reactions. This number is equal to 1 mol of electronic charge, which is called **Faraday's constant**, $9.6 \times 10^4$ C/mol.

## 1.3 Wave Equations

A wave is generally represented as

$$\psi(x,t) = \psi_0 \exp\left[2\pi i\left(\frac{x}{\lambda} - \nu t\right)\right] \tag{1.4}$$

by using Euler's relation $e^{i\theta} = \cos\theta + i\sin\theta$. $\psi(x, t)$ (psi) is a complex number, but in the circuit theory, the real part is recognized as the actual current or voltage. Then, the real part of this equation at $t = 0$ is

$$\text{Re}(\psi) = \psi_0 \cos 2\pi\frac{x}{\lambda}$$

which is depicted in Fig. 1.3a. Similarly, the real part at $x = 0$ is

$$\text{Re}(\psi) = \psi_0 \cos 2\pi\nu t$$

so at a certain point the wave oscillates as shown in Fig. 1.3b. At the next moment, the wave moves as represented by the dashed curve in Fig. 1.3a, (where the wave moves to the right). Equation 1.4 represents a propagating wave that moves to the right. Velocity of this propagating wave is obtained from the condition of the constant phase, for instance zero.

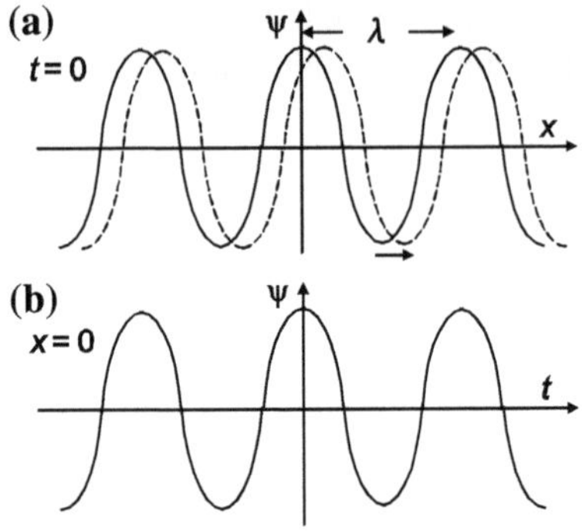


**Fig. 1.3** Real part of $\psi(x, t)$ **a** at $t = 0$, and **b** at $x = 0$

$$\frac{\Delta x}{\Delta t} = \lambda \nu \tag{1.5}$$

This is a generalization of the relation for light, $c = \lambda\nu$. When we use **wave number**, $k = 2\pi/\lambda$, and **angular frequency**, $\omega = 2\pi\nu$, Eq. 1.4 is converted to

$$\psi(x,t) = \psi_0 \exp[i(kx - \omega t)]. \tag{1.6}$$

Wave number $k$ represents the number of waves in the $2\pi$ unit, namely 6.28 m, where the unit of $k$ is $m^{-1}$. $\omega$ designates the number of waves oscillating in 6.28 s, where the unit is $s^{-1}$.

When Eq. 1.4 is differentiated by $x$

$$\frac{\partial \psi}{\partial x} = \frac{2\pi i}{\lambda}\psi = \frac{i}{\hbar}p\psi \text{ leads to } p\psi = \frac{\hbar}{i}\frac{\partial \psi}{\partial x} \tag{1.7}$$

where $\lambda = h/p$ is used. Similarly, differentiation of Eq. 1.4 by $t$ affords

$$\frac{\partial \psi}{\partial t} = -2\pi i \nu \psi = -\frac{i}{\hbar}E\psi \text{ leads to } E\psi = i\hbar\frac{\partial \psi}{\partial t} \tag{1.8}$$

where $E = h\nu$ is used. If we know the wave form $\psi(x, t)$, and the wave form does not change when $\psi(x, t)$ is differentiated with respect to $x$ or $t$, we can obtain particle's $p$ and $E$. In other words, observables such as $p$ and $E$ are represented by the operators

$$p \to \frac{\hbar}{i}\frac{\partial}{\partial x} \qquad E \to i\hbar\frac{\partial}{\partial t}.$$

Similarly, when we differentiate $\psi$ by $x$ twice, we obtain

$$\frac{\partial^2 \psi}{\partial x^2} = -\frac{4\pi^2}{\lambda^2}\psi = -\frac{p^2}{\hbar^2}\psi \text{ leads to } p^2\psi = -\hbar^2\frac{\partial^2 \psi}{\partial x^2} \tag{1.9}$$

In classical mechanics, the total energy of a particle is represented by a sum of the kinetic energy and the potential energy $V(x)$.

$$E = \frac{p^2}{2m} + V(x)$$

When $\psi$ is multiplied from the right, and using Eq. 1.9, we obtain

$$\left(-\frac{\hbar^2}{2m}\frac{\partial^2}{\partial x^2} + V(x)\right)\psi = E\psi. \tag{1.10}$$

This is the **Schödinger equation**. The left-hand side of the Schödinger equation is an operator representing the total energy, which is called **Hamiltonian**.

Equations 1.7, 1.8, and 1.10 are represented by a general form, $\mathbf{F}\psi = f\psi$. When an operator **F** is applied to a wave function $\psi$, the wave shape does not change and becomes $f$ times of the original function. This is a differential equation for $\psi$, which has solutions for some definite $f$ values. In this case, $f$ is called the **eigenvalue**, and $\psi$ is called the corresponding **eigenfunction**.

For example, consider a free electron with $V(x) = 0$. The Schödinger equation

$$-\frac{\hbar^2}{2m}\frac{\partial^2}{\partial x^2}\psi = E\psi$$

has solutions (eigenfunctions) represented by $\psi(x) = e^{ikx}$. By putting this wave function in the above equation, we can easily see the corresponding energy (eigenvalue) is $E = \hbar^2k^2/2m$. These wave functions are solutions of Eq. 1.7 at the same time. By putting this wave function in Eq. 1.7, we see the corresponding momentum is $p = \hbar k$. The plane wave $\psi(x) = e^{ikx}$ represents a propagating wave with the momentum $p = \hbar k$ and energy $E = \hbar^2k^2/2m$. Another state $\psi(x) = e^{-ikx}$ has a momentum $p = -\hbar k$, and energy $E = \hbar^2k^2/2m$, so this is a propagating wave moving to the left (Fig. 1.4).

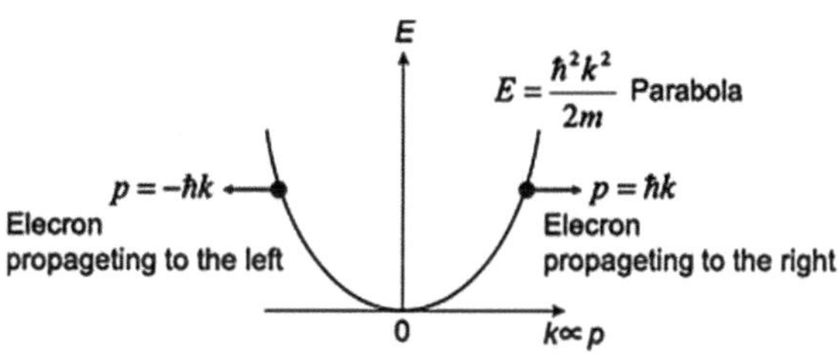


**Fig. 1.4** Energy and momentum of a free electron

## 1.4 Atomic Orbitals

We can obtain atomic orbitals by solving the Schödinger equation of atoms. Through use of the three-dimensional momentum, the kinetic energy is represented by

$$E = \frac{1}{2m}\left(p_x^2 + p_y^2 + p_z^2\right) \rightarrow -\frac{\hbar^2}{2m}\left(\frac{\partial^2}{\partial x^2} + \frac{\partial^2}{\partial y^2} + \frac{\partial^2}{\partial z^2}\right) = -\frac{\hbar^2}{2m}\nabla^2$$

so the three-dimensional Schödinger equation is

$$\left[-\frac{\hbar^2}{2m}\nabla^2 + V\right]\psi = E\psi. \tag{1.11}$$

Since a hydrogen-like atom with the atomic number $Z$ has the Coulomb attraction potential from the nucleus, $V = -\frac{Ze^2}{4\pi\varepsilon_0 r}$, the resulting Schödinger equation is

$$\left[-\frac{\hbar^2}{2m}\nabla^2 - \frac{Ze^2}{4\pi\varepsilon_0 r}\right]\psi = E\psi. \tag{1.12}$$

Transformation to the polar coordinates ($x = r \sin\theta \sin\varphi$, $y = r \sin\theta \cos\varphi$, and $z = r \cos\theta$) leads to the conversion of the derivatives

$$\nabla^2 \rightarrow \frac{\partial^2}{\partial r^2} + \frac{2}{r}\frac{\partial}{\partial r} + \frac{1}{r^2}\left[\frac{1}{\sin\theta}\frac{\partial}{\partial\theta}\left(\sin\theta\frac{\partial}{\partial\theta}\right) + \frac{1}{\sin^2\theta}\frac{\partial^2}{\partial\varphi^2}\right] \tag{1.13}$$

so the differential equation is separated with respect to the variables, $r$, $\theta$, and $\varphi$.

$$\psi = R(r)Y_{lm}(\theta,\varphi) = R(r)\Theta_{lm}(\theta)\Phi_m(\varphi) \tag{1.14}$$

This equation is put in Eqs. 1.12, and Eq. 1.13 is used. Then, divided by $R\Theta\Phi/r^2$ we obtain

$$r^2\left[\frac{1}{R}\left(\frac{\partial^2 R}{\partial r^2} + \frac{2}{r}\frac{\partial R}{\partial r}\right) + \frac{2m}{\hbar^2}\left(E + \frac{Ze^2}{4\pi\varepsilon_0 r}\right)\right] + \left[\frac{1}{\Theta\sin\theta}\frac{\partial}{\partial\theta}\left(\sin\theta\frac{\partial\Theta}{\partial\theta}\right) + \frac{1}{\Phi\sin^2\theta}\frac{\partial^2\Phi}{\partial\varphi^2}\right] = 0. \tag{1.15}$$

Since the first term contains only $r$, and the second term contains only $\theta$ and $\varphi$, these terms are constant independent of $r$ and $\theta$. Further calculation proves that these terms are represented by $l(l + 1)$ by using an integer $l$. Accordingly, the **radial part** $R(r)$ follows a differential equation.

$$\frac{\partial^2 R}{\partial r^2} + \frac{2}{r}\frac{\partial R}{\partial r} + \left[\frac{2m}{\hbar^2}\left(E + \frac{Ze^2}{4\pi\varepsilon_0 r}\right) - \frac{l(l+1)}{r^2}\right]R = 0 \tag{1.16}$$

*Example 1.3* Show that $R(r) = e^{-\zeta r}$ is a solution of Eq. 1.16 for $l = 0$. Obtain the corresponding $\zeta$ (zeta) and $E$.

For $l = 0$, $R(r) = e^{-\zeta r}$ is put in Eq. 1.16.

$$\zeta^2 - \frac{2\zeta}{r} + \left[\frac{2m}{\hbar^2}\left(E + \frac{Ze^2}{4\pi\varepsilon_0 r}\right)\right] = 0$$

Since this relation is valid for any $r$, the coefficient of the $1/r$ term is zero, leading to

$$\zeta = \frac{Zme^2}{4\pi\hbar^2\varepsilon_0} = \frac{Z}{a_0}. \tag{1.17}$$

Next, zero of the constant ($r$-independent) term affords, using the above $\zeta$

$$E = -\frac{\hbar^2}{2m}\zeta^2 = -\frac{Z^2me^4}{32\pi^2\varepsilon_0^2\hbar^2} = -Z^2R_0. \tag{1.18}$$

Remembering that $\hbar = h/2\pi$, Eqs. 1.17 and 1.18 are, respectively, equivalent to Eqs. 1.2 and 1.3. Equation 1.17 means inverse $\zeta$ is $a_0$, the Bohr radius. When we use $Z = 1$, the resulting wave function $R(r) = e^{-r/a_0}$ represents the 1s orbital of a hydrogen atom. Atomic orbitals have complicated normalization factors, but we do not discuss them because we do not use them in practical situations. Equation 1.18 is equal to Eq. 1.3 for $n = 1$, so the energy is −13.6 eV for $Z = 1$.

We can easily show that $R(r) = re^{-r/a_0}$ is another solution of Eq. 1.16 for $l = 1$. More generally, $R(r) = r^l e^{-\zeta r}$ is a solution of Eq. 1.16. Similarly to the above Example 1.3, we can show

$$\zeta = \frac{Z}{na_0} \quad \text{and } E = -\frac{\hbar^2}{2m}\zeta^2 = -\frac{Z^2R_0}{n^2}$$

in which $l + 1$ is replaced by $n$.

Since $re^{-\zeta r}$ is not orthogonal to $e^{-\zeta r}$, the former is orthogonalized to

$$R(r) \propto \left(2 - \frac{\rho}{2}\right)e^{-\rho/4}$$

by using $\rho = 2\zeta r = \frac{2Z}{a_0}r$. This is the usual 2s orbital. This equation has a spherical node at $\rho = 4$. The 3s orbital has a factor of $r$ square.

$$R(r) \propto \left(6 - 2\rho + \frac{\rho^2}{9}\right)e^{-\rho/6}$$

This wave function has two radial nodes, and is orthogonal to the 1s and 2s orbitals. Factors such as $6 - 2\rho + \rho^2/9$ are called **Laguerre's polynomials**, which appear in order to orthogonalize the orbitals.

In the actual molecular orbital calculations, we usually consider only the valence orbitals, and the orthogonality to the core orbitals is not necessary. In addition, the nodes appear much inside the usual chemical bonds. Accordingly, Laguerre's polynomial part is neglected, and the **Slater orbital** $R = e^{-\zeta r}$ is used. Here, $\zeta$ is called the **Slater exponent**, and generally represented by

$$\zeta = \frac{Z^*}{na_0}.$$

Note that $\zeta$ increases as $Z$ increases. In actual atoms, however, $\zeta$ suffers from screening of other electrons. If we suppose that the outermost electron is perfectly screened by the other inner electrons, the $+Z$ nucleus charge is screened by $Z - 1$ negative charges of the other electrons, so that the outermost electron always feels $Z = 1$ charge. Actually, the other electrons are not always located inside the particular electron, and the screening is imperfect, so we assume $Z^* = Z - S$. Slater assumed that for an orbital with the principal quantum number $n$, the screening constant $S$ is 0.35 for the other electrons with the same $n$, 0.85 for $n - 1$, and 1.0 for $n - 2$ and further inside electrons [1]. For example, 2s and 2p orbitals of carbon with $(1s)^2(2s)^2(2p)^2$ lead to $Z^* = Z - S = 6 - 3 \times 0.35 - 2 \times 0.85 = 3.25$, resulting in $\zeta = 3.25/2 = 1.625$.

The obtained $\zeta$ values are depicted in Fig. 1.5. Since $\zeta$ is inversely proportional to the size of the atomic orbital, $1/\zeta$ shown in Fig. 1.5b represents the size of the atomic orbital. As Eqs. 1.2 and 1.17 show, when we trace the periodic table from the left to the right, $\zeta$ increases approximately in proportion to Z, and the atomic radius shrinks in proportion to 1/Z (Appendix Table 1.1). Therefore, electronically negative atoms like oxygen and fluorine are smaller atoms than those earlier in their row. The same tendency holds for covalent radii and ionic radii. An atom is a many-electron system, but we can calculate the atomic orbitals precisely by an approximate method. As the starting point of molecular orbital calculations, we use

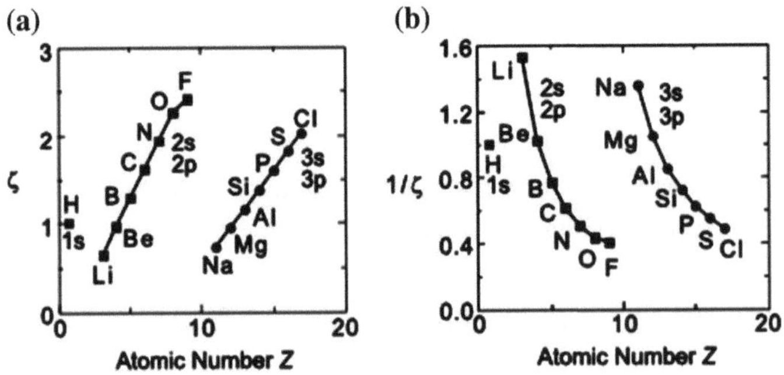


**Fig. 1.5** **a** $\zeta$ and **b** the inverse $1/\zeta$ for various atoms (Appendix Table 1)

atomic orbitals such as STO-3G, which means a Slater-type orbital represented by three Gaussian functions. The information of atomic orbitals is included in $\zeta$, which is incorporated in the ab initio programs as the basis set. Once $\zeta$ is given, we can calculate the energy from $-\zeta^2$.

Next, we shall investigate solutions of the **angular part** $\Phi_m(\varphi)$. The second term of Eq. 1.15 is separated into two terms, respectively, related to $\theta$ and $\varphi$. The $\varphi$ part satisfies $\frac{1}{\Phi}\frac{\partial^2\Phi}{\partial\varphi^2}$ = (constant). We can write (constant) = $m^2$, and the differential equation of $\Phi_m(\varphi)$ is

$$\frac{\partial^2\Phi}{\partial\varphi^2} = m^2\Phi. \tag{1.19}$$

The solutions are easily obtained as $\Phi(\phi) = e^{im\varphi}$. Here, $m$ is an integer because, similarly to Fig. 1.1a, the phase has to be integer times of $2\pi$ when the electron rotates around the orbital. A constant $\Phi$ is one of the solutions of Eq. 1.19, corresponding to $m = 0$.

Returning to Eq. 1.15, after the separation of variables, the $R$ term becomes a constant $l(l + 1)$, and the $\Phi$ term becomes a constant $m^2$. Then, the angular part $\Theta(\theta)$ satisfies

$$\frac{1}{\sin\theta}\frac{\mathrm{d}}{\mathrm{d}\theta}\left(\sin\theta\frac{\mathrm{d}\Theta}{\mathrm{d}\theta}\right) + \left[l(l+1) - \frac{m^2}{\sin^2\theta}\right]\Theta = 0. \tag{1.20}$$

Note that $\Theta$ = (constant) is a solution of this differential equation for $l = 0$ and $m = 0$. This corresponds to an s orbital obtained in Example 1.3.

*Example 1.4* Show that $\Theta_{10}(\theta) = \cos\theta$ is a solution of Eq. 1.20 for $l = 1$ and $m = 0$.

Putting $l = 1$, $m = 0$, and $\Theta = \cos\theta$, Eq. 1.20 leads to

$$-\frac{1}{\sin\theta}\frac{\mathrm{d}}{\mathrm{d}\theta}(\sin\theta\sin\theta) + 2\cos\theta = 0 \text{ namely } -\frac{2\sin\theta\cos\theta}{\sin\theta} + 2\cos\theta = 0.$$

This proves $\cos\theta$ is a solution of Eq. 1.20.

Since Example 1.3 shows that the radial part for $l = 1$ ($n = 2$) is $re^{-r/2a_0}$, we obtain

$$\psi = R(r)\Theta_{lm}(\theta)\Phi_m(\varphi) = re^{-r/2a_0}\cos\theta = ze^{-r/2a_0}.$$

This is a $2p_z$ orbital, which is shown in Fig. 1.6a. This wave function is zero at $z = 0$, and $\Psi$ is maximum at a finite $z$. $\Psi$ is negative at $z < 0$, but the probability distribution $\Psi^*\Psi$ is positive. When we rotate this function around the $z$ axis, we obtain a function as shown in Fig. 1.6b.

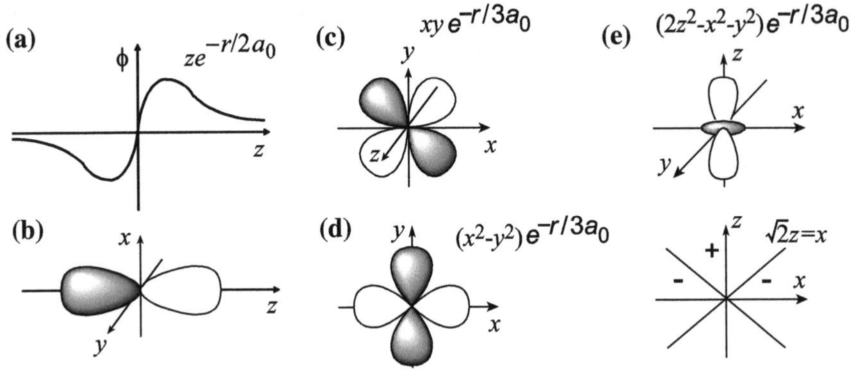


**Fig. 1.6** Shape of p and d orbitals

We can easily show that $\Theta_{11}(\theta) = \sin\theta$ is another solution of Eq. 1.20 for $l = 1$ and $m = \pm 1$. This leads to two functions corresponding to $m = \pm 1$

$$\psi_1 = re^{-r/2a_0}\sin\theta\, e^{i\varphi} \quad \text{and} \quad \psi_{-1} = re^{-r/2a_0}\sin\theta\, e^{-i\varphi}.$$

More conveniently, the complex functions are converted to two real functions by taking $\Psi_1 - \Psi_{-1}$ and $\Psi_1 + \Psi_{-1}$, which are the $2p_x$ orbital $xe^{-r/2a_0}$ and the $2p_y$ orbital $ye^{-r/2a_0}$. Thus, the $x$, $y$, and $z$ linear terms afford the $p$ orbitals. In general, Eq. 1.20 has solutions for $|m| \leq l$, although it is not directly obvious from this equation. Accordingly, for a given $l$, possible $m$ values are $m = -l, -l + 1, \ldots l - 1$, and $l$, so there are $2l + 1$ solutions.

Similarly to the $p$ orbitals, the $d$ orbitals are obtained from the $x$, $y$, and $z$ square terms, namely $x^2$, $y^2$, $z^2$, $xy$, $yz$, and $zx$. Among these six, only five are independent, because $l = 2$ leads to $2l + 1 = 5$ solutions corresponding to $m = 0, \pm 1$, and $\pm 2$. The shape of $d$ orbitals is depicted in Fig. 1.6c–e. The square terms afford the $xy$, $yz$, $zx$, and $x^2 - y^2$ orbitals, and the final one, orthogonalized to these four, is the $2z^2 - x^2 - y^2$ orbital, which has a cone-like node obtained by rotating a $\sqrt{2}z = x$ node around the $z$ axis. The combined angular part $Y_{lm}(\theta, \varphi) = \Theta_{lm}(\theta)\Phi_m(\varphi)$ is called **spherical harmonics**.

Thus obtained atomic orbitals are labeled by **principal quantum number** $n$, **azimuthal quantum number** $l$, and **magnetic quantum number** $m$.

| | $l = 0$ | 1 | 2 |
|---|---|---|---|
| $n = 1$ | 1s | | |
| $n = 2$ | 2s | 2p | |
| $n = 3$ | 3s | 3p | 3d |
| | ............ | | |

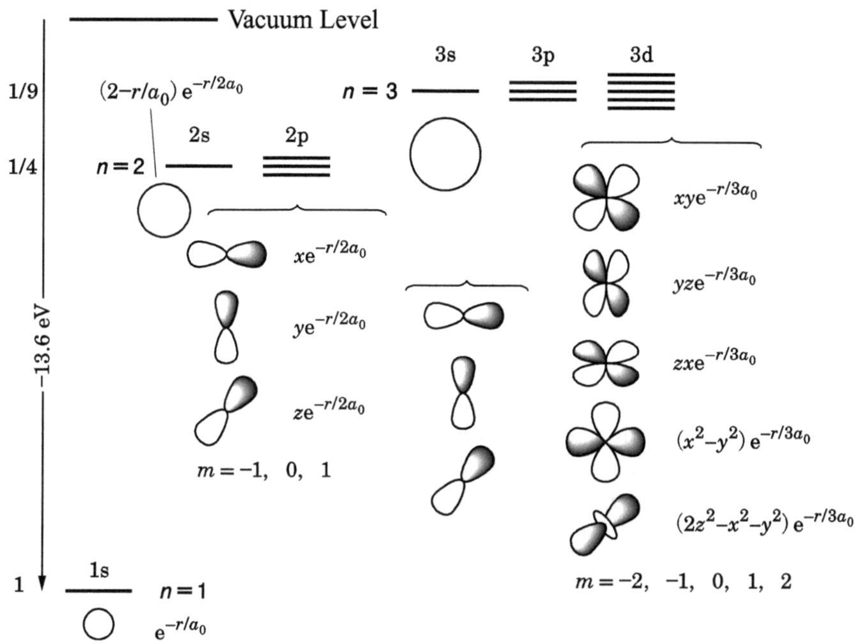


**Fig. 1.7** Energy and shape of hydrogen atomic orbitals

These orbitals are depicted in Fig. 1.7. For a given $n$, we can make different functions of $r$ from $l = 0$ to $n - 1$. For instance, 2s ($l = 0$) and 2$p$ ($l = 1$) for $n = 2$. Accordingly, the number of different $l$ is equal to $n$.

In quantum chemical calculations, **atomic unit** (a.u.) is sometimes used. We put $Z = 1$ and $n = 1$ in Eqs. 1.2 and 1.3, and the distance is converted to the unit of $r' = r/a_0$ ($a_0 = \varepsilon_0 h^2/\pi m e^2 = 0.529$ Å = 1 Bohr). The energy is measured in the unit of $E' = E/R_0$ ($R_0 = me^4/8\varepsilon_0^2 h^2 = 13.6$ eV = 1 Rydberg). The Schrödinger equation (Eq. 1.12) for $r'$ and $E'$ is

$$\left[-\nabla^2 - \frac{2}{r'}\right]\psi = E'\psi.$$

Physical constants such as $m$ and $h$ are eliminated from this equation. Such a unit is used inside the program of quantum chemical calculations. Sometimes, the energy unit is doubled to $E'' = E/2R_0$ ($2R_0 = me^4/4\varepsilon_0^2 h^2 = 27.2$ eV = 1 Hartree), and another form of the Schrödinger equation is used.

$$\left[-\frac{1}{2}\nabla^2 - \frac{1}{r'}\right]\psi = E''\psi.$$

## 1.5 Molecular Orbitals

The Schrödinger equation of a polyatomic molecule is

$$\left[\sum_i \left(-\frac{\hbar^2}{2m}\nabla_i^2 - \frac{1}{4\pi\varepsilon_0}\sum_n \frac{Z_n e^2}{r_{ni}}\right) + \frac{1}{4\pi\varepsilon_0}\sum_{i\neq j}\frac{e^2}{r_{ij}}\right]\Psi = E\Psi. \tag{1.21}$$

The first term represents kinetic energy of the $i$th electron, and the second term is the Coulomb attraction from the $n$th atoms. The final term is the electron–electron repulsion. This term is an origin of many intriguing phenomena (Chap. 5), but usually the average is included as a modified potential (Sect. 5.1). Now we shall neglect the electron–electron repulsion. The resulting Hamiltonian is a simple sum of the one-electron Hamiltonian related to the $i$th electrons, $H = H_1 + H_2 + H_3 + \cdots$. The overall wave function is represented by a product of one-electron wave functions, $\Psi = \psi_1\psi_2\psi_3\ldots$ Then, the variables are separated (Example 1.5), and the total energy is represented by a sum of the one-electron energies, $E = E_1 + E_2 + E_2 + \cdots$, where $H_i\psi_i = E_i\psi_i$ is the one-electron Schrödinger equation.

$$\left[-\frac{\hbar^2}{2m}\nabla^2 - \frac{e^2}{4\pi\varepsilon_0}\sum_n \frac{Z_n}{r_n}\right]\psi = E\psi. \tag{1.22}$$

*Example 1.5* Show the variables are separated as described above.

Note that $H_i$ works only on $\psi_i$. Using $H_i\psi_i = E_i\psi_i$, the overall Hamiltonian is converted.

$$\begin{aligned}(H_1 + H_2 + H_3 + \cdots)\psi_1\psi_2\psi_3\cdots &= (H_1\psi_1)\psi_2\psi_3\cdots + \psi_1(H_2\psi_2)\psi_3\cdots + \cdots \\ = E_1\psi_1\psi_2\psi_3\cdots + E_2\psi_1\psi_2\psi_3\cdots + \cdots &= (E_1 + E_2 + \cdots)\psi_1\psi_2\psi_3\ldots\end{aligned}$$

Accordingly, the overall $H\Psi = E\Psi$ is satisfied, if the individual $H_i\psi_i = E_i\psi_i$ is satisfied.

The above $H_i$ is the same equation for all $i$. It is the spirit of the molecular orbital theory that we solve the one-electron Schrödinger equation where $n$ nuclei are arranged in the appropriate positions and one electron is put in the potential. From this one-electron Schrödinger equation, the energy levels are obtained, and electrons are placed in the energy levels according to Pauli's exclusion principle.

Molecular orbitals are represented by linear combination of atomic orbitals (LCAO) constructed by multiplying a coefficient $c_j$ to each atomic orbital $\chi_j$.

$$\psi = \sum_{j}^{N} c_j \chi_j \tag{1.23}$$

From the left of Eq. 1.22, $\psi^*$ is multiplied, and this equation is integrated over the whole space, $\mathrm{d}\tau = \mathrm{d}x\,\mathrm{d}y\,\mathrm{d}z$, where the integration is performed from $-\infty$ to $\infty$ for $x$, $y$, and $z$, respectively. The resulting total energy is represented by

$$E = \frac{\int \psi^* H \psi \mathrm{d}\tau}{\int \psi^* \psi \mathrm{d}\tau}. \tag{1.24}$$

$\psi^*$ is a complex conjugate of $\psi$. However, in the molecular orbital theory, $\psi$ is always real, so practically $\psi^* = \psi$. In contrast, wave functions of energy bands in solids are complex numbers, and the complex conjugate needs to be explicitly considered (Chap. 2). Inserting Eq. 1.23 in Eq. 1.24, we obtain

$$E = \frac{\int (\sum_i^N c_i^* \chi_i^*) H (\sum_j^N c_j \chi_j) \mathrm{d}\tau}{\int (\sum_i^N c_i^* \chi_i^*)(\sum_j^N c_j \chi_j) \mathrm{d}\tau} = \frac{\sum_i^N \sum_j^N c_i^* c_j \int \chi_i^* H \chi_j \mathrm{d}\tau}{\sum_i^N \sum_j^N c_i^* c_j \int \chi_i^* \chi_j \mathrm{d}\tau}.$$

This equation is simplified by defining

$$\begin{aligned} &\textbf{Coulomb integral}: \alpha_i = \int \chi_i^* H \chi_i \mathrm{d}\tau \\ &\textbf{Resonance integral}: \beta_{ij} = \int \chi_i^* H \chi_j \mathrm{d}\tau \\ &\textbf{Overlap integral}: S_{ij} = \int \chi_i^* \chi_j \mathrm{d}\tau. \end{aligned} \tag{1.25}$$

The result is

$$E = \frac{\sum_i^N c_i^* c_i \alpha_i + \sum_{i \neq j}^N c_i^* c_j \beta_{ij}}{\sum_i^N \sum_j^N c_i^* c_j S_{ij}}.$$

$c_j$ is determined so as to minimize this energy. For this purpose, $E$ is differentiated by $c_j$, and put zero.

$$\frac{\partial E}{\partial c_i^*} = \frac{\left(c_i \alpha_i + \sum_{j(\neq i)}^N c_j \beta_{ij}\right)(\text{denominator}) - (\text{numerator}) \sum_j^N c_j S_{ij}}{(\text{denominator})^2} = 0$$

by using $E$ = (numerator)/(denominator), we obtain

$$(\alpha_i - ES_{ii})c_i + \sum_{j(\neq i)}^{N} (\beta_{ij} - ES_{ij})c_j = 0.$$

If the atomic orbitals $\chi_j$ and $\chi_i$ are different orbitals located on the same atom, these orbitals are orthogonal ($S_{ij} = 0$). We extend this relation to all combinations, and all orbitals are approximated to be orthogonal, $S_{ij} = 0$. Since an overlap integral for the same atomic orbital is normalized as $S_{ii}$=1, we assume that the atomic orbitals are orthonormal, $S_{ij} = \delta_{ij}$. Then, the above equation is reduced to

$$(\alpha_i - E)c_i + \sum_{j(\neq i)}^{N} \beta_{ij}c_j = 0.$$

We obtain similar equations for respective $c_i$, and the resulting simultaneous equations are expressed in a matrix form.

$$\begin{pmatrix} \alpha_1 - E & \beta_{12} & \beta_{13} & \cdots \\ \beta_{21} & \alpha_2 - E & \beta_{23} & \\ \beta_{31} & \beta_{32} & \alpha_3 - E & \\ \vdots & & & \end{pmatrix} \begin{pmatrix} c_1 \\ c_2 \\ c_3 \\ \vdots \end{pmatrix} = 0 \tag{1.26}$$

These simultaneous equations are satisfied when all $c_i$ are zero, but meaningful solutions are obtained for nonzero $c_i$. This condition is represented by zero of the coefficient determinant.

$$\begin{vmatrix} \alpha_1 - E & \beta_{12} & \beta_{13} & \cdots \\ \beta_{21} & \alpha_2 - E & \beta_{23} & \\ \beta_{31} & \beta_{32} & \alpha_3 - E & \\ \vdots & & & \end{vmatrix} = 0 \tag{1.27}$$

This is called **secular equation**. Since every diagonal element contains $E$, this determinant is an $N$th order equation of $E$, where $N$ is the number of atomic orbitals in Eq. 1.23. This equation always has $N$ real solutions as proved in Sect. 1.7. The solutions of this $N$th order equation afford $N$ energy levels (eigenvalues). When we adopt a solution $E$ obtained from Eq. 1.26, the simultaneous equations afford a set of $c_i$, which determine the shape of the molecular orbital according to Eq. 1.23.

In the secular equation, the diagonal element $\alpha_i$ is the energy of the atomic orbital $\chi_i$ (Eq. 1.25). Note that $\alpha_i$ is negative in general. In the **extended Hückel method**, $\alpha_i$ is approximated by the experimentally obtained ionization potential [2]. The ionization potentials are plotted in Fig. 1.8. The value for the hydrogen 1 s orbital is −13.6 eV, which is equal to the Rydberg constant. As Eqs. 1.3 and 1.18 show, $\alpha_i$ is approximately proportional to $-Z^2$. When we trace the periodic table, $|\alpha_i|$ increases gradually (Fig. 1.8). $\alpha_i$ for an alkali metal is about −5 eV, and that for C

2p is −11.4 eV, whereas that for F 2p is as large as −18.1 eV. Because the 2p level deepens continuously, the 2p level of Na becomes deeper, but the 2p of Na is a core level. The $\alpha_i$ values are also listed in Appendix Table 1.

In the extended Hückel method [3], the nondiagonal elements $\beta_{ij}$ are approximated by

$$\beta_{ij} = \frac{1.75}{2}(\alpha_i + \alpha_j)S_{ij}. \quad (1.28)$$

This is an average of $\alpha_i$ and $\alpha_j$, multiplied by 1.75. An approximate relation $H\chi_j = E_j\chi_j \sim \alpha_j\chi_j$ is put in Eq. 1.25, and we obtain $\beta_{ij} \approx \int \chi_i^* \alpha_j \chi_j \mathrm{d}\tau = \alpha_j S_{ij}$. Exchanging $\chi_i$ and $\chi_j$, we obtain $\beta_{ij} \approx \alpha_i S_{ij}$. This is the reason why we use Eq. 1.28. We can calculate overlap integrals $S_{ij}$ from the Slater orbitals. Note that $\beta$ is negative generally because $\alpha$ is negative. In the following, we investigate a few examples of molecular orbitals.

### *1.5.1 Polar Bonds*

Consider an LCAO molecular orbital $\Psi = c_1\chi_1 + c_2\chi_2$ consisting of two different atomic orbitals. For instance, we consider HCl, where $\chi_1$ is H 1s, and $\chi_2$ is Cl 3p$_z$ (Fig. 1.9a). Because these two are principal atomic orbitals to form the H–Cl bond, other Cl atomic orbitals are neglected. From Fig. 1.8, $\alpha_1$ is −13.6 eV, and $\alpha_2$ is −15.0 eV. The corresponding secular equation is

$$\begin{vmatrix} \alpha_1 - E & \beta \\ \beta & \alpha_2 - E \end{vmatrix} = 0. \quad (1.29)$$

This equation affords a square equation

$$E^2 - (\alpha_1 + \alpha_2)E + (\alpha_1\alpha_2 - \beta^2) = 0.$$

This equation is solved to

$$E_\pm = \frac{\alpha_1 + \alpha_2}{2} \pm \frac{1}{2}\sqrt{(\alpha_1 - \alpha_2)^2 + 4\beta^2}. \quad (1.30)$$

Note that when $\beta = 0$, these solutions are reduced to $\alpha_1$ and $\alpha_2$, implying that there is no chemical bond. $E_+$ and $E_-$ are separated from the average of $\alpha_1$ and $\alpha_2$ by the square root, which is always larger than $\alpha_1 - \alpha_2$. As a result, $E_+$ is lower than $\alpha_2$, and $E_-$ is higher than $\alpha_1$ (Fig. 1.9b). When we put two electrons into the $E_+$ level, the energy is lower than the perfectly ionic state $H^+Cl^-$ in which two electrons are put in the $\chi_2$ orbital. However, since the $E_+$ level is more populated on Cl than on H,

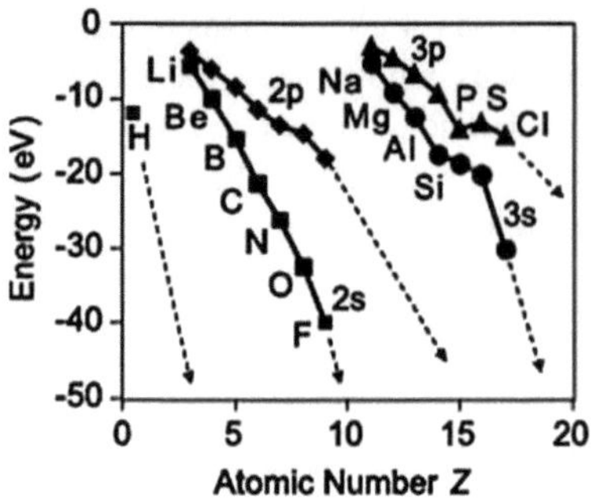


**Fig. 1.8** Ionization energy (Appendix Table 1)

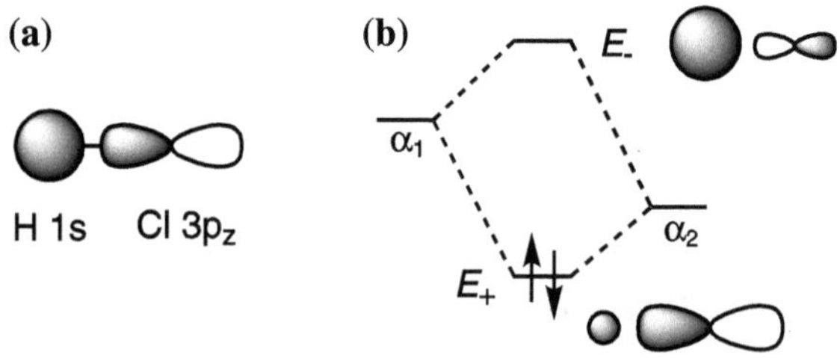


**Fig. 1.9** Polar bond

the electron is more distributed on Cl. The result is represented by $H^{\delta+}Cl^{\delta-}$; this is a polar bond. In a polar bond, electron is more populated on an atom with deeper $\alpha_i$. Such an atom is more electronegative. Accordingly, $\alpha$ is a measure of the electronegativity.

Putting $\alpha_1 = \alpha_2$ in Eq. 1.30, we obtain $E_+ = \alpha + \beta$ and $E_- = \alpha - \beta$. Then, we go back to a nonpolar molecule like $H_2$. In the $|\alpha_1 - \alpha_2| \gg |\beta|$ limit, Eq. 1.30 is rewritten as

$$E_{\pm} = \frac{\alpha_1 + \alpha_2}{2} \pm \frac{1}{2}|\alpha_1 - \alpha_2|\sqrt{1 + \frac{4\beta^2}{(\alpha_1 - \alpha_2)^2}}.$$

Approximation $\sqrt{1+x} \sim 1 + \frac{1}{2}x$ for $|x| \ll 1$ leads to

$$E_{\pm} = \frac{\alpha_1 + \alpha_2}{2} \pm \frac{1}{2}|\alpha_1 - \alpha_2|\left(1 + \frac{2\beta^2}{(\alpha_1 - \alpha_2)^2}\right)$$

which is further approximated to

$$E_+ \approx \alpha_2 - \frac{\beta^2}{|\alpha_1 - \alpha_2|} \quad \text{and} \quad E_- \approx \alpha_1 + \frac{\beta^2}{|\alpha_1 - \alpha_2|}.$$

Since stabilization coming from the bond formation is $\beta/|\alpha_1 - \alpha_2|$ times $\beta$, a larger value of $|\alpha_1 - \alpha_2|$ results in a weaker chemical bond. As a consequence, orbitals with largely different energies do not form an efficient chemical bond.

### 1.5.2 Diatomic Molecules

We investigate the molecular orbitals of $O_2$ as an example of a diatomic molecule, in which each atom has more than two atomic orbitals. The electron configuration of an oxygen atom is $(1s)^2(2s)^2(2p)^4$, and the total number of electrons is eight. The LCAO molecular orbitals are constructed from the atomic orbitals, 1s, 2s, $2p_x$, $2p_y$, and $2p_z$, coming from the two atoms. Labeling the two atoms as A and B, the molecular orbital is represented by the linear combination of ten atomic orbitals.

$$\begin{aligned}\psi = {} & c_1\chi_{A1s} + c_2\chi_{B1s} + c_3\chi_{A2s} + c_4\chi_{B2s} + c_5\chi_{A2p_z} + c_6\chi_{B2p_z} \\ & + c_7\chi_{A2p_x} + c_8\chi_{B2p_x} + c_9\chi_{A2p_y} + c_{10}\chi_{B2p_y}\end{aligned}$$

The secular equation is $10 \times 10$.

$$\begin{vmatrix} \alpha_{1s}-E & \beta_{1s} & & & & & 0 & 0 & 0 & 0 \\ \beta_{1s} & \alpha_{1s}-E & & & & & 0 & 0 & 0 & 0 \\ & & \alpha_{2s}-E & \beta_{2s} & & & 0 & 0 & 0 & 0 \\ & & \beta_{2s} & \alpha_{2s}-E & & & 0 & 0 & 0 & 0 \\ & & & & \alpha_{2p}-E & \beta_\sigma & 0 & 0 & 0 & 0 \\ & & & & \beta_\sigma & \alpha_{2p}-E & 0 & 0 & 0 & 0 \\ 0 & 0 & 0 & 0 & 0 & 0 & \alpha_{2p}-E & \beta_\pi & 0 & 0 \\ 0 & 0 & 0 & 0 & 0 & 0 & \beta_\pi & \alpha_{2p}-E & 0 & 0 \\ 0 & 0 & 0 & 0 & 0 & 0 & 0 & 0 & \alpha_{2p}-E & \beta_\pi \\ 0 & 0 & 0 & 0 & 0 & 0 & 0 & 0 & \beta_\pi & \alpha_{2p}-E \end{vmatrix} = 0$$

$\alpha_{1s}$, $\alpha_{2s}$, and $\alpha_{2p}$ are energies of 1s, 2s, and 2p orbitals, and $\beta_{1s}$ and $\beta_{2s}$ are resonance integrals between them. The bonding direction is the $z$ axis. There are two kinds of 2p $\beta$: $\beta_\sigma$ for the $\sigma$ bond and $\beta_\pi$ for the $\pi$ bond. The last four $\pi$ orbitals are orthogonal to the other $\sigma$ orbitals, so that all the nondiagonal elements are zero. Then, the seventh to the tenth lines are reduced to a two-dimensional determinant.

$$\begin{vmatrix} \alpha_{2p}-E & \beta_\pi \\ \beta_\pi & \alpha_{2p}-E \end{vmatrix} = 0$$

The same matrix appears twice, and both lead to $E = \alpha_{2p} \pm \beta_\pi$.

As for the first and second lines, $\alpha_{1s}$ is located at a deep position energetically much different from the other levels.

$$\begin{vmatrix} \alpha_{1s} - E & \beta_{1s} \\ \beta_{1s} & \alpha_{1s} - E \end{vmatrix} = 0$$

This determinant leads to $E = \alpha_{1p} \pm \beta_{1s}$. The 1s orbital of oxygen is so small that the overlap $\beta_{1s}$ with the adjacent atom is nearly zero. In addition, the 1s orbital does not contribute to the chemical bond because both the bonding and antibonding orbitals are occupied. These levels are called **core levels**, and usually not considered. The other levels contributing to the chemical bonds are called **valence levels**.

We have to solve the remaining $\sigma$ parts, but here we approximate that the third to the fourth lines are independent of the fifth to sixth lines. Then, we obtain $E = \alpha_{2s} \pm \beta_{2s}$ and $E = \alpha_{2p} \pm \beta_{\sigma}$. Thus, the molecular orbitals of $O_2$ are summarized in Fig. 1.10. Since generally $|\beta_{\sigma}|$ is larger than $|\beta_{\pi}|$, $\alpha_{2s} + \beta_{\sigma}$ is located lower than $\alpha_{2s} + \beta_{\pi}$, and $\alpha_{2s} - \beta_{\sigma}$ is located higher than $\alpha_{2s} - \beta_{\pi}$. Namely, the $\sigma$ levels are placed at a more outer position than the $\pi$ levels. Since an oxygen atom has eight electrons, totally sixteen electrons are placed, and the final two electrons enter the $\pi$ antibonding ($\pi^*$) levels.

The 2p $\sigma$, $\pi$, and $\pi^*$ levels contribute to the chemical bond, which are, respectively, occupied by two, four, and two electrons. The bonding energy is

$$2(\alpha_{2p} + \beta_{\sigma}) + 4(\alpha_{2p} + \beta_{\pi}) + 2(\alpha_{2p} + \beta_{\sigma}) - 8\alpha_{2p} = 2\beta_{\sigma} + 2\beta_{\pi}.$$

We subtract $8\alpha_{2p}$ in order to compare the molecular energy with the atomic energy. The bond order is defined by

$$(\text{Bond order}) = (N - N^*) \times 1/2$$

where $N$ is the number of electrons in the bonding orbitals and $N^*$ is the number of electrons in the antibonding orbitals. For $O_2$, this number is,

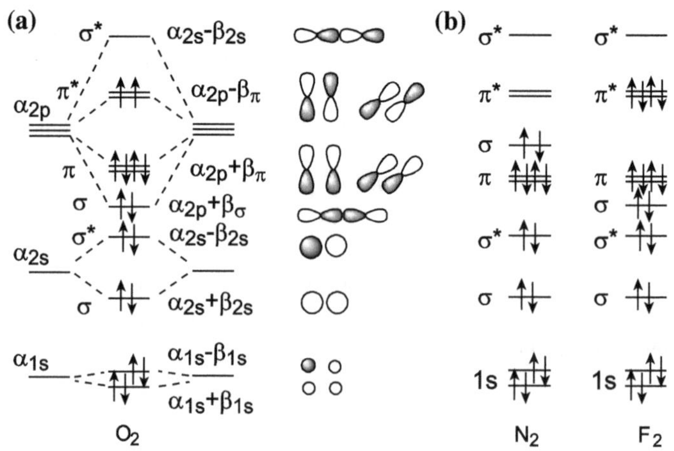


**Fig. 1.10** Molecular orbitals of **a** $O_2$, and **b** $N_2$ and $F_2$

$$[6 - 2] \times 1/2 = 2.$$

This means that an O=O molecule has a double bond consisting of one $\sigma$ bond and one $\pi$ bond. However, the above discussion indicates that one $\pi^*$ bond should be subtracted from two $\pi$ bonds, resulting in one net $\pi$ bond.

The total electrons of $N_2$ and $F_2$ are fourteen and eighteen, and the molecular orbital levels are depicted in Fig. 1.10. In $N_2$, on account of the mixing of the third–fourth and fifth–sixth lines of the secular equation, the 2s $\sigma$ level pushes up the 2p $\sigma$ level above the 2p $\pi$ level. This does not change the electron numbers; the 2s $\sigma$ has two electrons, and the 2p $\pi$ levels have four electrons. The bond orders are

$$N_2\text{: } [6 - 0] \times 1/2 = 3\sigma + 2\pi$$
$$F_2\text{: } [6 - 4] \times 1/2 = 1\sigma.$$

This means that N≡N has a triple bond, and F–F has a single bond.

The two $\pi^*$ levels of $O_2$ have two electrons. According to **Hund's rule** (Sect. 5.1), these two electrons have parallel spins, so the $O_2$ molecule is **paramagnetic**. By contrast, other molecules like $N_2$ and $F_2$ do not have unpaired electrons and are **diamagnetic**. Bond orders are estimated from the naive octet rule, but paramagnetism of $O_2$ is only derived from the molecular orbital theory. When we estimate $2|\beta|$ from the dissociation energy, $\beta$ is −2.2 eV for $H_2$, and $\beta_\sigma$ = −1.8 eV and $\beta_\pi$ = −1.5 eV for a C–C bond.

*Example 1.6* Consider the molecular orbital of NO, and estimate the bond order. Is NO magnetic?

Although N and O are different atoms, the energy levels are represented by Fig. 1.11 similarly to a homonuclear molecule. Fifteen electrons enter these levels, and the $\pi^*$ level has one electron. The bond order is

$$O_2\text{: } [6 - 1] \times 1/2 = 5/2.$$

Namely the bond order is 2.5. The molecule shows paramagnetism due to the unpaired electron.

*Example 1.7* Calculate the number of valence levels of the molecular orbitals in BEDT-TTF $C_{10}S_8H_8$ (Fig. 7.5). Obtain the number of the HOMO from the bottom. The valence levels are 2s and 2p for *C*, 3s and 3p for S, and 1s for H.

We use four atomic orbitals: 2s, $2p_x$, $2p_y$, and $2p_z$ for C, in which four electrons are incorporated. Similarly, S has four orbitals and six electrons, and *H* has one orbital and one electron. From Table 1.1, the total orbital number is 80 and the electron number is 96, so that the HOMO is the 48-th level from the bottom.

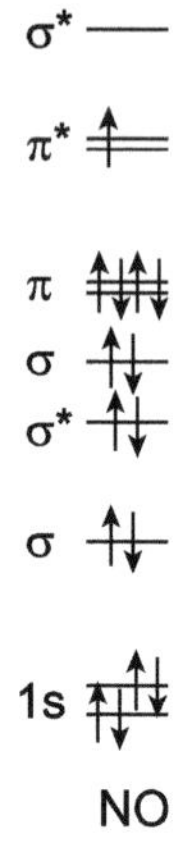


Fig. 1.11 Molecular orbital levels of NO

Table 1.1 Numbers of the orbitals and electrons in a BEDT-TTF molecule

| Atoms | | Orbitals | | Electrons | |
|---|---|---|---|---|---|
| C | 10 | 4 | 40 | 4 | 40 |
| S | 8 | 4 | 32 | 6 | 48 |
| H | 8 | 1 | 8 | 1 | 8 |
| | 26 | | 80 | | 96 |

## *1.5.3 Hückel Approximation*

Molecular orbital calculations considering only π-orbitals are called **Hückel approximation**. Since only single kinds of $\alpha$ and $\beta$ are incorporated, this is the simplest molecular orbital theory. Nonetheless, this is a powerful tool to understand π-conjugated molecules. In Chap. 2, energy bands of solids are introduced as an extension of the Hückel approximation. Here, we shall investigate a variety of molecules using the Hückel approximation [4].

Consider the molecular orbitals of butadiene (Fig. 1.12). We do not need to consider the $\sigma$ orbitals because they are orthogonal to the π orbitals. The molecular plane is in the $xy$ plane, and the LCAO is constructed from the carbon $2p_z$ orbitals $\chi_i$

$$\psi = c_1\chi_1 + c_2\chi_2 + c_3\chi_3 + c_4\chi_4.$$

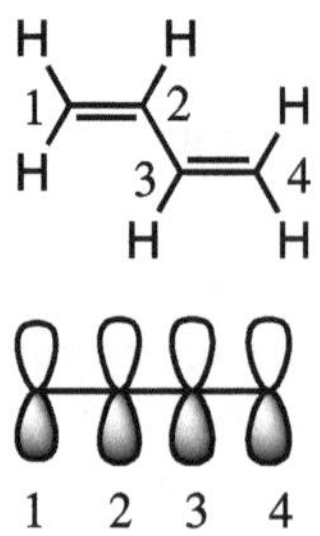


Fig. 1.12 π orbitals of butadiene

The carbon atoms are numbered from 1 to 4 as shown in Fig. 1.12. The secular equation is obtained similarly to Eq. 1.27

$$\begin{vmatrix} \alpha - E & \beta & 0 & 0 \\ \beta & \alpha - E & \beta & 0 \\ 0 & \beta & \alpha - E & \beta \\ 0 & 0 & \beta & \alpha - E \end{vmatrix} = 0$$

where $\alpha$ is for carbon 2p, and $\beta$ is for the π-bond, which is equivalent to $\beta_\pi$ appearing in Fig. 1.10. Strictly speaking, $\beta$ for the single bond between 2 and 3 is different from $\beta$ for the double bonds, 1–2 and 3–4. However, in the Hückel approximation, the conjugated bond $\beta$ is assumed to be the same as the double bond $\beta$. All lines (columns) are divided by $\beta$, and putting $x = (\alpha - E)/\beta$ this determinant is reduced to

$$\begin{vmatrix} x & 1 & 0 & 0 \\ 1 & x & 1 & 0 \\ 0 & 1 & x & 1 \\ 0 & 0 & 1 & x \end{vmatrix} = x\begin{vmatrix} x & 1 & 0 \\ 1 & x & 1 \\ 0 & 1 & x \end{vmatrix} - \begin{vmatrix} 1 & 1 & 0 \\ 0 & x & 1 \\ 0 & 1 & x \end{vmatrix} = x(x^3 - 2x) - (x^2 - 1)$$
$$= x^4 - 3x^2 + 1.$$

This is a quadratic equation of $x^2$ to afford

$$x^2 = \frac{3 \pm \sqrt{5}}{2}.$$

Consequently, $x$ is

$$x = \pm\sqrt{\frac{3 \pm \sqrt{5}}{2}} = \pm\frac{\sqrt{6 \pm 2\sqrt{5}}}{2} = \pm\frac{\sqrt{(1 \pm \sqrt{5})^2}}{2} = \pm\frac{1 \pm \sqrt{5}}{2}$$
$$= \pm 1.618, \pm 0.618.$$

The energy levels are depicted in Fig. 1.13. We can calculate $c_1 \sim c_4$ by putting these energies in the simultaneous equations (Eq. 1.26). However, we can more easily sketch the molecular orbitals (Fig. 1.13) because the bottom level does not have any node, and the number of nodes increases one by one. The lower two levels have bonding (π) nature, and the upper two levels have antibonding (π*) nature. The second level is called the **highest occupied molecular orbital** (**HOMO**), and the third level is the **lowest unoccupied molecular orbital** (**LUMO**). The HOMO and LUMO are sometimes called **frontier orbitals**.

Since one carbon atom gives up one π electron, four electrons are put into these levels. Note that $\alpha$ and $\beta$ are always negative. The electrons in these two levels have energies of $\alpha$ + 1.618$\beta$ and $\alpha$ + 0.618$\beta$, so the bonding energy from these electrons is

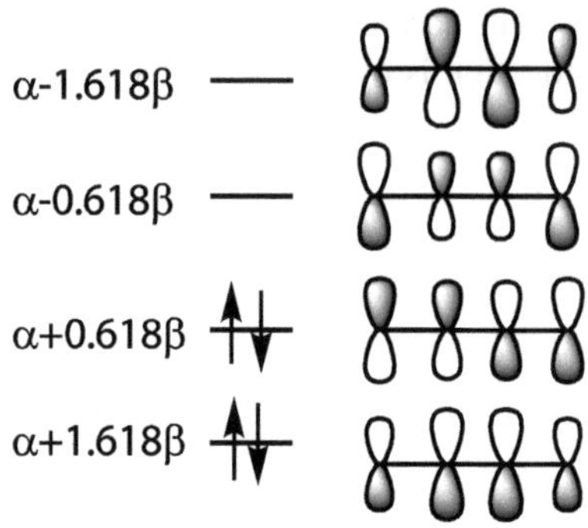


**Fig. 1.13** Molecular orbitals of butadiene

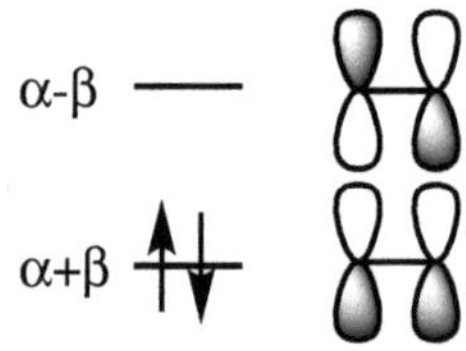


**Fig. 1.14** Molecular orbitals of ethylene

$$2(\alpha + 1.618\beta) + 2(\alpha + 0.618\beta) - 4\alpha = 4.48\beta.$$

Here, $4\alpha$ is subtracted because each 2p orbital has the energy of $\alpha$ in the atomic state.

Similarly the secular equation of ethylene is

$$\begin{vmatrix} \alpha - E & \beta \\ \beta & \alpha - E \end{vmatrix} = 0.$$

This leads to the bonding level $\alpha + \beta$ and the antibonding level $\alpha - \beta$ (Fig. 1.14). The bonding energy is

$$2(\alpha + \beta) - 2\alpha = 2\beta.$$

This is the energy of a double bond. Then, the energy of two double bonds is $4\beta$. The π bonding energy of butadiene is by $0.48\beta$ larger than $4\beta$. This is a gain of the conjugated double bonds, which is called **delocalization energy**.

The procedure of the Hückel approximation is summarized as follows.

1) Number the carbon atoms having π electrons. The total number is $N$.
2) In an $N \times N$ determinant, all diagonal elements are $\alpha - E$.
3) Nondiagonal elements are $\beta$ for bonded carbon pairs. $\beta$ is the same for double bonds and single bonds between the conjugated double bonds. Nondiagonal elements are zero for nonbonded carbon pairs.

4) This secular equation is reduced to an $N$th order equation of $E$. Solving this equation, we obtain $N$ energy levels.
5) Since a carbon atom gives up one electron, we put $N$ electrons. Summing the energies of the occupied levels, the $\pi$ bond energy is obtained.

*Example 1.8* Calculate energy levels and the delocalization energy of benzene according to the Hückel approximation.

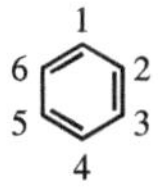


The secular equation is obtained according to the above rule.

$$\begin{vmatrix} \alpha - E & \beta & 0 & 0 & 0 & \beta \\ \beta & \alpha - E & \beta & 0 & 0 & 0 \\ 0 & \beta & \alpha - E & \beta & 0 & 0 \\ 0 & 0 & \beta & \alpha - E & \beta & 0 \\ 0 & 0 & 0 & \beta & \alpha - E & \beta \\ \beta & 0 & 0 & 0 & \beta & \alpha - E \end{vmatrix} = 0$$

The diagonal elements are $\alpha - E$, and the adjacent positions to the diagonal elements are $\beta$. The other nondiagonal elements are zero except for the 1–6 element. Because this determinant is represented by $3 \times 3$ determinants A and B in the following form, this relation is reduced to

$$\begin{vmatrix} \mathrm{A} & \mathrm{B} \\ \mathrm{B} & \mathrm{A} \end{vmatrix} = \begin{vmatrix} \mathrm{A}+\mathrm{B} & 0 \\ \mathrm{B} & \mathrm{A}-\mathrm{B} \end{vmatrix} = |\mathrm{A}+\mathrm{B}||\mathrm{A}-\mathrm{B}| = 0.$$

This relation is obtained by adding the lower three lines to the upper three lines, and subtracting the left three lines from the right three lines. By applying this relation to the above $6 \times 6$ determinant, and using $x = (\alpha - E)/\beta$, we obtain

$$\begin{vmatrix} x & 1 & 1 \\ 1 & x & 1 \\ 1 & 1 & x \end{vmatrix} \begin{vmatrix} x & 1 & -1 \\ 1 & x & 1 \\ -1 & 1 & x \end{vmatrix} = 0.$$

This equation is reduced to two cubic equations.

$$x^3 - 3x + 2 = 0 \quad \text{and } x^3 - 3x - 2 = 0$$

These equations are, respectively, zero at $x = 1$ and $x = -1$, so the equations are easily factorized.

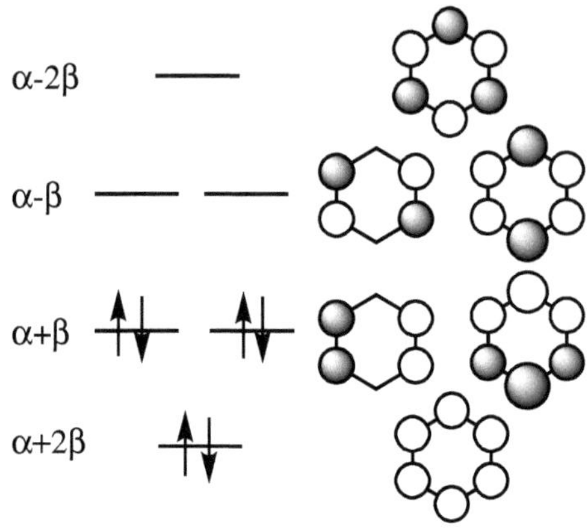


**Fig. 1.15** Molecular orbitals of benzene

$$(x-1)^2(x+2)=0 \quad \text{and } (x+1)^2(x-2)=0$$

The solutions are

$$x=1,1,\ -2,\ -1,\ -1,\ \text{and } 2.$$

The resulting energy levels are shown in Fig. 1.15. The bonding energy is

$$2(\alpha+2\beta)+4(\alpha+\beta)-6\alpha=8\beta.$$

If a benzene molecule is composed of three double bonds, the resulting bonding energy is $3 \times 2\beta = 6\beta$. Therefore, the additional delocalization energy is $8\beta - 6\beta = 2\beta$. The delocalization energy is as large as the double bond energy.

*Example 1.9* Calculate energy levels and the delocalization energy of cyclobutadiene according to the Hückel approximation.

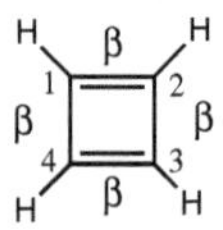


According to the atom numbering, the secular equation is

$$\begin{vmatrix} \alpha-E & \beta & 0 & \beta \\ \beta & \alpha-E & \beta & 0 \\ 0 & \beta & \alpha-E & \beta \\ \beta & 0 & \beta & \alpha-E \end{vmatrix} = 0.$$

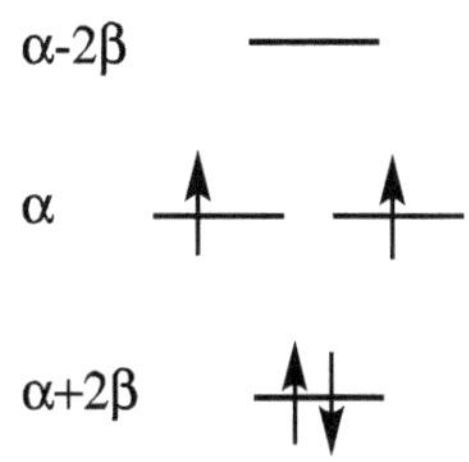


**Fig. 1.16** Energy levels of cyclobutadiene

Since this is again in the form $\begin{vmatrix} A & B \\ B & A \end{vmatrix} = 0$ for 2 × 2 determinants A and B, this is reduced to

$$\begin{vmatrix} \alpha - E & 2\beta \\ 2\beta & \alpha - E \end{vmatrix} \begin{vmatrix} \alpha - E & 0 \\ 0 & \alpha - E \end{vmatrix} = 0.$$

The solutions are $E = \alpha \pm 2\beta$, $\alpha$, and $\alpha$ (Fig. 1.16). The π-bonding energy is

$$2(\alpha + 2\beta) + 2\alpha - 4\alpha = 4\beta.$$

Since the bonding energy of two double bonds is $2 \times 2\beta = 4\beta$, the delocalization energy is zero.

Thus, cyclobutadiene does not have delocalization energy. This is surprising because linear butadiene has delocalization energy. In general, an annulene consisting of $4n$ carbon atoms does not show delocalization, whereas a $4n + 2$ annulene exhibits delocalization. This is called the **Hückel rule**. Four, ten, and fourteen membered rings are stabilized by the Hückel rule, and called **aromatic**. For example, naphthalene composed of ten carbon atoms is aromatic, though there are two rings. Eight-membered rings such as cyclooctatetraene are **antiaromatic**.

The Hückel rule is derived from the form of the secular equation, namely from the geometry of the molecules. In Chap. 2 we will generally prove why the delocalization takes place only in the $4n + 2$ membered rings.

When dark red bromine is reacted with a molecule having a double bond, an addition reaction occurs. This reaction is usually done in a carbon tetrachloride solution of bromine, and the decolorization of the bromine solution indicates the presence of a double bond (the brominated molecule is nearly colorless), which is used as a qualitative analysis of organic compounds. Cyclobutadiene reacts with bromine in this manner, but benzene and other aromatic compounds do not decolorize bromine (Fig. 1.17). Instead, benzene reacts with $FeBr_3$ to give the substitution product, bromobenzene. Substitution occurs due to the extra stabilization of the aromatic ring.

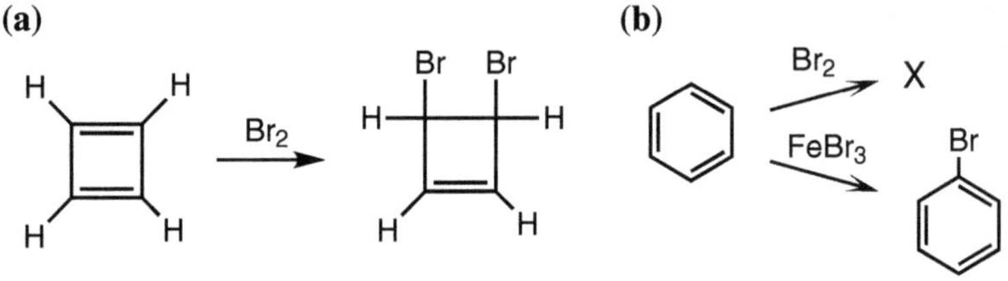


Fig. 1.17 Bromination of **a** cyclobuthadiene, and **b** benzene

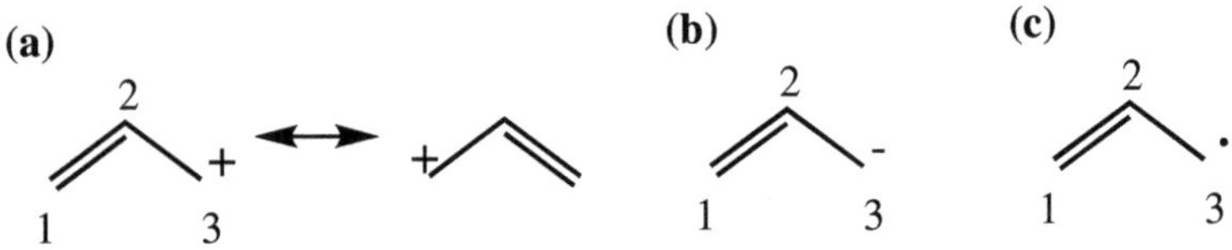


Fig. 1.18 **a** Allyl cation, **b** anion, and **c** radical

Organic chemistry deals with **carbocation** and **carbanion** as reaction intermediates, where + and − ions are, respectively, generated on a carbon atom. These ions are located on a carbon atom with three $\sigma$ bonds, and have $sp^2$ orbitals. Thus, these ionic carbons are stabilized by delocalization when a double bond or a benzene ring is bonded to the ionic carbon. For instance, in an allyl cation (Fig. 1.18a), the cation is delocalized not only on the original carbon 3 but also on the carbon 1. An ordinary carbon atom has four valence electrons, and in an $sp^2$ carbon, three of which are used for the $\sigma$ bonds. Thus, a radical has one $\pi$ electron. Similarly, a carbocation has three valence electrons, and the $\pi$ electron is zero. A carbanion has five valence electrons, in which two are $\pi$ electrons.

*Example 1.10* Calculate energy levels and the delocalization energy of allyl cation, radical, and anion (Fig. 1.18) according to the Hückel approximation.

Atom numbering is shown in Fig. 1.18. The secular equation is

$$\begin{vmatrix} \alpha - E & \beta & 0 \\ \beta & \alpha - E & \beta \\ 0 & \beta & \alpha - E \end{vmatrix} = 0.$$

The usual substitution affords the equation

$$x^3 - 2x = 0 \ \text{leading to} \ (x^2 - 2)x = 0$$

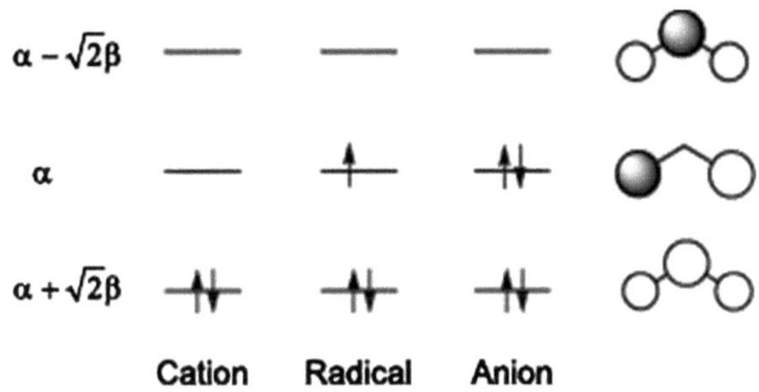


**Fig. 1.19** Energy levels and electron configuration in allyl cation, radical, and anion

so the solutions are

$$x = \sqrt{2},\ 0,\ \text{and}\ -\sqrt{2}.$$

The resulting energy levels are shown in Fig. 1.19. The cation, radical, and anion have, respectively, two, three, and four π electrons. The bonding energy is

$$\begin{array}{ll} \text{cation} & 2(\alpha+\sqrt{2}\beta) - 2\alpha = 2\sqrt{2}\beta \\ \text{radical} & 2(\alpha+\sqrt{2}\beta) + \alpha - 3\alpha = 2\sqrt{2}\beta \\ \text{anion} & 2(\alpha+\sqrt{2}\beta) + 2\alpha - 4\alpha = 2\sqrt{2}\beta. \end{array}$$

Accordingly, the stability is the same within this approximation.

Putting the lowest energy level $E = \alpha + \sqrt{2}\beta$ in the simultaneous equation (Eq. 1.26), the first and third lines are

$$-\sqrt{2}\beta c_1 + \beta c_2 = 0$$
$$\beta c_2 - \sqrt{2}\beta c_3 = 0.$$

These relations lead to $c_1{:}c_2{:}c_3 = 1{:}\sqrt{2}{:}1$. After the normalization, the molecular orbital corresponding to the $E = \alpha + \sqrt{2}\beta$ level is obtained as

$$\psi_1 = \frac{1}{2}\chi_1 + \frac{1}{\sqrt{2}}\chi_2 + \frac{1}{2}\chi_3.$$

In the same way, the second ($E = \alpha$) and third ($E = \alpha - \sqrt{2}\beta$) molecular orbitals are

$$\psi_2 = \frac{\chi_1 - \chi_3}{\sqrt{2}} \quad \text{and} \quad \psi_3 = \frac{1}{2}\chi_1 - \frac{1}{\sqrt{2}}\chi_2 + \frac{1}{2}\chi_3.$$

The shape is depicted in Fig. 1.19.

*Example 1.11* Calculate energy levels and the delocalization energy of cyclopropenyl cation according to the Hückel approximation. Compare the delocalization energy with those of the radical and the anion.

The positive charge is delocalized to the three carbon atoms. The secular equation

$$\begin{vmatrix} x & 1 & 1 \\ 1 & x & 1 \\ 1 & 1 & x \end{vmatrix} = 0$$

affords

$$x^3 - 3x + 2 = 0 \text{ leading to } (x - 1)^2(x + 2) = 0.$$

The solutions are

$$x = 1, 1, \text{ and } -2.$$

The energy levels are shown in Fig. 1.20. The cation, radical, and anion have, respectively, two, three, and four $\pi$ electrons. The bonding energy is

$$\begin{aligned} &\text{cation} \quad 2(\alpha + 2\beta) - 2\alpha = 4\beta \\ &\text{radical} \quad 2(\alpha + 2\beta) + (\alpha + \beta) - 3\alpha = 3\beta \\ &\text{anion} \quad 2(\alpha + 2\beta) + 2(\alpha - \beta) - 4\alpha = 2\beta. \end{aligned}$$

Compared to the energy of a double bond $2\beta$, the cation has $2\beta$ stabilization energy, while it reduces to $\beta$ in the radical, and it is zero in the anion. When we put $N = 0$ in the Hückel $4n + 2$ rule, we know 2 satisfies the Hückel rule. A two-membered ring is impossible, but the above example demonstrates that the Hückel rule holds for the electron number instead of the carbon number. The $2\pi$ electron system is aromatic, as the cyclopropenyl cation exhibits extraordinary stability. We will see in Chap. 7 that it is an important strategy in designing organic conductors to use molecules whose cationic or anionic form has extra stability due to the aromaticity.

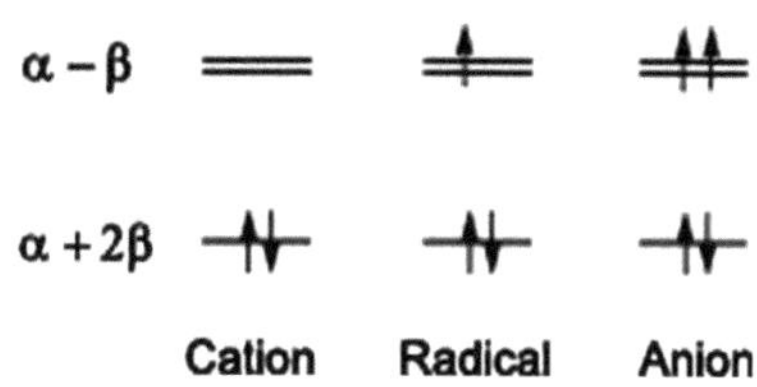


**Fig. 1.20** Energy levels and electron configuration of cyclopropenyl cation, radical, and anion

*Example 1.12* In cyclobutadiene, the π electrons are not entirely delocalized like benzene, but the double bonds and the single bonds are different (**bond alternation**). Assuming that the resonance integrals of the double and single bonds are, respectively, β and β′, calculate the energy levels and the delocalization energy according to the Hückel approximation.

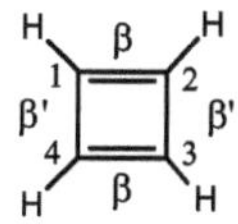


Following the atom numbering, the secular equation is

$$\begin{vmatrix} \alpha - E & \beta & 0 & \beta' \\ \beta & \alpha - E & \beta' & 0 \\ 0 & \beta' & \alpha - E & \beta \\ \beta' & 0 & \beta & \alpha - E \end{vmatrix} = 0.$$

Since this is again in the form $\begin{vmatrix} \mathrm{A} & \mathrm{B} \\ \mathrm{B} & \mathrm{A} \end{vmatrix} = 0$ for 2 × 2 determinants A and B, this is reduced to

$$\begin{vmatrix} \alpha - E & \beta + \beta' \\ \beta + \beta' & \alpha - E \end{vmatrix} \begin{vmatrix} \alpha - E & \beta - \beta' \\ \beta - \beta' & \alpha - E \end{vmatrix} = 0.$$

The solutions are $E = \alpha \pm (\beta \pm \beta')$ (Fig. 1.21). Since a double bond is shorter than a single bond, we have assumed $|\beta| > |\beta'|$. Because the degeneracy of $\alpha + \beta - \beta'$ and $\alpha - \beta + \beta'$ is lost, the spins are paired and the molecule is thus diamagnetic. The bonding energy is

$$2(\alpha + \beta + \beta') + 2(\alpha + \beta - \beta') - 4\alpha = 4\beta.$$

The HOMO energy is $|\beta - \beta'|$ lower than $\alpha$. Stabilization due to the bond alternation is $2|\beta - \beta'|$.

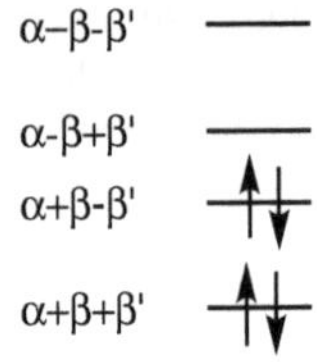


**Fig. 1.21** Energy levels of bond-alternating cyclobutadiene

We shall see in Chap. 2 that bond alternation corresponds to the Peierls transition. An $N = 4n + 2$ system shows stabilization due to the delocalization and the aromaticity. In contrast, an $N = 4n$ system is stabilized by the bond alternation.

## 1.6 Symmetry Orbital

Molecular orbital calculation is simplified considering the molecular symmetry. Here, we see how the molecular symmetry is used to calculate the molecular orbitals.

An ammonia molecule looks the same after a rotation of 120° around the central axis (Fig. 1.22a). Similarly, a benzene molecule has a 60° rotation, and a water molecule has a 180° rotation. When a molecule looks the same after a rotation through (360/*n*)°, the symmetry is called an ***n*-fold rotation**. An ammonia molecule has a threefold rotation axis ($C_3$), a benzene molecule has a sixfold rotation axis ($C_6$), and a water molecule has a twofold rotation axis ($C_2$).

A water molecule has another symmetry: a **reflection** in a **mirror** plane ($\sigma$ in Fig. 1.22(b)). The difference from the twofold axis is defined by seeing how the hydrogen coordinates are converted.

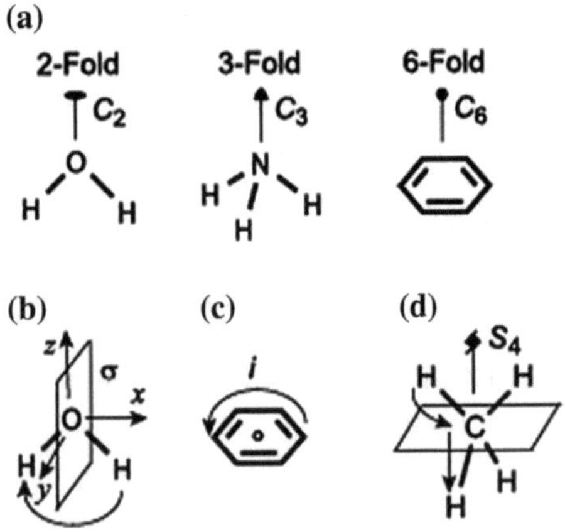


**Fig. 1.22** Symmetry elements. **a** Rotation axis **b** Mirror **c** Inversion **d** Improper rotation

$$C_2\,(x, y, z) \to (-x,\ -y, z)$$
$$\sigma\,(x, y, z) \to (-x, y, z)$$

A water molecule has another mirror plane.

$$\sigma'\,(x, y, z)\ \to\ (x, -y, z)$$

However, all atoms are located on this mirror plane, and do not move by the reflection. Similarly, a benzene molecule has a mirror plane on which all atoms are placed.

$$\sigma\,(x, y, z)\ \to\ (x, y, -z)$$

Here, $C_6$ is defined as the $z$ axis. A mirror is represented by $\sigma$ based on the German word Spiegel.

In the above symmetry operations, signs of one or two coordinates are changed. An operation to change three signs

$$i(x, y, z)\ \to\ (-x, -y, -z)$$

is called an **inversion** (Fig. 1.22c). A benzene molecule has an inversion center.

A tetrahedral molecule, for instance methane, has $C_2$ and $C_3$, but does not have $C_4$. However, after a 90° rotation around $C_2$, a reflection is performed through a mirror plane perpendicular to the rotation axis, and these successive transformations move the upper hydrogen to a lower hydrogen as shown in Fig. 1.22d. Such a combined symmetry operation is called a **fourfold improper rotation** ($S_4$). Successive transformations are necessary to cover all possibilities to move a hydrogen atom to the positions of all other equivalent hydrogen atoms. In an $n$-fold improper rotation axis ($S_n$), $n$ is even.

Figure 1.23a is a top view of an ammonia molecule. A threefold axis generates two rotations: a 120° rotation and a −120° rotation, which are designated as $C_3^+$ and $C_3^-$. These $C_3^+$ and $C_3^-$ are called **symmetry operations**, and the $C_3$ axis is called a **symmetry element**. Two successive $C_3^+$ operations result in another symmetry operation identical to $C_3^-$. This is designated as $C_3^+C_3^+ = C_3^-$. After three successive

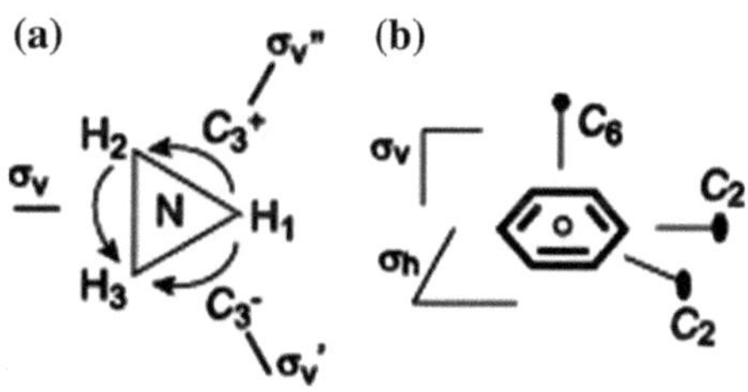


**Fig. 1.23** Symmetry in **a** an ammonia, and **b** a benzene molecule

$C_3^+$ operations, a hydrogen atom returns to the original position. An operation to keep all atoms at the original position is represented by $E$. Then, we can write $C_3^+C_3^+$ $C_3^+ = E$. An ammonia molecule has mirror planes $\sigma_v$ parallel to $C_3$, where v means vertical. We can easily see $\sigma_v\sigma_v = E$. Like this, successive symmetry operations lead to another symmetry operation. Thus, the symmetry operations make a **group**. A mathematical group satisfies the following criteria.

(1) Successive operations result in another element of the group.
(2) The associative law $a(bc) = (ab)c$ holds.
(3) The group includes the identity element $E$.
(4) Any element $a$ has the inverse element $a^{-1}$ satisfying $aa^{-1} = a^{-1}\,a = E$.

A collection of the symmetry operations of an ammonia molecule, $E$, $C_3^+$, $C_3^-$, $\sigma_v$, $\sigma_v'$, and $\sigma_v''$, makes a group. A group of symmetry operations of a molecule is called **point group**, because the symmetry operations do not move at least one point. For instance, a benzene molecule have $C_6$, $\sigma_h$ (h means horizontal), three $C_2$ passing carbon atoms, three $C_2'$ as perpendicular bisectors of the bonds, three $\sigma_v$ involving $C_2$, three $\sigma_v'$ involving $C_2'$, $E$, and $i$ (Fig. 1.23b).

The point group of an ammonia molecule is different from that of a benzene molecule. Different point groups are labeled by the Schönflies system, which are determined according to Fig. 1.24. First, the rotation axis with maximum $n$ is

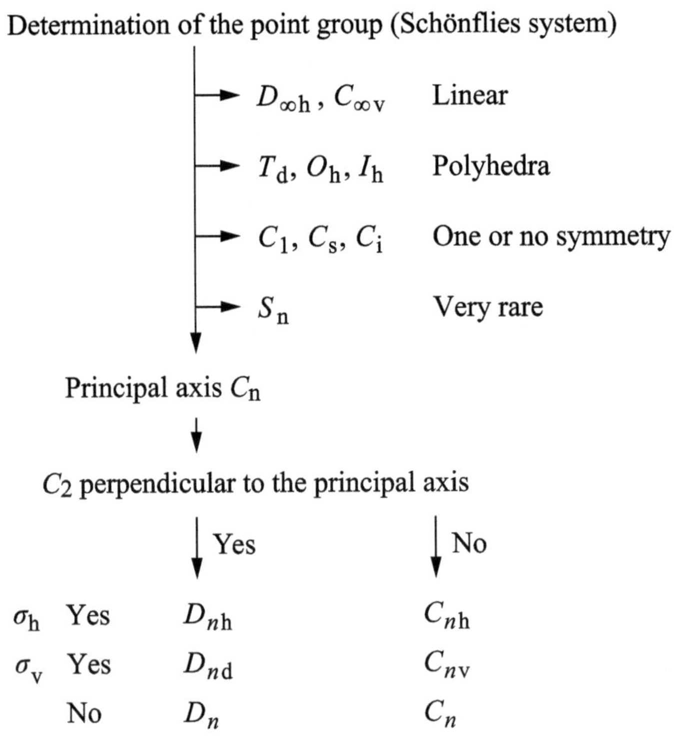


**Fig. 1.24** Determination of the Schönflies system

defined as the **principal axis**. Second, the existence of $C_2$ perpendicular to the principal axis is investigated. Because a benzene molecule has such $C_2$, benzene temporarily falls into the category of $D_6$. Third, the existence of $\sigma_h$ is examined. Since a benzene molecule has $\sigma_h$, the point group of benzene is determined to be $D_{6h}$. When $\sigma_h$ does not exist, the existence of $\sigma_v$ is examined. For instance, an ammonia molecule has $C_3 \rightarrow C_2$ no $\rightarrow \sigma_h$ no $\rightarrow \sigma_v$ yes, to lead to $C_{3v}$. Similarly, a water molecule belongs to the group $C_{2v}$.

Some molecules are classified independent of such a principal axis criterion. Linear molecules have a $\infty$-fold rotation axis, because it can be rotated by an infinitesimally small angle. Depending on the existence of $\sigma_h$, linear molecules are classified to $D_{\infty h}$ or $C_{\infty v}$. For example, $CO_2$ belongs to $D_{\infty h}$ and CO belongs to $C_{\infty v}$. Polyhedrons are classified to $T_d$ for tetrahedron, $O_h$ for octahedron and cube, and $I_h$ for dodecahedron and icosahedron. Some molecules do not have any principal axis. When a mirror is the only symmetry, it is $C_s$, and when an inversion is the only element, it is $C_i$. A group without any symmetry element is designated as $C_1$.

*Example 1.13* Identify the point groups to which the following molecules belong. Assume the skeleton to be a square.

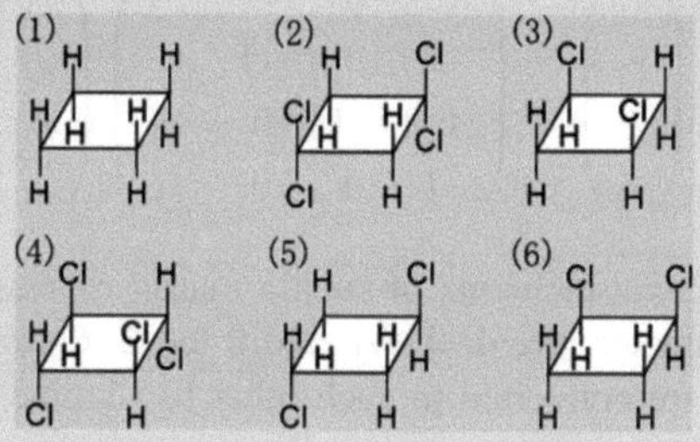


(1) $D_{4h}$, (2) $D_{2h}$, (3) $C_{2v}$, (4) $D_{2d}$, (5) $C_{2h}$, (6) $C_s$.

A butadiene molecule belongs to the $C_{2h}$ point group (Fig. 1.25), which has symmetry elements: $E$, $C_2$, $\sigma_h$, and $i$. Here, $C_2$ is perpendicular to the molecular plane, and $\sigma_h$ includes the molecular plane. The four π orbitals $\chi_1 \sim \chi_4$ transform as follows.

$$E\begin{pmatrix}\chi_1\\\chi_2\\\chi_3\\\chi_4\end{pmatrix}=\begin{pmatrix}1&0&0&0\\0&1&0&0\\0&0&1&0\\0&0&0&1\end{pmatrix}\begin{pmatrix}\chi_1\\\chi_2\\\chi_3\\\chi_4\end{pmatrix},\quad C_2\begin{pmatrix}\chi_1\\\chi_2\\\chi_3\\\chi_4\end{pmatrix}=\begin{pmatrix}0&0&0&1\\0&0&1&0\\0&1&0&0\\1&0&0&0\end{pmatrix}\begin{pmatrix}\chi_1\\\chi_2\\\chi_3\\\chi_4\end{pmatrix}$$

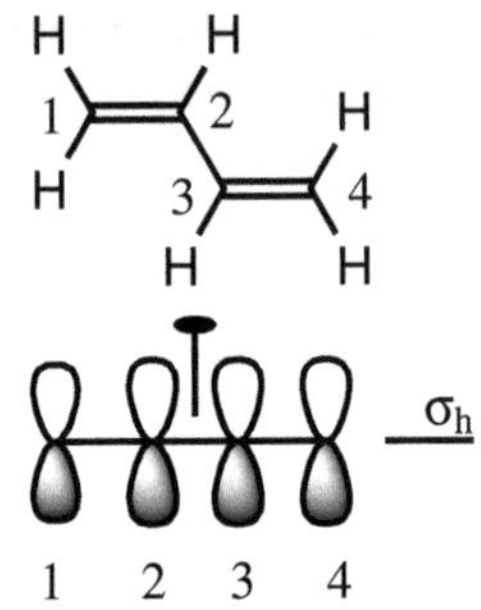


**Fig. 1.25** π orbitals of a butadiene molecule

Such a **matrix representation** implies the $C_2$ rotation moves $\chi_1$ to $\chi_4$. Other operations are similarly represented.

$$\sigma_h\begin{pmatrix}\chi_1\\\chi_2\\\chi_3\\\chi_4\end{pmatrix}=\begin{pmatrix}-1&0&0&0\\0&-1&0&0\\0&0&-1&0\\0&0&0&-1\end{pmatrix}\begin{pmatrix}\chi_1\\\chi_2\\\chi_3\\\chi_4\end{pmatrix},$$

$$i\begin{pmatrix}\chi_1\\\chi_2\\\chi_3\\\chi_4\end{pmatrix}=\begin{pmatrix}0&0&0&-1\\0&0&-1&0\\0&-1&0&0\\-1&0&0&0\end{pmatrix}\begin{pmatrix}\chi_1\\\chi_2\\\chi_3\\\chi_4\end{pmatrix}.$$

The sum of the diagonal elements of such a matrix representation is called **character**. Here, the characters are 4, 0, −4, and 0 for $E$, $C_2$, $\sigma_h$, and $i$.

Since $\chi_1$ and $\chi_4$ are converted to each other by symmetry operations, we shall consider two functions, $\psi_1 = (\chi_1 + \chi_4)/\sqrt{2}$ and $\psi_2 = (\chi_1 - \chi_4)/\sqrt{2}$. These functions are transformed as follows.

$$E\begin{pmatrix}\psi_1\\\psi_2\end{pmatrix}=\begin{pmatrix}1&0\\0&1\end{pmatrix}\begin{pmatrix}\psi_1\\\psi_2\end{pmatrix},\quad C_2\begin{pmatrix}\psi_1\\\psi_2\end{pmatrix}=\begin{pmatrix}1&0\\0&1\end{pmatrix}\begin{pmatrix}\psi_1\\\psi_2\end{pmatrix},$$

$$\sigma_h\begin{pmatrix}\psi_1\\\psi_2\end{pmatrix}=\begin{pmatrix}-1&0\\0&-1\end{pmatrix}\begin{pmatrix}\psi_1\\\psi_2\end{pmatrix},\quad i\begin{pmatrix}\psi_1\\\psi_2\end{pmatrix}=\begin{pmatrix}-1&0\\0&-1\end{pmatrix}\begin{pmatrix}\psi_1\\\psi_2\end{pmatrix}.$$

$\psi_1$ and $\psi_2$ do not mix with each other. Therefore, the matrix element is represented by a one-dimensional form: $i\,\psi_1 = -\psi_1$. So the characters 1, 1, −1, and −1 indicate whether $\psi_1$ is an even or odd function of the symmetry operations. Because $\psi_1$ cannot be simplified, it is called an **irreducible representation**. By contrast, a function such as $\chi_1 \sim \chi_4$, which transforms to each other by symmetry operations, is called a **reducible representation**. A reducible representation is **reduced** to irreducible representations.

**Table 1.2** Character table of $C_{2h}$

| $C_{2h}$ | $E$ | $C_2$ | $\sigma_h$ | $i$ |
|---|---|---|---|---|
| $A_g$ | 1 | 1 | 1 | 1 |
| $A_u$ | 1 | 1 | -1 | -1 |
| $B_g$ | 1 | -1 | -1 | 1 |
| $B_u$ | 1 | -1 | 1 | -1 |

To investigate an irreducible representation of a point group, we use a character table. For example, the character table of the $C_{2h}$ point group is shown in Table 1.2 [5]. The character table shows whether a given function is even or odd for the symmetry operations. For example, a function belonging to the irreducible representation $A_g$ is even for all symmetry operations. An irreducible representation provides a classification of the symmetry. Irreducible representations are named according to the following rule:

| | | |
|---|---|---|
| A | Even for the principal rotation | one-dimensional |
| B | Odd for the principal rotation | one-dimensional |
| E | two-dimensional | |
| T | three-dimensional | |

A two-dimensional irreducible representation appears when the matrix is not simplified lower than two-dimensional. Such a case happens when the molecule has a rotation of more than threefold. The subscript is identified as follows:

| | | |
|---|---|---|
| g | gerade | even for the inversion |
| u | ungerade | odd for the inversion |

where gerade and ungerade are German words. Parity for the symmetry operation with the next priority to the principal axis is designated by the subscript as $A_1$ and $A_2$.

A reducible representation like $\chi_1 \sim \chi_4$ is reduced in the following way.

The number of the irreducible representations $a_i$ included in a reducible representation with characters $\chi(R)$, is given by

$$a_i = \frac{1}{h} \sum_R \underline{\chi}(R) \underline{\chi}_i(R) \tag{1.31}$$

where $R$ represents the symmetry operations. $\underline{\chi}_i(R)$ is the characters of the irreducible representation $i$ taken from the character table, and $\chi(R)$ is the characters of the reducible representation. $h$ is the number of the symmetry operations. Here, characters are underlined as $\underline{\chi}$ in order to be distinguished from atomic orbitals.

For instance, characters of $\chi_1 \sim \chi_4$ are 4, 0, −4, and 0. The "inner products" with the characters in Table 1.1 afford

$$\begin{aligned}
&A_g \quad \frac{1}{4}(4\times 1 + 0\times 1 + (-4)\times 1 + 0\times 1) = 0\\
&A_u \quad \frac{1}{4}(4\times 1 + 0\times 1 + (-4)\times(-1) + 0\times(-1)) = 2\\
&B_g \quad \frac{1}{4}(4\times 1 + 0\times(-1) + (-4)\times(-1) + 0\times 1) = 2\\
&B_u \quad \frac{1}{4}(4\times 1 + 0\times(-1) + (-4)\times 1 + 0\times(-1)) = 0.
\end{aligned}$$

Accordingly, $\chi_1 \sim \chi_4$ are reduced to $2A_u + 2B_g$.

Irreducible orbitals like $\psi_1$ and $\psi_2$ are called **symmetry orbitals**. We can obtain symmetry orbitals as follows.

$$\psi_i = \sum_R \underline{\chi}_i(R) R\chi_j \tag{1.32}$$

First, the transformations of $\chi_1 \sim \chi_4$ are listed as shown in Table 1.3. This table represents $R\chi_j$. The "inner products" with the character table afford

$$\begin{aligned}
A_u \quad & 1\times\chi_1 + 1\times\chi_4 + (-1)\times\chi_1 + (-1)\times\chi_4 \to \psi_1 = (\chi_1 + \chi_4)/\sqrt{2}\\
& 1\times\chi_2 + 1\times\chi_3 + (-1)\times\chi_2 + (-1)\times\chi_3 \to \psi_3 = (\chi_2 + \chi_3)/\sqrt{2}
\end{aligned}$$

$$\begin{aligned}
B_g \quad & 1\times\chi_1 + (-1)\times\chi_4 + (-1)\times\chi_1 + 1\times\chi_4 \to \psi_2 = (\chi_1 - \chi_4)/\sqrt{2}\\
& 1\times\chi_2 + (-1)\times\chi_3 + (-1)\times\chi_2 + 1\times\chi_3 \to \psi_4 = (\chi_2 - \chi_3)/\sqrt{2}.
\end{aligned}$$

These symmetry orbitals are normalized by multiplying $1/\sqrt{2}$, because they are composed of two atomic orbitals.

Symmetry of a product $\psi_1\psi_2$ is given by the products of the characters. For instance

| | | | | | |
|---|---|---|---|---|---|
| $\psi_1$ | $A_u$ | 1 | 1 | −1 | −1 |
| $\psi_2$ | $B_g$ | 1 | −1 | −1 | 1 |

**Table 1.3** Transformations of the atomic orbitals in butadiene

| $C_{2h}$ | $E$ | $C_2$ | $\sigma_h$ | $i$ |
|---|---|---|---|---|
| $\chi_1$ | $\chi_1$ | $\chi_4$ | $-\chi_1$ | $-\chi_4$ |
| $\chi_2$ | $\chi_2$ | $\chi_3$ | $-\chi_2$ | $-\chi_3$ |
| $\chi_3$ | $\chi_3$ | $\chi_2$ | $-\chi_3$ | $-\chi_2$ |
| $\chi_4$ | $\chi_4$ | $\chi_1$ | $-\chi_4$ | $-\chi_1$ |

lead to

| | | | | | |
|---|---|---|---|---|---|
| $\psi_1\psi_2$ | | 1 | −1 | 1 | −1 → $B_u$ |

so $\psi_1\psi_2$ belongs to $B_u$. This is obvious because 1 and −1 represent even and odd functions.

An integral over the whole space $\int \psi_1\psi_2 d\tau$ is nonzero only when $\psi_1\psi_2$ involves the total symmetry irreducible representation $A_g$. This is obvious because a single odd direction makes the integral zero. When $\psi_1\psi_2$ is reducible and contains $A_g$ with other irreducible representations, it gives a nonzero integral. This is expressed as "$\psi_1\psi_2$ involves $A_g$."

An integral over the whole space $\int \psi_1\psi_2 d\tau$ is zero when $\psi_1$ and $\psi_2$ belong to different irreducible representations. This is because an odd direction makes the integral zero. In contrast, an integral over the whole space $\int \psi_1\psi_1' d\tau$ for $\psi_1$ and $\psi_1'$ belonging to the same irreducible representation is nonzero. This is because 1 × 1 and (−1) × (−1) are 1, and the irreducible representation of $\psi_1\psi_1'$ is $A_g$.

For the molecular orbitals of butadiene, instead of the LCAO $\Phi = c_1\chi_1 + c_2\chi_2 + c_3\chi_3 + c_4\chi_4$, we shall adopt a linear combination of the symmetry orbitals $\Phi = a_1\psi_1 + a_2\psi_2 + a_3\psi_3 + a_4\psi_4$. Since integrals between different irreducible representations are zero, the secular equation is divided to two 2 × 2 blocks; one is for the $A_u$ part, and another is for the $B_g$ part. The $A_u$ part concerning to $\psi_1$ and $\psi_3$ is

$$\begin{vmatrix} \int \frac{\chi_1+\chi_4}{\sqrt{2}} H \frac{\chi_1+\chi_4}{\sqrt{2}} d\tau - E & \int \frac{\chi_1+\chi_4}{\sqrt{2}} H \frac{\chi_2+\chi_3}{\sqrt{2}} d\tau \\ \int \frac{\chi_1+\chi_4}{\sqrt{2}} H \frac{\chi_2+\chi_3}{\sqrt{2}} d\tau & \int \frac{\chi_2+\chi_3}{\sqrt{2}} H \frac{\chi_2+\chi_3}{\sqrt{2}} d\tau - E \end{vmatrix} = 0.$$

The respective matrix elements are calculated as follows.

Line 1 Column 1 $\frac{1}{2}(\int \chi_1 H\chi_1 d\tau + \int \chi_4 H\chi_4 d\tau) - E = \alpha - E$

Line 1 Column 2 $\frac{1}{2}(\int \chi_1 H\chi_2 d\tau + \int \chi_3 H\chi_4 d\tau) = \beta$

Line 2 Column 2 $\frac{1}{2}(\int \chi_2 H\chi_2 d\tau + \int \chi_3 H\chi_3 d\tau + 2\int \chi_2 H\chi_3 d\tau) - E = \alpha + \beta - E$

The resulting secular equation is

$$\begin{vmatrix} \alpha - E & \beta \\ \beta & \alpha + \beta - E \end{vmatrix} = 0.$$

Substituting $x = (\alpha - E)/\beta$, this leads to

$$\begin{vmatrix} x & 1 \\ 1 & x+1 \end{vmatrix} = 0$$

which is reduced to

$$x^2 + x - 1 = 0.$$

The solutions are

$$x = \frac{-1 \pm \sqrt{5}}{2}.$$

For $B_g$, we similarly obtain

$$\begin{vmatrix} \alpha - E & \beta \\ \beta & \alpha - \beta - E \end{vmatrix} = 0.$$

This leads to

$$\begin{vmatrix} x & 1 \\ 1 & x - 1 \end{vmatrix} = 0$$

which is reduced to

$$x^2 - x - 1 = 0.$$

The solutions are

$$x = \frac{1 \pm \sqrt{5}}{2}.$$

The resulting molecular orbitals are depicted in Fig. 1.26. This is identical to Fig. 1.13, but the molecular orbitals are labeled by the irreducible representations. In general, molecular orbitals belong to an irreducible representation. Alternatively, the irreducible representation of a molecular orbital is determined by examining the even/odd properties of the molecular orbitals, and comparing the obtained characters with the character table (Table 1.1).

In butadiene, the solution of the 4 × 4 matrix (Sect. 1.5.3) looks simpler than the solution of the symmetry orbitals. However, symmetry orbitals are necessary to calculate more complicated molecular orbitals.

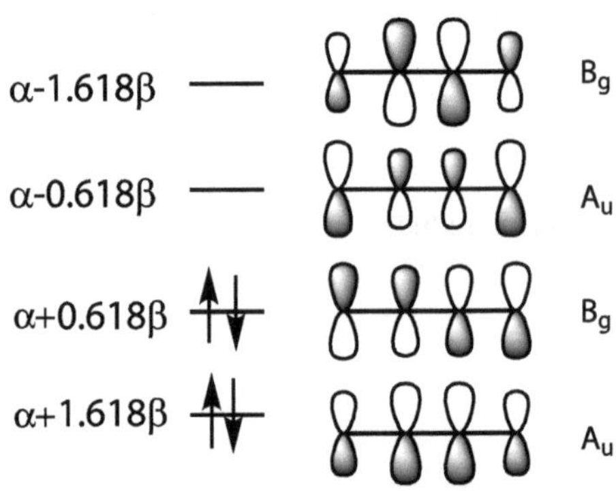


**Fig. 1.26** Molecular orbitals of butadiene

*Example 1.14* Using the symmetry orbitals, calculate molecular orbitals of naphthalene according to the Hückel approximation. Since the decomposition is simplified by considering a part of the symmetry operations, we consider a partial point group $D_2$ instead of the real point group $D_{2h}$. The character table is shown in Table 1.4.

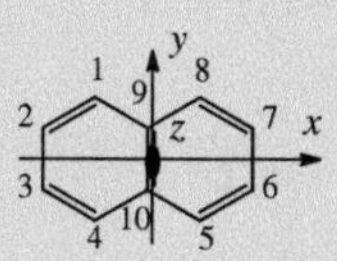


**Table 1.4** Character table of $D_2$

| $D_2$ | $E$ | $C_2(x)$ | $C_2(y)$ | $C_2(z)$ |
|---|---|---|---|---|
| $A$ | 1 | 1 | 1 | 1 |
| $B_1$ | 1 | −1 | −1 | 1 |
| $B_2$ | 1 | −1 | 1 | −1 |
| $B_3$ | 1 | 1 | −1 | −1 |

(1) It is difficult to solve a 10 × 10 secular equation, but the calculation is remarkably simplified by considering the symmetry. The atom numbering and the $x$, $y$, and $z$ axes are defined in the figure. First, we investigate how $\chi_i$ moves under the symmetry operations $C_2$, as shown in Table 1.5. The number of staying atoms affords the characters $\chi$ as 10, 0, −2, and 0.

**Table 1.5** Transformation of atomic orbitals in naphthalene

| | $E$ | $C_2(x)$ | $C_2(y)$ | $C_2(y)$ |
|---|---|---|---|---|
| | $\chi_1$ | $-\chi_4$ | $-\chi_8$ | $\chi_5$ |
| | $\chi_2$ | $-\chi_3$ | $-\chi_7$ | $\chi_6$ |
| | $\chi_3$ | $-\chi_2$ | $-\chi_6$ | $\chi_7$ |
| | $\chi_4$ | $-\chi_1$ | $-\chi_5$ | $\chi_8$ |
| | $\chi_5$ | $-\chi_8$ | $-\chi_4$ | $\chi_1$ |
| | $\chi_6$ | $-\chi_7$ | $-\chi_3$ | $\chi_2$ |
| | $\chi_7$ | $-\chi_6$ | $-\chi_2$ | $\chi_3$ |
| | $\chi_8$ | $-\chi_5$ | $-\chi_1$ | $\chi_4$ |
| | $\chi_9$ | $-\chi_{10}$ | $-\chi_9$ | $\chi_{10}$ |
| | $\chi_{10}$ | $-\chi_9$ | $-\chi_{10}$ | $\chi_9$ |
| $\chi$ | 10 | 0 | −2 | 0 |

(2) Eq. 1.31 is calculated from these characters and the character Table 1.4.

$$A \quad \frac{1}{4}(10\times 1+(-2)\times 1)=2$$

$$B_1 \quad \frac{1}{4}(10\times 1+(-2)\times(-1))=3$$

$$B_2 \quad \frac{1}{4}(10\times 1+(-2)\times 1)=2$$

$$B_3 \quad \frac{1}{4}(10\times 1+(-2)\times(-1))=3.$$

Then, the orbitals are reduced to 2 $A$ + 3 $B_1$ + 2 $B_2$ + 3 $B_3$.

(3) The symmetry orbitals are obtained by multiplying the characters in Table 1.4 to the transformed orbitals $R\chi_j$ in Table 1.5.

$$A \quad \frac{1}{2}(\chi_1-\chi_4+\chi_5-\chi_8),\ \frac{1}{2}(\chi_2-\chi_3+\chi_6-\chi_7)$$

$$B_1 \quad \frac{1}{2}(\chi_1+\chi_4+\chi_5+\chi_8),\ \frac{1}{2}(\chi_2+\chi_3+\chi_6+\chi_7),\ \frac{1}{\sqrt{2}}(\chi_9+\chi_{10})$$

$$B_2 \quad \frac{1}{2}(\chi_1+\chi_4-\chi_5-\chi_8),\ \frac{1}{2}(\chi_2+\chi_3-\chi_6-\chi_7)$$

$$B_3 \quad \frac{1}{2}(\chi_1-\chi_4-\chi_5+\chi_8),\ \frac{1}{2}(\chi_2-\chi_3-\chi_6+\chi_7),\ \frac{1}{\sqrt{2}}(\chi_9-\chi_{10})$$

The normalization factor derived from

$$\int(\chi_1+\chi_4+\chi_5+\chi_8)(\chi_1+\chi_4+\chi_5+\chi_8)\mathrm{d}\tau=$$
$$\int\chi_1\chi_1\mathrm{d}\tau+\int\chi_4\chi_4\mathrm{d}\tau+\int\chi_5\chi_5\mathrm{d}\tau+\int\chi_8\chi_{81}\mathrm{d}\tau=4$$

is $\frac{1}{2}$ (or $\frac{1}{\sqrt{2}}$ for an orbital composed of two atomic orbitals). According to the ordinary Hückel approximation, the overlap integrals $S = \int \chi_1 \chi_2 \mathrm{d}\tau$ are assumed to be zero. The symmetry orbitals are constructed by summing the equivalent orbitals: (1, 4, 5, 8), (2, 3, 6, 7), and (9, 10) by multiplying $\pm 1$.

(4) For instance, the secular equation for A is

$$\begin{vmatrix} \alpha - E & \beta \\ \beta & \alpha - \beta - E \end{vmatrix} = 0.$$

This relation is obtained as follows.

$$\begin{aligned} M_{11} &= \int \frac{1}{2}(\chi_1 - \chi_4 + \chi_5 - \chi_8) H \frac{1}{2}(\chi_1 - \chi_4 + \chi_5 - \chi_8)\mathrm{d}\tau \\ &= \frac{1}{4}\left[\int \chi_1 H \chi_1 \mathrm{d}\tau + \int \chi_4 H \chi_4 \mathrm{d}\tau + \int \chi_5 H \chi_5 \mathrm{d}\tau + \int \chi_8 H \chi_8 \mathrm{d}\tau\right] = \frac{4\alpha}{4} = \alpha \end{aligned}$$

$$\begin{aligned} M_{12} &= \int \frac{1}{2}(\chi_1 - \chi_4 + \chi_5 - \chi_8) H \frac{1}{2}(\chi_2 - \chi_3 + \chi_6 - \chi_7)\mathrm{d}\tau \\ &= \frac{1}{4}\left[\int \chi_1 H \chi_2 \mathrm{d}\tau + \int \chi_4 H \chi_3 \mathrm{d}\tau + \int \chi_5 H \chi_6 \mathrm{d}\tau + \int \chi_8 H \chi_7 \mathrm{d}\tau\right] = \frac{4\beta}{4} = \beta \end{aligned}$$

$$\begin{aligned} M_{22} &= \int \frac{1}{2}(\chi_2 - \chi_3 + \chi_6 - \chi_7) H \frac{1}{2}(\chi_2 - \chi_3 + \chi_6 - \chi_7)\mathrm{d}\tau \\ &= \frac{1}{4}\left[\int \chi_2 H \chi_2 \mathrm{d}\tau + \int \chi_3 H \chi_3 \mathrm{d}\tau + \int \chi_6 H \chi_6 \mathrm{d}\tau \right.\\ &\quad \left. + \int \chi_7 H \chi_7 \mathrm{d}\tau - 2\int \chi_2 H \chi_3 \mathrm{d}\tau - 2\int \chi_6 H \chi_7 \mathrm{d}\tau\right] \\ &= \frac{4\alpha - 4\beta}{4} = \alpha - \beta \end{aligned}$$

Here, we put $x = (\alpha - E)/\beta$ to obtain $\begin{vmatrix} x & 1 \\ 1 & x-1 \end{vmatrix} = 0$ leading to $x(x-1) - 1 = 0$.

By solving $x^2 - x - 1 = 0$, we obtain $x = \frac{1 \pm \sqrt{5}}{2} = -0.618,\ 1.618$.

In the same way

$B_1$ $\begin{vmatrix} x & 1 & \sqrt{2} \\ 1 & x+1 & 0 \\ \sqrt{2} & 0 & x+1 \end{vmatrix} = 0$ leads to $x(x+1)^2 - 2(x+1) - (x+1) = 0$.

$(x^2 + x - 3)\ (x+1) = 0$ is solved as $x = -1, \frac{-1 \pm \sqrt{13}}{2} = -2.303, -1, 1.303$.

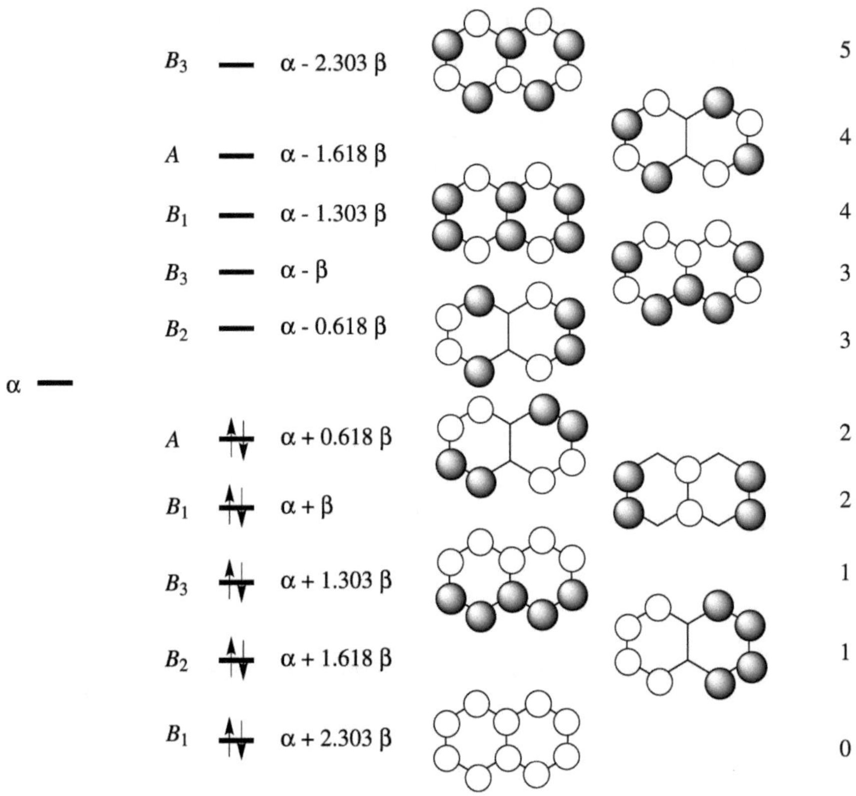


**Fig. 1.27** Molecular orbitals of naphthalene

$B_2$ $\begin{vmatrix} x & 1 \\ 1 & x+1 \end{vmatrix} = 0$ leads to $x(x+1) - 1 = 0$.

$x^2 + x - 1 = 0$ is solved as $x = \frac{-1 \pm \sqrt{5}}{2} = -1.618, 0.618$.

$B_3$ $\begin{vmatrix} x & 1 & \sqrt{2} \\ 1 & x-1 & 0 \\ \sqrt{2} & 0 & x-1 \end{vmatrix} = 0$ leads to $x(x-1)^2 - 2(x-1) - (x-1) = 0$.

$(x^2 - x - 3)(x - 1) = 0$ is solved as $x = 1, \frac{1 \pm \sqrt{13}}{2} = -1.303, 1, 2.303$.

(5) In summary, the energy levels are

$$
\begin{array}{lllll}
E = \alpha + 2.303\beta, & \alpha + 1.618\beta, & \alpha + 1.303\beta, & \alpha + \beta, & \alpha + 0.618\beta, \\
B_1 & B_2 & B_3 & B_1 & A \\
\alpha - 0.618\beta, & \alpha - \beta, & \alpha - 1.303\beta, & \alpha - 1.618\beta, & \alpha - 2.303\beta \\
B_2 & B_3 & B_1 & A & B_3
\end{array}
$$

(Figure 1.27). The π bonding energy is

$$2(2.303 + 1.618 + 1.303 + 1 + 0.618)\beta = 13.664\beta.$$

Because five double bonds have $10\beta$ bonding energy, the stabilization by delocalization is $3.664\beta$.

(6) In order to obtain the shape of the molecular orbitals, we have to solve the original simultaneous equations. However, the molecular orbitals are derived from the symmetry orbitals, and the number of nodes increases one by one. We can therefore depict the molecular orbitals as shown in Fig. 1.27. Two orbitals ($B_2$ and $B_3$) have one node, but one is vertical and the other is horizontal. Similarly, there are, respectively, two ways to introduce two, three, and four nodes.

*Example 1.15* A cyclopentadienyl anion, $C_5H_5^-$, has a pentagonal shape (Figure (a)), because the negative charge is delocalized to five carbon atoms. Calculate molecular orbitals according to the Hückel approximation using the symmetry orbitals. This molecule has $D_{5h}$ symmetry, but the $C_2$ symmetry is sufficient to simplify the orbitals. The character table is given in Table 1.6.

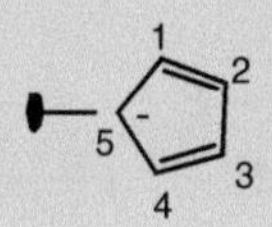


**Table 1.6** Character table of $C_2$

| $C_2$ | $E$ | $C_2$ |
|---|---|---|
| A | 1 | 1 |
| B | 1 | -1 |

(1) It is a tough calculation to solve a $5 \times 5$ secular equation without using the symmetry. Each atomic orbital moves by $C_2$ as shown in Table 1.7. The characters are 5 and $-1$.

**Table 1.7** Transformation of atomic orbitals in a cyclopentadienyl anion

| $E$ | $C_2$ |
|---|---|
| $\chi_1$ | $-\chi_4$ |
| $\chi_2$ | $-\chi_3$ |
| $\chi_3$ | $-\chi_2$ |
| $\chi_4$ | $-\chi_1$ |
| $\chi_5$ | $-\chi_5$ |
| 5 | -1 |

(2) Eq. 1.31 affords the irreducible representations.

$$\text{A}\quad \frac{1}{2}(1 \times 5 + 1 \times (-1)) = 2$$
$$\text{B}\quad \frac{1}{2}(1 \times 5 + (-1) \times (-1)) = 3$$

Then, the matrix is reduced to 2A + 3B.

(3) From the inner products of Tables (b) and (c), the symmetry orbitals are obtained.

$$\text{A}\quad \frac{\chi_1 - \chi_4}{\sqrt{2}}, \frac{\chi_2 - \chi_3}{\sqrt{2}}$$
$$\text{B}\quad \frac{\chi_1 + \chi_4}{\sqrt{2}}, \frac{\chi_2 + \chi_3}{\sqrt{2}}, \chi_5.$$

(4) The secular equation for A is

$$\begin{vmatrix} \int \frac{\chi_1-\chi_4}{\sqrt{2}} H \frac{\chi_1-\chi_4}{\sqrt{2}} \mathrm{d}\tau - E & \int \frac{\chi_1-\chi_4}{\sqrt{2}} H \frac{\chi_2-\chi_3}{\sqrt{2}} \mathrm{d}\tau \\ \int \frac{\chi_1-\chi_4}{\sqrt{2}} H \frac{\chi_2-\chi_3}{\sqrt{2}} \mathrm{d}\tau & \int \frac{\chi_2-\chi_3}{\sqrt{2}} H \frac{\chi_2-\chi_3}{\sqrt{2}} \mathrm{d}\tau - E \end{vmatrix} = 0$$

which leads to

$$\begin{vmatrix} \alpha - E & \beta \\ \beta & \alpha - \beta - E \end{vmatrix} = 0.$$

Substituting $x = (\alpha - E)/\beta$, we obtain

$$\begin{vmatrix} x & 1 \\ 1 & x - 1 \end{vmatrix} = 0$$

which gives

$$x^2 - x - 1 = 0.$$

The solutions are

$$x = \frac{1 \pm \sqrt{5}}{2}.$$

Similarly, the secular equation for B is

$$\begin{vmatrix} \int \frac{\chi_1+\chi_4}{\sqrt{2}} H \frac{\chi_1+\chi_4}{\sqrt{2}} \mathrm{d}\tau - E & \int \frac{\chi_1+\chi_4}{\sqrt{2}} H \frac{\chi_2+\chi_3}{\sqrt{2}} \mathrm{d}\tau & \int \frac{\chi_1+\chi_4}{\sqrt{2}} H \chi_5 \mathrm{d}\tau \\ \int \frac{\chi_1+\chi_4}{\sqrt{2}} H \frac{\chi_2+\chi_3}{\sqrt{2}} \mathrm{d}\tau & \int \frac{\chi_2+\chi_3}{\sqrt{2}} H \frac{\chi_2+\chi_3}{\sqrt{2}} \mathrm{d}\tau - E & \int \frac{\chi_2+\chi_3}{\sqrt{2}} H \chi_5 \mathrm{d}\tau \\ \int \frac{\chi_1+\chi_4}{\sqrt{2}} H \chi_5 \mathrm{d}\tau & \int \frac{\chi_2+\chi_3}{\sqrt{2}} H \chi_5 \mathrm{d}\tau & \int \chi_5 H \chi_5 \mathrm{d}\tau - E \end{vmatrix} = 0$$

which leads to

$$\begin{vmatrix} \alpha - E & \beta & \sqrt{2}\beta \\ \beta & \alpha + \beta - E & 0 \\ \sqrt{2}\beta & 0 & \alpha - E \end{vmatrix} = 0.$$

Substituting $x = (\alpha - E)/\beta$, we obtain

$$\begin{vmatrix} x & 1 & \sqrt{2} \\ 1 & x+1 & 0 \\ \sqrt{2} & 0 & x \end{vmatrix} = 0$$

which gives

$$x^3 + x^2 - 3x - 2 = 0.$$

Since $-2$ is a solution, this equation is easily factorized.

$$(x+2)(x^2 - x - 1) = 0.$$

The solutions are

$$x = -2 \quad \text{and} \quad \frac{1 \pm \sqrt{5}}{2}.$$

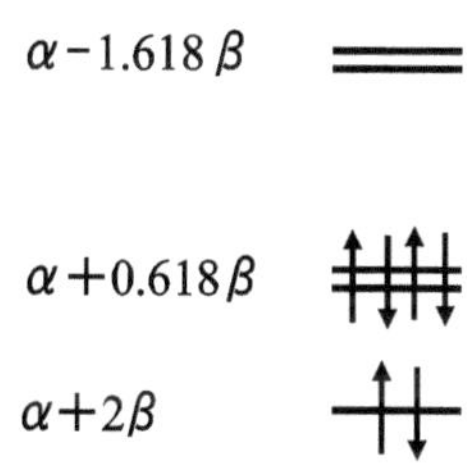


**Fig. 1.28** Energy levels of a cyclopentadienyl anion

(5) The energy levels are shown in Fig. 1.28. The anion has six π electrons. The π bonding energy is

$$2(\alpha + 2\beta) + 4(\alpha + 0.618\beta) - 6\alpha = 6.472\beta.$$

Since this is larger than the energy of three double bonds $6\beta$, the molecule is stabilized by the delocalization. This is attributed to the aromaticity of the six π electrons. This anion is discussed again in Example 2.1.

## 1.7 Properties of Eigenvalues and Eigenfunctions

We shall discuss the mathematical background of quantum mechanics. First, we shall investigate the basic relations of eigenvalues and eigenfunctions. In quantum mechanics, physical quantities such as energy and momentum are represented by operators. Such operators contain differentiation with respect to $x$ and $t$. For instance, the kinetic energy involves second-order differentiation for $x$. A little more complicated example is angular momentum, which is $\mathbf{L} = \mathbf{r} \times \mathbf{p}$ in the classical mechanics. Accordingly, the $z$ component in quantum mechanics is

$$L_z = xp_y - p_x y = \frac{\hbar}{i}\left(x\frac{\partial}{\partial y} - y\frac{\partial}{\partial x}\right).$$

These are all **linear operators F** that satisfy the following relations.

$$\mathbf{F}(\psi_1 + \psi_2) = \mathbf{F}\psi_1 + \mathbf{F}\psi_2$$

$$\mathbf{F}(c\psi_1) = c\mathbf{F}\psi_1.$$

Therefore, we can make a new state by superposition of the waves $\psi = c_1\psi_1 + c_2\psi_2 + c_3\psi_3 + \cdots$

When differentiated, the imaginary unit $i$ is multiplied. So the operators **F** appearing in quantum mechanics are the **Hermitian operators** satisfying

$$\int \psi_1^* \mathbf{F}\psi_2 \mathrm{d}\tau = \int \psi_2(\mathbf{F}^*\psi_1^*)\mathrm{d}\tau$$

where $\mathbf{F}^*$ and $\psi^*$ are complex conjugates of **F** and $\psi$, respectively. It is characteristic of the Hermitian operators that the eigenvalues are real. In order to prove this, we note

$$\mathbf{F}\psi = f\psi, \quad \mathbf{F}^*\psi^* = f^*\psi^*$$

which leads to

$$\int \psi^*\mathbf{F}\psi \mathrm{d}\tau = f \int \psi^*\psi \mathrm{d}\tau = f.$$

However, since **F** is a Hermitian, this is equal to

$$\int \psi(\mathbf{F}^*\psi^*)\mathrm{d}\tau = \int \psi(f^*\psi^*)\mathrm{d}\tau = f^*.$$

This leads to $f = f^*$, and the eigenvalue $f$ is real. The Hermitian operators guarantee that the eigenvalues representing the physical quantities are real.

In the Hermitian operators, eigenfunctions belonging to different eigenvalues are orthogonal. In order to prove this, we suppose

$$\mathrm{F}_1\psi_1 = f_1\psi_1 \quad \text{and} \quad F_2\psi_2 = f_2\psi_2 \quad \text{where} \quad f_1 \neq f_2.$$

This leads to

$$\int \psi_2^*\mathbf{F}\psi_1 \mathrm{d}\tau = \int \psi_2^* f_1 \psi_1 \mathrm{d}\tau = f_1 \int \psi_2^*\psi_1 \mathrm{d}\tau.$$

Since **F** is a Hermitian, this is equal to

$$\int \psi_1\mathbf{F}^*\psi_2^* \mathrm{d}\tau = \int \psi_1 f_2 \psi_2^* \mathrm{d}\tau = f_2 \int \psi_1\psi_2^* \mathrm{d}\tau.$$

Subtraction of these two relations leads to

$$(f_1 - f_2)\int \psi_2^*\psi_1 \mathrm{d}\tau = 0$$

but from $f_1 \neq f_2$, we conclude $\int \psi_2^*\psi_1 \mathrm{d}\tau = 0$. Then, $\psi_1$ and $\psi_2$ are orthogonal. For instance, waves with different $k$ satisfy

$$\int e^{-ik_1x}e^{ik_2x}\mathrm{d}x = \int e^{i(k_2-k_1)x}\mathrm{d}\tau = 0 \,\text{because of}\; k_1 \neq k_2.$$

Two degenerate eigenfunctions with the same eigenvalue are also made orthogonal. Suppose

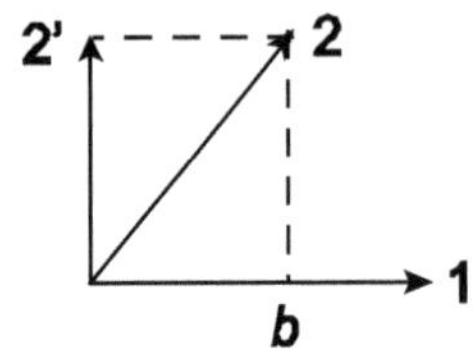


**Fig. 1.29** Schumidt's orthogonalization

$$\mathbf{F}\psi_1 = f\psi_1, \quad \mathbf{F}\psi_2 = f\psi_2$$

and the inner product is defined by $\int \psi_2 * \psi_1 \mathrm{d}\tau = b$. As shown in Fig. 1.29, we shall make $\psi_2' = \psi_2 - b^*\psi_1$ to orthogonalize $\psi_1$ and $\psi_2'$. This function satisfies

$$\int \psi_2'^* \psi_1 \mathrm{d}\tau = \int (\psi_2 - b^*\psi_1)^* \psi_1 \mathrm{d}\tau = \int \psi_2^* \psi_1 \mathrm{d}\tau - b \int \psi_1^* \psi_1 \mathrm{d}\tau = 0$$

and $\psi_2'$ is orthogonal to $\psi_1$. This is called **Schmidt's orthogonalization**. We see

$$\mathbf{F}\psi_2' = \mathbf{F}\psi_2 - b^*\mathbf{F}\psi_1 = f\psi_2 - b^* f\psi_1 = f\psi_2'.$$

This implies $\psi_2'$ is another eigenfunction that has the same eigenvalue $f$ as $\psi_1$.

In general, when functions $\psi_1$ and $\psi_2$ are eigenfunctions of **F** with the same eigenvalue $f$, the linear combination $\psi = c_1\psi_1 + c_2\psi_2$ is another eigenfunction of **F** with the same eigenvalue $f$. This condition is represented by

$$\mathbf{F}\psi_1 = f\psi_1, \mathbf{F}\psi_2 = f\psi_2$$

which leads to

$$\mathbf{F}\psi = \mathbf{F}(c_1\psi_1 + c_2\psi_2) = c_1\mathbf{F}\psi_1 + c_2\mathbf{F}\psi_2 = c_1 f\psi_1 + c_2 f\psi_2 = f\psi.$$

For example, $\mathrm{p}_x$ and $\mathrm{p}_y$ orbitals are represented by $xe^{-r/2a_0}$ and $ye^{-r/2a_0}$, but

$$\begin{pmatrix} \psi_1 \\ \psi_2 \end{pmatrix} = \begin{pmatrix} \cos\varphi & \sin\varphi \\ -\sin\varphi & \cos\varphi \end{pmatrix} \begin{pmatrix} xe^{-r/2a_0} \\ ye^{-r/2a_0} \end{pmatrix}$$

generates another p orbital which is rotated by $\varphi$ in the $xy$ plane (Fig. 1.30). For instance, we put $\varphi = 45°$ to obtain $\psi_1 = \frac{x+y}{\sqrt{2}} e^{-r/2a_0}$. Since this p orbital has a node

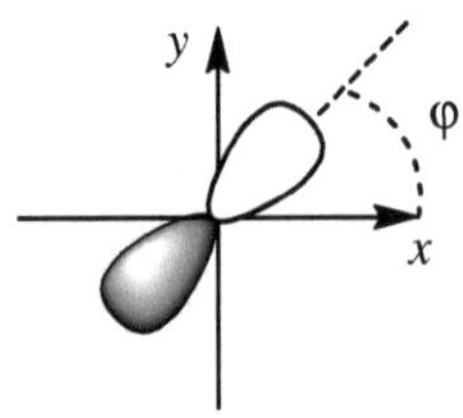


**Fig. 1.30** Rotation of a p orbital

on the $x+y = 0$ line, this is a p orbital extending in the $x = y$ direction. We can carry out the three-dimensional rotation including the $p_z$ orbital. Since an atom has a spherical symmetry, the choice of the $x$ and $y$ axes is artificial. So the $p_x$ and $p_y$ orbitals are no more fixed, and a p orbital may be oriented in any arbitrary direction.

The above discussion guarantees all eigenfunctions are orthogonal. A collection of eigenfunctions of an operator produces an orthonormal system.

$$\int \psi_m^* \psi_n \mathrm{d}\tau = \delta_{mn}$$

When we define the inner product of functions by the integral like $b$, the functions are regarded as vectors. Here, the linear combination $\psi = c_1\psi_1 + c_2\psi_2 + \cdots$ makes a vector space called the **Hilbert space**. Any point in the three-dimensional space is represented as $\boldsymbol{r} = c_1\,\boldsymbol{x} + c_2\,\boldsymbol{y} + c_3\,\boldsymbol{z}$ by using the unit vectors, $\boldsymbol{x}$, $\boldsymbol{y}$, and $\boldsymbol{z}$ and the Cartesian coordinates ($c_1$, $c_2$, $c_3$). In the functional space, any function is represented as a linear combination $\psi = c_1\psi_1 + c_2\psi_2 + \cdots$ by using the unit vectors $\psi_1$, $\psi_2$ … For $\psi_1$, $\psi_2$ …, we may use eigenfunctions of an operator **F**. Different **F** constitute different Hilbert spaces. A Hilbert space is sometimes composed of infinite functions.

Any function $\psi$ is expanded by a linear combination of an orthonormal system $\psi_i$.

$$\psi = \sum_i c_i \psi_i.$$

Multiplying $\psi_j^*$ from the left, and integration in the whole space gives

$$\int \psi_i^* \psi \mathrm{d}\tau = \int \psi_i^* \sum_j c_j \psi_j \mathrm{d}\tau = \sum_j c_j \delta_{ij} = c_i.$$

Thus, the expansion coefficient $c_i = \int \psi_i^* \psi \mathrm{d}\tau = \langle \psi_i | \psi \rangle$ is obtained. According to Dirac, such an integral is represented by using a bra: $\langle$and a ket:$\rangle$. When a bra and a ket appear at the same time, a "bracket" is formed, where the integration is performed. In contrast, a single $|\psi\rangle$ means a wave function $\psi$. $\langle \psi_1 | \psi_2 \rangle$ is an inner product of $\psi_1$ and $\psi_2$, and $\langle \psi_1 | \mathrm{H} | \psi_2 \rangle$ is an integral $\int \psi_1^* \mathrm{H} \psi_2 \mathrm{d}\tau$. The function in the bra is the complex conjugate, though a star is not designated.

For example, an arbitrary function $\psi(x)$ is expanded by an orthonormal system $e^{ikx}$

$$\psi(x) = \sum_k f(k) e^{ikx} = \int f(k) e^{ikx} \mathrm{d}x.$$

The expansion coefficient is given by

$$f(k) = \int \psi(x)e^{-ikx}\mathrm{d}x.$$

This is the Fourier expansion. Expansion by an orthonormal system is a generalization of the Fourier expansion to a general set of functions.

When an arbitrary function $\psi(x)$ is expanded in this way, the integral is

$$\int \psi^* \psi \mathrm{d}\tau = \int \left(\sum_i c_i^* \psi_i^*\right)\left(\sum_j c_j \psi_j\right)\mathrm{d}\tau = \sum_i \sum_j c_i^* c_j \int \psi_i^* \psi_j \mathrm{d}\tau$$

However, the final integral is reduced to $\delta_{ij}$. When $\psi$ is normalized, this leads to

$$\sum_i |c_i|^2 = 1.$$

This equation implies that the state $\psi$ is a convolution of the states $\psi_i$, in which the respective states have probabilities $|c_i|^2$. Assume $\psi_i$ is not the eigenfunctions of **F**, and consider the following integral.

$$\begin{aligned}\langle \mathrm{F}\rangle &= \int \psi^* \mathrm{F}\psi \mathrm{d}\tau = \int \left(\sum_i c_i^* \psi_i^*\right)\mathrm{F}\left(\sum_j c_j \psi_j\right)\mathrm{d}\tau \\ &= \sum_i \sum_j c_i^* c_j f_j \delta_{ij} = \sum_i |c_i|^2 f_i\end{aligned}$$

Since the probability of the state $i$ is $|c_i|^2$, the value $f_i$ is obtained in the probability of $|c_i|^2$ when we measure the physical quantity **F**. Accordingly, $\langle \mathbf{F} \rangle$ is an average value of **F**, and called **expectation value**. For instance, as for the function $\psi = e^{ikx}$, the expectation value of the momentum

$$p \rightarrow \frac{\hbar}{i}\frac{\partial}{\partial x}$$

is

$$\langle \psi | \frac{\hbar}{i}\frac{\partial}{\partial x} | \psi \rangle = \hbar k,$$

and the expectation value of the kinetic energy

$$E \rightarrow -\frac{\hbar^2}{2m}\frac{\partial^2}{\partial x^2}$$

is

$$\langle E\rangle = \left\langle \psi\right| - \frac{\hbar^2}{2m}\frac{\partial^2}{\partial x^2}\left|\psi\right\rangle = \frac{\hbar^2 k^2}{2m}.$$

In this particular case, the given function is an eigenfunction of the two operators at the same time. If the function is not the eigenfunction, the expansion $\psi = c_1\psi_1 + c_2\psi_2 + \cdots$ provides the expectation value in the form of a sum of $|c_i|^2$ multiplied by the eigenvalues $f_i$.

Here, we introduce the commutation relation. For $x$ and $p_x$, the following relations hold for any function $\psi$.

$$xp_x\psi = x\frac{\hbar}{i}\frac{\partial\psi}{\partial x}$$

$$p_x x\psi = \frac{\hbar}{i}\frac{\partial}{\partial x}(x\psi) = \frac{\hbar}{i}\psi + x\frac{\hbar}{i}\frac{\partial\psi}{\partial x}$$

Subtraction of these relations gives

$$(xp_x - p_x x)\psi = i\hbar\psi$$

so $x$ and $p_x$ are not commutable. Since this relation holds for any $\psi$, the operator part is written in the form

$$xp_x - p_x x \rightarrow [x, p_x] = i\hbar.$$

This is the definition of a **commutator**, where $x$ and $p_x$ are conjugated. Non-conjugated variables are commutable

$$[x, y] = 0, \ [x, p_y] = 0, \text{ and } \ [p_x, p_y] = 0.$$

In the same way, the definition of the energy operator $E$ leads to $[E, t] = i\hbar$.

If operators $\mathbf{F}_1$ and $\mathbf{F}_2$ are commutable, $\mathbf{F}_1$ and $\mathbf{F}_2$ have the same set of eigenfunctions. Suppose $\mathbf{F}_1\psi_1 = f_1\psi_1$, and we obtain

$$\mathbf{F}_2\mathbf{F}_1\psi_1 = f_1(\mathbf{F}_2\psi_1).$$

However, this equals $\mathbf{F}_1\mathbf{F}_2\psi_1$ if $\mathbf{F}_1$ and $\mathbf{F}_2$ are commutable. Accordingly, both $\mathbf{F}_2\psi_1$ and $\mathbf{F}_1$ are eigenfunctions corresponding to the same eigenvalue $f_1$, but we have not assumed that $\psi_1$ is degenerate. This is resolved if $\mathbf{F}_2\psi_1$ is a multiple of $\psi_1$, which leads to $\mathbf{F}_2\psi_1 = f_2\psi_1$. This relation means $\psi_1$ is an eigenfunction of $\mathbf{F}_2$ as well.

The inverse is also true. If $\mathbf{F}_1$ and $\mathbf{F}_2$ have a common set of eigenfunctions $\psi_i$, $\mathbf{F}_1$ and $\mathbf{F}_2$ are commutable. In order to show this, any function $\psi$ is expanded like $\psi = \sum_i c_i\psi_i$, and we show

$$(\mathbf{F}_1\mathbf{F}_2 - \mathbf{F}_2\mathbf{F}_1)\psi = \sum_i c_i(\mathbf{F}_1\mathbf{F}_2 - \mathbf{F}_2\mathbf{F}_1)\psi_i = \sum_i c_i(f_1f_2 - f_2f_1)\psi_i = 0.$$

Since this relation holds for any $\psi$, we can generally say $\mathbf{F}_1\mathbf{F}_2 - \mathbf{F}_2\mathbf{F}_1 = 0$. We have previously shown substitution to the operators $p \rightarrow \frac{\hbar}{i}\frac{\partial}{\partial x}$ and $E \rightarrow -\frac{\hbar^2}{2m}\frac{\partial^2}{\partial x^2}$, but these two operators are obviously commutable because they are first- and second-order derivatives. Therefore, $p$ and $E$ have a common set of eigenfunctions $\psi = e^{ikx}$.

From these properties, an operator $\mathbf{F}$ is represented by a matrix using an orthonormal system $\psi_i$ as a basis set. We define a matrix

$$\mathbf{F} = \begin{pmatrix} F_{11} & F_{12} & \cdots \\ F_{21} & F_{22} & \cdots \\ \vdots & \vdots & \ddots \end{pmatrix} \quad \text{where } F_{ij} = \int \psi_i * F\psi_j \mathrm{d}\tau. \tag{1.33}$$

Here, the orthonormal system $\psi_i$ is not necessarily the eigenfunctions of $\mathbf{F}$. When the orthonormal system $\psi_i$ consists of $N$ functions, the matrix is $N$th order. If an eigenfunction of $\mathbf{F}$ is expanded as $\psi = \sum_i c_i\psi_i$, we see

$$\begin{pmatrix} F_{11} & F_{12} & \cdots \\ F_{21} & F_{22} & \cdots \\ \vdots & \vdots & \ddots \end{pmatrix}\begin{pmatrix} c_1 \\ c_2 \\ \vdots \end{pmatrix} = f\begin{pmatrix} c_1 \\ c_2 \\ \vdots \end{pmatrix} \tag{1.34}$$

In order to show this, note $\psi$ is an eigenfunction with the eigenvalue $f$, and we obtain $\mathbf{F}\psi = f\psi$. To this equation, $\psi_i^*$ is multiplied from the left. After integrated in the whole space, this equation affords

$$(\text{Left-hand side}) = \langle\psi_i|\mathbf{F}|\psi\rangle = \langle\psi_i|\mathbf{F}|\sum_j c_j\psi_j\rangle = \sum_j c_j\langle\psi_i|\mathbf{F}|\psi_j\rangle = \sum_j c_jF_{ij}$$

$$(\text{Right-hand side}) = \langle\psi_i|f|\sum_j c_j\psi_j\rangle = f\sum_j c_j\langle\psi_i|\psi_j\rangle = fc_i.$$

These relations are summarized in a matrix form as shown in Eq. 1.34. In a special case where $\psi_i$ is eigenfunctions of $\mathbf{F}$, the matrix $\mathbf{F}$ is reduced to

$$\begin{pmatrix} f_1 & 0 & \\ 0 & f_2 & \\ & & \ddots \end{pmatrix}.$$

**Fig. 1.31** Orbital made by convolution of $p_z$ and $p_{xz}$ orbitals

This is a diagonal matrix in which the diagonal elements are the eigenvalues $f_i$. A product of operators is a product of matrices, so it is not commutable.

Since the operator **F** is a Hermitian, the matrix **F** satisfies $F_{ji} = F_{ij}^*$. Thus, the matrix is in the form.

$$\begin{pmatrix} f_{11} & f_{12} + if'_{12} & \\ f_{12} - if'_{12} & f_{22} & \\ & & \ddots \end{pmatrix}$$

A nondiagonal element is a complex conjugate of the transposed element. Such a matrix is called **Hermitian matrix**. Diagonal elements of a Hermitian matrix are real, and in particular, the eigenvalues are real.

Equation 1.34 is an equation to obtain the eigenvalues of the matrix.

$$\begin{pmatrix} F_{11} - f & F_{12} & \cdots \\ F_{21} & F_{22} - f & \cdots \\ \vdots & \vdots & \ddots \end{pmatrix} \begin{pmatrix} c_1 \\ c_2 \\ \vdots \end{pmatrix} = 0 \tag{1.35}$$

This is simultaneous equations for the variables $c_1, c_2, \ldots$, and the condition to obtain nonzero solutions of $c_1, c_2, \ldots$ is given by zero of the coefficient determinant.

$$\begin{vmatrix} F_{11} - f & F_{12} & \cdots \\ F_{21} & F_{22} - f & \cdots \\ \vdots & \vdots & \ddots \end{vmatrix} = 0$$

When the determinant is $N$th order, this is an $N$th order equation for $f$. Then, we obtain $N$ real solutions (energy levels) as the eigenvalues. Putting these $N$ eigenvalues in Eq. 1.35, we obtain a set of solutions ($c_1, c_2, \ldots$), which are coefficients of the expansion using the basis set $\psi_i$. Then, the molecular orbitals are obtained as the eigenfunctions.

In Sect. 1.4, we solved a differential equation (the Schrödinger equation) of a hydrogen atom, but it was a laborious work. If we start from a basis set $\psi_i$, an operator is converted to a matrix, and the solution is reduced to an eigenvalue problem represented by the secular equation. The above matrix-form calculation has close resemblance to the molecular orbital calculation in Sect. 1.5, because the molecular orbital calculation is an eigenvalue problem for the Hamiltonian of a molecule. In the secular equation, $\alpha$ and $\beta$ correspond to the diagonal and nondiagonal elements of the Hamiltonian. The basis set is the atomic orbitals. Although the atomic orbitals on different atoms are not strictly orthonormal, the LCAO approximation affords quite satisfactory results.

Any function on an atom is generally represented by a linear combination of the atomic orbitals. Even a considerably distorted orbital is sufficiently expanded when we include not only the valence orbitals but also higher energy orbitals such as 3d, and 4f. For example, we can make such an orbital as Fig. 1.31 by adding p and d orbitals. When we attempt to improve molecular orbital calculations, we actually use this kind of orbital. We usually use only the valence orbitals, but the results are satisfactory in spite of the limitations.

A transformation from an orthonormal system $|\psi\rangle$ to another orthonormal system $|\psi'\rangle$ is generally represented by $|\psi'\rangle = \mathbf{U}\,|\psi\rangle$ using a **unitary matrix U**. In a unitary matrix, the transposed complex conjugate $(u^+)_{ij} = (u^*)_{ji}$ is the inverse matrix. In order to show this, $|\psi'\rangle = \mathbf{U}\,|\psi\rangle$ is written as $\psi'_n = \sum_m c_{nm}\psi_m$, to which $\psi'^*$ is multiplied from the left, and integrated.

$$\int \psi_k'^* \psi_n' \mathrm{d}\tau = \int \left(\sum_l c_{kl}^* \psi_l^*\right)\left(\sum_m c_{nm}\psi_m\right)\mathrm{d}\tau = \sum_l \sum_m c_{kl}^* c_{nm}\delta_{lm}$$
$$= \sum_m c_{km}^* c_{nm} = \sum_m c_{nm} c_{mk}^+$$

The new set $|\psi'\rangle$ is orthonormal when the matrix $\mathbf{U}$ composed of $c_{nm}$ satisfies $\mathbf{UU}^+ = \mathbf{E}$. This is the definition of the unitary matrix. At the **unitary transformation**, the eigenvalue equation $\mathbf{F}|\psi\rangle = f|\psi\rangle$ is converted to $\mathbf{FU^+U}|\psi\rangle = f\,\mathbf{U^+U}|\psi\rangle$. When $\mathbf{U}$ is multiplied from the left, this is converted to $\mathbf{UFU}^+\,|\psi'\rangle = f\,|\psi'\rangle$, so the matrix is transformed to $\mathbf{F} \rightarrow \mathbf{F}' = \mathbf{UFU}^+$. If $\mathbf{F}$ is converted to a diagonal matrix $\mathbf{F}'$ by a unitary transformation, the eigenvalues are aligned on the diagonal elements. Then, $|\psi'\rangle = \mathbf{U}|\psi\rangle$ affords the eigenfunctions. The problem to obtain eigenvalues is substituted by the problem to **diagonalize** a matrix by a unitary transformation. It is the central issue of quantum mechanics to obtain eigenvalues of the Hamiltonian, but this is practically reduced to the diagonalization of the Hamiltonian. Diagonalization is easily performed in a computer program, and matrices with hundreds of lines are diagonalized in molecular orbital calculations. Block diagonalization using symmetry orbitals, described in the last section, is another example of the unitary transformation.

## 1.8 Uncertainty Principle

As another basic concept of quantum mechanics, we shall investigate the uncertainty principle. Although the mysterious aspect has been sometimes emphasized, the uncertainty principle is a fundamental property of waves. In the last section, we described the commutators

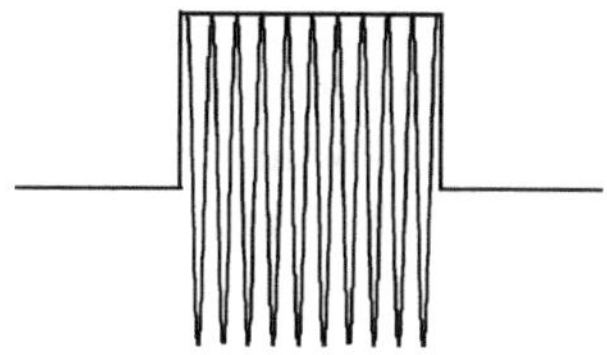

**Fig. 1.32** A pulse of light

$$[x,p] = i\hbar,\ [E,t] = i\hbar$$

where $p_x$ is simply written as $p$. We assume that the state is not an eigenstate of $x$ and $p$. As described in the last section, when we measure $x$ and $p$, different $x$ and $p$ values are observed with the probabilities of $|c_i|^2$. The deviations from the expectation values $\langle x\rangle$ and $\langle p\rangle$ are defined as

$$\Delta x = x - \langle x\rangle$$

$$\Delta p = p - \langle p\rangle.$$

We consider

$$\begin{aligned}(\Delta x)^2(\Delta p)^2 &= \int \psi^*(\Delta x)^2\psi \mathrm{d}x \int \psi^*(\Delta p)^2\psi \mathrm{d}x \\ &= \int (\Delta x\psi)^*(\Delta x\psi)\mathrm{d}x \int (\Delta p\psi)^*(\Delta p\psi)\mathrm{d}x\end{aligned}$$

where we used that $\Delta x$ and $\Delta p$ are Hermitian. In general, the Schwarz inequality holds for any functions $f$ and g.

$$\int |f|^2\mathrm{d}x \int |g|^2\mathrm{d}x \geq |\int f^* g\mathrm{d}x|^2$$

Putting $\Delta x\psi$ in $f$, and $\Delta p\psi$ in g, the above equation is converted to

$$\begin{aligned}(\Delta x)^2(\Delta p)^2 &\geq \left|\int \psi^* \Delta x\Delta p\psi \mathrm{d}x\right|^2 = \left|\int \psi^*\left[\frac{1}{2}(\Delta x\Delta p - \Delta p\Delta x) + \frac{1}{2}(\Delta x\Delta p + \Delta p\Delta x)\right]\psi \mathrm{d}x\right|^2. \\ &= \frac{1}{4}\left|\int (\Delta x\Delta p - \Delta p\Delta x)\mathrm{d}x\right|^2 + \frac{1}{4}\left|\int (\Delta x\Delta p + \Delta p\Delta x)\mathrm{d}x\right|^2.\end{aligned}$$

The first equality is easily obtained by simplifying the right-hand side. The former term of the last relation is $i\hbar$ from the commutation relation. Even if the second term is zero, the following inequality holds.

$$(\Delta x)^2(\Delta p)^2 \geq \frac{\hbar^2}{4} \quad \text{namely} \quad (\Delta x)(\Delta p) \geq \frac{\hbar}{2}$$

This is **Heisenberg's uncertainty principle**, which states that position and momentum of a particle are not determined at the same time in the accuracy smaller than $\hbar/2$. For instance, the momentum of a plane wave $\psi = e^{ikx}$ is $\hbar k$, implying $\Delta p = 0$, which leads to $\Delta x \to \infty$. Then, the position is entirely uncertain.

A similar relation

$$(\Delta E)(\Delta t) \geq \frac{\hbar}{2}$$

is obtained from the commutation relation between $E$ and $t$. We need a long-time observation to determine the energy with sufficient accuracy. Putting $E = h\nu$ in this relation, we obtain

$$(\Delta \nu)(\Delta t) \geq \frac{1}{4\pi}.$$

Through use of a laser, we can make a pulse as short as femto ($10^{-15}$) to atto ($10^{-18}$) second. Frequency of 1 eV light is about $10^{14}$ Hz, but we cannot make a pulse shorter than a half wave of the electromagnetic wave (Fig. 1.32). This is the meaning of the above equation. We can prove this relation from the property of classical waves without using quantum mechanics. However, through use of fundamental relations like $E = h\nu$, we can show that such quantities as position and momentum are uncertain. The interpretation of the uncertainty principle has developed extensive philosophical discussion, but everything stems from the fact that there is no pulse shorter than the wave.

## 1.9 Analytical Mechanics and Phase

The final comment is related to the phase in quantum mechanics. Inserting $p = \hbar k$ and $E = \hbar\omega$ in the plane wave equation, we obtain

$$e^{i(kx-\omega t)} = e^{i\frac{px-Et}{\hbar}} = e^{i\frac{S}{\hbar}}. \tag{1.36}$$

The quantity $S = px - Et$ is called **action**, which equals to the integration of the difference between the kinetic energy $T$ and the potential energy $V$ when a particle moves from time $t_1$ to $t_2$.

$$S = \int_{t_1}^{t_2} L(t)\mathrm{d}t = \int_{t_1}^{t_2} (T - V)\mathrm{d}t$$

$L = T - V$ is called **Lagrangian**. In analytical mechanics [6], motion minimizing the action $S$ is realized; this is called **the principle of least action**. From this principle, all kinds of equations of motion are derived. For instance, suppose the position and velocity $(x, v)$ of a particle are slightly deviated from the actually realized motion. In such a case, $S$ increases. The actual motion is obtained by minimizing $S$.

$$\delta S = \int_{t_1}^{t_2} \left( \frac{\partial L}{\partial v} \mathrm{d}v + \frac{\partial L}{\partial x} \mathrm{d}x \right) \mathrm{d}t = 0$$

This equation represents the principle of least action. The integrand has to be zero for any $t$. Through use of $v = \mathrm{d}x/\mathrm{d}t$, this equation leads to

$$\frac{\mathrm{d}}{\mathrm{d}t} \frac{\partial L}{\partial v} - \frac{\partial L}{\partial x} = 0.$$

The negative sign comes from the partial integration of the first term. When we put $T = mv^2/2$ and $V$, this relation leads to

$$m \frac{\mathrm{d}v}{\mathrm{d}t} = -\frac{\partial V}{\partial x} = F.$$

This is Newton's equation of motion. The principle of least action tells us the motion minimizing $T - V$ is realized. This is a mathematical expression that nature does not make a useless motion.

The definition of Hamiltonian in analytical mechanics

$$H = p \frac{\partial x}{\partial t} - L$$

is integrated with respect to $t$.

$$Et = H\delta t = px - \int L\mathrm{d}t = px - S$$

This leads to Eq. 1.36.

$$S = px - Et$$

Action is an important concept in analytical mechanics. Equation 1.36 tells us 1/$h$ times of the action works as phase in quantum mechanical. This relation is valid not only for electrons but also for any other particles. Therefore, $h$ is the minimum unit of phase for all particles. When we try to divide a particle smaller than this unit, the particle is obscured and it behaves as a wave. Once the last relation in Eq. 1.36 is obtained from analytical mechanics, the general wave equation (Eq. 1.4) gives the relations $E = \hbar\omega$ and $p = \hbar k$. Accordingly, the two relations $E = h\nu$ and $\lambda = h/p$, which we assumed as theorems in the beginning of this chapter, are derived from a single concept that $S/h$ gives the phase.

## References

1. J. Slater, Phys. Rev. **36**, 57 (1930)
2. R. Hoffmann, J. Chem. Phys. **40**, 2474, 2480, 2745 (1964)
3. R.S. Mulliken, C.A. Rieke, D. Orloff, H. Orloff, J. Chem. Phys. **17**, 1248 (1949)
4. J.C. Roberts, *Notes on Molecular Orbital Calculations*. Benjamin (1962)
5. F.A. Cotton, *Chemical Applications of Group Theory*. Wiley (1963)
6. L.D. Landau, E.M. Lifshitz, *Mechanics*. Elsevier (1960)

# Chapter 2
# Quantum Chemistry of Solids

**Abstract** Energy band theory is introduced as an extension of the molecular orbital theory, and applied to organic conductors. From this, we can discuss the Fermi surface of organic conductors.

**Keywords** Energy band · Tight-binding approximation · Bloch condition · Bandwidth · Hückel rule · Periodical boundary condition · Density of states · Free-electron model · Fermi-Dirac distribution · Bose-Einstein distribution · Boltzmann distribution · Transfer integral · Hole · Fermi surface · Brillouine zone · Peierls transition

## 2.1 Tight-Binding Approximation

In the last chapter, we have investigated molecular orbitals of π-conjugated rings with the carbon number from $N = 3$ (cyclopropenyl cation) to $N = 6$ (benzene). In this chapter, we shall investigate energy levels of π-conjugated rings for general $N$ [1, 2]. The solution affords energy levels of large π-systems in which the secular equation is not easily solved. The solution also provides a general proof of the Hückel $4n + 2$ rule. Large $N$ leads to a large π-conjugated ring (Fig. 2.1a). The linear chain corresponds to polyacetylene (Fig. 2.1b), while linear polyacetylene has terminal states (Fig. 2.1c). The ordinary carbon atom is bonded to one hydrogen and two carbon atoms, whereas the terminal carbon is connected to two hydrogen and one carbon atoms. Then, the terminal has a different electronic state. In order to avoid such a terminal state, a ring is considered (Fig. 2.1a). When $N$ is as large as the Avogadro number, this leads to energy bands in a bulk solid. Hereafter, $a$ is the distance between the adjacent carbon atoms.

Since this is a natural extension of the Hückel approximation, we consider each carbon atom to have one π orbital. The LCAO-MO is constructed from $N$ orbitals.

$$\psi = \sum_{n}^{N} c_n \chi_n \tag{2.1}$$



T. Mori, *Electronic Properties of Organic Conductors*,
DOI 10.1007/978-4-431-55264-2_2

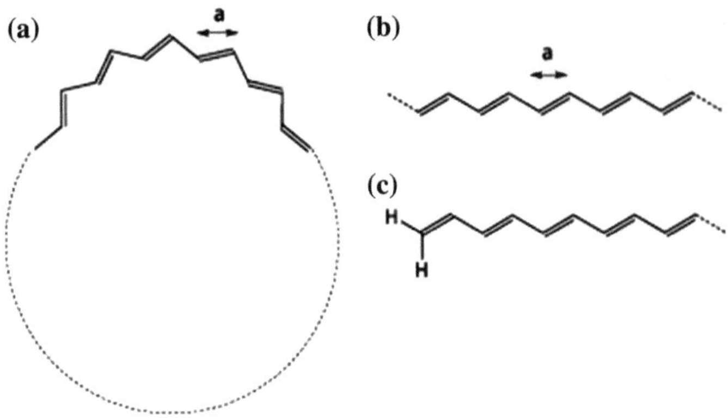


**Fig. 2.1** (**a**) A π-conjugated ring with the carbon number *N*, (**b**) a linear π-system, polyacetylene, and (**c**) edge states of the polyacetylene

The secular equation is obtained in the form of an $N \times N$ determinant.

$$\begin{vmatrix} \alpha - E & \beta & 0 & \cdots \\ \beta & \alpha - E & \beta & \cdots \\ 0 & \beta & \alpha - E & \cdots \\ \vdots & \vdots & \vdots & \vdots \end{vmatrix} = 0 \tag{2.2}$$

It is possible to solve this equation for general *N*, but we will adopt another easier way.

The system has a periodicity of *a*. Then, after the translation of $r \to r + a$, the resulting function $\psi(r + a)$ is again the eigenfunction. The Hamiltonian also has the periodicity, so the translation $r \to r + a$ does not change the energy. Accordingly, $\psi(r)$ and $\psi(r + a)$ are eigenfunctions with the same energy. If there is no degeneracy,[1] these two functions are related to each other like $\psi(r + a) = c(a)\psi(r)$ using a constant $c(a)$. One more translation results in $\psi(r + 2a) = c(2a)\psi(r)$, which leads to $\psi(r + 2a) = c(a)\psi(r + a) = c(a)c(a)\psi(r)$. Thus, we obtain $c(2a) = c(a)c(a)$. When $c(a) = \mathrm{e}^{ika}$ is assumed, this $c(a)$ satisfies the above relation. As a consequence, we obtain $\psi(r + a) = \mathrm{e}^{ika}\,\psi(r)$; this condition is known as the **Bloch condition**.

This is a general relation derived from the periodicity. The LCAO (Eq. 2.1) satisfying this condition is represented as

$$\begin{aligned} \psi &= c_0\left[\chi_0 + \mathrm{e}^{ika}\chi_1 + \mathrm{e}^{i2ka}\chi_2 + \mathrm{e}^{i3ka}\chi_3 + \mathrm{e}^{i4ka}\chi_4 + \cdots\right] \\ &= c_0 \sum_n \mathrm{e}^{inka}\chi_n. \end{aligned} \tag{2.3}$$

[1]This assumption is not strictly correct because the states at *k* and −*k* are generally degenerate. Nonetheless, this assumption gives the correct conclusion if the unnecessary hybridization of the degenerate states is appropriately avoided.

This form of wave function will be called a **crystal orbital** in contrast to the molecular orbital. Here, $e^{ika}$ is a complex whose absolute value is one, and the phase is nonzero. After the translation of *a*, the amplitude does not change, but the phase changes. Here, *k* is introduced as a coefficient which determines the phase shift *ka*. Since *ka* is dimensionless, *k* has a unit of inverse length. This quantity is identical to the wave number introduced in the last chapter.

Usually, we solve the secular equation and then obtain the wave function. In contrast, here the wave function has been mostly determined from the periodicity. Once the wave function $\psi$ is known, the energy is obtained by multiplying $\psi^*$ to $H\psi = E\psi$ from the left and integrating it in the whole space.

$$E = \frac{\int \psi^* H \psi \mathrm{d}\tau}{\int \psi^* \psi \mathrm{d}\tau} \tag{2.4}$$

When Eq. 2.3 is substituted in Eq. 2.4, we obtain

$$E = \frac{\int \left(\sum_m \mathrm{e}^{-imka} \chi_m^*\right) H \left(\sum_n \mathrm{e}^{inka} \chi_n\right) \mathrm{d}\tau}{\int \left(\sum_m \mathrm{e}^{-imka} \chi_m^*\right) \left(\sum_n \mathrm{e}^{inka} \chi_n\right) \mathrm{d}\tau} = \frac{\sum_n \sum_m \mathrm{e}^{i(n-m)ka} \int \chi_m^* H \chi_n \mathrm{d}\tau}{\sum_n \sum_m \mathrm{e}^{i(n-m)ka} \int \chi_m^* \chi_n \mathrm{d}\tau}. \tag{2.5}$$

We focus on a given *n* (Fig. 2.2). According to the usual definition of the Hückel approximation, $\int \chi_m^* H \chi_n \mathrm{d}\tau$ is $\alpha$ for $m = n$, and $\beta$ for $m = n \pm 1$, namely for the adjacent atoms. This integral is zero for all other combinations, because they are not directly bonded. Since the factors $e^{ika}$ and $e^{-ika}$ appear for $m = n \pm 1$, we obtain

$$E = \frac{N\left(\mathrm{e}^{ika}\beta + \alpha + \mathrm{e}^{-ika}\beta\right)}{N}. \tag{2.6}$$

The denominator is *N* because the overlap integral is one for $m = n$ and zero for all others. Namely, for a given *n*, a rotation for *m* affords one. When added for *n*, this gives *N*. For a given *n*, the numerator affords $\beta e^{ika}$, $\alpha$, and $\beta e^{-ik}$ corresponding to $m = n - 1$, *n*, and $n + 1$, respectively. When added for a rotation of *n*, the same terms appear *N* times. Thus, *N* cancels with the denominator. The resulting energy is

$$E = \alpha + 2\beta \cos ka \tag{2.7}$$

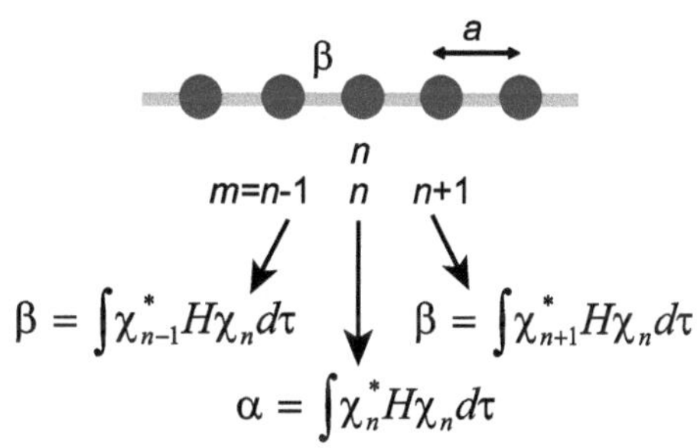


**Fig. 2.2** Integrals appearing in the tight-bind approximation

where $\cos ka = \frac{e^{ika} + e^{-ika}}{2}$ is used. Equation 2.7 is an important relation which represents the energy of the **tight-binding approximation**. This name comes from the starting atomic orbitals, which are tightly bound to the atoms. In solid-state physics, this name is used in contrast to the free-electron approximation, but it is practically equivalent to the LCAO approximation. Therefore, the tight-binding approximation is the solid state version of the molecular orbital theory. The original Hamiltonian **H** includes the potential energy of $N$ atoms. However, following the standard Hückel approximation, the energy is represented by a simple function of α and β. We shall investigate the meaning of this equation as follows.

Equation 2.7 affords a cosine curve as shown in Fig. 2.3a. The cosine curve is upside down because β is negative. The energy is $E = \alpha + 2\beta$ at $k = 0$, where the energy takes minimum. The maximum energy is $E = \alpha - 2\beta$, which appears at $ka = \pm\pi$, namely at $k = \pm\pi/a$. Although this is a periodical function, it is sufficient to consider the region of $-\pi/a < k < \pi/a$. In Fig. 2.3, only this region is plotted.

Since the atoms form a ring (Fig. 2.1a), the atom at $N + 1$ is identical to the first atom, and the $N$-th atom corresponds to the zero-th atom (Fig. 2.3b). The condition to equate the phases of the $N$-th atom and the zero-th atom is satisfied when $e^{iNka} = 1$. This is fulfilled when $Nka = 2\pi n$, where $n$ is an integer. This leads to $k = 2\pi n/Na$. The restriction on $k$ is called the **periodical boundary condition**. When $N$ is large, $2\pi/Na$ is very small, though $k$ is restricted to the integer times of this quantity. Figure 2.3a shows the discrete energy levels aligned with this $k$ interval. In a solid, $N$ is as large as the Avogadro number, and $k$ is practically continuous. Since the energy levels are almost continuous, this is called an **energy band**. Since the energy band exists in the range of $\alpha + 2\beta < E < \alpha - 2\beta$, the **bandwidth** is $4|\beta|$.

Even though the energy band is practically continuous, each energy level has the wave function. For example, at $k = 0$, $e^{inka}$ is always one, so the corresponding wave function (Eq. 2.3) is

$$\varphi = \chi_0 + \chi_1 + \chi_2 + \chi_3 + \cdots \tag{2.8}$$

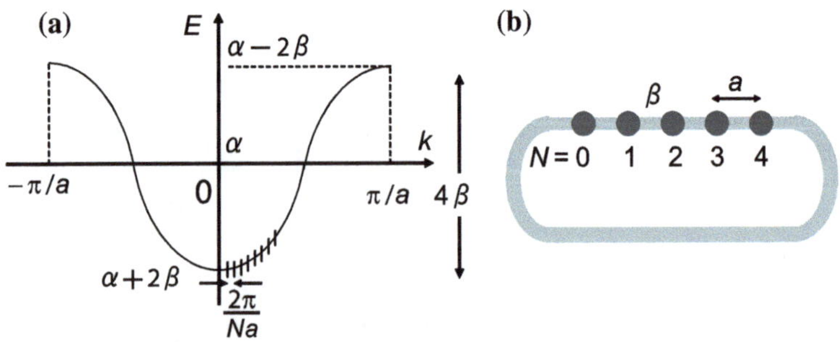


**Fig. 2.3** (**a**) $E(k)$ in the one-dimensional tight-binding approximation, and (**b**) the periodical boundary condition

This is a bonding orbital spreading over the whole solid (Fig. 2.4). For $k = \pm\pi/a$, $nka = n\pi$ leads to $e^{inka} = 1, -1, 1, -1$, so the wave function is

$$\varphi = \chi_0 - \chi_1 + \chi_2 - \chi_3 + \cdots \tag{2.9}$$

This is an antibonding orbital where a node is located on every carbon–carbon bond (Fig. 2.4). This is the reason that this state has the highest energy. In the energy level next to $k = 0$, the phase of the successive carbon increases by $e^{i2\pi/N}$, and the whole rotation leads to the $2\pi$ phase shift. The corresponding wave function has two nodes during the rotation. Due to the small number of nodes, we have to gradually increase the phase, and the coefficient of a crystal orbital has to be a complex number. In general, a coefficient of a molecular orbital is real, whereas a coefficient of an orbital in a solid is complex. In the succeeding energy levels, the phase shift increases as $4\pi$, $6\pi$ ... during the rotation, and the node number increases as 4, 6 ... The final level at $k = \pi/a$ has $N$ nodes, and a node is located on every carbon–carbon bond.

We shall investigate the number of energy levels in the energy band. Since the interval of the energy levels is $2\pi/Na$, the number of the energy levels existing between $k = -\pi/a$ and $k = \pi/a$ is

$$\frac{2 \times \frac{\pi}{a}}{\frac{2\pi}{Na}} = N. \tag{2.10}$$

The carbon number is $N$, and the number of the atomic orbitals is $N$. Therefore, it is quite reasonable that the number of the energy levels is $N$ in analogy with the number of solutions obtained from a secular equation (Eq. 2.2). This justifies that we consider only the region of $-\pi/a < k < \pi/a$. This region is called the **first Brillouin zone**.

In polyacetylene, each carbon atom has one $\pi$ electron. Thus, the total number of the $\pi$ electrons is $N$. The occupation of the energy band is shown in Fig. 2.5a. When we place electrons from the lowest levels, the energy levels are occupied up to a certain energy, above which the levels are unoccupied. The border energy is called

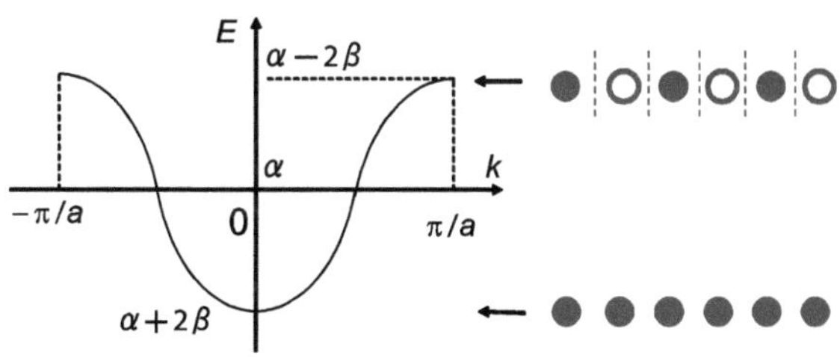


**Fig. 2.4** Orbitals in a one-dimensional tight-binding band

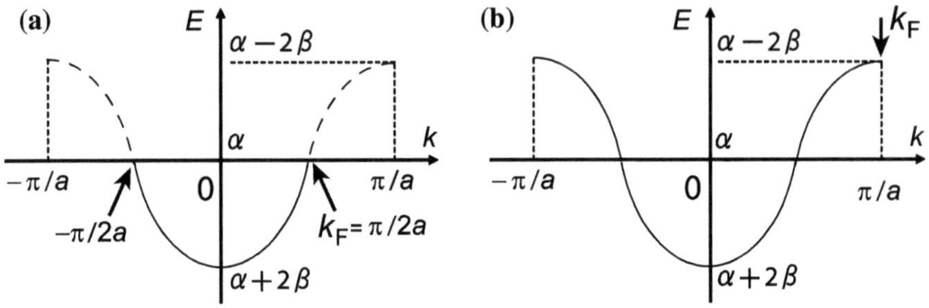


**Fig. 2.5** (**a**) A half-filled one-dimensional tight-binding band, and (**b**) an entirely filled one-dimensional tight-binding band

the **Fermi energy** $E_F$, and the corresponding $k$ is the **Fermi wave number** $k_F$. Since the $N$ electrons enter the $-k_F < k < k_F$ range, we obtain

$$2\frac{2k_F}{\frac{2\pi}{Na}} = N \tag{2.11}$$

where two before the fraction comes from the upspin and downspin electrons entering an energy level. Solving this relation, we obtain

$$k_F = \frac{\pi}{2a} \tag{2.12}$$

so, $k_F$ is half of $\pi/a$. Therefore, the Fermi energy appears at the band center at $E_F = \alpha$. This situation is called **half-filled**. Since the energy band is continuous around $E_F$, a half-filled energy band is a metallic band. However, actual polyacetylene is an insulator unless doped. This discrepancy will be resolved in the later part of this chapter.

When the total electron number is $2N$, we obtain similarly

$$2\frac{2k_F}{\frac{2\pi}{Na}} = 2N. \tag{2.13}$$

This is reduced to $k_F = \pi/a$. Here, the energy band is entirely occupied (Fig. 2.5b). In this situation, each energy level has two electrons. This is the entirely occupied state. Most organic compounds are closed shell molecules, and the HOMO is occupied like this. Ordinary molecular crystals have such a state and they are insulators.

In the above discussion, the tight-binding approximation is applied to an energy band in a solid. However, Eq. 2.7 is also valid for small $N$. For example, $N = 6$ corresponds to benzene, where $k$ is restricted to $k = \frac{2\pi n}{Na} = \frac{2\pi n}{6a}$. Then, the phase in the cosine function rotates by an interval of $ka = \pi/3$. Substituting this in Eq. 2.7, we obtain

$$E = \alpha + 2\beta \cos \frac{2\pi}{6} n. \tag{2.14}$$

As shown in Fig. 2.6, the energy for $n = 0, \pm 1, \pm 2$, and 3 leads to

$$E = \alpha + 2\beta, \alpha + \beta, \alpha + \beta, \alpha - \beta, \alpha - \beta, \alpha - 2\beta. \tag{2.15}$$

These results are the same as Example 1.8.

In Fig. 2.7, energy levels of π-conjugated rings are depicted for $N = 4$, 6, and 8. Since a cosine function is a real part of $e^{ika}$, it is convenient to suppose a complex plane of $x + iy$, whose real axis $x$ is taken downwards (Fig. 2.7b). Starting from the $x$ direction corresponding to $n = 0$, the circle is equally divided by $N$, and the height affords the energy levels. For $N = 4$, the energy levels appear at the successive π/2 positions, so two energy levels appear at $E = \alpha$. These levels are half-filled, but due

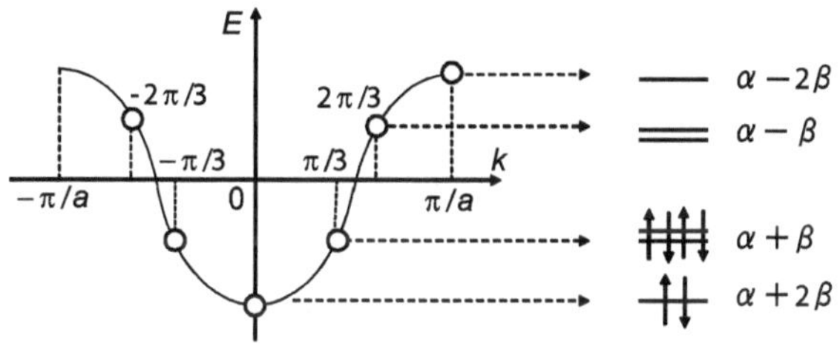


**Fig. 2.6** Molecular orbital levels of benzene obtained from the tight-binding approximation

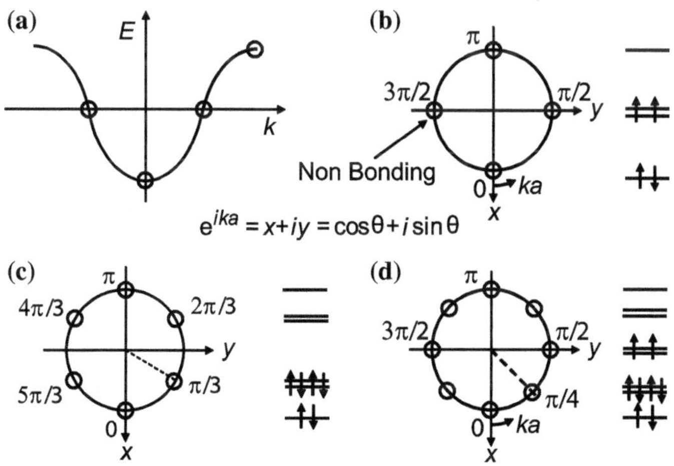


**Fig. 2.7** Energy levels of (**a**, **b**) cyclobutadiene ($N = 4$), (**c**) benzene ($N = 6$), and (**d**) cyclooctatetraene ($N = 8$) obtained from the tight-binding approximation

to the nonbonding character, the delocalization energy becomes zero. For $N = 6$, an energy gap appears at $E = \alpha$ (Fig. 2.7c). Then, all electrons enter the bonding levels below $E = \alpha$, to give rise to the delocalization energy. For $N = 8$, the energy levels appear at the successive $\pi/4$ positions (Fig. 2.7d), and two energy levels appear at $E = \alpha$. Therefore, the delocalization energy is zero. Zero delocalization energy occurs at $N = 4n$, while the delocalization energy appears at $N = 4n + 2$ in the aromatic rings. This is the most general proof of the Hückel rule.

*Example 2.1* Calculate the energy levels of cyclopentadienyl anion using the tight-binding approximation. Calculate the bonding energy and compare it with those of the radical and the cation. The same molecule has been previously discussed in Example 1.15.

For $N = 5$, $k$ is limited to $k = \frac{2\pi n}{5a}$, so the energy is $E = \alpha + 2\beta\cos\frac{2\pi}{5}n$. The values for $n = 0, \pm1$, and $\pm2$ are shown in Fig. 2.8.

In a radical, every carbon atom gives up one $\pi$ electron, and the total number of the $\pi$ electrons is five. An anion has one additional electron, and the $\pi$ electron number is six, whereas a cation has four electrons. These four to six electrons occupy the $\pi$ levels as shown in Fig. 2.8. The bonding energy is estimated by summing the energy of the occupied levels.

$$\begin{array}{lll} \text{Anion} & 6\pi & 2(\alpha+2\beta)+4(\alpha+0.618\beta)-6\alpha=6.472\beta \\ \text{Radical} & 5\pi & 2(\alpha+2\beta)+3(\alpha+0.618\beta)-5\alpha=5.854\beta \\ \text{Cation} & 4\pi & 2(\alpha+2\beta)+2(\alpha+0.618\beta)-4\alpha=5.236\beta \end{array} \tag{2.16}$$

Since bonding energy of three double bonds is $6\beta$, the anion has larger bonding energy than this, showing large delocalization energy.

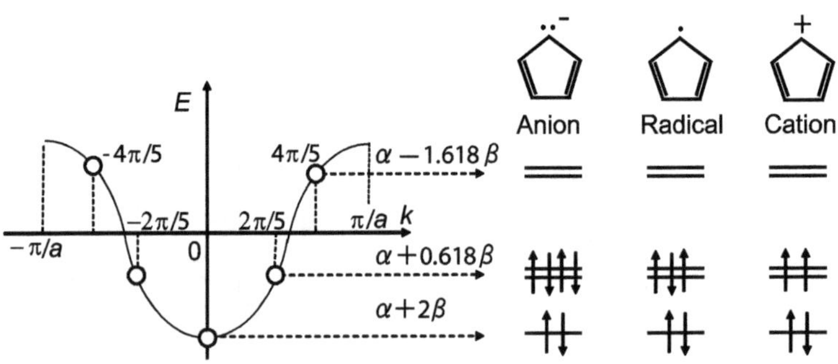


**Fig. 2.8** Energy levels in the cyclopentadienyl anion, radical, and cation

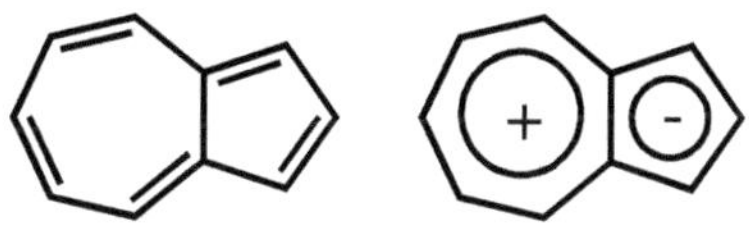

**Fig. 2.9** Azulene

A cyclopentadienyl anion shows aromaticity owing to the six π electrons. Actually, this is an extraordinarily stable organic anion. It forms an ionic compound with $Fe^{2+}$ with the composition of $Fe(C_5H_5)_2$. This entirely air-stable orange solid is known as ferrocene. Ferrocene is used as a standard material of electrochemistry (Sect. 7.2). Ferrocene dissolves in organic solvents, and $Fe^{2+}$ is electrochemically oxidized to $Fe^{3+}$.

The original Hückel rule tells us that π-conjugated systems with the carbon number $4n + 2$ show aromaticity. The above example shows that π-conjugated systems with the π electron number $4n + 2$ are stable in a similar way. As another example, in azulene (Fig. 2.9), the five-membered ring tends to have six π electrons like a cyclopentadienyl anion, and the seven-membered ring is stabilized in the cationic 6π form. Azulene is a hydrocarbon with the formula of $C_{10}H_8$, which is an isomer of naphthalene containing only carbon and hydrogen. Nonetheless, azulene is a very polar molecule. A five-membered ring makes a stable anion with 6π electrons, and a seven-membered ring makes a stable cation with 6π electrons. Accordingly, we can make stable organic anions and cations. Design of organic electron donors and acceptors based on this principle is investigated in Chap. 7.

## 2.2 Free-Electron Model

The one-dimensional free-electron model was discussed in Chap. 1. Here, we investigate the three-dimensional free-electron model [3–7]. Total energy of a three-dimensional electron is

$$E = \frac{p_x^2 + p_y^2 + p_z^2}{2m} + V \tag{2.17}$$

and the Schrödinger equation is

$$\left[ -\frac{\hbar^2}{2m} \left( \frac{\partial^2}{\partial x^2} + \frac{\partial^2}{\partial y^2} + \frac{\partial^2}{\partial z^2} \right) + V \right] \psi = E\psi. \tag{2.18}$$

Electrons in solids are bound to the nucleus by the Coulomb attraction. In a metal electron, however, the attraction is largely shielded by other core electrons. As a result, a metal electron is approximated by a free electron in a constant potential

$V$. $V$ is a negative constant, but we take the $V$ value as energy zero, and use $V = 0$ in Eq. 2.18. The eigenfunction of the free electron is

$$\psi(x, y, z) = \mathrm{e}^{i\left(k_x x + k_y y + k_z z\right)}. \tag{2.19}$$

The eigenvalue is

$$E = \frac{\hbar^2}{2m}\left(k_x^2 + k_y^2 + k_z^2\right). \tag{2.20}$$

As shown in Fig. 2.10, the energy is represented by a parabola along the $k_x$, $k_y$, and $k_z$ axes. When we include $E$, we could not plot the function in the four-dimensional space, but we imagine that the respective point ($k_x$, $k_y$, $k_z$) in the three-dimensional $k$-space has the characteristic energy proportional to the distance from the origin (Eq. 2.20). This is a discussion within fundamental quantum mechanics, but consideration of $N$ electrons introduces solid-state physics.

Here, we suppose the electron is not in an infinite space, but enclosed in a box with the length $L$. The boundary at $x = L$ is not a wall but connected to $x = 0$ as shown in Fig. 2.11a, b. This is the periodical boundary condition, which is necessary to avoid the surface state. This condition requires $\psi(x + L, y, z) = \psi(x, y, z)$, which is satisfied when $\mathrm{e}^{ik_x L} = 1$. This leads to $k_x = \frac{2\pi}{L} n$, where $n$ is an integer. This is equivalent to the previous periodical boundary condition by noting $L = Na$. We apply similar conditions to $k_y$ and $k_z$, and obtain

$$k_x = \frac{2\pi}{L} n_x, k_y = \frac{2\pi}{L} n_y, \text{ and } k_z = \frac{2\pi}{L} n_z \tag{2.21}$$

using three integers, $n_x$, $n_y$, and $n_z$. We assumed the same $L$ for three directions, so we suppose a cube with the edge $L$.

The periodical boundary condition in three directions is defined like this. However, it is difficult to imagine a cube whose edges are circularly connected in three directions. In a two-dimensional space, we can rotate a donut surface in two different directions (Fig. 2.11c). Then, we imagine a similar surface in the

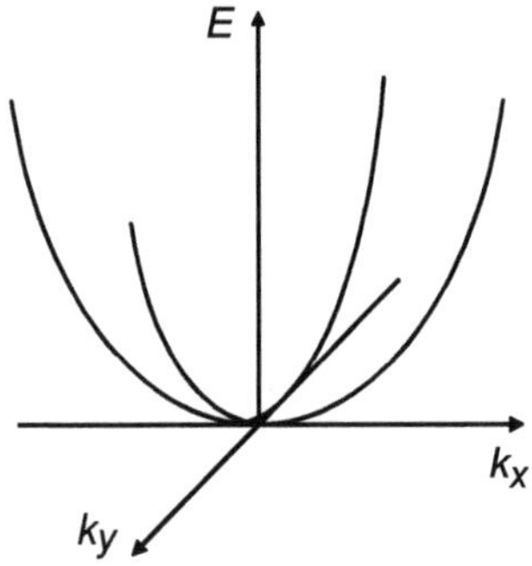


**Fig. 2.10** Free electron energy

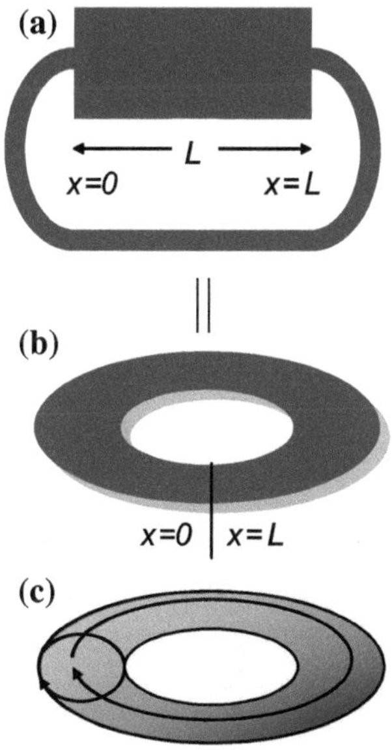


**Fig. 2.11** Periodical boundary condition

three-dimensional space. In general, we implicitly believe we live in a Euclidean space spreading to infinity. However, we can trace a spherical surface along a great circle in any directions to come back to the starting position. A spherical surface is two dimensional, but the surface of a four-dimensional sphere is three-dimensional. On this surface, we go to the right and return from the left, going to the front and returning from the back, and going up and returning from underneath. Einstein imagined such a space in his general theory of relativity.

Returning to the original subject, Eq. 2.21 implies that the $k$-space is restricted to discrete points with the interval of $2\pi/L$ (Fig. 2.12a). In the three-dimensional $k$-space, the energy level exists only on the lattice points ($k_x$, $k_y$, $k_z$) with the interval

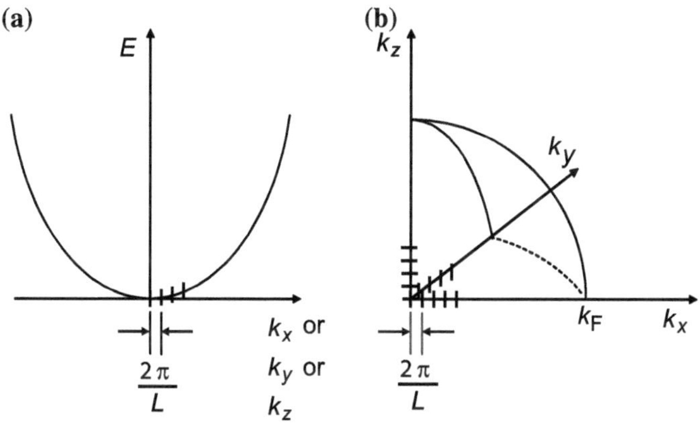


**Fig. 2.12** Lattice points and the Fermi surface in the three-dimensional $k$-space

of $2\pi/L$ (Fig. 2.12b). The energy of the respective point is given by Eq. 2.20, which is proportional to the distance from the origin. When $N$ electrons are incorporated, starting from the origin, each level is occupied by two electrons according to the Pauli exclusion principle. The resulting occupied states have a spherical shape. The boundary between the occupied and unoccupied states is called the **Fermi surface**. The Fermi surface of the three-dimensional free-electron model is a sphere. When the radius of this sphere is $k_F$, the volume is $\frac{4\pi}{3}k_F^3$. Since each $\left(\frac{2\pi}{L}\right)^3$ volume contains an energy level, the volume of the Fermi surface is related to the electron number $N$.

$$2\frac{\frac{4\pi}{3}k_F^3}{\left(\frac{2\pi}{L}\right)^3} = N \tag{2.22}$$

The factor two before the fraction comes from the upspin and downspin electrons entering into an energy level. This relation is reduced to

$$\frac{V}{3\pi^2}k_F^3 = N \tag{2.23}$$

where $V = L^3$ is the volume of the solid. Equation 2.20 affords $E_F = \frac{\hbar^2}{2m}k_F^2$. Through the use of this relation, $k_F$ is eliminated.

$$E_F = \frac{\hbar^2}{2m}\left(\frac{3\pi^2 N}{V}\right)^{\frac{2}{3}} \tag{2.24}$$

This relation implies that $E_F$ is determined by the electron density $N/V$. This equation is solved for $N$, and we obtain

$$N = \frac{V}{3\pi^2} \times \left(\frac{2mE_F}{\hbar^2}\right)^{\frac{3}{2}}. \tag{2.25}$$

When $E$ is slightly increased by d$E$ (Fig. 2.13), the electron number $N$ increases by d$N$. The change is obtained by differentiating $N$ by $E$.

$$D(E) = \frac{dN}{dE} = \frac{V}{2\pi^2}\left(\frac{2m}{\hbar^2}\right)^{\frac{3}{2}} E^{\frac{1}{2}} \tag{2.26}$$

From now on, the subscript F is omitted from $E_F$ and $k_F$ because the following relations hold more generally. $D(E)$ is the number of energy levels existing between $E$ and d$E$, and called the **density of states**. Density of states in the three-dimensional free-electron model is proportional to $\sqrt{E}$ (Fig. 2.13b). This plot looks like the parabola of $E(k)$ (Fig. 2.13a), but the horizontal axis represents a different quantity. The shaded region in Fig. 2.13b corresponds to the occupied region. In Fig. 2.13a, however, we cannot shade the region below $E_F$, because the occupied energy levels exist only on the energy band.

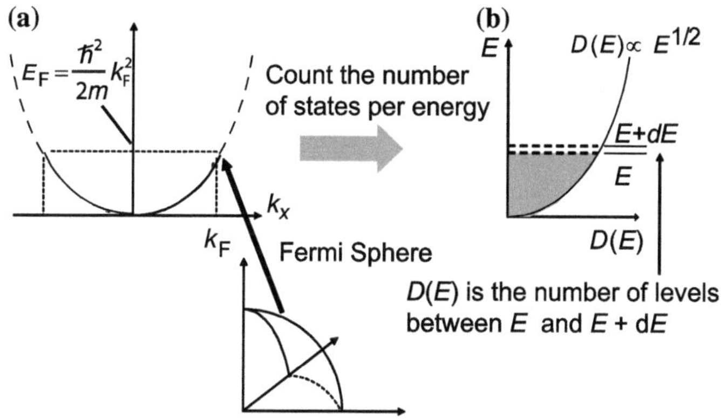


**Fig. 2.13** Density of states $D(E)$ at the Fermi level

We shall extract the density of states in a different way. When the energy changes from $E$ to $E$ + d$E$ (Fig. 2.14), the spherical surface is $4\pi k^2$, and the volume of the spherical shell is $4\pi k^2 \mathrm{d}k$. This is divided by $\left(\frac{2\pi}{L}\right)^3$ to give

$$2\frac{4\pi k^2 \mathrm{d}k}{\left(\frac{2\pi}{L}\right)^3} = \mathrm{d}N. \tag{2.27}$$

$E = \frac{\hbar^2 k^2}{2m}$ gives $\mathrm{d}E = \frac{\hbar^2 k}{m}\mathrm{d}k$. These d$N$ and d$E$ are substituted in the definition of $D(E)$

$$D(E) = \frac{\mathrm{d}N}{\mathrm{d}E} = 2\frac{4\pi k^2}{\left(\frac{2\pi}{L}\right)^3}\frac{\mathrm{d}k}{\frac{\hbar^2 k}{m}\mathrm{d}k} = \frac{V}{2\pi^2}\frac{2m}{\hbar^2}k = \frac{V}{2\pi^2}\left(\frac{2m}{\hbar^2}\right)^{\frac{3}{2}} E^{\frac{1}{2}} \tag{2.28}$$

where $k = \left(\frac{2mE}{\hbar^2}\right)^{\frac{1}{2}}$ is used. This relation is identical to Eq. 2.26.

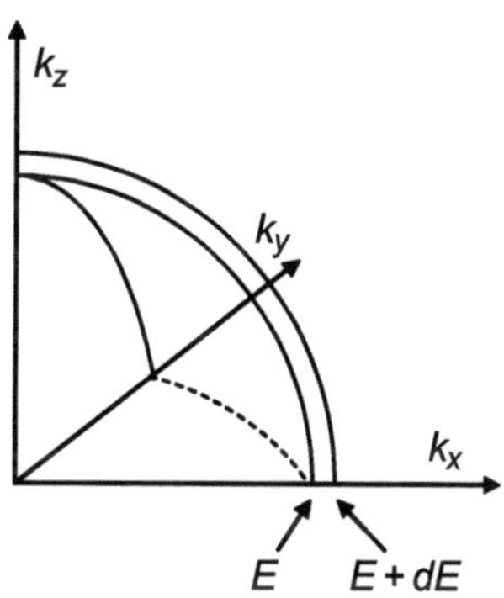


**Fig. 2.14** Density of states at the Fermi spherical shell

Since the energy is represented by a parabola (Fig. 2.13a), the slope d$E$/d$k$ increases with increasing $E$. Then, the number of states per energy decreases. However, the spherical surface increases in proportion to $k^2$, and these two factors make $D(E)$ increase proportionally to $\sqrt{E}$. We will see that many important properties of metals such as specific heat (Sect. 2.3), conductivity (Sect. 3.1.3), and magnetic susceptibility (Sect. 4.4) increase in proportion to the density of states.

*Example 2.2* Energy in a two-dimensional metal is given by $E = \frac{\hbar^2}{2m}(k_x^2 + k_y^2)$. Describe the shape of the Fermi surface. Calculate the density of states.

The Fermi surface of a two-dimensional metal is obtained by $k_x^2 + k_y^2 = k_F^2$ = constant, which gives a circle in the $k_x$-$k_y$ plane. In the three-dimensional $k$-space, this gives an infinitely elongated cylinder (Fig. 2.15a). The number of the energy levels is calculated from the area of the circle $\pi k^2$.

$$2\frac{\pi k^2}{\left(\frac{2\pi}{L}\right)^2} = N \tag{2.29}$$

This is reduced to

$$N = \frac{L^2}{2\pi}k^2 = \frac{mL^2}{\hbar^2\pi}E. \tag{2.30}$$

This is differentiated by $E$ to give the density of states.

$$D(E) = \frac{\mathrm{d}N}{\mathrm{d}E} = \frac{mL^2}{\hbar^2\pi} \tag{2.31}$$

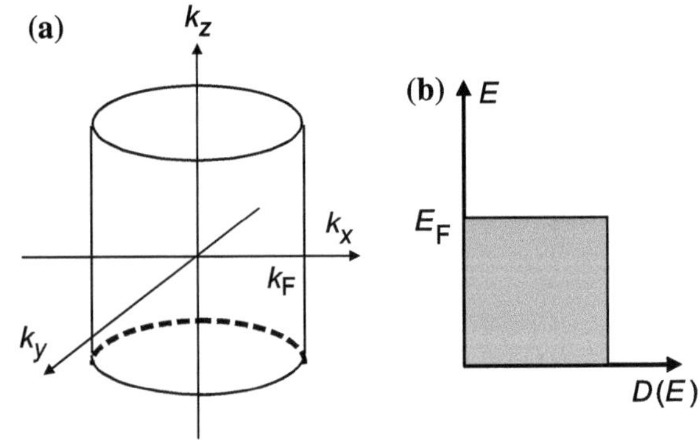


**Fig. 2.15** (**a**) Fermi surface and (**b**) density of states in a two-dimensional metal

The density of states of a two-dimensional metal does not depend on $E$ (Fig. 2.15b).

*Example 2.3* Energy in a one-dimensional metal is given by $E = \frac{\hbar^2}{2m}k_x^2$. Describe the shape of the Fermi surface. Calculate the density of states.

The Fermi surface of a one-dimensional metal is obtained by $k_x^2 = k_F^2 = \text{constant}$, which gives $k_x = \pm k_F$. In the three-dimensional $k$-space, this is a couple of planes (Fig. 2.15a), in which the $-k_F < k < k_F$ region is occupied, and the outside is unoccupied. The number of the energy levels is calculated from the length of the segment $2k_F$.

$$2\frac{2k}{\frac{2\pi}{L}} = N \tag{2.32}$$

Since $E = \frac{\hbar^2 k^2}{2m}$ leads to $k = \frac{\sqrt{2mE}}{\hbar}$, $k$ is deleted to give

$$N = \frac{2L\sqrt{2mE}}{\pi\hbar} \tag{2.33}$$

which is differentiated by $E$ to afford the density of states.

$$D(E) = \frac{dN}{dE} = \frac{L\sqrt{2m}}{\pi\hbar\sqrt{E}} \tag{2.34}$$

As shown in Fig. 2.16b, the density of states in a one-dimensional metal is proportional to $E^{-1/2}$, which diverges at the band edge ($E = 0$).

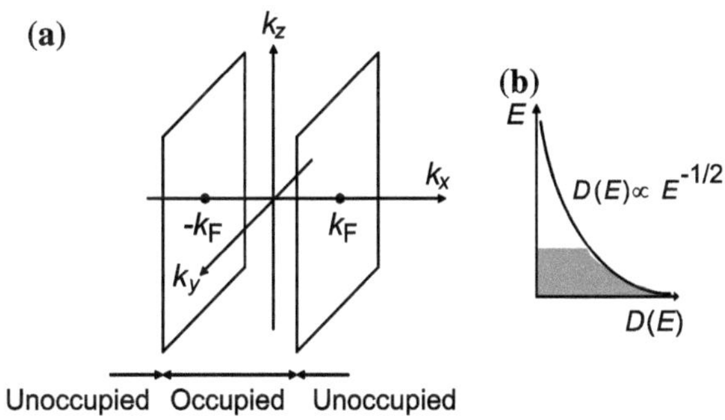


**Fig. 2.16** (**a**) Fermi surface and (**b**) density of states in a one-dimensional metal

## 2.3 Fermi Distribution

An isolated molecule has discrete energy levels, whereas a solid has continuous energy bands. A solid consisting of a large number of atoms ($N$) has many energy levels, which are densely distributed with the energy interval proportional to $1/N$. In a molecule, energy levels below the HOMO are occupied, and those above the LUMO are unoccupied. In a solid, the energy bands below the Fermi energy are occupied. However, since the energy band is continuous, some energy levels are located only slightly above the Fermi energy, and some electrons are thermally excited to these energy levels (Fig. 2.17). This happens because the energy interval in the energy band is smaller than the thermal energy. Therefore, we have to consider a **distribution function** $f(E)$ representing the probability of the occupancy. $f(E)$ is one when 100 % occupied, and zero when entirely vacant. The distribution of metallic electrons is represented by the **Fermi-Dirac distribution function**. Here, we shall extract the Fermi-Dirac distribution function from statistical mechanics.

Suppose the number of energy levels with the same energy $E$ is $C_i$. In the three-dimensional free-electron model, the $C_i$ levels are located on the same spherical Fermi surface. Here, the states with different energies are labeled by the subscript $i$. Suppose $N_i$ electrons are incorporated in these levels (Fig. 2.17). Since an electron is a Fermi particle, we cannot distinguish these $N_i$ electrons. Then, $N_i$ occupied states are chosen from the $C_i$ states, and the statistical weight is

$$W_i = \frac{C_i!}{N_i!(C_i - N_i)!}. \tag{2.35}$$

The entropy is obtained from the definition in statistical mechanics.

$$S = k_B \ln W = k_B \ln \prod_i W_i = k_B \sum_i \ln W_i = k_B \sum_i \ln \frac{C_i!}{N_i!(C_i - N_i)!} \tag{2.36}$$

This is simplified by using Stirling's approximation $\ln N! = N \ln N - N$.

$$S = k_B \sum_i (C_i \ln C_i - N_i \ln N_i - (C_i - N_i) \ln(C_i - N_i)). \tag{2.37}$$

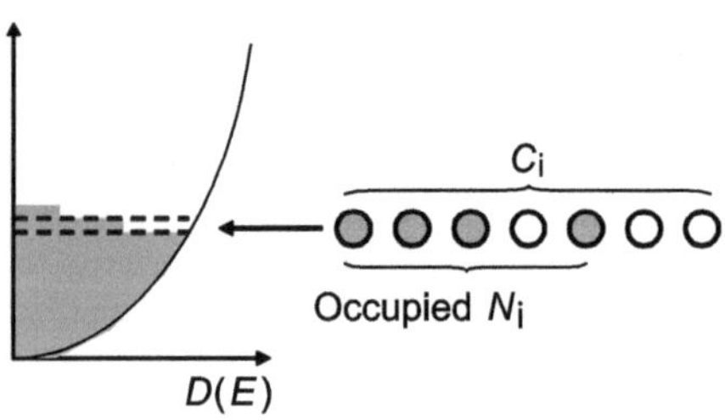


**Fig. 2.17** Fermi-Dirac distribution function

The Gibbs free energy is represented by $F = E - TS - \mu N$, where μ is the chemical potential. Note that the total energy is $E = \Sigma N_i E_i$ and the total electron number is $N = \Sigma N_i$. Here, the distribution of $N_i$ which minimizes $F$ is realized. This is obtained by placing the $N_i$ derivative of $F$ to be zero.

$$\frac{\partial F}{\partial N_i} = E_i + k_B T(\ln N_i - \ln(C_i - N_i)) - \mu = 0 \tag{2.38}$$

This leads to

$$\frac{C_i - N_i}{N_i} = e^{\frac{E_i-\mu}{k_B T}}. \tag{2.39}$$

As shown in Fig. 2.18, the distribution function corresponds to $f(E) = N_i/C_i$. This is obtained from the above equation.

$$f(E_i) = \frac{N_i}{C_i} = \frac{1}{e^{\frac{E_i-\mu}{k_B T}} + 1} \tag{2.40}$$

This is the Fermi distribution function. In this notation, the density of states is $D(E) = C_i$, and the distribution function is $f(E) = N_i/C_i$. Then, as shown in Fig. 2.18, $D(E) \times f(E)$ gives the electron number $N_i$.

The Fermi distribution function is depicted in Fig. 2.19. At $T = 0$ K, Eq. 2.40 gives

$$\begin{aligned} f(E) &= \frac{1}{e^{-\infty}+1} = 1 \quad \text{for } E < \mu \\ f(E) &= \frac{1}{e^{+\infty}+1} = 0 \quad \text{for } E > \mu \end{aligned} \tag{2.41}$$

which affords a step function (Fig. 2.19). At a finite temperature, $e^{\frac{E-\mu}{k_B T}}$ changes gradually, and $f(E)$ drops smoothly (Fig. 2.19). At $E = \mu$, $f(E)$ is always 1/2. Since

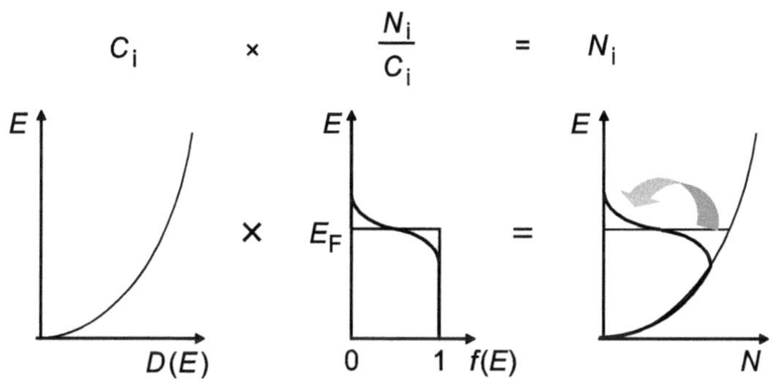


**Fig. 2.18** Density of states × Distribution function = Number of electrons

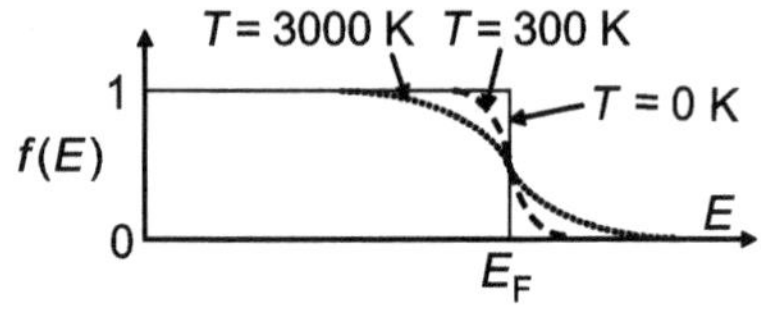


**Fig. 2.19** Fermi distribution functions at finite temperatures

$e^{\frac{E-\mu}{k_B T}}$ changes only around the $E \sim \mu \pm k_B T$ region, $f(E)$ is mostly either one or zero except for the very neighbor of $E \sim \mu$. $\mu$ of an inorganic metal is in the same order as the bandwidth, which is as large as 5 eV ~ 50,000 K. At 300 K, the step is smoothed in the small region of 300 K/50,000 K ~ 0.6 %, and the excited electrons are less than 1 % of the total electrons.

In order to demonstrate this point more clearly, we shall calculate the specific heat of metallic electrons. Since the electron number is represented by $D(E)$ $f(E)$ (Fig. 2.18), the internal energy is obtained by multiplying $E$ to the electron number.

$$U(T) = \int_0^\infty (E - E_F) D(E) f(E) \mathrm{d}E \tag{2.42}$$

Here, $E - E_F$ instead of $E$ is multiplied in order to adopt $E_F$ as the standard. The specific heat at constant volume is obtained by differentiating the internal energy by temperature.

$$C_V = \frac{\partial U}{\partial T} = \int_0^\infty (E - E_F) D(E) \frac{\partial f(E)}{\partial T} \mathrm{d}E \tag{2.43}$$

Note that $T$ appears only in $f(E)$. The $T$ derivative of $f(E)$ is obtained by using $x = \frac{E-\mu}{k_B T}$ in

$$f(E_i) = \frac{1}{e^{\frac{E_i-\mu}{k_B T}} + 1} = \frac{1}{e^x + 1}. \tag{2.44}$$

This affords

$$\frac{\partial f}{\partial T} = \frac{E-\mu}{k_B T^2} \frac{e^x}{(e^x+1)^2}. \tag{2.45}$$

Since $f(E)$ is either one or zero except for the neighbor of $E \sim \mu$, the derivative is mostly zero. This function is only non zero around the neighbor of $E_F$ (Fig. 2.20b).

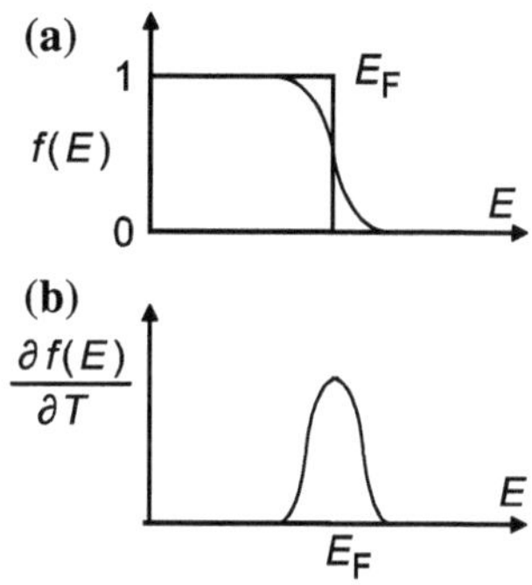


**Fig. 2.20** (**a**) Fermi distribution function and (**b**) the derivative

The integral in Eq. 2.43 is finite only around $E_F$, and $D(E)$ is approximated by $D(E_F)$. Then, $D(E_F)$ is taken out of the integral to afford

$$C_V = D(E_F)\int_0^{\infty}(E-E_F)\frac{\partial f(E)}{\partial T}\mathrm{d}E = D(E_F)\int_0^{\infty}(k_B Tx)\frac{x}{T}\frac{\mathrm{e}^x}{(\mathrm{e}^x+1)^2}k_B T\mathrm{d}x$$
$$= k_B^2 TD(E_F)\int_0^{\infty}x^2\frac{\mathrm{e}^x}{(\mathrm{e}^x+1)^2}\mathrm{d}x. \quad (2.46)$$

Since this integral is $\pi^{2/3}$, the specific heat is

$$C_V = \frac{\pi^2}{3}D(E_F)k_B^2 T. \quad (2.47)$$

The specific heat is proportional to $T$, and represented as $C_v = \gamma T$. The coefficient $\gamma = \frac{\pi^2}{3}D(E_F)k_B^2$ contains only $D(E_F)$ as a characteristic property of a material. As another notation, the density of states in a three-dimensional metal is obtained from Eq. 2.25 as

$$D(E_F) = \frac{3}{2}\frac{N}{E_F} = \frac{3}{2}\frac{N}{k_B T_F} \quad (2.48)$$

so the specific heat is

$$C_v = \frac{\pi^2}{3}\frac{3}{2}\frac{N}{k_B T_F}k_B^2 T = \frac{\pi^2}{2}Nk_B\frac{T}{T_F} = \frac{\pi^2}{2}nR\frac{T}{T_F}. \quad (2.49)$$

Here, the **Fermi temperature** $T_F$ is defined by $E_F = k_B T_F$. This is in the order of 50,000 K, implying only $T/T_F \sim$ 300 K/50,000 K $\sim$ 0.6 % electrons are thermally excited. If electrons follow the classical distribution instead of the Fermi distribution, the electron specific heat should be $C_v \sim 3R$ according to **Dulong–Petit's law**, where $R$ is the gas constant and the number three appears from the $x$, $y$,

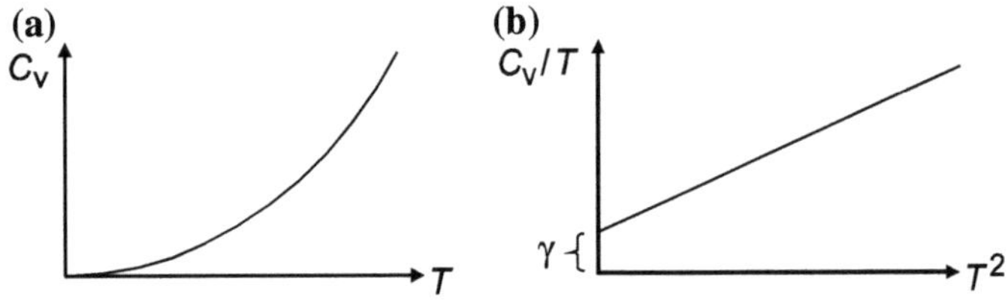


**Fig. 2.21** Temperature dependence of low-temperature specific heat at low temperatures

and $z$ directions. Equation 2.49 becomes less than 1 % of this value because the electrons are Fermi particles, reflecting the quantum mechanical nature of the metal electrons. The states below $E_F$ are occupied and those above $E_F$ are unoccupied, and this situation is practically unchanged at a finite temperature. Then, the Fermi surface is obscured only very slightly by the thermal excitation. Even the metal electrons are occupied similarly to the ordinary molecular orbitals, where the states below the HOMO are occupied, and those above the LUMO are unoccupied.

When the electron-specific heat is added to the lattice-specific heat from the thermal vibration following $T^3$ (Fig. 2.21a), we obtain

$$C_v = \gamma T + \beta T^3. \tag{2.50}$$

This is divided by $T$ to lead to

$$C_v/T = \gamma + \beta T^2 \tag{2.51}$$

so the plot of $C_v/T$ versus $T^2$ gives a straight line (Fig. 2.21b). The intercept of this plot affords γ. The second term $\beta T^3$ is overwhelming at high temperatures, but becomes small at liquid helium temperatures, where the $\gamma T$ term is more important. Then, low-temperature specific heat measurement below liquid helium temperatures gives γ from the plot like Fig. 2.21b. Since γ is determined only from $D(E_F)$ (Eq. 2.47), this affords an estimation of $D(E_F)$. In an insulator having no metallic electrons, a similar plot affords a straight line passing zero.

*Example 2.4* In **Bose–Einstein statistics**, each state can be occupied by more than two particles. In this condition, we shall consider a case where $N_i$ particles enter the $C_i$ states with the same energy. Figure 2.22 represents an example in which these states are occupied by 3, 1, 2, 0, 2 … particles. Suppose the white balls represent the boarders between the states, and the black balls are particles incorporated as 3, 1, 2, 0, 2 … The statistical weight is given by the orders of $N_i$ black balls and $C_i - 1$ white balls.

$$W_i = \frac{(C_i + N_i - 1)!}{N_i!(C_i - 1)!} \tag{2.52}$$

From this relation, calculate the Bose–Einstein distribution by minimizing the free energy. For simplicity, replace $C_i - 1$ with $C_i$.

Assuming $C_i - 1 \to C_i$, and using Stirling's approximation, we obtain

$$\ln W_i = (C_i + N_i)\ln(C_i + N_i) - N_i \ln N_i - C_i \ln C_i. \tag{2.53}$$

This is substituted in $F = E - TS - \mu N$, and differentiated by $N_i$ to give

$$\frac{\partial F}{\partial N_i} = E_i + k_B T(\ln(N_i + C_i) - \ln N_i) - \mu = 0. \tag{2.54}$$

This gives

$$\frac{C_i + N_i}{N_i} = \exp\left(\frac{E_i - \mu}{k_B T}\right). \tag{2.55}$$

The distribution function is

$$f(E_i) = \frac{N_i}{C_i} = \frac{1}{\exp\left(\frac{E_i - \mu}{k_B T}\right) - 1}. \tag{2.56}$$

At $T = 0$ K, this equation leads to

$$f(E) = \frac{1}{e^{+\infty} - 1} = 0 \quad \text{for } E > \mu Z$$

$$e^0 = 1 \text{ affords} f(E) \to \infty \quad \text{for } E = \mu \tag{2.57}$$

and all particles fall into the lowest state (Fig. 2.23). This situation is called the Bose–Einstein condensation.

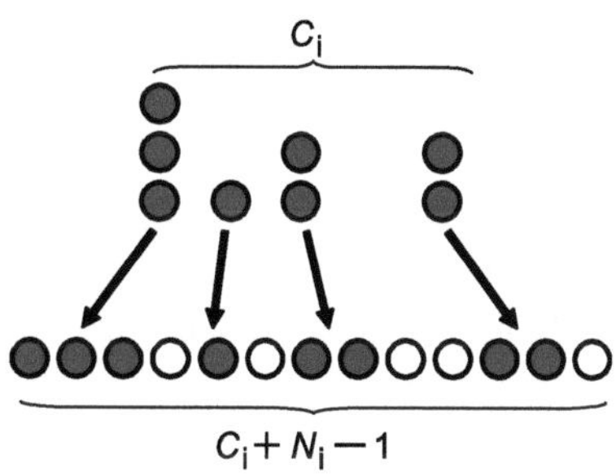


**Fig. 2.22** Bose distribution

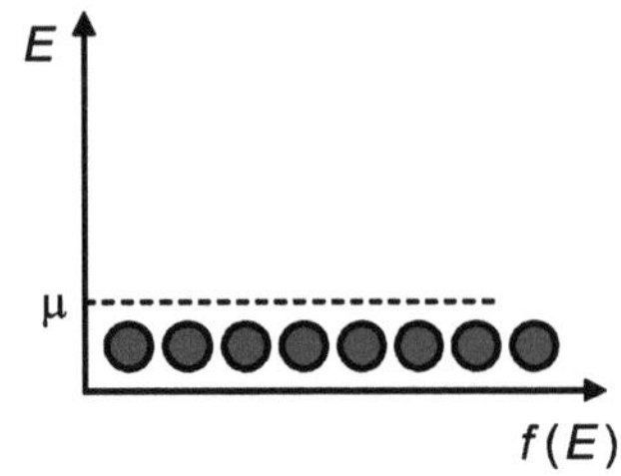


**Fig. 2.23** Bose-Einstein condensation at $T = 0$ K

When $E_i - \mu$ is replaced by $\hbar\omega$ in Eq. 2.56, the Planck distribution

$$f(\omega) = \frac{1}{\exp\left(\frac{\hbar\omega}{k_B T}\right) - 1} \tag{2.58}$$

is obtained. Since a photon is a Bose particle, the black body radiation (Sect. 1.1) follows the Planck distribution. A phonon is a representation of the lattice vibration, which is another Bose particle. The lattice vibration also follows this relation.

In quantum statistical mechanics, the Fermi particles are particles with half-integer spins. The examples are electrons, protons, neutrons, and $^3$He. These particles follow the Fermi distribution. The Bose particles are particles with integer spins. The examples are photons, phonons, and $^4$He. These particles follow the Bose distribution. These distributions are generally represented by

$$f(E_i) = \frac{1}{\exp\left(\frac{E_i - \mu}{k_B T}\right) \pm 1} \tag{2.59}$$

where the + sign gives the Fermi distribution, and the – sign gives the Bose distribution. In the large energy limit ($E - \mu \gg k_B T$), the exponential is larger than one, and $\pm 1$ is neglected to give the classical **Boltzmann distribution**.

$$f(E_i) = \exp\left(-\frac{E_i - \mu}{k_B T}\right) \tag{2.60}$$

*Example 2.5* In the classical Boltzmann distribution, $n_i$ particles of the total $N = \Sigma n_i$ particles are distributed into the $i$-th state. The statistical weight is (Fig. 2.24)

$$W = \frac{N!}{n_1! n_2! n_3! \cdots}. \tag{2.61}$$

From this relation, calculate the Boltzmann distribution by minimizing the free energy.

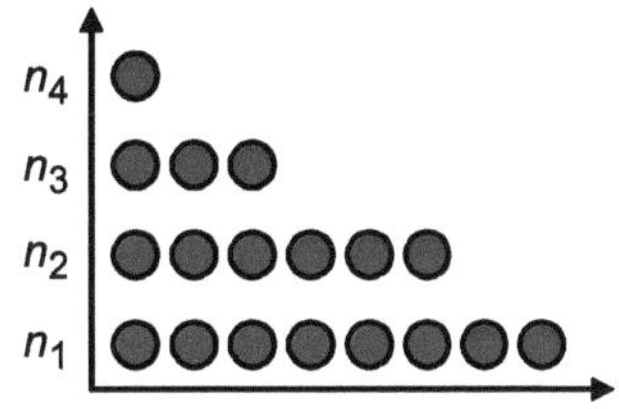


**Fig. 2.24** Boltzmann distribution

Stirling's approximation affords

$$\ln W = \ln \frac{N!}{n_1!n_2!n_3!\cdots} = N \ln N - \sum_i n_i \ln n_i. \tag{2.62}$$

This is placed in $F = E - TS - \mu N$ to give

$$F = \sum_i E_i n_i - k_B T \left( N \ln N - \sum_i n_i \ln n_i \right) - \mu \sum_i n_i. \tag{2.63}$$

This is differentiated by $n_i$ and equated zero to afford

$$\frac{\partial F}{\partial n_i} = E_i + k_B T (\ln n_i + 1) - \mu = 0. \tag{2.64}$$

This gives the Boltzmann distribution.

$$f(E_i) = n_i = \exp\left( -\frac{E_i - \mu}{k_B T} \right) \tag{2.65}$$

The Fermi, Bose, and Boltzmann distributions are summarized in Fig. 2.25. At $T = 0$ K, the Fermi distribution is step-like, and in the Bose distribution all particles condense into the $E = \mu$ state. These ground states appear, respectively, when a state is occupied only by a single particle, and by multiple particles. At $E - \mu \gg k_B T$, the exponential becomes much larger than one, and all distributions converge to the classical Boltzmann distribution.

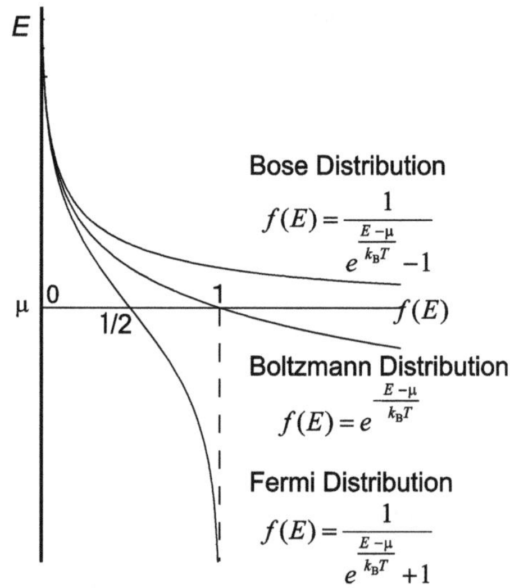


**Fig. 2.25** Various kinds of distributions

## 2.4 Relation Between the Tight-Binding Approximation and the Free-Electron Model

The tight-binding and the free-electron models are two different views of solid electrons. We can expand the tight-binding energy band (Eq. 2.7) around $k = 0$.

$$E(k) = E(0) + bk^2 + \cdots \tag{2.66}$$

Here, the $k$-linear term for the cosine function is zero. When $bk^2$ is regarded as $\frac{\hbar^2 k^2}{2m*}$, the free-electron model is an approximation of the tight-binding band (Fig. 2.26). Therefore, the **effective mass** $\frac{1}{m*} = \frac{1}{\hbar^2}\frac{\partial^2 E(k)}{\partial k^2}$ is not equal to the electron mass.

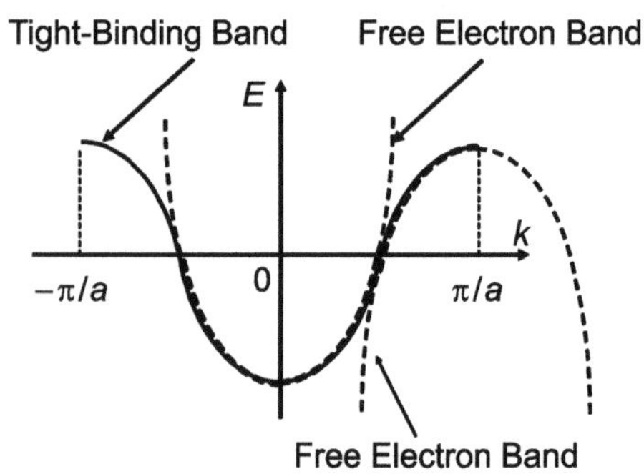


**Fig. 2.26** Relation between the tight-binding and free electron models

*Example 2.6* Calculate the relation of the effective mass $m^*$ in the free-electron model to β in the tight-binding approximation.

Equation 2.7 is differentiated by $k$ twice.

$$\frac{\partial E}{\partial k} = -2\beta a \sin ka$$
$$\frac{\partial^2 E}{\partial k^2} = -2\beta a^2 \cos ka. \tag{2.67}$$

This is substituted in the definition of the effective mass $\frac{1}{m^*} = \frac{1}{\hbar^2}\frac{\partial^2 E(k)}{\partial k^2}$, and $k$ is assumed to be zero.

$$m^* = -\frac{\hbar^2}{2\beta a^2 \cos ka} \rightarrow -\frac{\hbar^2}{2\beta a^2} \tag{2.68}$$

The resonance integral β in the molecular orbital theory is called a **transfer integral** in solid-state physics, and represented by $t$. Then, Eq. 2.7 is rewritten as

$$E = \alpha + 2t \cos ka. \tag{2.69}$$

As shown in Fig. 2.26, the difference between the tight-binding band and the free-electron model increases at large $k$. In particular, the free-electron band increases without a limit, whereas the tight-binding band reaches to a limit at $k = \pi/a$. A free-electron band does not have a bandwidth. However, we can expand the tight-binding band at $k = \pi/a$, and a downward parabola is obtained (Fig. 2.26). Here, $b < 0$ leads to $m^* < 0$, indicating the existence of holes with positive charges. Accordingly, the lower half band is approximated by electrons, while the upper half band has holes (Fig. 2.27).

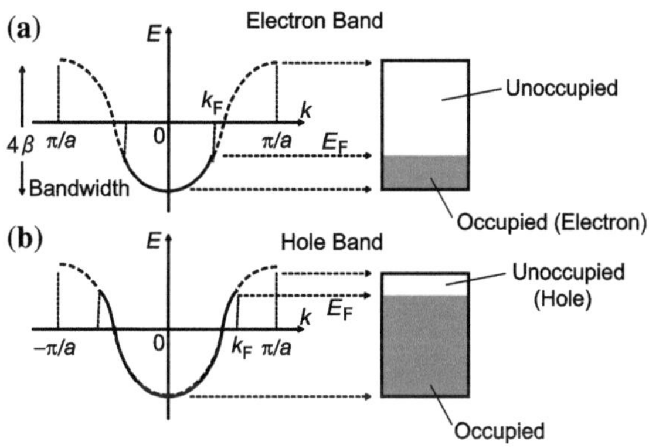


**Fig. 2.27** (**a**) Electron band and (**b**) hole band

*Example 2.7* Calculate the density of states at the band center (in the half-filled case) of a one-dimensional tight-binding band with the bandwidth 1 eV.

Equation 2.67 leads to

$$D(E)=\frac{\partial N}{\partial E}=\frac{\partial N}{\partial k}\frac{\partial k}{\partial E}=-\frac{a}{2\pi}\frac{1}{2\beta a\sin ka}=-\frac{1}{4\beta\pi\sin ka}.$$

Here, sin $ka$ is one at the band center, so $4\beta = 1$ eV affords $D(E) = 1/\pi = 0.318$ states/eV. This is minimum in the band, and $D(E)$ increases and diverges towards the band edge similarly to the one-dimensional free-electron band (Example 2.3).

## 2.5 Two-Dimensional Energy Band

Next, we shall investigate the two-dimensional tight-binding band. As shown in Fig. 2.28, consider a square lattice with a transfer integral β between the neighboring atoms. Similarly to Eq. 2.3, a crystal orbital that satisfies the Bloch condition is represented by

$$\psi=\sum_n\sum_m e^{ink_xa}e^{imk_ya}\chi_{nm}. \tag{2.70}$$

This is substituted in Eq. 2.4 to give the energy.

$$E\propto\sum_n\sum_m\sum_p\sum_q e^{i(n-p)k_xa}e^{i(m-q)k_ya}\int\chi_{np}^{*}H\chi_{mq}\mathrm{d}\tau \tag{2.71}$$

For a given $(n, m)$, nonzero integrals are $\beta e^{ikxa}$ for $(p, q) = (n-1, m)$, $\alpha$ for $(n, m)$, $\beta e^{-ikxa}$ for $(n+1, m)$, $\beta e^{ikya}$ for $(n, m-1)$, and $\beta e^{-ikya}$ for $(n, m+1)$. These terms are collected to

$$E\propto e^{ik_xa}\beta+\alpha+e^{-ik_xa}\beta+e^{ik_ya}\beta+e^{-ik_ya}\beta. \tag{2.72}$$

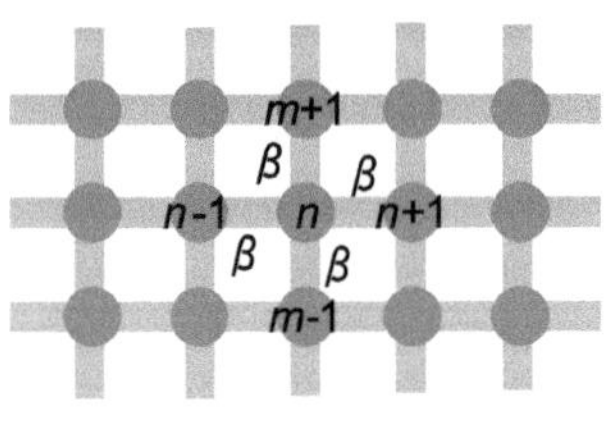


**Fig. 2.28** Two-dimensional square lattice

This affords

$$E = \alpha + 2\beta \cos k_x a + 2\beta \cos k_y a. \quad (2.73)$$

We consider the range of $k$ to be $-\pi/a < k_x < \pi/a$ and $-\pi/a < k_y < \pi/a$, so the first Brillouin zone is represented by a square shown in Fig. 2.29a. Energy given by Eq. 2.73 is depicted in Fig. 2.29b. Energy at the representative points in the $k$ space is

$$\begin{aligned}(k_x, k_y) = {} & (0,0) && \alpha + 4\beta \\ & (\pi/a, 0) && \alpha \\ & (0, \pi/a) && \alpha \\ & (\pi/a, \pi/a) && \alpha - 4\beta.\end{aligned}$$

The minimum energy is $\alpha + 4\beta$, and the maximum energy is $\alpha - 4\beta$, so the bandwidth is $8|\beta|$. The bandwidth is generally given as $2z|\beta|$, where $z$ is the number of the neighboring atoms (coordination number). In a one-dimensional band, $z = 2$ affords $4|\beta|$. In the present two-dimensional square lattice, $z = 4$ affords $8|\beta|$. Similarly, the tight-binding energy of a three-dimensional cubic lattice is

$$E = \alpha + 2\beta \cos k_x a + 2\beta \cos k_y a + 2\beta \cos k_z a. \quad (2.74)$$

This affords the bandwidth $12|\beta|$.

Suppose the two-dimensional square band is half-filled. The Fermi energy is $E_F = \alpha$, and the Fermi surface is a square shown in Fig. 2.30. It seems surprising that the Fermi surface is surrounded by straight lines, but applying the sum equation of cosine functions to Eq. 2.73, we obtain the straight lines because the sine becomes zero at $\pm k_x \pm k_y = \pi/a$. Inside of the square is occupied, whereas the outside squares are unoccupied. The outside squares in Fig. 2.30b are regarded as hole pockets.

Copper-oxide high-temperature superconductors have a CuO square lattice (Fig. 2.31a). The copper atoms are octahedrally coordinated, and oxygen atoms exist in between the Cu atoms. The axial interaction is less important, so the CuO network is regarded as a square lattice. The Fermi energy is located in the Cu $d_{x^2-y^2}$

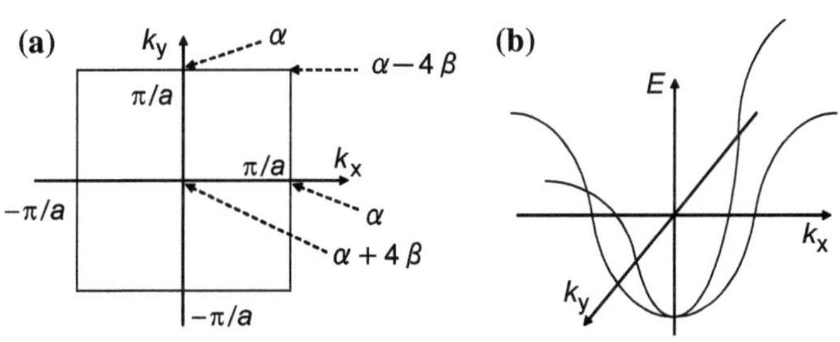


**Fig. 2.29** (**a**) First Brillouin zone and (**b**) energy band in the two-dimensional square lattice

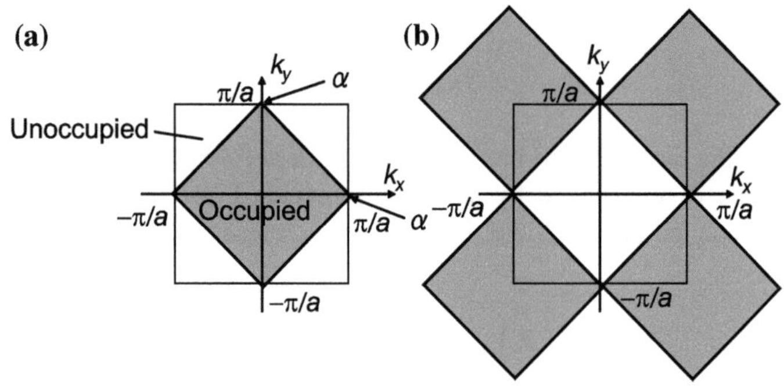


**Fig. 2.30** Fermi surface in the half-filled two-dimensional square lattice

**Fig. 2.31** (**a**) Structure and (**b**) Fermi surface of a cupper oxide superconductor

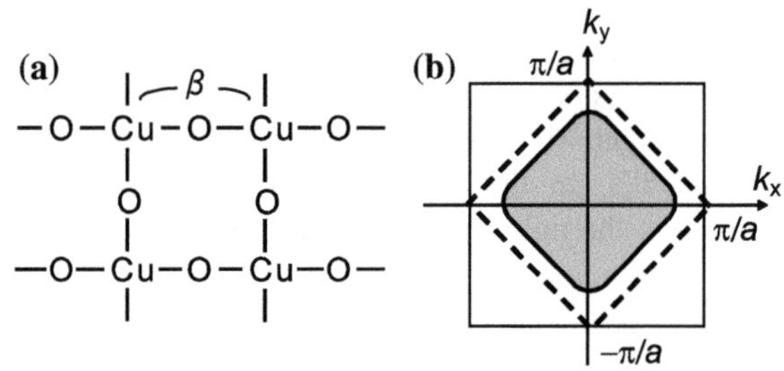


band represented by Eq. 2.73. The non superconducting parent compound $La_2CuO_4$ has $Cu^{2+}$ and the Cu $d_{x^2-y^2}$ band is half-filled. The Fermi surface is represented by Fig. 2.30, and this compound is a Mott insulator (Chap. 5). Superconductors have such compositions as $(La_{1-x}Sr_x)_2CuO_4$; because $Sr^{2+}$ has a smaller charge than $La^{3+}$, $Cu^{2+}$ is partly converted to $Cu^{3+}$. This compound has less electrons than the parent compound, and the CuO network is hole doped. The resulting superconducting material has smaller (electron) Fermi surface than that of the parent compound (Fig. 2.31b), and shows metallic conductivity.

*Example 2.8* Calculate the energy band of a triangular lattice, in which the three directions have different transfers, $\beta_a$, $\beta_b$, and $\beta_p$. Calculate the bandwidth.

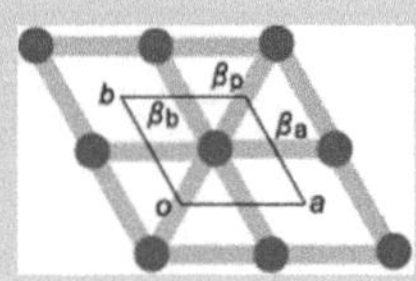


The atomic orbital in a unit cell is one, and the transfers are $\beta_a$ for $\pm a$, $\beta_b$ for $\pm b$, and $\beta_p$ for $\pm(a + b)$. Since the neighboring atoms are six, the energy band is obtained similarly to Eq. 2.72.

$$E = \alpha + \beta_a e^{ik_a a} + \beta_a e^{-ik_a a} + \beta_b e^{ik_b b} + \beta_b e^{-ik_b b} + \beta_p e^{i(k_a a + k_b b)} + \beta_b e^{-i(k_a a + k_b b)}$$
$$= \alpha + 2\beta_a \cos k_a a + 2\beta_b \cos k_b b + 2\beta_p \cos(k_a a + k_b b)$$

Since each cosine function changes over the $\pm 1$ range, the bandwidth is $4|\beta_a| + 4|\beta_b| + 4|\beta_p|$.

## 2.6 Tight-Binding Approximation for General Crystals

We shall consider a unit cell which contains more than two atoms. For example, a unit cell in Fig. 2.32 involves two atoms, 1 and 2. In analogy with the LCAO in the molecular orbital theory, the whole orbital is represented by using the atomic orbitals $\chi_1$ and $\chi_2$

$$\psi = \sum_{i=1}^{2} c_i \chi_i \tag{2.75}$$

where $i$ = 1 and 2. In order to take account of $\chi_i$ involved in different cells, $\chi_i$ is replaced by the crystal orbital.

$$\chi_i \to \sum_{n}^{N} e^{inka} \chi_i(n). \tag{2.76}$$

Combining Eqs. 2.75 and 2.76, we obtain

$$\psi = \sum_{i=1}^{2} \sum_{n}^{N} c_i e^{inka} \chi_i(n). \tag{2.77}$$

In general, $i$ is a sum for orbitals in a unit cell, and $n$ is a sum for different cells.

Similarly, to the usual molecular orbital theory, the energy is obtained by using this equation in Eq. 2.4. The resulting equation contains $c_1$ and $c_2$, so we calculate $\frac{\partial E}{\partial c_i} = 0$, leading to the secular equation.

$$\begin{vmatrix} \alpha_{11}(k) - E & \beta_{12}(k) \\ \beta_{21}(k) & \alpha_{22}(k) - E \end{vmatrix} = 0 \tag{2.78}$$

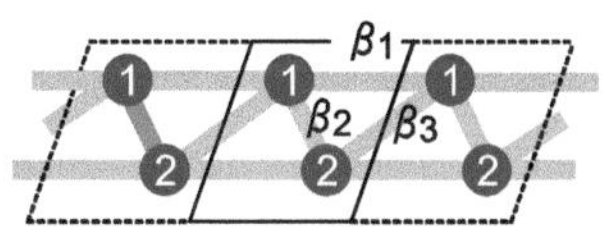


**Fig. 2.32** Two atoms in a unit cell

The diagonal term $\alpha_{ii}(k)$ is an integral where the Hamiltonian is sandwiched by the same crystal orbital $\sum e^{inka}\chi_i(n)$

$$\begin{aligned}\alpha_{ii}(k) &= \int \left(\sum_m e^{-imka}\chi_i^*(m)\right) H \left(\sum_n e^{inka}\chi_i(n)\right) d\tau \\ &= \alpha_i + \sum_{n'} \beta_{ii}(n')e^{in'ka}.\end{aligned} \tag{2.79}$$

which includes not only $\alpha_i$ coming from the same $\chi_i$ in the same unit cell, but also $\beta_{ii}(k)$ coming from different $\chi_i$ belonging to the neighboring cells. The latter is $k$ dependent. For example, atom 1 in Fig. 2.32 includes $\beta_1$ to the neighboring cells.

$$\alpha_{11} = \alpha_1 + \beta_1 e^{ika} + \beta_1 e^{-ika} = \alpha_1 + 2\beta_1 \cos ka \tag{2.80}$$

The nondiagonal term is similarly obtained.

$$\begin{aligned}\beta_{ij}(k) &= \int \left(\sum_m e^{-imka}\chi_i^*(m)\right) H \left(\sum_n e^{inka}\chi_j(n)\right) d\tau \\ &= \sum_n \beta_{ij}(n')e^{in'ka}\end{aligned} \tag{2.81}$$

For example, the term 12 in Fig. 2.31 is

$$\beta_{12} = \beta_2 + \beta_3 e^{-ika}. \tag{2.82}$$

This consists of the intra-cell $\beta_2$ and the inter-cell $\beta_3$ multiplied by $e^{-ika}$. In general, when the transfer $\beta$ exists in the $r$ direction between the atoms 1 and 2, the matrix element is obtained by summing $\beta e^{ikr}$. The following example illustrates a practical case.

*Example 2.9* Calculate the energy band of a one-dimensional alternating chain, where $\beta$ is alternately $\beta_1$ and $\beta_2$ as shown in Fig. 2.33. Estimate the overall bandwidth and the energy gap.

A unit cell contains two atoms (1 and 2), so the secular equation is 2 × 2, where the diagonal terms are $\int \chi_1^* H \chi_1 d\tau = \int \chi_2^* H \chi_2 d\tau = \alpha$ and the nondiagonal terms are $\beta_1 + \beta_2 e^{i2ka}$ and $\beta_1 + \beta_2 e^{-i2ka}$.

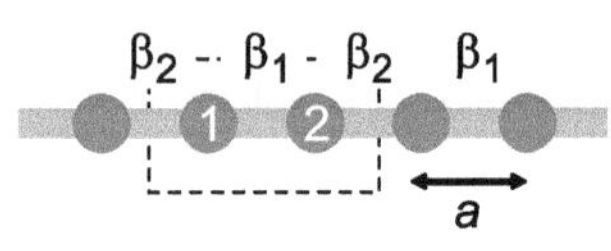


**Fig. 2.33** One-dimensional alternating chain

The secular equation is

$$\begin{vmatrix} \alpha - E & \beta_1 + \beta_2 e^{i2ka} \\ \beta_1 + \beta_2 e^{-i2ka} & \alpha - E \end{vmatrix} = 0. \tag{2.83}$$

This is reduced to

$$(\alpha - E)^2 = \left(\beta_1 + \beta_2 e^{i2ka}\right)\left(\beta_1 + \beta_2 e^{-i2ka}\right) \tag{2.84}$$

which gives

$$E = \alpha \pm \sqrt{\beta_1^2 + \beta_2^2 + 2\beta_1\beta_2 \cos 2ka}. \tag{2.85}$$

This is depicted in Fig. 2.34. Since the energy at $k = 0$ is $E = \pm|\beta_1 + \beta_2|$, the overall bandwidth is $W = 2\,|\beta_1 + \beta_2|$. The energy at $k = \pi/a$ is $E = \pm|\beta_1 - \beta_2|$, so the energy gap is $E_g = 2|\beta_1 - \beta_2|$.

When we assume $\beta_1 = \beta_2 = \beta$ in Eq. 2.85, we obtain

$$\begin{aligned} E &= \alpha \pm \sqrt{\beta^2 + \beta^2 + 2\beta^2 \cos 2ka} = \alpha \pm \beta\sqrt{2(1 + \cos 2ka)} \\ &= \alpha + 2\beta \cos ka. \end{aligned} \tag{2.86}$$

Here, we used the double-angle formula, $1 + \cos 2x = 2\cos^2 x$. This is identical to Eq. 2.7. When $\beta_1 = \beta_2$, the energy gap $2|\beta_1 - \beta_2|$ is zero, and the upper and the lower bands are connected (Fig. 2.35a). We have defined the unit cell as $2a$ (Fig. 2.33), so the uniform chain has a periodicity of $a$. When the upper band is moved to the $k > \pi/2a$ region, the band is depicted in the $-\pi/a < k < \pi/a$ region (Fig. 2.25b). This is the original cosine band. The band scheme like (a) is called a **reduced zone**, and the scheme (b) is called an **extended zone**. These two represent the same band. The energy band at $\beta_1 \neq \beta_2$ is depicted (c) in the reduced zone and (d) in the extended zone. If this band is half-filled, the lower band is fully occupied, but the upper band is empty. Figure 2.35a is equivalent to Fig. 2.5a, where polyacetylene is a metal. In the actual polyacetylene, the single and double bonds have different $\beta_1 \neq \beta_2$ (bond alternation), and an energy gap emerges as shown in Fig. 2.35c, d. Note the lattice periodicity is doubled in this case.

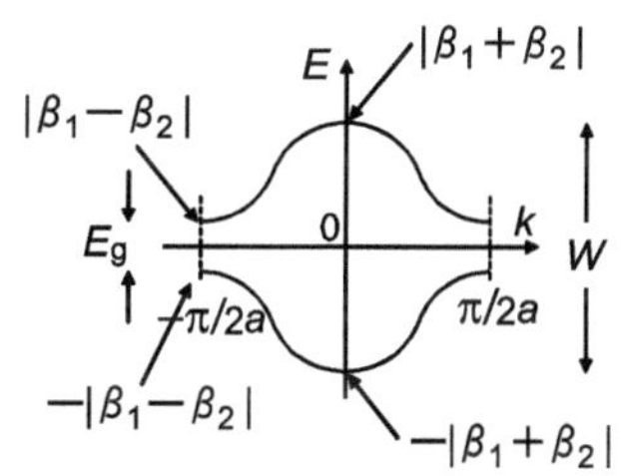


**Fig. 2.34** Energy band in a one-dimensional alternating chain

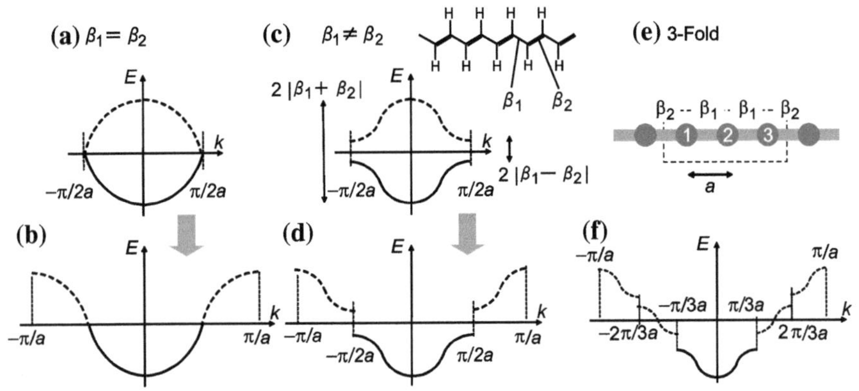


**Fig. 2.35** (**a, b**) Uniform one-dimensional band and (**c, d**) alternating one-dimensional band. (**e**) Three-fold chain and (**f**) the energy band

Similarly, when the periodicity of the chain is $3a$ (Fig. 2.35e), energy gaps appear at $k = \pi/3a$ as show in Fig. 2.35f. Figure 2.35e has only two kinds of $\beta_1$ and $\beta_2$, but the periodicity $\beta_1\beta_1\beta_2$ is three-fold. In general, when the lattice periodicity is $n$ times the original lattice, the repeating unit in the $k$ space becomes $1/n$, and an energy gap emerges at $\pi/na$. When $k_F$ is located at $\pi/na$, the material becomes an insulator. Consequently, a $1/n$-filled one-dimensional band becomes insulating when the lattice periodicity becomes $na$. This is called the **Peierls transition**. Polyacetylene is regarded as a Peierls insulator. Bond alternation in antiaromatic cyclobutadiene (Example 1.12) is also regarded as a Peierls insulator. In Chap. 7, we discuss organic conductors with noninteger periodicity such as 3.39 molecules (Fig. 7.14) and 2.29 molecules (Fig. 7.43). Even in these cases, a gap appears at $k_F$ corresponding to the inverse of the periodicity.

In general, when a unit cell contains $N$ atomic orbitals, the LCAO of the crystal orbital leads to a secular equation

$$\begin{vmatrix} \alpha_{11}(k) - E & \beta_{12}(k) & \cdots \\ \beta_{21}(k) & \alpha_{22}(k) - E & \cdots \\ \vdots & \vdots & \vdots \end{vmatrix} = 0. \tag{2.87}$$

The matrix elements depend on $\boldsymbol{k} = (k_x, k_y, k_z)$. At a certain $\boldsymbol{k}$, however, the matrix elements are complex numbers, and the solution gives $N$ energy levels. Note the solutions are real numbers because the matrix is a Hermetian. When we use a slightly different $\boldsymbol{k}$, we obtain slightly different matrix elements. The energy levels are slightly different. By connecting these energy levels, we obtain the energy bands (Fig. 2.36). Even when $N$ is very large, the number of the energy bands is the same as the number of the starting orbitals $N$.

Figure 2.36 is the energy bands of germanium. Germanium forms a diamond lattice from the Ge-Ge covalent bonds, where the original $4s4p^3$ atomic orbitals construct the bonding and antibonding orbitals. The occupied bonding orbitals form

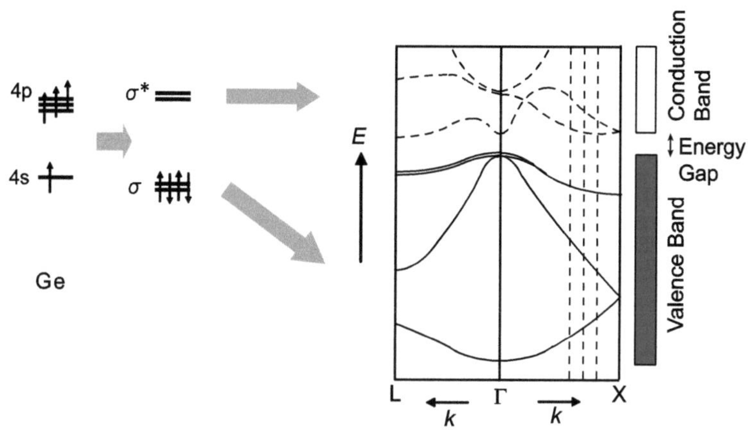


**Fig. 2.36** An example of a three-dimensional energy band: chemical bonds and energy bands in germanium

the valence band, and the unoccupied antibonding orbitals form the conduction band. An energy gap emerges between the valence and conduction bands, so germanium is a semiconductor. The energy gap is 0.66 eV for germanium, which increases to 1.12 eV in silicon and 5.47 eV in diamond.

## 2.7 Brillouin Zone

The unit cell in the *k* space is called the Brillouin zone. The Brillouin zone of the two-dimensional square lattice (Sect. 2.5) is shown in Fig. 2.37. In crystallography, lattice points in the inverse space are located on the points such as (1, 0) and (0, 1). In solid-state physics, these points are on $2\pi/a$, and regarded as equivalent to the origin Γ, when we discuss the electronic state. In solid-state physics, the Fourier transform is defined by $e^{ikr}$, so the lattice point appears at $2\pi/a$. In contrast, the

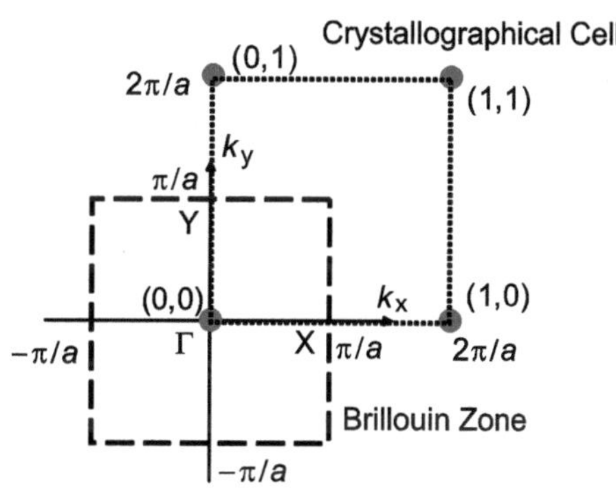


**Fig. 2.37** The first Brillouin zone (*dashed*) and the crystallographical cell (*dotted*) in the two-dimensional square lattice

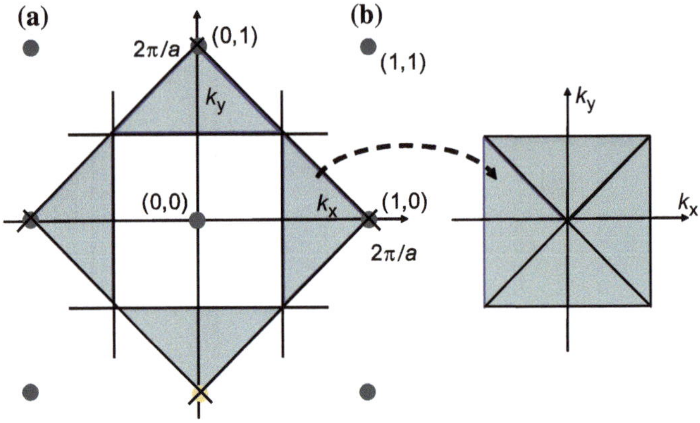


**Fig. 2.38** The second Brillouin zone in the two-dimensional square lattice

Fourier transform in crystallography is defined by $e^{2\pi ikr}$, and the lattice point is located on $k = 1/a$. Note the definition of the $k$ space is different by 2π.

The boundary of the Brillouin zone is defined by the midpoint to the adjacent lattice point. Therefore, the boundary appears at $\pi/a$. In general, the zone boundary is obtained by the perpendicular bisector between the original and adjacent lattice points. The resulting Brillouin zone is a square (Fig. 2.37). A crystallographical unit cell is surrounded by the lattice points, which is different from the Brillouin zone (Fig. 2.37). However, these two squares have the same area, and represent the equivalent unit cell. The origin of the Brillouin zone (0, 0) is conventionally called the Γ point, and the crossing points of the $k_x$ and $k_y$ axes to the zone boundary are, respectively, called the X and Y points.

Since the Brillouin zone is surrounded by perpendicular bisectors to the adjacent lattice points, the perpendicular bisector to the (1, 1) point defines an oblique boarder (Fig. 2.38). Accordingly, shaded regions surrounded by the new perpendicular bisectors are generated out of the first Brillouin zone. The right triangle is located left of the (1, 0) point, so moved to the left of the (0, 0) point. Similarly, the left triangle is moved to the right and so on to generate a new square that is identical to the first Brillouin zone. This square is a region originally located out of the first Brillouin zone, and called the second Brillouin zone. In the same way, perpendicular bisectors to more distant lattice points define the third and fourth Brillouin zones.

Next, we investigate non orthogonal lattices. As shown in Fig. 2.39, the perpendicular bisectors define an irregular hexagon, which is the first Brillouin zone. In an oblique lattice, perpendicular bisectors to diagonal lattice points constitute the boarder. In crystallography, a unit cell is a parallelogram, but for example, the

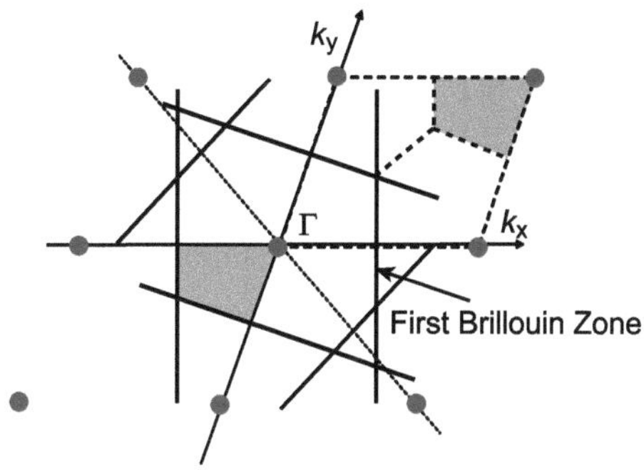


**Fig. 2.39** The Brillouin zone for a non orthogonal lattice

shaded regions in Fig. 2.39 are equivalent to each other. We can construct the first Brillouin zone by moving four separate panels from the parallelogram.

Examples of three-dimensional Brillouin zones are shown in Fig. 2.40. Brillouin zones of primitive cubic and orthorhombic lattices are again primitive cubic and orthorhombic lattices (Fig. 2.40b), because the lattice is orthogonal. However, similar to Fig. 2.39, a C-center orthogonal system leads to a hexagonal face (Fig. 2.40c). The reciprocal cell of a body-center cubic lattice is a face-center cubic, and that of a face-center cubic lattice is a body-center cubic. The corresponding Brillouin zones are shown in Fig. 2.40d, e. The polyhedron in Fig. 2.40d is surrounded by rhombuses with equivalent edges. This modified polyhedron is called a rhombus dodecahedron [8]. The polyhedron in Fig. 2.40e is produced by cutting corners of a cube (Fig. 2.40f) or an octahedron (Fig. 2.40g), and called a truncated cube or truncated octahedron. This polyhedron is surrounded by squares and hexagons with equivalent edges. Note that we can fill the whole three-dimensional space by stacking these polyhedra attaching a hexagon face to a hexagon and a square face to a square. Not only this polyhedron but also all Brillouin zones fulfill this condition. The Brillouin zone of a triclinic system has a complicated shape (Fig. 2.40a), but can fill the whole three-dimensional space. This is obvious because these polyhedra are surrounded by perpendicular bisectors of lattice points. The body-center and face-center cubic lattices involve more than two lattice points, but the Brillouin zone contains only a single Γ point. In general, the Brillouin zone is a minimal repeating unit centered at the Γ point.

*Example 2.10* Identify the point groups of Fig. 2.40d, e.

Both are $O_h$.

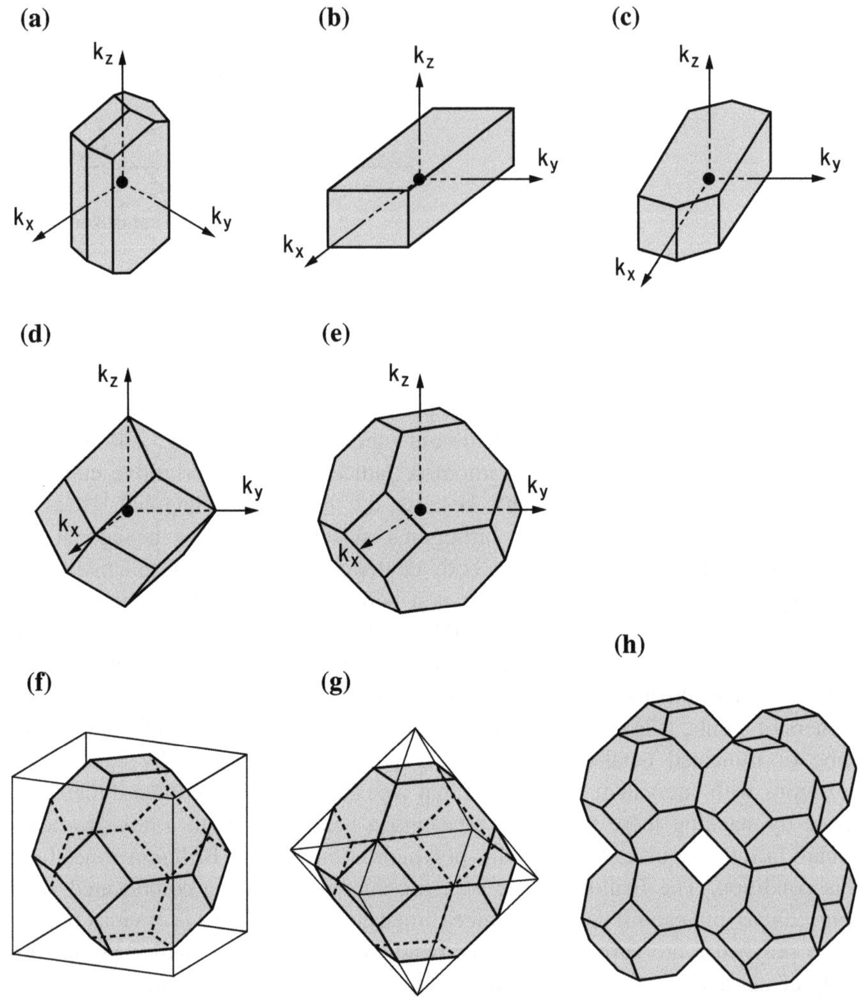


**Fig. 2.40** Three-dimensional Brillouin zones. **a** Triclinic. **b** Orthorhombic P. **c** Orthorhombic C. **d** Body-center cubic. **e** Face-center cubic. **f** Truncated cubic. **g** Truncated octahedron. **h** Face-center cubic

## 2.8 Fermi Surface of Metals

Next, we discuss the Fermi surface of metals [9]. Alkali metals such as sodium and potassium have a body-center lattice, and the Brillouin zone is represented by Fig. 2.40d. When one electron enters the energy band, the volume of the Fermi surface is half the first Brillouin zone, because the s band receives up to two electrons. So the Fermi surface is a sphere and does not touch the zone boundary

**Fig. 2.41** Spherical Fermi surface of an alkali metal

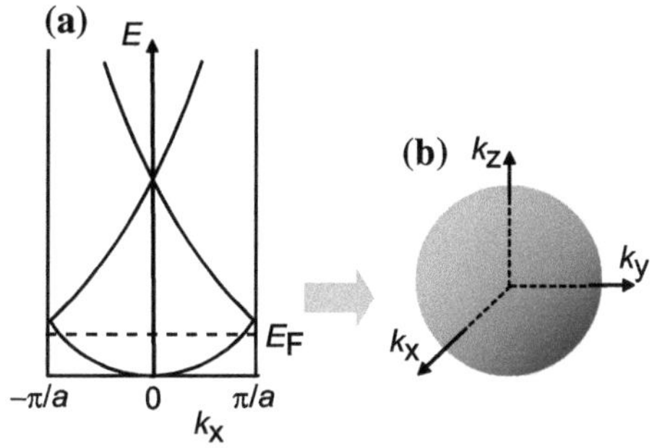


(Fig. 2.41). Experimental methods described in Sect. 3.4 have proved the deviation from the perfect sphere is less than 1 %.

Alkali earth metals such as magnesium and calcium have two valence electrons, and the volume of the Fermi surface is equal to the Brillouin zone. Accordingly, the Fermi surface spread out of the first Brillouin zone (Fig. 2.42). These metals have a hexagonal or body-center lattice, but a square lattice is schematically depicted in Fig. 2.42. In the first Brillouin zone, the shaded region is occupied as shown in Fig. 2.42c. Holes exist in the white region to make a star-like hole pocket at the corner. In the second Brillouin zone, the shaded region in Fig. 2.42d is occupied to make an elliptical electron pocket. Like this, starting from a naive free-electron model, we can predict fairly complicated electron and hole pockets. Such a Fermi surface has been proved by actual experiments. The group 3 and 4 elements have a similar Fermi surface.

The Fermi surface of a transition metal is similarly composed of many electron and hole pockets. However, we shall investigate the global feature (Fig. 2.43a). For example, electron configuration of a titanium atom is $4s^2 3d^2$. In the solid metal, the s band is larger than the d band, and the s band spreads over the d band. Then, it is appropriate to approximate that the s band has one electron, and other electrons enter the d band like $4s^1 3d^3$. In elemental metals, the electron configuration is generally represented by $s^1 d^{n-1}$. Since the five d orbitals can receive up to ten electrons, six-electron systems such as chromium, molybdenum, and tungsten have

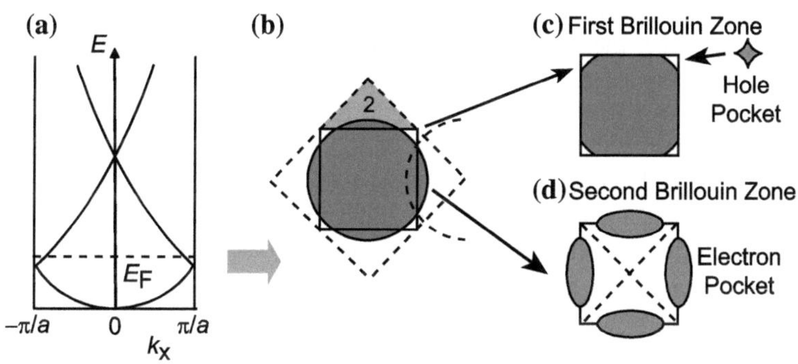


**Fig. 2.42** Fermi surface of an alkali earth metal

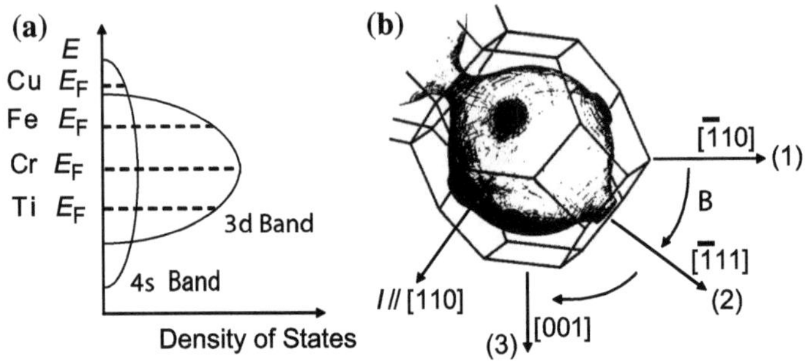


**Fig. 2.43** **a** Energy bands of transition metals. **b** Fermi surface of copper

$s^1d^5$ configuration, and the Fermi energy is located at the center of the d band. The d band is composed of many energy bands, but the density of states is maximum at the band center. Since the metal electrons constitute a kind of covalent bonds, these metals form strong chemical bonds with the surrounding atoms. This is the reason that these metals exhibit the highest melting points among all elements.

Iron metal has $4s^13d^7$ configuration, and the d band is mostly occupied up to the upper end. Copper metal has $4s^13d^{10}$ configuration, and the d band is entirely occupied. Since the s band has one electron, the Fermi surface is almost spherical similarly to alkali metals (Fig. 2.43b), where the volume of the Fermi sphere is half the Brillouin zone. Copper has a face-center cubic lattice, and the Brillouin zone is represented by Fig. 2.40e. However, the Fermi surface of copper touches the zone boundary, and is connected to the adjacent Fermi sphere (Fig. 2.43b). The presence of such a connected Fermi surface has been proved experimentally (Sect. 3.5).

## 2.9 Fermi Surface of Low-Dimensional Metals

There are examples of one- and two-dimensional metals. A platinum complex with the composition of $K_2[Pt(CN)_4]Br_{0.3}xH_2O$ forms a one-dimensional conductor called KCP (Fig. 2.44a). Platinum forms a square-planar complex like $[Pt(CN)_4]^{2-}$, which is stacked to form a one-dimensional Pt–Pt chain (//*c*). When partially oxidized with bromine, the one-dimensional band composed of the Pt $d_{z^2}$ orbitals is partially filled. This complex shows metallic conductivity around room temperature, but undergoes a Peierls transition at 250 K, below which the platinum chain is distorted according to the $2k_F = 0.30\ c^*$ periodicity.

One-dimensional columns of $NbSe_3$ are composed of trigonal prisms (Fig. 2.44b). Since a unit cell contains three different columns, there are three different sheets of the Fermi surface. Therefore, $NbSe_3$ undergo the Peierls transitions twice at 143 and 53 K. However, the interchain interaction is too large to make the columns entirely insulating. Then, after the resistance once goes up, a metallic decrease is restored again.

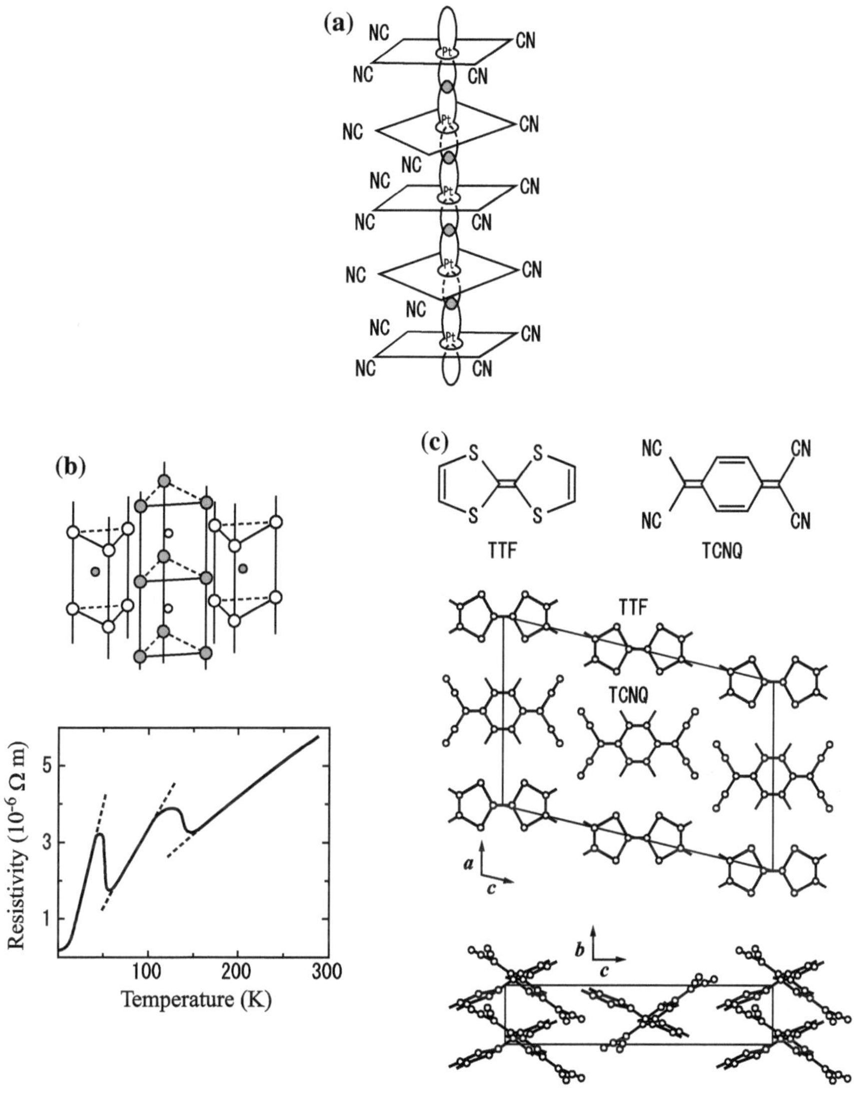


**Fig. 2.44** One-dimensional metals: (**a**) KCP, (**b**) structure and resistivity of $NbSe_3$, and (**c**) (TTF) (TCNQ)

Many organic conductors consist of planar molecules with an extended π-electron system, and tend to form one-dimensional conductors (Sect. 7.4). Figure 2.44c shows a typical one-dimensional organic charge-transfer complex composed of tetrathiafulvalene (TTF) and tetracyanoquinodimethane (TCNQ). The conductivity is highest along the stacking ($b$) axis.

Organic conductors are usually one-dimensional. Therefore, ordinary organic conductors undergo a Peierls transition and become insulating at low temperatures.

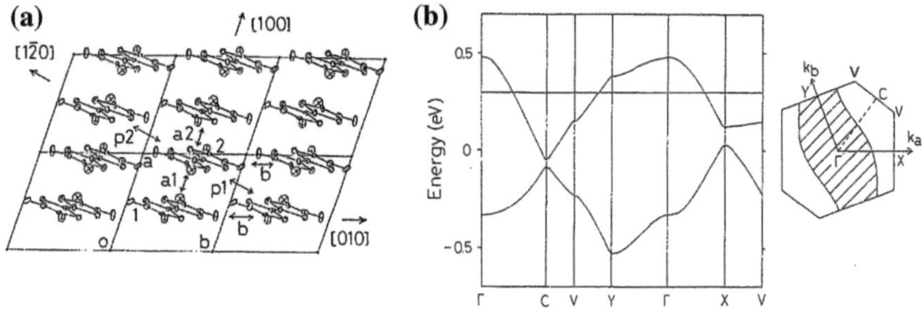


**Fig. 2.45** (**a**) Crystal structure and (**b**) energy band of $(TMTSF)_2PF_6$. The transfer integrals are $\beta_{a1}$ = 200, $\beta_{a2}$ = 230, $\beta_b$ = 35, $\beta_{p1}$ = 20, and $\beta_{a1}$ = 7 meV

In order to make organic superconductors, it is necessary to suppress the Peierls transition and to maintain the metallic conductivity down to low temperatures. For this purpose, two-dimensional organic conductors have been explored. Figure 2.45 shows the crystal structure of the first organic superconductor, $(TMTSF)_2PF_6$ (TMTSF: tetramethyltetraselenafulvalene in Fig. 7.16). TMTSF has many molecular orbitals, but to investigate the electronic properties, it is sufficient to consider the molecular orbital in which the Fermi energy is located. From the composition, TMTSF has 1/2+ charge, so we only consider the partially filled HOMO. A unit cell contains two molecules designated as 1 and 2 (Fig. 2.45a), and the secular equation is derived similar to Eqs. 2.80 and 2.82.

$$\begin{vmatrix} \beta_{11} - E & \beta_{12} \\ \beta_{21} & \beta_{22} - E \end{vmatrix} = 0$$

$$\beta_{11} = \beta_{22} = \beta_b e^{ik_b b} + \beta_b e^{-ik_b b} = 2\beta_b \cos k_b b$$

$$\beta_{12} = \beta_{21}^* = \beta_{a1} + \beta_{a2} e^{-ik_a a} + \beta_{p1} e^{-ik_b b} + \beta_{p2} e^{-i(k_a a - k_b b)} \tag{2.88}$$

Here, $\alpha$ is omitted. Molecule 1 is sandwiched by two Molecules 1 located at $\pm b$, to which the central molecule is connected by $\beta_b$. This forms $\beta_{11}$. Molecule 2 is similarly related to Molecules 2 in the adjacent $\pm b$ cells, and $\beta_{22}$ has the same form. The interaction between Molecule 1 and Molecule 2 ($\beta_{12}$) consists of $\beta_{a1}$ and $\beta_{a2}$ in the $\pm a$ directions, in addition to $\beta_{p1}$ and $\beta_{p2}$ in the oblique direction. The secular equation is reduced to

$$(\beta_{11} - E)^2 = \beta_{12}\beta_{12}^*. \tag{2.89}$$

The insertion of $\beta_{ij}$ affords the solutions. However, it is more convenient to multiply $e^{ik_a a/2}$ to $\beta_{12}$.

$$\beta_{12}e^{ik_a a/2} = \beta_{a1}e^{ik_a a/2} + \beta_{a2}e^{-ik_a a/2} + \beta_{p1}e^{i(k_a a/2 - k_b b)} + \beta_{p2}e^{-i(k_a a/2 - k_b b)} \quad (2.90)$$

and $e^{-ik_a a/2}$ is multiplied to $\beta_{12}$*. It is generally possible to multiply this kind of an arbitrary phase. Equation 2.88 is written assuming both Molecules 1 and 2 are located in the same cell, but Eq. 2.90 is obtained supposing Molecule 2 is placed at ***a***/2. The energy band is represented by using $\Delta = \beta_{12}\,\beta_{12}^*$.

$$E(k) = 2\beta_b \cos(k_b b) \pm \sqrt{\Delta}$$

$$\begin{aligned}\Delta = &\left[(\beta_{a1} + \beta_{a2})\cos\left(\frac{k_a a}{2}\right) + (\beta_{p1} + \beta_{p2})\cos\left(\frac{k_a a}{2} - k_b b\right)\right]^2 \\ &+ \left[(\beta_{a1} - \beta_{a2})\sin\left(\frac{k_a a}{2}\right) + (\beta_{p1} - \beta_{p2})\sin\left(\frac{k_a a}{2} - k_b b\right)\right]^2.\end{aligned} \quad (2.91)$$

The transfer integrals are estimated from the molecular orbital calculation as listed in the caption of Fig. 2.45. Through use of these values, the energy band is calculated as shown in Fig. 2.45b. The Brillouin zone is represented by a hexagon because the unit cell is triclinic (Figs. 2.39 and 2.40a). The energy bands are plotted starting from the Γ point to the upper right C point, followed by the anticlockwise rotation to V, Y, Γ, and X, to go to another V point. Since a unit cell contains two molecules, Eq. 2.89 is a quadratic equation and affords two energy bands. The 1/2+ charge of the TMTSF molecule implies the HOMO band is 3/4-filled, and the upper 1/4 is empty. The hole part corresponds to the shaded region in the Brillouin zone. The shaded parts are chosen from the top of the energy band until the shaded area becomes half the Brillouin zone. The boarder of the shaded region is a winding curve extending in the $b$* direction, so the warping Fermi surface extending in the $b$*$c$* direction indicates a one-dimensional metal band along the $a$ axis. This is associated with $\beta_{a1}$ and $\beta_{a2}$, which are ten times larger than the other transfers. The interchain transfers, however, make the Fermi surface considerably warping. As a

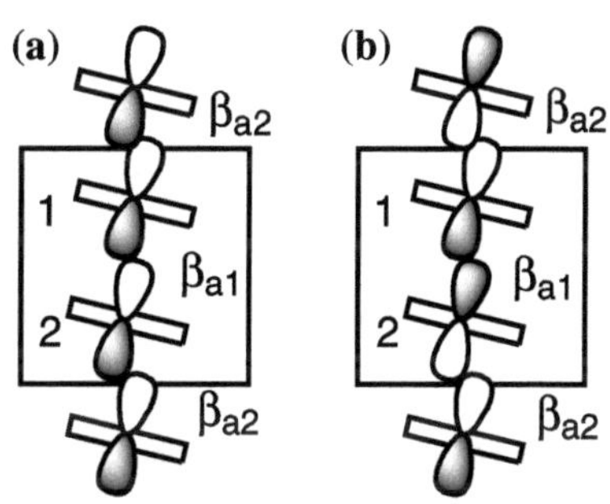


**Fig. 2.46** π orbitals of $(TMTSF)_2PF_6$

result, this complex maintains the metallic conductivity down to 12 K and exhibits superconductivity at 0.9 K under pressure.

All transfer values are positive (Fig. 2.45 caption), but this is accidental. If the π orbitals are defined as Fig. 2.46a, the overlap integrals are negative, and the transfer integrals are positive. However, if the sign of the molecular orbital on Molecule 2 is taken opposite (Fig. 2.46b), the overlap integrals are positive, and the transfer integrals are negative. It is always possible to chose the signs of the basis functions arbitrarily. Since the transfer $\beta_b$ connects the same kind of molecules, the sign is uniquely defined. In Eq. 2.91, all transfers except for $\beta_b$ appear as the squares, so even if all these transfers are taken negative, the resulting energy band is unchanged. Like this, the signs of the transfer integrals are changeable as far as the signs are chosen systematically according to the basis functions.

*Example 2.11* Figure 2.47 is a popular structure of organic crystals, known as the θ-phase or a herringbone structure. Obtain the equation of the energy band.

For simplicity, we assume $\alpha = 0$. A unit cell contains two shaded molecules, where $\beta_a$ connects the same molecules (1-1 or 2-2) in the $\pm a$ direction, and $\beta_p$ connects the different molecules (1-2) in the diagonal directions ($\pm a/2$, $\pm b/2$). Obtain the element $\beta_{11}$ and $\beta_{12}$, and calculate the formula of the energy band $E(k_a, k_b)$.

1. Viewed from the shaded Molecule 1, Molecules 1 in the adjacent $+a$ and $-a$ cells are connected by the transfer integral $\beta_a$. Then, we obtain the diagonal element.

$$\beta_{11} = \beta_a e^{ik_a a} + \beta_a e^{-ik_a a} = 2\beta_a \cos k_a a \quad (2.92)$$

Note that $\beta_{22}$ has the same form.

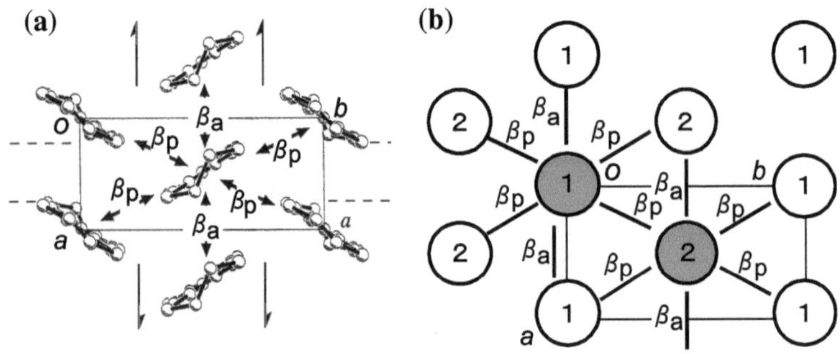


**Fig. 2.47** Structure of the θ-phase. The transfer integrals are $\beta_a = 25$ and $\beta_p = 101$ meV

2. Viewed from the central shaded Molecule 1, four Molecules 2 in the ($\pm a/2$, $\pm b/2$) directions are connected by the transfer integral $\beta_p$. Then, we obtain the non diagonal element.

$$\begin{aligned}\beta_{12} &= \beta_p e^{i\frac{k_a a + k_b b}{2}} + \beta_p e^{-i\frac{k_a a + k_b b}{2}} + \beta_p e^{i\frac{k_a a - k_b b}{2}} + \beta_p e^{-i\frac{k_a a - k_b b}{2}} \\ &= 2\beta_p\left(\cos\frac{k_a a + k_b b}{2} + \cos\frac{k_a a - k_b b}{2}\right)\end{aligned} \quad (2.93)$$

3. $\begin{vmatrix} \beta_{11} - E & \beta_{12} \\ \beta_{12} & \beta_{11} - E \end{vmatrix} = 0$ leads to $E = \beta_{11} \pm \beta_{12}$. The results of (1) and (2) are substituted in this formula.

$$\begin{aligned}E &= 2\beta_a \cos k_a a \pm 2\beta_p\left(\cos\frac{k_a a + k_b b}{2} + \cos\frac{k_a a - k_b b}{2}\right) \\ &= 2\beta_a \cos k_a a \pm 4\beta_p \cos\frac{k_a a}{2}\cos\frac{k_b b}{2}\end{aligned} \quad (2.94)$$

When the transfer integrals obtained from the molecular orbital calculation (Fig. 2.47 caption) are placed in this equation, the energy band is obtained as shown in Fig. 2.48. The two-dimensional network affords the elliptical Fermi surface.

Since a unit cell contains two molecules, there are two energy bands in Fig. 2.48. However, on the CX and CY zone boundaries, the energy bands are degenerate to one. This is due to the two-fold screw axis along the *a* axis and the glide plane along the *b* axis (Fig. 2.47a). A molecule has symmetry elements such as rotation, inversion, and mirror, but a space group in a crystal has additional symmetry elements such as a screw axis and a glide plane, in which translation is combined

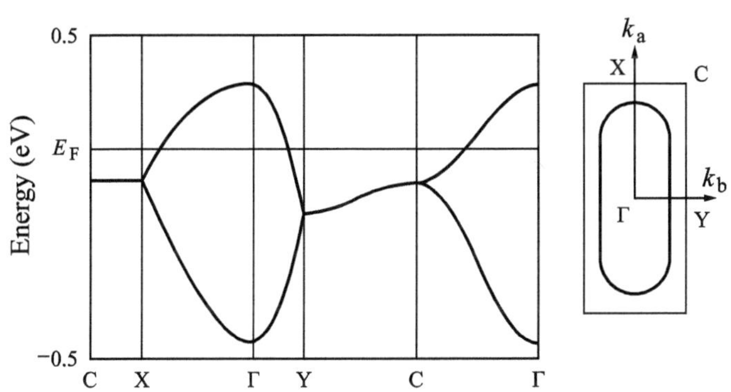


**Fig. 2.48** Energy band in the θ-phase

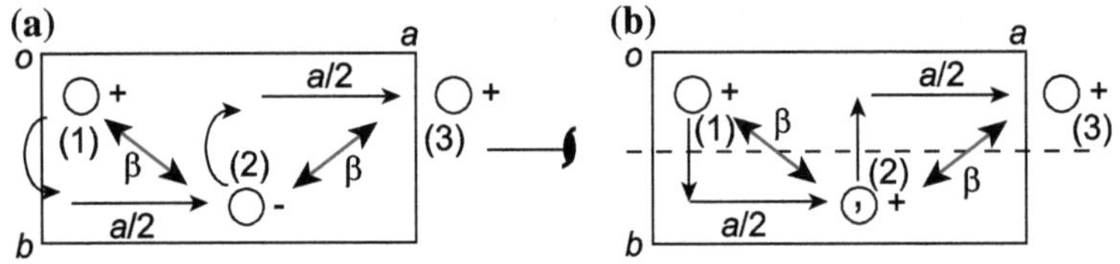


**Fig. 2.49** (**a**) A screw axis, and (**b**) a glide plane

with rotation and mirror, respectively. A **two-fold screw axis** is a combination of two-fold rotation and translation, where after the 180° rotation, the atom is translated by ***a***/2 (Fig. 2.49a). The symbols + and − represent an atom located above and below the sheet. Then, the atom (1) moves to (2). One more operation moves (2) to (3), which corresponds to a one-unit-cell (***a***) translation of (1). In a **glide plane** (b), the original atom (1) is reflected, and translated by ***a***/2 to (2). The comma implies, if the original atom has right-hand chirality, the atom with comma has left-hand chirality. One more operation moves (2) to (3), which is again the ***a*** translation.

The transfer integral between (1) and (2) is equivalent to the transfer integral between (2) and (3), because they are generated by the same symmetry operation applied to the crystallographically equivalent molecules. Accordingly, although the periodicity of the crystal is ***a***, the periodicity of β is ***a***/2. This is the origin of the degenerate zone boundary in Fig. 2.48. This is just opposite to the Peierls transition (Fig. 2.35a), where at $\beta_1 = \beta_2$, the energy band is similarly degenerated at 2π/*a*. Space groups including symmetry operations with translation such as a screw axis and a glide plane are called **non-symmorphic** [10]. In general, energy bands of non-symmorphic space groups are degenerate at the zone boundary in the direction of the translation. The band structure of the θ-phase (Fig. 2.48) is degenerate both along the *a* and *b* axes, respectively, owing to the screw axis and the glide plane. This is obvious because two $\beta_p$ are repeated in the unit cell both along the *a* and *b* axes.

In a molecule, the secular equation is reduced to separate blocks according to the symmetry, and energy levels belonging to different symmetry are independently obtained. In a crystal, such block diagonalization does not work because the symmetry disappears at the general *k* point even when the crystal has symmetry. Then, the degeneracy at the zone boundary in a non-symmorphic space group is an important conclusion derived from the symmetry. Even in a triclinic lattice, special points such as Γ, X, Y, and C have inversion symmetry. At these points, all elements in the secular equation become real, and the energy levels are classified to symmetrical (gerade) and antisymmetrical (ungerade). These special points are important to discuss the electronic structure of graphene.

C-center cells have two lattice points in a unit cell. For the band calculation, the C-center cell is transformed to a primitive cell (Fig. 2.50). The resulting cell has oblique axes, where a monoclinic cell is transformed to a triclinic cell, and orthorhombic turns to monoclinic. The resulting cell contains a single lattice point, and the volume is half of the original cell. The volume of the Brillouin zone

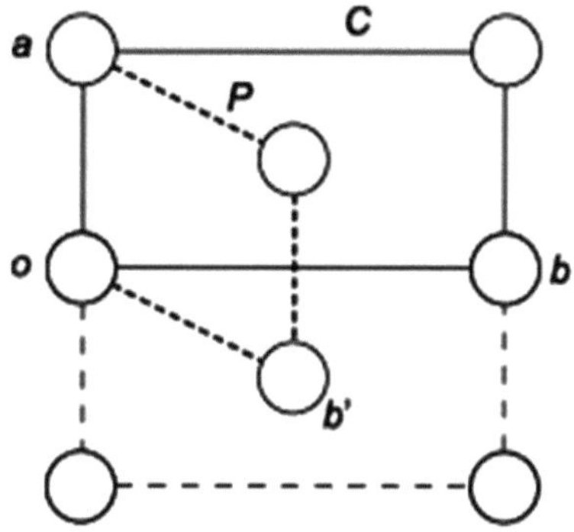


**Fig. 2.50** Transformation from a C-center lattice to a primitive lattice

corresponding to the primitive cell is twice of the volume assuming the C-center cell. Then, the energy bands are represented in the extended zone scheme. When we use the Brillouin zone of the C-center cell, the energy bands are folded into the reduce zone, and the number of the energy bands are twice larger, corresponding to the original two lattice points.

The transfer integrals are calculated on the basis of molecular orbitals of the single molecule. For an electron donor, we calculate the transfer integrals between the HOMO. According to the spirit of the extended Hückel approximation, the transfer integrals are obtained from Eq. 1.28.

$$\beta_{kl} \approx \int \psi_k H \psi_l \mathrm{d}\tau = ES_{kl}$$

Then, the transfer integrals $\beta_{kl}$ are estimated from the overlap integrals $S_{kl}$ by multiplying the HOMO energy $E$. In general, $E = -10$ eV is used because this value affords good agreement with the experiments. Since the HOMO is represented by LCAO (Eq. 1.23) of the atomic orbitals $\chi_i$, the overlap integral $S_{kl}$ is

$$S_{kl} = \sum_i \sum_j c_i c_j \int \chi_i \chi_j \mathrm{d}\tau.$$

First, we evaluate the overlap integrals between $\chi_i$ and $\chi_j$. Then, $c_i$ and $c_j$ are multiplied, and $S_{kl}$ is obtained. The values of $c_i$ and $c_j$ are taken from the molecular orbital calculation. The transfer integrals listed in the captions of Figs. 2.45 and 2.48 are obtained like this.

The overlap integral $S_{kl}$ between two BEDT-TTF molecules is plotted in Fig. 2.51 as a function of the angle $\varphi$ between the molecular plane and the intermolecular vector. Along the stack ($\varphi = 90°$), $S_{kl}$ is about $-20 \times 10^{-3}$. By multiplying $E = -10$ eV, the corresponding transfer integral is 0.2 eV. Then, the one-dimensional stack gives the bandwidth of $4\beta = 0.8$ eV. The overlap integral has additional peaks at 60°, 30°, and 0°. The sign of the overlap integral is unimportant when the energy band is formed. Between these peaks, the overlap

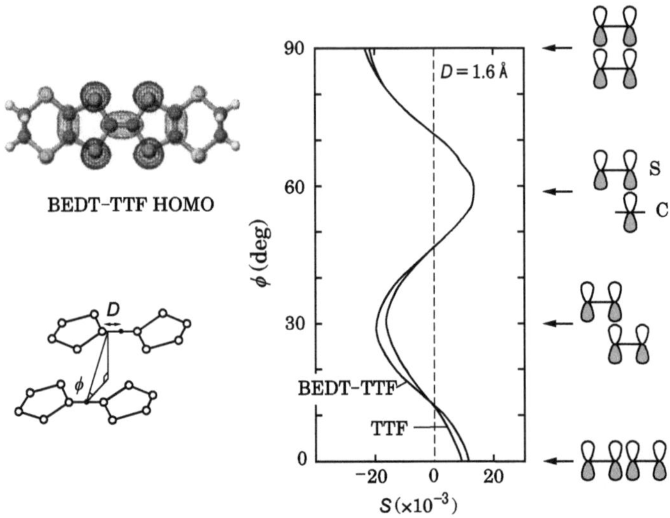


**Fig. 2.51** Orientation dependence of the overlap integral between the HOMO of two BEDT-TTF molecules [11]

integral crosses zero. It is not certain whether the β zero coming from the HOMO phase is related to some physical properties. However, the angle dependence of the overlap integral is sometimes very important in determining the systematic change of the properties of organic conductors (Sect. 7.7.4). Instead of the direct calculation of the overlap integrals, molecular orbital calculation is sometimes carried out for two molecules. The HOMO splitting corresponds to the bonding and antibonding combinations, and gives 2β. Since the molecular orbitals change depending on the dimer geometry, this method does not afford $\beta = 0$.

Bandwidths of the representative organic conductors are listed in Table 2.1 [12–18]. It is obvious from Eq. 2.92 that the bandwidth of the θ-phase is $4|\beta_a| + 8|\beta_p|$.

**Table 2.1** Bandwidths of the representative organic conductors

| Compound | Bandwidth/eV | References |
|---|---|---|
| $(TMTSF)_2X$ | 1.0 | |
| $(BEDT\text{-}TTF)_2X$ | 1.0 | |
| Pentacene HOMO | 0.2 | [12, 13] |
| Rubrene HOMO | 0.34 | [14] |
| Picene HOMO | 0.27 | [13] |
| Sexithiophene HOMO | 0.2 | [13] |
| M(TCNQ) | 1.0 | |
| $C_{60}$ LUMO | 0.5 | [15, 16] |
| $La_{2-x}Sr_xCuO_4$ | 2 | [17] |
| Graphene | 10 | [18] |

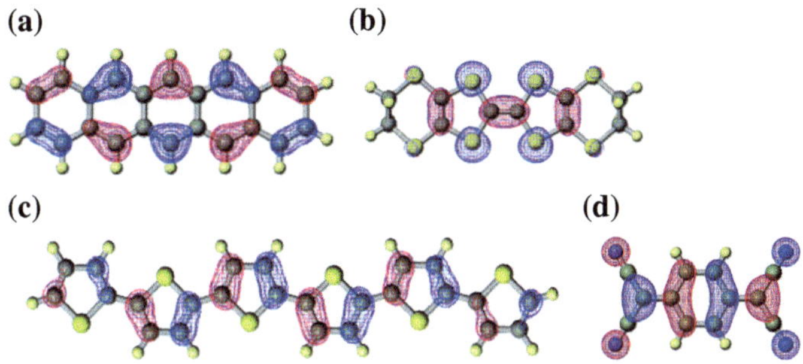


**Fig. 2.52** Molecular orbitals of the representative organic conducting molecules. The *shaded regions* depend on the molecular orbital signs. **a** Pentacene HOMO. **b** BEDT-TTF HOMO. **c** Sexithiophene HOMO. **d** TCNQ LUMO

Note that in organic charge-transfer complexes like $(\text{BEDT-TTF})_2X$, the energy band is quarter-filled, but in organic semiconductors like pentacene and thiophene, the HOMO is entirely occupied. Bandwidths of TTF family charge-transfer complexes are usually about 1.0 eV.

The bandwidths are determined by the overlap of the atomic orbitals. The exponent is $\zeta = 1.625$ for C 2p, whereas $\zeta = 1.817$ for S 3p (Appendix 2). As shown in Fig. 1.5, the smaller is $\zeta$, the larger is the orbital. For the same $\zeta$, however, 3p is larger than 2p. The van der Waals distance is 1.70 Å for C, while 1.85 Å for S [19]. Therefore, the atomic orbital of S affords much larger overlap than C. 1 eV bandwidth of the TTF family mainly comes from this S-S overlap.

Molecular orbitals of the representative organic conductors are shown in Fig. 2.52. Four sulfur atoms in TTF have the same sign; this is advantageous to give a large overlap without largely depending on the orientation [11, 13]. In pentacene, the HOMO is mostly located on the carbon atoms, and the side hydrogen atoms prohibit the transverse interaction. However, the herringbone structure realizes the two-dimensional band. The HOMO of thiophene has nodes on the sulfur atoms, and the coefficients are zero. Then, the S-S overlap does not contribute to the bandwidth. The calculated bandwidth of oligothiophene is as small as that of pentacene (0.2 eV). However, the next HOMO has a finite coefficient on sulfur, and this contribution enlarges the bandwidth as large as 0.4 eV. TCNQ does not contain sulfur atoms, but forms an energy band as large as 1 eV along the stack. The LUMO bandwidth of $C_{60}$ is around 0.4 $\sim$ 0.5 eV, and the density of states is enhanced due to the three-fold degeneracy. Bandwidth of inorganic compounds with connected chemical bonds is large; for example, 2 eV for the $d_{x^2-y^2}$ band in the CuO plane of the cupper oxide superconductors (Fig. 2.31) [11]. The bandwidth of graphite and graphene comes from the C = C covalent bonds, which is as large as 10 eV.

## References


1. R. Hoffmann, *Solids and Surfaces: A Chemist's View of Bonding in Extended Structures* (Wiley, Hoboken, 1988)
2. J.-M. André, J. Delhalle, J.-L. Brédas, *Quatum Chemistry Aided Design of Organic Polymers* (World Scientific, 1991)
3. C. Kittel, *Introduction to Solid State Physics*, 8th edn. (Wiley, Hoboken, 2004)
4. N.W. Ashcroft, N.D. Mermin, *Solid State Physics* (Thomson, 1976)
5. H. Ibach, H. Luth, *Solid State Physics* (Springer, Berlin, 1993)
6. N.F. Mott, H. Jones, *The Theory of Properties of Metals and Alloys* (Dover, 1958)
7. R.E. Peierls, *Quantum Theory of Solids* (Clarendon, 1956)
8. P.R. Cromwell, *Polyhedra* (Springer, Berlin, 1997)
9. A.P. Cracknell, K.C. Wong, *The Fermi Surface* (Clarendon, 1973)
10. G. Burns, *Introduction to Group Theory with Applications* (Academic Press, Cambridge, 1977)
11. T. Mori, A. Kobayashi, Y. Sasaki, H. Kobayashi, G. Saito, H. Inokuchi, Bull. Chem. Soc. Jpn. **57**, 627 (1984)
12. R.C. Haddon, T. Siegrist, R.M. Fleming, P.M. Bridenbaugh, R.A. Laudise, J. Mater. Chem. **5**, 1719 (1995)
13. H. Kojima, T. Mori, Bull. Chem. Soc. Jpn. **84**, 1049 (2011)
14. D.A. de Silva Filho, E.G. Kim, J.L. Bredas, Adv. Mater. **17**, 1072 (2005)
15. R.C. Haddon, X. Chi, M.E. Itkis, J.E. Anthony, D.L. Eaton, T. Siegrist, C.C. Mattheus, T.T.M. Pastra, J. Phys. Chem. **16**, 433 (2002)
16. M.S. Dresselhaus, G. Dresselhaus, P.C. Eklund, *Science of Fullerenes and Carbon Nanotubes* (Academic Press, Cambridge, 1996)
17. C.P. Poole, H.A. Farach, R.J. Creswick, *Superconductivity* (Academic Press, Cambridge, 1995)
18. A. Zunger, Phys. Rev. B **17**, 626 (1978)
19. F.A. Cotton, G. Wilkinson, *Inorganic Chemistry* (Interscience, 1962)

# Chapter 3
# Transport Properties

**Abstract** Energy band theory is useful to understand various conducting phenomena of organic conductors. Based on the low-temperature magnetoresistance, we can investigate the Fermi surface of organic conductors.

**Keywords** Transport properties · Group velocity · Effective mass · Relaxation time · Conductivity · Hall effect · Thermal conductivity · Plasma frequency · Cyclotron resonance · Landau level · Shubnikov–de Haas oscillation · Magnetic breakdown · Lifshitz–Kosevich equation · Fermi surface · Magnetoresistance · Fermiology · Boltzmann equation · Thermoelectric power · Seebeck coefficient · Peltier effect · Figure of merit · Power factor · Quantum hall effect

## 3.1 Energy Band and Transport Properties

### *3.1.1 Group Velocity*

In Chap. 2, we have described how energy levels of molecules form energy bands in a solid. In the energy band, different levels are distinguished by $k$, and the Fermi surface appears in a metallic solid. Since $\hbar k$ is the momentum, the Fermi surface represents the momentum distribution of the electrons with energy $E_F$. In this chapter, we shall discuss **transport properties** such as electrical and thermal conductivity on the basis of the energy band. From the transport properties under a magnetic field, we can experimentally determine the Fermi surface.

As described in Sect. 1.3, a wave is generally represented as $U = U_0 e^{i(kx-\omega t)}$. Consider convolution of two waves with slightly different frequencies, $\omega \pm \Delta\omega$, and wave numbers, $k \pm \Delta k$.

$$\begin{aligned} U &= U_0\left[e^{i[(k+\Delta k)x-(\omega+\Delta\omega)t]} + e^{i[(k-\Delta k)x-(\omega-\Delta\omega)t]}\right] \\ &= U_0 e^{i(kx-\omega t)}\left[e^{i(\Delta kx-\Delta\omega t)} + e^{-i(\Delta kx-\Delta\omega t)}\right] \\ &= 2U_0 e^{i(kx-\omega t)}\cos(x\Delta k - t\Delta\omega) \end{aligned} \tag{3.1}$$



T. Mori, *Electronic Properties of Organic Conductors*,
DOI 10.1007/978-4-431-55264-2_3

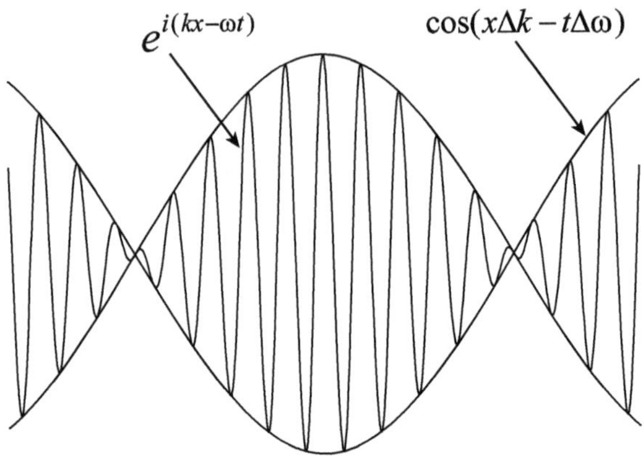


**Fig. 3.1** Waves generated by a convolution of waves with slightly different frequencies

A slow beat appears in addition to the original wave $e^{i(kx-\omega t)}$, where the envelop of the beat is represented by $\cos(x\Delta k - t\Delta\omega)$ (Fig. 3.1). The velocity of the original wave is obtained by the condition in which $e^{i(kx-\omega t)}$ shows a constant phase, for example zero.

$$v = \frac{\Delta x}{\Delta t} = \frac{\omega}{k}. \tag{3.2}$$

In the same way, the constant phase of $\cos(x\Delta k - t\Delta\omega)$ gives the velocity with which the envelop moves.

$$v_g = \frac{\Delta\omega}{\Delta k} \tag{3.3}$$

The former is called **phase velocity**, and the latter is called **group velocity**. In general, when the dispersion relation of a wave $\omega(k)$ is known, the slope of the line connecting $\omega(k)$ and the origin gives the phase velocity $\omega/k$, while the gradient of $\omega(k)$ gives the group velocity.

An electron with definite energy (frequency) and momentum (wave number) is represented by $e^{i(kx-\omega t)}$. This is an infinitely spreading plane wave, where the position $x$ and time $t$ as a particle are entirely indefinite according to the uncertainty principle (Sect. 1.8). To consider an electron as a particle, we have to imagine a wave with a finite extent (Fig. 3.2). Such a wave is called **wave packet**, where, according to the uncertainty principle, not only $x$ and $t$, but also $k$ and $\omega$ are uncertain. The velocity of the particle is given by the velocity of the envelop, namely by the group velocity. Then, the group velocity affords the velocity of the particle.

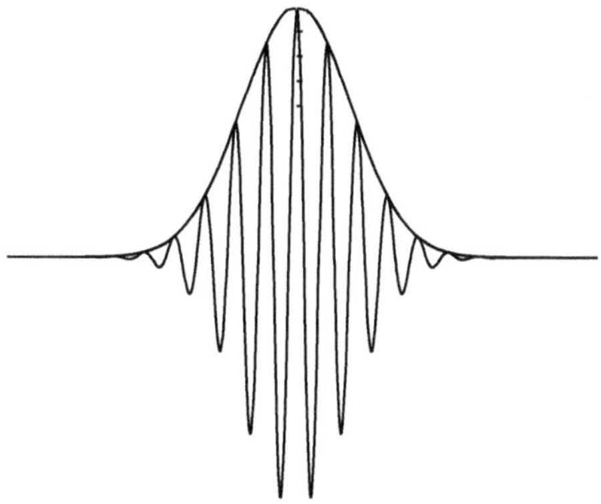

**Fig. 3.2** Wave packet

### 3.1.2 Effective Mass

When an energy band $E(k)$ is given, the group velocity is obtained using $E = \hbar\omega$.

$$v_{\mathrm{g}} = \frac{\partial \omega}{\partial k} = \frac{1}{\hbar}\frac{\partial E(k)}{\partial k} \tag{3.4}$$

The electron velocity is proportional to the gradient of the energy band (Fig. 3.3a). In the three-dimensional space, this leads to

$$\mathbf{v}_{\mathrm{g}} = \frac{1}{\hbar}\left(\frac{\partial E(k)}{\partial k_x}, \frac{\partial E(k)}{\partial k_y}, \frac{\partial E(k)}{\partial k_z}\right) \tag{3.5}$$

so the velocity vector is perpendicular to the Fermi surface (Fig. 3.3b).

When Eq. 3.4 is differentiated by $t$, we obtain

$$\frac{\partial v_{\mathrm{g}}}{\partial t} = \frac{1}{\hbar}\frac{\partial^2 E(k)}{\partial k^2}\frac{\partial k}{\partial t}. \tag{3.6}$$

Suppose an external electric field $\varepsilon$ is applied. Here, the electric field is designated as $\varepsilon$ to be distinguished from the energy $E$. Owing to the applied force, $-e\varepsilon$, the electron moves with a distance of $v_{\mathrm{g}}\delta t$, and the resulting work (impulse) is $\delta E = -e\varepsilon v_{\mathrm{g}}\delta t$. Using $\delta E = \left(\frac{\partial E}{\partial k}\right)\delta k = \hbar v_{\mathrm{g}}\delta k$, we obtain $-e\varepsilon v_{\mathrm{g}}\delta t = \hbar v_{\mathrm{g}}\delta k$, and

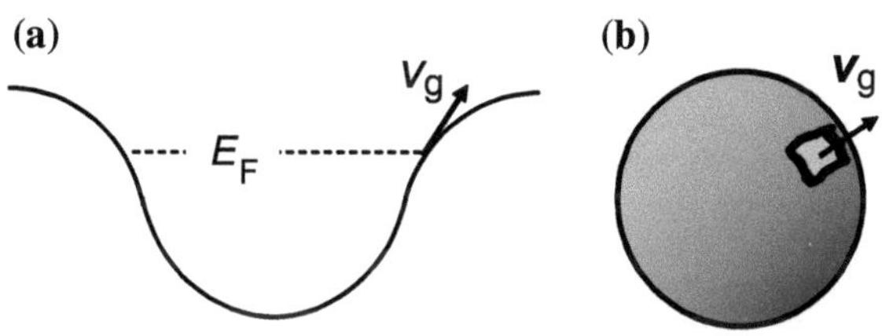


**Fig. 3.3** **a** Group velocity $v_{\mathrm{g}}$ in an energy band, and **b** the velocity vector at the Fermi surface

$$\hbar \frac{\partial k}{\partial t} = -e\varepsilon. \tag{3.7}$$

Alternatively, the same relation is obtained by differentiating $\hbar k$ by $t$ in the equation of motion.

$$\frac{\partial p}{\partial t} = \hbar \frac{\partial k}{\partial t} = F = -e\varepsilon \tag{3.8}$$

Substituting Eq. 3.7 in Eq. 3.6, we obtain

$$\frac{\partial v_g}{\partial t} = \frac{1}{\hbar^2} \frac{\partial^2 E(k)}{\partial k^2} (-e\varepsilon). \tag{3.9}$$

The left-hand side is the acceleration, and the right-hand side, $F = -e\varepsilon$, is the force. Then, the equation of motion $m^*a = F$ affords

$$\frac{1}{m^*} = \frac{1}{\hbar^2} \frac{\partial^2 E(k)}{\partial k^2}. \tag{3.10}$$

This relation gives the **effective mass** $m^*$, which determines how the electron is accelerated when an electric field is applied. In general, the effective mass $m^*$ is different from the free electron mass. The effective mass is the coefficient when the energy band is approximated by a parabola in the free-electron model (Fig. 2.26). Since Eq. 2.68 gives the relation $m^* = -\hbar^2/2\beta a^2$, a band with a large bandwidth $4\beta$ has a small effective mass. In such a case, the gradient of the energy band is large, and the velocity is large (Fig. 3.4). The large band width is associated with the large overlap of the adjacent atomic orbitals, then the electron moves easily. In contrast, an energy band with a small $\beta$ has a large effective mass. Here, the gradient of the energy band is small, and the velocity is small.

From Eq. 2.67, the second derivative of the one-dimensional tight-binding band gives the effective mass.

$$m^* = -\frac{\hbar^2}{2\beta a^2 \cos ka} \tag{3.11}$$

This equation demonstrates $m^*$ is positive in the lower half of the energy band, but negative in the upper half of the energy band (Fig. 3.5). An electron in an electric field moves attracted by the positive electrode. In Fig. 3.5c, however, the

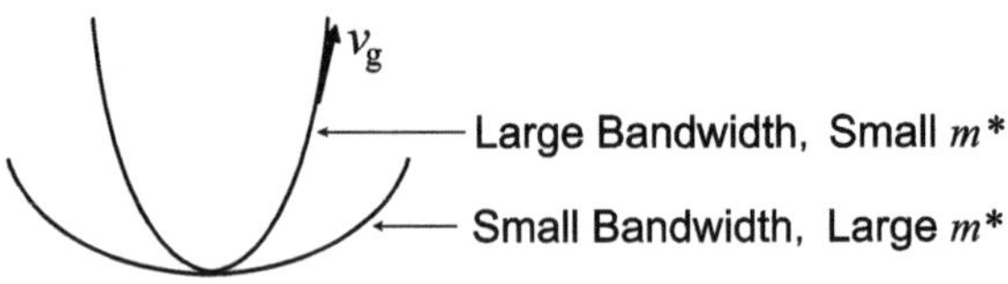


**Fig. 3.4** Effective masses in energy bands with large and small band widths

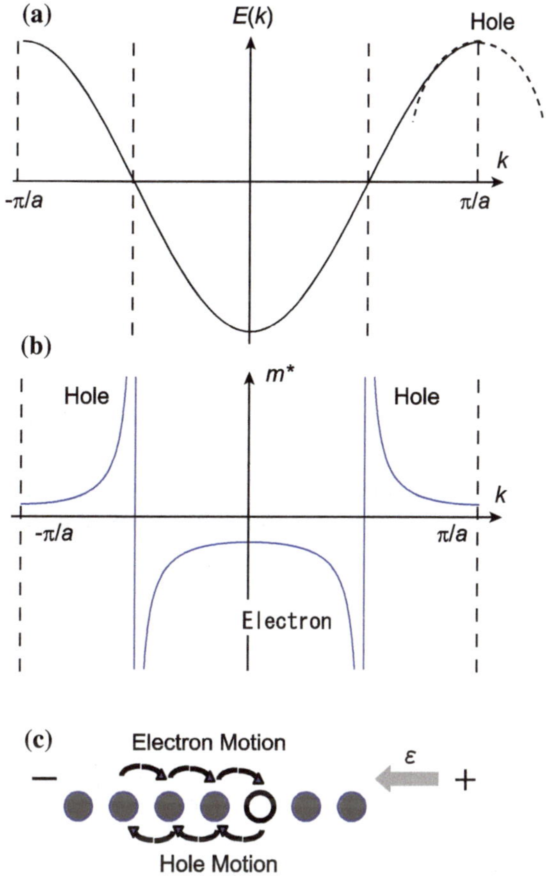


**Fig. 3.5** **a** $E(k)$, **b** $m^*$ in the one-dimensional tight-binding band and **c** the movement of a hole

successive movement of the black balls to the right leads to the movement of the white ball to the left attracted by the negative electrode. Therefore, the white ball is regarded as a particle with a positive charge, and called hole. In the lower half of the energy band, the carrier is an electron, whereas in the upper half of the energy band the carrier is a hole. The energy of the hole is

$$E(k) = \frac{\hbar^2 k^2}{2m^*} = -\frac{\hbar^2 k^2}{2|m^*|} < 0. \tag{3.12}$$

In contrast to an electron, a hole is most stable at the top of the energy band. A hole behaves like a balloon, going up to the top of the band. Then, the holes come to the upper end of the energy band.

### 3.1.3 Relaxation Time

Electrical conductivity is mediated by transport of electrons and holes. Classical equation of motion in an electric field $\varepsilon$ is

$$m\frac{\mathrm{d}v}{\mathrm{d}t} + \frac{m}{\tau}v = -e\varepsilon \tag{3.13}$$

where the effective mass is simply written as $m$. The second term of the left-hand side represents the friction, where $m/\tau$ corresponds to the friction coefficient. In order to investigate the meaning of this equation, suppose the force is removed at $t = 0$ by putting $\varepsilon = 0$, then

$$\frac{\mathrm{d}v}{\mathrm{d}t} = -\frac{v}{\tau} \text{ leads to } v = v_0 e^{-\frac{t}{\tau}}. \tag{3.14}$$

This equation means the electrons are scattered within the characteristic time constant $\tau$, which is therefore called **relaxation time**. Next, we suppose an equilibrium state at $\varepsilon \neq 0$, where $\frac{\mathrm{d}v}{\mathrm{d}t} = 0$ leads to $v = -\frac{e\tau}{m}\varepsilon$. Since the current density is represented by

$$J = -nev = \frac{ne^2\tau}{m}\varepsilon \tag{3.15}$$

the **conductivity** is

$$\sigma = \frac{J}{\varepsilon} = \frac{ne^2\tau}{m}. \tag{3.16}$$

The conductivity increases in proportion to the carrier density $n = N/V$, and decreases with increasing the effective mass $m$. The conductivity increases when the relaxation time $\tau$ is large. The conductivity is proportional to $e^2$, so does not depend on the polarity of the carriers. Therefore, from the conductivity we cannot determine whether the carriers are electron or hole. This equation is valid both for metals and for semiconductors. Using the **mobility** $\mu = \frac{e\tau}{m}$, the conductivity is represented as

$$\sigma = ne\mu. \tag{3.17}$$

In a metal, the temperature dependence of the electrical resistivity mostly comes from $\tau$ by the relation $\rho = \frac{1}{\sigma} \propto \frac{1}{\tau}$. Here, $\tau$ is derived from the lattice vibration (phonon) and impurity scattering

$$\frac{1}{\tau} = \frac{1}{\tau_{\mathrm{ph}}} + \frac{1}{\tau_{\mathrm{i}}} \tag{3.18}$$

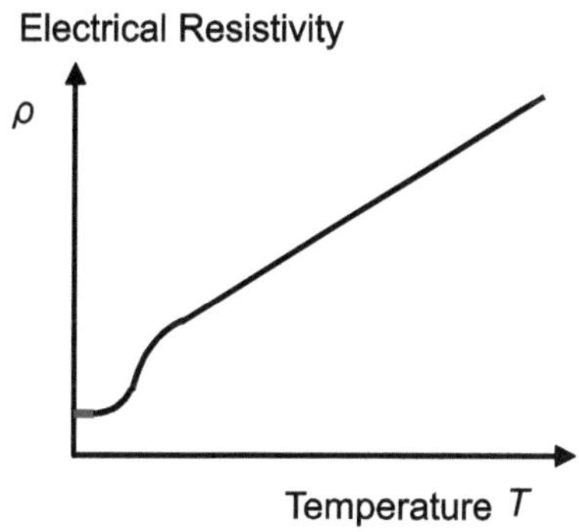


**Fig. 3.6** Temperature dependence of the resistivity of a metal

because these two scattering factors contribute independently like a parallel circuit. Scattering by phonon is given by $\frac{1}{\tau_{\text{ph}}} = C_v \cdot T$. Since the phonon specific heat $C_v$ is constant at high temperatures according to Dulong–Petit's law, we can conclude a linear temperature dependence $\frac{1}{\tau_{\text{ph}}} \propto T$. In contrast, the impurity scattering does not depend on the temperature. As a result, the resistivity of a metal shows temperature dependence as shown in Fig. 3.6. At high temperatures above 100 K, the resistivity is proportional to the temperature as $\rho \propto \frac{1}{\tau_{\text{ph}}} \propto T$ owing to the phonon scattering. At low temperatures around 50 K, the resistivity exhibits a more steep temperature dependence like $\rho \propto T^n$, but impure metals do not show such an obviously steep region. At very low temperatures below 10 K, the phonon scattering is negligible, and the resistivity mainly comes from the temperature-independent impurity scattering. Then, the low-temperature resistivity is proportional to the impurity concentration, and called **residual resistivity**. As a whole, the resistivity of a metal decreases upon cooling, and becomes constant at very low temperatures.

### *3.1.4 Hall Effect*

Consider motion of an electron in a magnetic field. As shown in Fig. 3.7, we apply an electric current $j_x$ in the $x$ direction, and a magnetic field $B$ in the $z$ direction, and

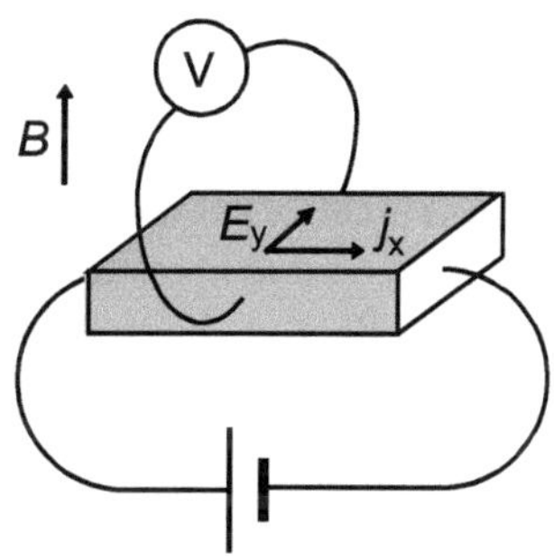


**Fig. 3.7** Hall effect

measure the electromotive force $E_y$ in the $y$ direction. An electron in an electric field $E$ and a magnetic field $B$, feels the Lorentz force.

$$\mathbf{F} = -e(\mathbf{v} \times \mathbf{B} + \mathbf{E}) \tag{3.19}$$

The condition that current does not flow in the $y$ direction is

$$F_y = -ev_xB + eE_y = 0 \text{ which leads to } E_y = v_xB. \tag{3.20}$$

Using $J_x = -nev_x$, we obtain

$$E_y = -\frac{1}{ne}J_xB \tag{3.21}$$

then the electromotive force is proportional to the applied current and magnetic field. The coefficient $R_{\mathrm{H}} = -\frac{1}{ne}$ is called **Hall coefficient**. Since $R_{\mathrm{H}}$ contains $e$, $R_{\mathrm{H}}$ depends on the carrier polarity. Since the magnitude of $R_{\mathrm{H}}$ is determined only by the carrier concentration $n$, the value of $n$ is derived from $R_{\mathrm{H}}$. Then from the conductivity $\sigma = ne\mu$, we can estimate the Hall mobility $\mu$. Hall is the name of the discoverer, and differs from "hole."

*Example 3.1* In an organic conductor with the composition (BEDT-TTF)$_2X$, a hole exists in a volume of approximately 800 Å$^3$ in general. When the conductivity is 200 S/cm, calculate the drift mobility.

From Eq. 3.17, the mobility is

$$\mu = \frac{\sigma}{ne} = \frac{(200\ \mathrm{S/cm}) \times (800 \times 10^{-24}\ \mathrm{cm}^3)}{(1.6 \times 10^{-19}\ \mathrm{C})} = 1.0\ \mathrm{cm^2/Vs}.$$

Note the unit is S = $\Omega^{-1}$ = C/sV. Here, the carrier concentration is $n = 1.25 \times 10^{21}$ cm$^{-3}$.

The Hall coefficients have been reported to be about $R_{\mathrm{H}} = 10 \times 10^{-3}$ cm$^3$/C in β-(BEDT- TTF)$_2$I$_3$ and κ-(BEDT-TTF)$_2$Cu(NCS)$_2$ [1, 2], which lead to the carrier concentration.

$$n = \frac{1}{R_{\mathrm{H}}e} = \frac{1}{(10 \times 10^{-3}\ \mathrm{cm^3/C})(1.6 \times 10^{-19}\ \mathrm{C})} = 0.625 \times 10^{21}\ \mathrm{cm}^3$$

This is 50 % of the above value. In κ-(BEDT-TTF)$_2$Cu(NCS)$_2$, the conductivity is 10 S/cm at room temperature and 1000 S/cm at 20 K (Fig. 7.36b), and using Eq. 3.17 and $\mu = R_{\mathrm{H}} \times \sigma$, the Hall mobilities are 0.1 and 10 cm$^2$/Vs. These values are typical in a correlated metal bordering on a semiconductor.

The Hall coefficient of $(TMTSF)_2PF_6$ has been reported to be $R_H = 6 \times 10^{-3}$ cm$^3$/C [3]. This leads to $n = 1.0 \times 10^{-21}$ cm$^3$. The room-temperature conductivity, 500 S/cm (Fig. 7.17b), affords the Hall mobility, $\mu = R_H \times \sigma = (6 \times 10^{-3}$ cm$^3$/C) $\times$ (500 S/cm) = 3 cm$^2$/Vs.

### 3.1.5 *Thermal Conductivity*

In a metal, not only the electrical conduction but also the thermal conduction is mediated by the metal electrons. In analogy with the thermal conductivity mediated by phonons, thermal conductivity carried by metal electrons is

$$\kappa = \frac{1}{3} C_v v_F l_F. \tag{3.22}$$

This equation means that the heat with the electron heat capacity $C_v$ is carried by an electron with the velocity at the Fermi level, $v_F$, and the mean free path $l_F$. The factor 1/3 appears because one of the three different directions contributes to the thermal conductivity. Using $l_F = v_F \tau$ and Eq. 2.47 for $C_v$, we obtain

$$\kappa = \frac{1}{3}\frac{\pi^2}{3} D(E_F) k_B^2 T \cdot v_F \cdot v_F \tau. \tag{3.23}$$

In the three-dimensional free electron model, Eq. 2.48 affords $D(E_F) = \frac{3N}{2E_F} = \frac{3N}{mv_F^2}$, and

$$\kappa = \frac{\pi^2 N k_B^2 T \tau}{3m}. \tag{3.24}$$

This is compared with the conductivity given by Eq. 3.16, to afford

$$\frac{\kappa}{\sigma} = \frac{\frac{\pi^2 N k_B^2 T\tau}{3m}}{\frac{Ne^2\tau}{m}} = \frac{\pi^2}{3}\left(\frac{k_B}{e}\right)^2 T. \tag{3.25}$$

This equation affords the relation between the thermal conductivity and the electrical conductivity

$$\frac{\kappa}{\sigma T} = L. \tag{3.26}$$

This is called **Wiedemann–Franz law**, and $L$ is called **Lorenz number**.

*Example 3.2* Calculate the Lorenz number, $2.44 \times 10^{-8}$ WΩ/K$^2$. When the room-temperature conductivity is $\sigma = 100$ S/cm, estimate the thermal conductivity.

The Lorenz number is

$$\frac{\pi^2}{3}\left(\frac{k_B}{e}\right)^2 = \frac{1}{3}\left(\frac{\pi \times (1.3806 \times 10^{-23}\,\mathrm{J/K})}{1.6022 \times 10^{-19}\mathrm{C}}\right)^2 = 2.44 \times 10^{-8}\,\mathrm{W\Omega/K^2}.$$

Using this, the expected thermal conductivity is

$$\left(2.44 \times 10^{-8}\,\mathrm{W\Omega/K^2}\right) \times (100\,\mathrm{S/cm}) \times (300\,\mathrm{K}) = 7.3 \times 10^{-4}\,\mathrm{W/cm\,K}$$

or 0.073 W/m K. Since thermal conductivity of an organic polymer such as polyethylene is typically 0.2 W/m K [4], the lattice contribution is still larger than the electronic contribution. For example, room-temperature thermal conductivity is 1 W/m K for glass and 400 W/m K for copper.

Since electrical resistivity of a metal exhibits such a temperature dependence as Fig. 3.6, the thermal conductivity exhibits a temperature dependence as shown in Fig. 3.8. Around room temperature, $\sigma \propto 1/T$ results in basically temperature-independent thermal conductivity. Around 50 K, $\sigma \propto 1/T^n$ affords the increasing thermal conductivity, $\kappa \propto 1/T$, followed by the $\sigma =$ constant region, where $\kappa$ decreases in proportion to $T$. In various metals, thermal conductivity is proportional to the electrical conductivity; highly conducting metals like copper has large thermal conductivity, and comparatively low conducting metals like stainless steel show small thermal conductivity. In nonconducting materials such as glass and fiber, the thermal conductivity is mediated by the lattice vibration (phonon), and the thermal conductivity is in several orders smaller than metals. We can distinguish metals by touching it, because it is cool due to the large thermal conductivity. This is directly associated with the good electrical conductivity.

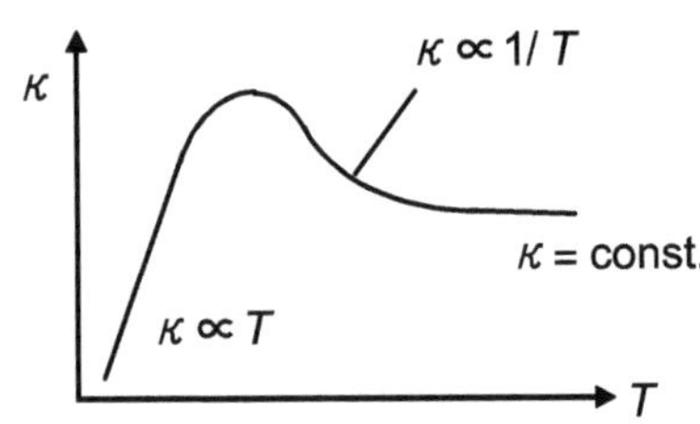


**Fig. 3.8** Temperature dependence of thermal conductivity in a metal

## 3.2 Optical Response of Metals

In addition to the transport properties, metallic luster is also related to the conduction electrons. Suppose a metal is irradiated by light (Fig. 3.9a), and consider how the oscillating electric field enters the metal. From electromagnetism, $D(\omega) = \varepsilon_0 E(\omega) + P$ and $D(\omega) = \varepsilon\varepsilon_0 E(\omega)$ afford

$$\varepsilon(\omega) = \frac{D(\omega)}{\varepsilon_0 E(\omega)} = 1 + \frac{P(\omega)}{\varepsilon_0 E(\omega)}. \tag{3.27}$$

Assume the light polarized linearly along $x$, and the metallic electron is vibrated by the electric field $E$. The equation of motion is

$$m\frac{d^2x}{dt^2} = -eE. \tag{3.28}$$

The light is an alternating electric field with frequency, $\omega$. Then, $E$ and $x$ are proportional to $e^{i\omega t}$. The above equation of motion is reduced to $-m\omega^2 x = -eE$ or $x = eE/m\omega^2$, which affords the polarization

$$P = -nex = -\frac{ne^2E}{m\omega^2}. \tag{3.29}$$

Substituting this relation in Eq. 3.27, we obtain

$$\varepsilon(\omega) = 1 - \frac{ne^2}{\varepsilon_0 m\omega^2} = 1 - \frac{\omega_p^2}{\omega^2} \text{ where } \omega_p = \frac{ne^2}{\varepsilon_0 m}. \tag{3.30}$$

Here, $\omega_p$ is called **plasma frequency**. The ac dielectric constant $\varepsilon(\omega)$ shows a frequency dependence as shown in Fig. 3.9b. To investigate the physical meaning, $E \propto e^{i(Kr-\omega t)}$ is substituted in the Maxwell equation, $\mu_0 \frac{\partial^2 D}{\partial t^2} = \nabla^2 E$, to give

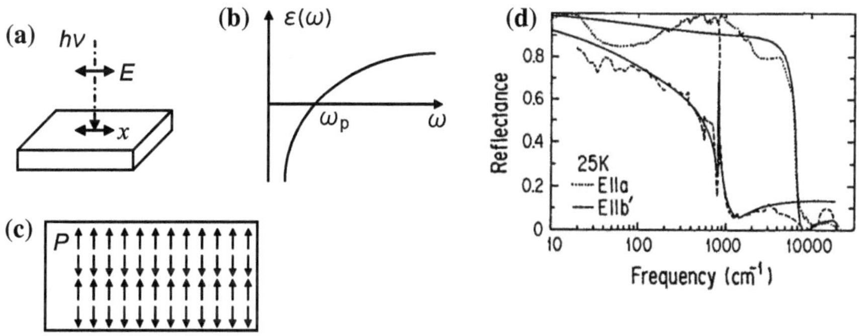


**Fig. 3.9** **a–c** Plasma frequency of a metal and **d** the reflectance of $(TMTSF)_2PF_6$ [5]

$\mu_0\varepsilon_0\varepsilon\omega^2 = K^2$. When $\mu_0\varepsilon_0=1/c^2$ is used, this leads to $\omega^2 - \omega_p^2 = c^2K^2$. For $\omega < \omega_p$, this equation means $K$ is imaginary, and light attenuates as $E \propto e^{-Kr}$ in the metal. Then, the light cannot enter the metal, and is totally reflected. Since $\omega_p$ of ordinary metals is located in the near ultraviolet region, the metal reflects visible light. This is the origin of the metallic luster. For $\omega > \omega_p$, the quantity $K$ is real, and light propagates as $E \propto e^{-iKr}$. Accordingly, the metal is transparent for the ultraviolet light and X-ray whose frequency is larger than $\omega_p$.

Reflectance of an organic superconductor, $(TMTSF)_2PF_6$, is shown in Fig. 3.9d. The plasma frequency of this material is located in the near infrared region, about 10,000 $cm^{-1}$, at which the reflectance drops steeply. Then, this material reflects infrared light, but does not reflect visible light. As a consequence, this material looks black but has metallic luster. From the plasma frequency, the bandwidth is estimated to be about 1 eV. In Chap. 5, we shall investigate the electron correlation, which enhances the effective mass and the density of states around $E_F$. The bandwidth obtained from the plasma frequency is not influenced by such a correlation effect, and basically equivalent to the bandwidth estimated from the band calculation (Fig. 2.45b). Since $\omega_p^2$ is proportional to the bandwidth, $\omega_p$ along the $b$ axis is by one order smaller than that along the $a$ axis (Fig. 3.9c).

In order to investigate the plasma frequency, suppose $\omega = \omega_p$. This leads to $D = 0$, and from Eq. 3.27,

$$E = -\frac{1}{\varepsilon_0}P = \frac{nex}{\varepsilon_0}.$$

The equation of motion is

$$m\frac{d^2x}{dt^2} = -\frac{ne^2}{\varepsilon_0}x$$

or

$$\frac{d^2x}{dt^2} + \omega_p^2 x = 0.$$

This equation means $\omega_p$ is a characteristic frequency. This is a longitudinal compression wave called plasma oscillation. When light is irradiated, the metallic electron oscillates according to the AC electric field. If the frequency is sufficiently low, the metallic electron entirely follows the electric field so that the light cannot enter the material, and the light is reflected. In such a case, $\varepsilon(\omega)$ is negative. When metallic electrons are regarded as free electrons, the free electron "sea" is equivalent to plasma, in which the positive and the negative charges are independent. A plasma has the characteristic plasma frequency. When the frequency of the light exceeds the plasma frequency, the metallic electron does not follow the oscillation, and the light is not reflected.

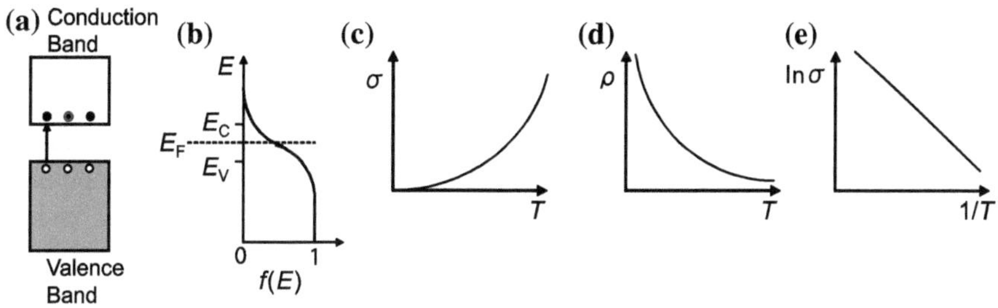


**Fig. 3.10** Energy band and conduction in a semiconductor. **a** Energy band, **b** Fermi distribution, **c** electrical conductivity, **d** electrical resistivity, and **e** the Arrhenius plot of the conductivity

## 3.3 Semiconductor

Semiconductors such as silicon and germanium have energy bands as shown in Fig. 2.36. Since the *k*-dependence is unimportant in the following discussion, we consider the energy bands like Fig. 3.10a. The valence band is completely occupied, and the conduction band is entirely unoccupied. The top of the valence band is designated as $E_v$, and the bottom of the conduction band is $E_c$. The conduction is mediated by electrons in the conduction band and holes in the valence band, generated by thermal activation from the valence band to the conduction band. The distribution is given by the Fermi distribution function (Eq. 2.40)

$$f(E) = \frac{1}{\exp\left(\frac{E-\mu}{k_B T}\right) + 1}. \quad (3.31)$$

Since the energy gap, $E_g = E_c – E_v > 1$ eV, is larger than the temperature, $k_B T \sim 0.03$ eV, we can suppose $E - \mu > E_c - E_F >> k_B T$. Then, the exponent is much larger than one, and the distribution function is approximated as $f(E) = \exp\left(-\frac{E-\mu}{k_B T}\right)$. Namely, in a semiconductor, we may assume the Boltzmann distribution instead of the Fermi distribution. Accordingly, the number of the activated electrons is

$$N_e = \int D(E) f(E) \mathrm{d}E = N_e^0 \exp\left(-\frac{E_C - E_F}{k_B T}\right) \quad (3.32)$$

and the number of the activated holes is

$$N_p = N_p^0 \exp\left(\frac{E_V - E_F}{k_B T}\right). \quad (3.33)$$

When we multiply these two, we obtain

$$N_e \cdot N_p = N_e^0 N_p^0 \exp\left(\frac{E_V - E_F}{k_B T} - \frac{E_C - E_F}{k_B T}\right) = N_e^0 N_p^0 \exp\left(-\frac{E_g}{k_B T}\right). \tag{3.34}$$

For given $E_g$ and $T$, this product is a constant. This rule is sometimes called **law of mass action**. Even when one of the carriers increases due to doping or applying a bias, another carrier decreases owing to the $E_F$ shift, and the product is kept constant.

If the effective masses of electron and hole are the same and as a result $N_{e0} = N_{p0}$ in an intrinsic semiconductor, $N_e = N_p$ leads to $E_F - E_V = E_C - E_F$ and

$$E_F = \frac{E_C + E_V}{2}. \tag{3.35}$$

Then, the Fermi level is located at the center of the energy gap. From Eq. 3.34, we obtain

$$N_e = N_p = \sqrt{N_e^0 N_p^0} \exp\left(-\frac{E_g}{2k_B T}\right). \tag{3.36}$$

From $\sigma = Ne\mu$ (Eq. 3.17), the conductivity is obtained by adding the electron and hole contributions

$$\sigma = N_e e \mu_e + N_p e \mu_p \propto \exp\left(-\frac{E_g}{2k_B T}\right). \tag{3.37}$$

As shown in Fig. 3.10c, conductivity of a semiconductor is determined by the carrier concentration, and decreases at low temperatures. The mobility $\mu$ also changes depending on the temperature, but the carrier concentration overwhelms $\mu$ because of the exponential dependence. This is contrasting with the conductivity of a metal, whose temperature dependence mainly comes from $\tau$ or $\mu$, and $n$ is regarded as approximately constant. The inverse of Eq. 3.37 affords the resistivity.

$$\rho = \frac{1}{\sigma} \propto \exp\left(\frac{E_g}{2k_B T}\right) \tag{3.38}$$

As shown in Fig. 3.10d, the resistivity increases exponentially upon cooling. This leads to

$$\ln \rho = \frac{E_g}{2k_B T} \tag{3.39}$$

so when the logarithm of $\sigma$ or $\rho$ is plotted against $1/T$, a straight line is obtained, and the slope gives $E_g$. Note that the slope in an intrinsic semiconductor corresponds to $E_a = E_g/2$, which is sometimes called activation energy.

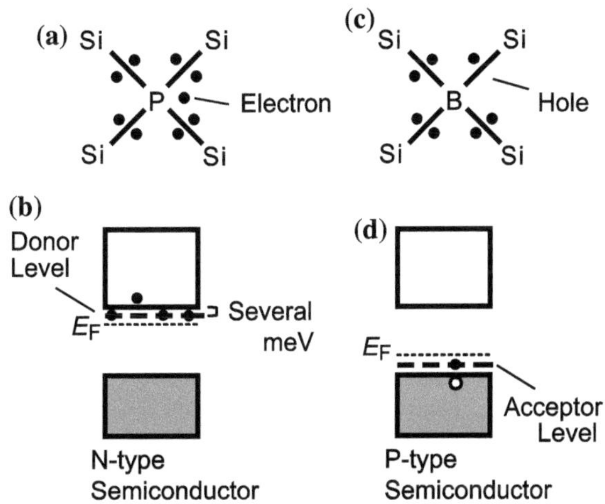


**Fig. 3.11** Doped semiconductors. **a** N-type semiconductor and **b** the energy level. **c** P-type semiconductor and **d** the energy level

Pure silicon and germanium show very high resistivity. When we add a small amount of phosphorus or arsenic with five valence electrons in silicon, these atoms enter the same site as silicon and give up the fifth electron to the conduction band; these elements work as an electron donor. Although silicon has a diamond structure, Fig. 3.11a is schematically depicted in a plane. Since a nucleus of the phosphorus and arsenic atom is balanced with the five electrons, the fifth electron is weakly bound to these atoms, which makes a donor level by a few meV below the conduction band (Fig. 3.10b). This energy difference is so small that considerable electrons move around the conduction band even at room temperature. Here, the Fermi level goes up just below the donor level, and Eq. 3.36 holds as well. In this case, the electrons are the **majority carriers** and the holes are the **minority carriers**. Since the majority carriers have negative charges, an electron-doped semiconductor is called N-type semiconductor. The conductivity of an N-type semiconductor follows an activated type, where the activation energy comes from the difference between the donor level and $E_c$.

In contrast, when we add a small amount of boron or aluminium with three valence electrons, a hole is generated. The unoccupied levels are bound to the nucleus originally corresponding to the three electrons, to generate an acceptor level just above the valence band (Fig. 3.11d). However, the electrons in the valence band are easily activated to the acceptor level, and the remaining holes move around the valence band. Here, the Fermi level drops just above the acceptor level, and according to Eq. 3.36, the majority carrier is the hole and the minority carrier is the electron. Since the majority carriers have positive charges, such a semiconductor is called P-type semiconductor. The activation energy reflects the difference between the acceptor level and $E_v$.

## 3.4 Quantum Oscillation and Observation of the Fermi Surface

Next, we describe motion of metal electrons in a strong magnetic field, from which we can investigate the Fermi surface [6]. Equation of motion for a free electron under a magnetic field is represented using the Lorentz force.

$$\frac{\mathrm{d}\mathbf{p}}{\mathrm{d}t} = m\frac{\mathrm{d}\mathbf{v}}{\mathrm{d}t} = -e\mathbf{v} \times \mathbf{B}. \tag{3.40}$$

When the magnetic field $B$ is parallel to $z$, the cross product in the right-hand side is written as

$$\frac{\mathrm{d}v_x}{\mathrm{d}t} = \frac{eB}{m}v_y = \omega_c v_y \text{ and } \frac{\mathrm{d}v_y}{\mathrm{d}t} = -\omega_c v_x. \tag{3.41}$$

for the $x$ and $y$ directions, respectively. Here, $\omega_c = \frac{eB}{m}$ is the **cyclotron frequency**. The solution of this equation is represented by

$$v_x = v_0 \sin \omega_c t$$
$$v_y = v_0 \cos \omega_c t \tag{3.42}$$

and one more integration affords

$$x = -\frac{v_0}{\omega_c} \cos \omega_c t$$
$$y = \frac{v_0}{\omega_c} \sin \omega_c t. \tag{3.43}$$

An electron in a magnetic field rotates around a circular orbit (Fig. 3.12). Tangent of the circle in the real space gives the velocity $v$. From $\hbar k = p = mv$, the trajectory in the $k$-space is also a circle, where $k$ is a vector in the radial direction. The position

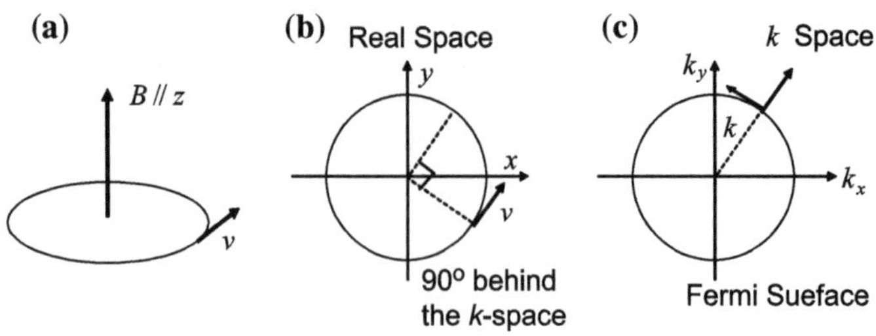


**Fig. 3.12** Rotational motion of an electron in a magnetic field

in the real space is 90° behind the position in the $k$-space. The angular frequency of the rotation is the cyclotron frequency. Comparing Eqs. 3.42 and 3.43, the radius in the real space $|r|$, is related to the radius in the $k$-space space $|p|$.

$$|r| = \frac{v_0}{\omega_c} = \frac{m}{eB} v_0 = \frac{|p|}{eB} \tag{3.44}$$

With increasing the magnetic field, $|r|$ becomes small, and at the same time $|p|$ decreases. By contrast, in a weak magnetic field $|r|$ becomes large. At the limit of $B = 0$, $|r|$ becomes infinity, and the electron moves straightly.

The change of the electron energy under a magnetic field is

$$\frac{\mathrm{d}E(\mathbf{k})}{\mathrm{d}t} = \frac{\mathrm{d}E}{\mathrm{d}\mathbf{p}} \frac{\mathrm{d}\mathbf{p}}{\mathrm{d}t} = -\mathbf{v} \cdot (e\mathbf{v} \times \mathbf{B}) = 0 \tag{3.45}$$

where Eqs. 3.5 and 3.40 are used. At the final step, a cross product of $v$ and $B$ is perpendicular to $v$, and its inner product with $v$ is zero. Then, the application of a magnetic field does not change the electron energy. This is because the Lorentz force affords force only in the direction perpendicular to the magnetic field. An electron started from a point on the Fermi surface moves only on the Fermi surface. On account of the Lorentz force, an electron moves in a plane perpendicular to the magnetic field. In the $k$-space, the cross section of the Fermi surface with a plane perpendicular to the magnetic field affords the trajectory (Fig. 3.13). The position of the perpendicular plane is given by the initial $k_z$. During the circular motion, the momentum $p_z = \hbar k_z$ is kept constant. The rotation in the $xy$-plane is given by Eq. 3.43, and the electron follows a screw motion.

*Example 3.3* Calculate the cyclotron frequency of a free electron under a magnetic field of 1 T.

$\omega_c = \frac{eB}{m} = \frac{(1.6\times10^{-19}\mathrm{C})(1\mathrm{T})}{(9.1\times10^{-31}\mathrm{kg})}$ = 1.76 x $10^{11}$ rad/s, or the frequency is $\nu = \frac{\omega_c}{2\pi}$ = 28 GHz.

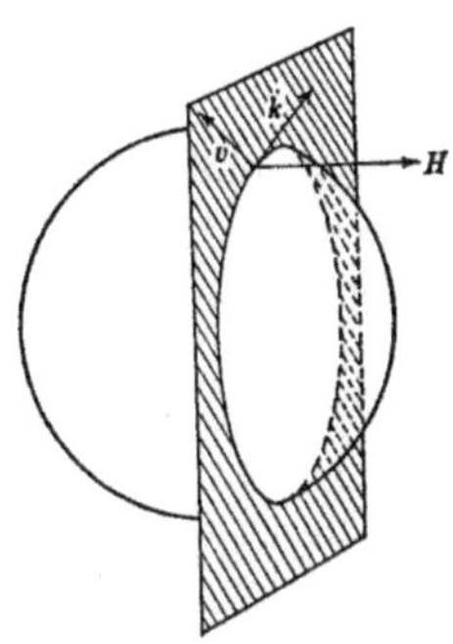


**Fig. 3.13** Rotation of an electron in the $k$-space

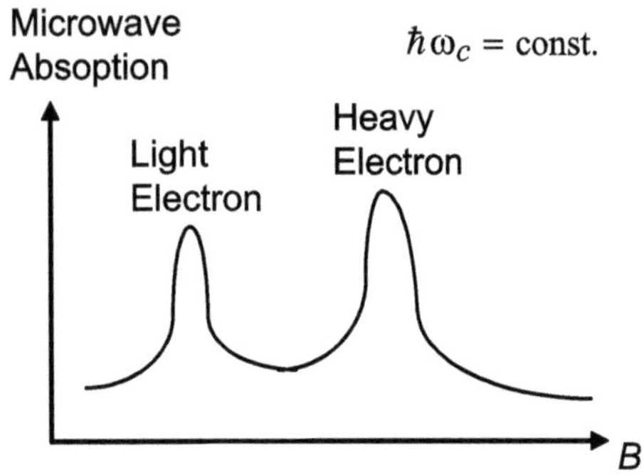


**Fig. 3.14** Microwave absorption due to the cyclotron resonance

This is in the microwave region. From Eq. 3.19, the unit of $B$ (T) is (force)/(Cm/s), which is therefore (kgm/s$^2$)/(Cm/s) = kg/Cs. Then, the unit of the above equation is s$^{-1}$.

Owing to such rotational motion, an electron in a magnetic field resonates with a electromagnetic wave with the frequency $\omega_c$. The frequency is in the microwave region, so when microwave is irradiated to a metal in a magnetic field, the absorption occurs at the resonating frequency (Fig. 3.14). This phenomenon is called **cyclotron resonance**. The effective mass is estimated by measuring the cyclotron resonance.

When the magnetic field is strong enough, the radius of the rotational motion becomes small, and the electron is not scattered during the rotation. This condition is represented by $\omega_c\tau \gg 1$. $B$ has to be strong enough to make $\omega_c$ large, and $\tau$ has to be sufficiently large. In such a case, after one rotation, the electron phase has to be the same as the original phase. In analogy with the orbit of a hydrogen atom (Sect. 1.1), the quantization is applied to (circumference) = (wavelength) x (integer) according to Bohr's condition

$$2\pi r = n\lambda = n\frac{h}{p} \quad \text{leads to} \quad r \cdot p = n\hbar. \tag{3.46}$$

Insertion of Eq. 3.44 gives $\frac{p}{eB} \cdot p = n\hbar$ or

$$k^2 = \frac{eB}{\hbar} n. \tag{3.47}$$

Bohr's condition applies to $x$ and $y$ components independently, so the above equation leads to $x \cdot p_x = n\hbar$ and $y \cdot p_y = n\hbar$. This is inserted in the $k_x^2 + k_y^2$ part of the three-dimensional free electron energy (Eq. 2.20) to give

$$E = \frac{\hbar^2}{2m} 2\frac{eB}{\hbar} n + \frac{\hbar^2}{2m} k_z^2 = \hbar\omega_c n + \frac{\hbar^2}{2m} k_z^2. \tag{3.48}$$

Rotational motion is quantized entirely in the same way as the harmonic oscillator (the spring motion). Similarly, the above relation is used in the area $S_k = \pi(k_x^2 + k_y^2)$ of the circle in the $k$-space to afford

$$S_k = \frac{2\pi eB}{\hbar} n. \tag{3.49}$$

This equation implies that the area of the rotation is quantized (Fig. 3.15a). The area of the orbit corresponding to $n = 2$ has twice as large as the area of the $n = 1$ orbit. Accordingly, the interval between the circles becomes gradually small. In the three-dimensional space, the motion is restricted on a cylindrical surface elongated

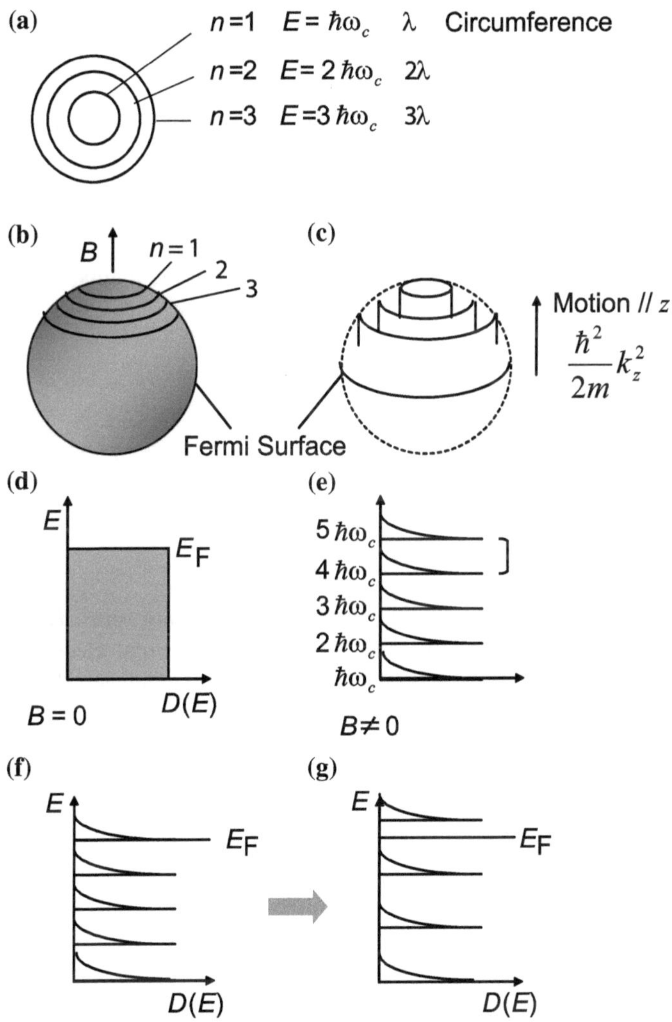


**Fig. 3.15** **a–c** Landau tubes and **d–g** Landau levels

parallel to $B$ (Figs. 3.15b, c). These cylinders are called **Landau tubes**. The corresponding energy is given by integer times of $\hbar\omega_c$ from Eq. 3.48. These are called **Landau levels**. At $B = 0$, the interval of the Landau levels is zero, and the state density is continuous (Fig. 3.15d). At nonzero $B$, the state density splits to discrete levels with the same interval $\hbar\omega_c$ (Fig. 3.15e). When the magnetic field is so small that the splitting of the Landau levels is smaller than the thermal energy $k_BT$, the splitting is negligible. However, the splitting is significant when $\hbar\omega_c$ is larger than $k_BT$.

Equation 3.48 contains $\frac{\hbar^2}{2m}k_z^2$, which is a straight motion along $z$. Then, the electron makes a spiral motion. The density of states coming from the one-dimensional electron is represented by $D(E) \propto E^{-1/2}$ similarly to Eq. 2.34. Such a one-dimensional band diverges at the band edge. Including the one-dimensional motion along $z$, smoothly decreasing tails are added to the discrete Landau levels (Fig. 3.15e).

We shall investigate the number of electrons entering in a Landau level. The periodical boundary condition $k = (2\pi/L)n$ is put in Eq. 3.44 to give

$$|r| = \frac{|p|}{eB} = \frac{\hbar k}{eB} = \frac{h}{eBL}n$$

which means the interval of the states is $h/eBL$. Then, the number of the states is obtained as

$$N = \frac{L}{h/eBL} = \frac{eB}{h}L^2 \quad (3.50)$$

implying a unit area has $eB/h$ states. Form this, the density of states is given by

$$D(E) = \frac{2eBL^2/h}{\hbar\omega_c} = \frac{2eBL^2}{h}\frac{m}{\hbar eB} = \frac{mL^2}{\hbar^2\pi}$$

where the factor two comes from the upward and downward spins. This is the same as Eq. 2.31 for two-dimensional free electrons, so the total electron number is unchanged. Thus, the splitting of the energy band to Landau levels does not change $E_F$.

In Fig. 3.15f, $E_F$ is situated at the peak of a Landau level. This Landau level corresponds to the largest Landau tube, and the orbit is the equator of the Fermi sphere (Fig. 3.15c). Here, $E_F$ is located at the bottom of the Landau level, corresponding to the maximum of the state density. When the magnetic field is increased slightly, the interval of the Landau levels increases, and the largest Landau tube goes out of the equator, to become unoccupied. Then, the state density jumps to nearly zero. With further increasing the magnetic field, $E_F$ moves to the next Landau level, and to the next maximum. Accordingly, by changing the magnetic field, the state density oscillates when $E_F$ crosses the Landau levels. Accordingly, many physical properties oscillate depending on $B$. Oscillation of the electrical resistivity is called **Shubnikov–de Haas oscillation**, and oscillation of the magnetic susceptibility is called **de Haas–van Alphene oscillation**. Similarly, many other properties such as heat capacity exhibit oscillation. In order to estimate the interval of the oscillation, the areas of the $n$-th and $n + 1$-th orbits are equated to the area of the equator $S_k$ in Eq. 3.49.

$$S_k = \frac{2\pi e}{\hbar} B_n n = \frac{2\pi e}{\hbar} B_{n+1}(n+1) \tag{3.51}$$

By deleting $n$ from these relations, we obtain

$$\frac{1}{B_{n+1}} - \frac{1}{B_n} = \Delta\left(\frac{1}{B}\right) = \frac{2\pi e}{\hbar S_k}. \tag{3.52}$$

Peaks appear in the regular interval of $1/B$, and the frequency affords the cross section of the Fermi surface. In a spheral Fermi surface, $S_k$ corresponds to the area surrounded by the equator. In general, the maximum or minimum cross section of the Fermi surface cut by a plane perpendicular to the magnetic field is obtained. For a large Fermi surface, fine oscillation is observed, and for a small Fermi surface, slowly varying oscillation is observed.

Figure 3.16 shows the Shubnikov–de Haas oscillation observed in an organic superconductor β-$(BEDT\text{-}TTF)_2I_3$. Owing to the elliptical Fermi surface (or cylindrical Fermi surface in the three-dimensional space, Fig. 3.16a), remarkable oscillation (Fig. 3.16b) has been observed. From the oscillation, $S_k$ is estimated to be half the first Brillouin zone, in excellent agreement with the band calculation in Fig. 3.16a, where the hatched area is half of the Brillouin zone.

Figure 3.17 shows the Shubnikov–de Haas oscillation observed in an organic superconductor κ-$(BEDT\text{-}TTF)_2Cu(NCS)_2$. The predominant oscillation in Fig. 3.17b corresponds to the rugby ball around the Z point designated as $\alpha$ in Fig. 3.17a, which amounts to 18 % of the Brillouin zone. Above 40 T, additional fine peaks are visible, which correspond to 100 % of the Brillouin zone. This comes from the β-orbit in Fig. 3.17a, where the electron jumps from the α-orbit to the global orbit across the small energy gap. Such a phenomenon is called **magnetic breakdown**.

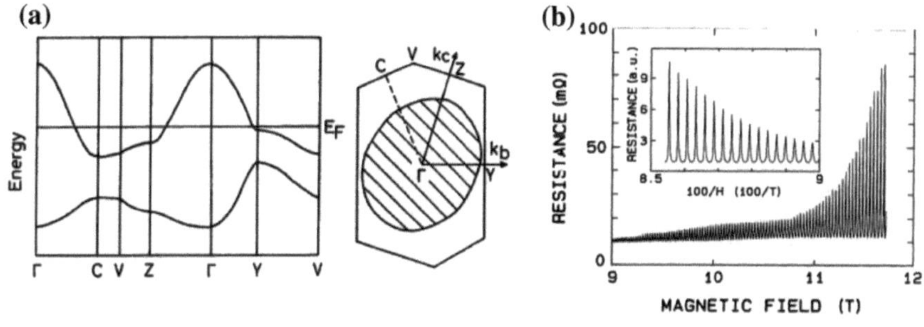


**Fig. 3.16** **a** Band structure, [7] and **b** the Shubnikov–de Haas oscillation of β-$(BEDT\text{-}TTF)_2I_3$. [8]

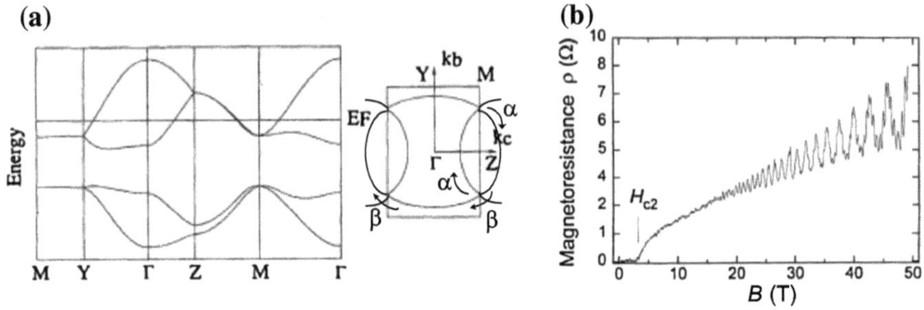


**Fig. 3.17** **a** Band structure, [9] and **b** the Shubnikov–de Haas oscillation of κ-(BEDT-TTF)$_2$Cu (NCS)$_2$ [10]

The Shubnikov–de Haas oscillation is observed only when the organic conductor has a closed cylindrical Fermi surface. The observation of the Shubnikov–de Haas oscillation is a clear evidence of the two-dimensional Fermi surface. Many ordinary one-dimensional conductors do not afford such an oscillation even if they are metallic down to low temperatures. To observe the quantum oscillation, the electron has to rotate the orbit more than once without scattered. In order to fulfill the condition $\omega_c\tau \gg 1$, the measurements in Figs. 3.15 and 3.16 are carried out at a temperature lower than 1 K and at a magnetic field higher than 10 T. The Shubnikov–de Haas oscillation in organic conductors is clearly observed because the Fermi surface consists of a cylinder with uniform cross section. In ordinary three-dimensional metals, the Fermi surface is investigated by the de Haas–van Alphene oscillation, where sensitive measurement of the magnetic susceptibility is possible using a modulated field.

*Example 3.4* In κ-(BEDT-TTF)$_2$Cu(NCS)$_2$, the Shubnikov–de Haas oscillations have been observed at 667 and 3900 T (magnetic breakdown). The area of a unit cell is 107.8 A$^3$ in the conducting layer. Calculate the areas of the Fermi surface as ratios to the Brillouin zone.

The area of the Brillouin zone is

$$S_{\mathrm{BZ}} = \frac{4\pi^2}{S_{\mathrm{cell}}} = \frac{4\pi^2}{(107.8 \times 10^{-20}\,\mathrm{m}^2)} = 3.662 \times 10^{19}\,\mathrm{m}^{-2}.$$

Equation 3.52 gives

$$S_k = \frac{2\pi e}{\hbar\Delta(1/B)} = \frac{2\pi \times (1.602 \times 10^{-19}\,\mathrm{C}) \times (667\,\mathrm{T})}{(1.0546 \times 10^{-34}\,\mathrm{Js})} = 6.366 \times 10^{19}\,\mathrm{m}^{-2}$$

so the ratio

$$\frac{S_k}{S_{\mathrm{BZ}}} = \frac{(0.6366 \times 10^{-19}\mathrm{m}^2)}{(3.662 \times 10^{-19}\mathrm{m}^2)} = 0.17$$

corresponds to the rugby ball-like α-orbit in Fig. 3.17a. For the breakdown β-orbit

$$S_k = \frac{2\pi e}{\hbar \Delta(1/B)} = \frac{2\pi \times (1.602 \times 10^{-19}\,\mathrm{C}) \times (3900\,\mathrm{T})}{(1.0546 \times 10^{-34}\,\mathrm{Js})} = 3.722 \times 10^{19}\,\mathrm{m}^{-2}$$

affords

$$\frac{S_k}{S_{\mathrm{BZ}}} = \frac{(3.722 \times 10^{-19}\mathrm{m}^2)}{(3.662 \times 10^{-19}\mathrm{m}^2)} = 1.016.$$

The 1.6 % error comes from the thermal expansion, because the cell area has been estimated at 104 K. The unit of T is kg/Cs from Example 3.3, and J is kg $\mathrm{m}^2/\mathrm{s}^2$, so the above unit is

$$\frac{(\mathrm{C}) \times (\mathrm{kg/Cs})}{(\mathrm{Js})} = \frac{(\mathrm{C}) \times (\mathrm{kg/Cs})}{(\mathrm{kgm}^2/\mathrm{s})} = (\mathrm{m}^{-2}).$$

Once the composition is determined to be $(\mathrm{BEDT\text{-}TTF})_2\mathrm{X}$, the cell area is approximately the same, and the globally connected Fermi surface affords about the same Shubnikov–de Haas frequency at 3900 T [11]. In such a case, we measure the carrier concentration by the Shubnikov–de Haas frequency. The corresponding two-dimensional carrier density is $n = 1.86 \times 10^{14}\ \mathrm{cm}^{-2}$ from $2/S_{\mathrm{cell}}$.

The amplitude $A$ of the Shubnikov–de Haas oscillation is given by the **Lifshitz–Kosevich** equation

$$A \propto \sqrt{B}\frac{K(m^*/m_0)T/B}{\sinh(K(m^*/m_0)T/B)}\exp\left(-\frac{K(m^*/m_0)T_D}{B}\right)$$

which exponentially decreases as elevating the temperature [12], where $K = 2\pi^2 k_B m_0/e\hbar$ = 14.96 T. The factor containing sinh is approximated to

$$\frac{2K(m^*/m_0)T}{B}\exp(-\frac{K(m^*/m_0)T}{B}).$$

Usually the Fourier transform of the oscillation is calculated, and the amplitude $A$ of the frequency is plotted like log $A/T$ versus $T$ as shown in Fig. 3.18. From the slope, we can estimate the effective cyclotron mass $m^*$. When there is only one kind of oscillation, we can more simply plot the amplitude similarly, and obtain $m^*$. Since the exponentials in the above equations have similar forms, we next plot the field

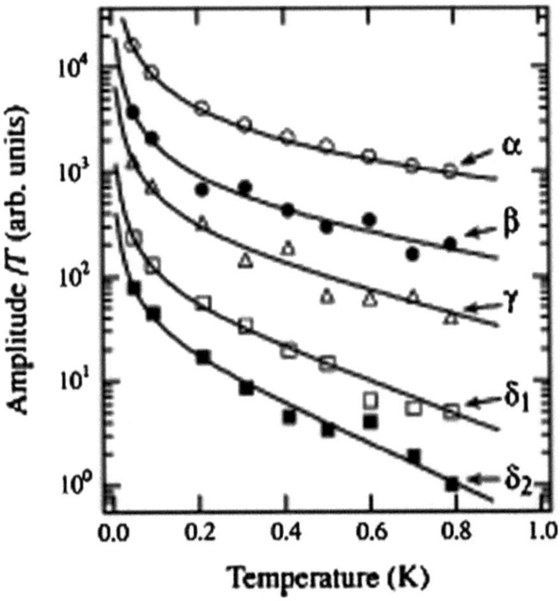


**Fig. 3.18** Mass plot of (MDT-TSF)$(AuI_2)_{0.436}$ (Figs. 7.40 and 7.43) [13]

dependence of $A$ like log $A\sqrt{B}$ versus $1/B$ at a definite temperature to obtain the **Dingle temperature**

$$T_D = \hbar/2\pi k_B \tau.$$

The Dingle temperature depends on τ, and is small for large τ, so depends on the sample quality. The parameters $m^*$ and $T_D$ represent how the oscillation amplitude attenuates depending on $T$ and $B$, respectively.

Effective masses obtained from the Shubnikov–de Haas oscillations in organic conductors are listed in Table 3.1. It is a general rule that the Fermi surface with large $S_k$ has large $m^*$. For instance, the β-orbit of the κ-phase has larger $m^*$ than the

**Table 3.1** Effective mass, $m^*$, and superconducting transition temperature, $T_c$, in organic conductors [14]

| Compound | $S_k/S_{BZ}$/% | $m^*/m_0$ | $T_c$/K |
|---|---|---|---|
| α-$(BEDT\text{-}TTF)_2KHg(SCN)_4$ | 16 | 1.4 | 1 (3 kbar uniaxial) |
| α-$(BEDT\text{-}TTF)_2NH_4Hg(SCN)_4$ | 13 | 2.1 | 0.8 |
| θ-$(BEDT\text{-}TTF)_2I_3$ | 19, 102 | 1.8, 3.5 | 3.6 |
| $\beta_H$-$(BEDT\text{-}TTF)_2I_3$ | 51 | 4.65 | 8 |
| κ-$(BEDT\text{-}TTF)_2I_3$ | α:15 β:102 | α:1.85 β:3.9 | 3.6 |
| κ-$(BEDT\text{-}TTF)_2Ag(CN)_2H_2O$ | α:17 | α:2.7 | 5.0 |
| κ-$(BEDT\text{-}TTF)_2Cu(NCS)_2$ | α:15.7 β:105 | α:3.5 β:6.5 | 10.4 |
| κ-$(BEDT\text{-}TTF)_2Cu[N(CN)_2]Br$ | β:106 | β:6.4 | 11.8 |

α-orbit. This also comes from the relation that the cyclotron mass is related to the derivative of $S_k$

$$m* = \frac{\hbar^2}{2\pi}\left(\frac{\partial S_k}{\partial E}\right)_{E=E_F}.$$

The superconducting transition temperatures $T_c$, listed in Table 3.1 tend to increase with increasing $m^*$. From Eq. 6.42, $T_c$ increases with increasing the state density, and accordingly with increasing $m^*$. Later we show the state density around $E_F$ is large for high $T_c$ compounds (Fig. 3.23). According to the Stoner model in Sect. 5. 7.1, we see that the state density is enhanced by the Coulomb interaction. The effective mass reflects all the influence.

*Example 3.5* Under the condition of $\omega_c\tau = \mu B = 1$ at 10 T, calculate the mobility μ in a unit of $cm^2/Vs$

The above equation is obtained from $\omega_c = eB/m$ and $\mu = e\tau/m$. The unit of T is kg/Cs from Example 3.3. The product of voltage and current affords the power $P = VI$, and the unit is W = J/s = $kgm^2/s^3$ from Example 3.4. Then, the unit of V is

$$[V] = \left[\frac{kgm^2}{s^3}\right]\left[\frac{s}{C}\right] = \left[\frac{kgm^2}{Cs^2}\right].$$

The unit of the mobility $\mu$ ($m^2/Vs$) is transformed to

$$\left[\frac{m^2}{Vs}\right] = \left[\frac{Cs^2}{kgm^2}\frac{m^2}{s}\right] = \left[\frac{Cs}{kg}\right]$$

which is identical to the inverse of T. The magnetic field of 10 T gives $\mu = 0.1$ $m^2/Vs$ = 1000 $cm^2/Vs$. Example 2.1 demonstrates that typical mobility of organic conductors is around 1–10 $cm^2/Vs$ at room temperature. The ratio of resistivity at room and helium temperatures is called the residual resistivity ratio (RRR), which is about $10^2$–$10^3$ as exemplified by Figs. 7.17b, 7.36b and 7.47c. As a result, the mobility is expected to be more than 1000 $cm^2/Vs$ at helium temperatures, which satisfies the condition $\omega_c\tau \gg 1$, to observe the quantum oscillation.

## 3.5 Magnetoresistance

When the magnetic field is too weak and the temperature is too high to observe quantum oscillations, the resistance simply increases with the magnetic field. Magnetoresistance gives a lot of information on the Fermi surface. To investigate magnetoresistance, we shall start from the equation of motion (Eq. 3.13) under a magnetic field.

$$m\frac{d\mathbf{v}}{dt}+\frac{m}{\tau}\mathbf{v}=-e(\mathbf{E}+\mathbf{v}\times\mathbf{B}) \tag{3.53}$$

Assuming $B//z$ and the steady states $d\mathbf{v}/dt = 0$, the respective components are

$$v_x=-\frac{e\tau}{m}E_x-\frac{e\tau}{m}Bv_y,\ v_y=-\frac{e\tau}{m}E_y+\frac{e\tau}{m}Bv_x,\ v_z=-\frac{e\tau}{m}E_z. \tag{3.54}$$

Substituting $v_x$ and $v_y$ in the other equations, we obtain

$$v_x=-\frac{e\tau}{m}E_x+\left(\frac{e\tau}{m}\right)^2BE_y-\left(\frac{e\tau}{m}\right)^2B^2v_x$$

$$v_y=-\frac{e\tau}{m}E_y-\left(\frac{e\tau}{m}\right)^2BE_x-\left(\frac{e\tau}{m}\right)^2B^2v_y. \tag{3.55}$$

These relations are simplified to the $x$ and $y$ components of the conductivity tensor using $J=-nev$, Eq. 3.16, $\sigma_0=\frac{ne^2\tau}{m}$, and $\omega_c=\frac{eB}{m}$.

$$\sigma=\frac{\sigma_0}{1+(\omega_c\tau)^2}\begin{pmatrix}1 & -\omega_c\tau\\ \omega_c\tau & 1\end{pmatrix} \tag{3.56}$$

The inverse matrix affords the resistivity tensor.

$$\rho=\rho_0\begin{pmatrix}1 & \omega_c\tau\\ -\omega_c\tau & 1\end{pmatrix} \tag{3.57}$$

The nondiagonal term $\rho_{yx}=-\rho_0\omega_c\tau=-B/ne$ represents the Hall effect. The diagonal term is $\rho_{xx}=\rho_0$, implying the **transverse magnetoresistance** is zero. This means the resistance $\rho_{xx}$ does not increase when $I//x$ and $B//z$ are applied. This comes from the cancellation of the Hall effect with the denominator of Eq. 3.56, because the current does not bent due to the perpendicular field. Similarly, the third relation of Eq. 3.54 implies that $B//z$ does not influence $I//z$, so the **longitudinal magnetoresistance** is also zero.

It is generally observed that the longitudinal magnetoresistance is practically zero, but the transverse magnetoresistance is nonzero in many occasions. To investigate the transverse magnetoresistance, suppose a two-carrier model. Then instead of Eq. 3.56, the current is represented by a sum of two contributions.

$$J_x=\left(\frac{\sigma_1}{1+\beta_1^2B^2}+\frac{\sigma_2}{1+\beta_2^2B^2}\right)E_x+\left(\frac{\sigma_1\beta_1}{1+\beta_1^2B^2}+\frac{\sigma_2\beta_2}{1+\beta_2^2B^2}\right)BE_y \tag{3.58}$$

Here, 1 and 2 designate two kinds of carriers, and we use $\omega_{c1}\tau_1 = \frac{e\tau_1}{m_1}B = \beta_1 B$. It is a lengthy calculation to convert this relation to ρ, but the result is

$$\frac{\rho(B) - \rho_0}{\rho_0} = \frac{\sigma_1\sigma_2(\beta_1 - \beta_2)B^2}{(\sigma_1 + \sigma_2)^2 + (\beta_2\sigma_1 + \beta_1\sigma_2)^2 B^2}. \tag{3.59}$$

As shown in Fig. 3.19a, the transverse magnetoresistance increases in proportion to $B^2$ at weak fields, and saturates at high fields.

There are several exceptional cases where the magnetoresistance does not saturate. When the electron and hole concentrations are, respectively, $n_1$ and $n_2$, the Hall conductivity is

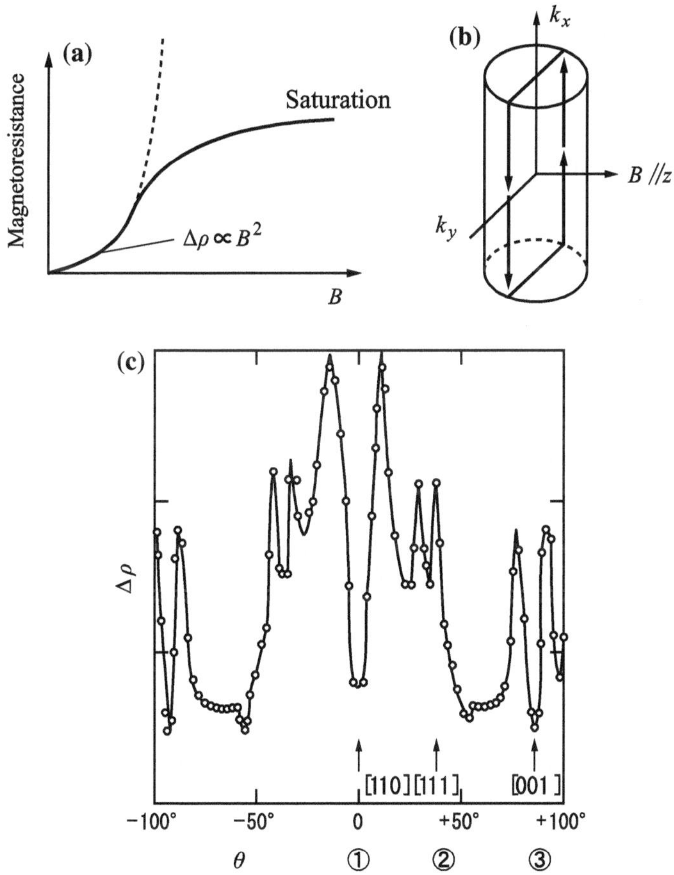


**Fig. 3.19** **a** Typical magnetoresistance. **b** An open orbit. **c** Field orientation dependence of the magnetoresistance for a single crystal gold [15]

$$\sigma_{xy} = -\frac{e(n_1 - n_2)}{B}. \tag{3.60}$$

This becomes zero at $n_1 = n_2$. In this case, Eq. 3.56 is converted to

$$\sigma = \frac{\sigma_0}{1+(\omega_c\tau)^2}\begin{pmatrix} 1 & 0 \\ 0 & 1 \end{pmatrix} \tag{3.61}$$

and the inverse matrix leads to $\rho_{xx} = \rho_0(1+(\omega_c\tau)^2)$. In such a compensated metal, the transverse magnetoresistance does not saturate even at a strong field ($\omega_c\tau \gg 1$), and increases in proportion to $B^2$ as shown in the dotted curve in Fig. 3.19a.

The second exception is derived from the open Fermi surface (Fig. 3.19b). Assume a cylindrical Fermi surface elongating along $x$. Under $B//z$, the electron orbit is the cross section perpendicular to $z$, and the trajectory runs along $x$. When the equation of motion given by Eq. 3.53

$$\hbar\frac{\mathrm{d}\mathbf{k}}{\mathrm{d}t} = -e\mathbf{v}\times\mathbf{B} \tag{3.62}$$

is averaged over a period of a closed orbit, $k$ oscillates, and the average is $< v_x >= 0$ and $< v_y > = 0$. For an open orbit elongating along $x$, $k_x$ increases continuously to give $< v_x > = 0$ and $< v_y > \neq 0$. Alternatively, this is due to the constant finite $\hbar k_y$. Substituting $< v_x > = 0$ in the first relation of Eq. 3.55, we obtain $E_x = -\omega_c\tau E_y$. This is put in the second relation to give

$$J_y = -nev_y = \frac{\sigma_0(E_y - \omega_c\tau E_x)}{1+(\omega_c\tau)^2} = \frac{\sigma_0(1+(\omega_c\tau)^2)E_y}{1+(\omega_c\tau)^2} = \sigma_0 E_y. \tag{3.63}$$

Accordingly, $\sigma_{yy}$ in Eq. 3.56 is converted to

$$\sigma = \sigma_0\begin{pmatrix} \frac{1}{1+(\omega_c\tau)^2} & -\frac{\omega_c\tau}{1+(\omega_c\tau)^2} \\ \frac{\omega_c\tau}{1+(\omega_c\tau)^2} & 1 \end{pmatrix} \rightarrow \sigma_0\begin{pmatrix} (\omega_c\tau)^{-2} & -(\omega_c\tau)^{-1} \\ (\omega_c\tau)^{-1} & 1 \end{pmatrix}. \tag{3.64}$$

Here, the final relation is obtained assuming a high magnetic field, $\omega_c\tau \gg 1$. The inverse matrix affords $\rho_{xx} = \rho(\omega_c\tau)^2 \propto B^2$, and the magnetoresistance for $I//x$ increases without saturation.

A variety of intriguing phenomena have been observed in magnetoresistance of elemental metals. Magnetoresistance of alkali metals saturates in all directions as shown in Fig. 3.19a, because these metals have a close spherical Fermi surface. Magnetoresistance of bismuth and molybdenum increases in all directions because these metals are compensated metals with the same number of electrons and holes. In several metals like copper, the magnetoresistance saturates depending on the orientations (Fig. 3.19c). The Fermi surface of copper and gold consists of connected spheres (Fig. 2.43). At a field along [001] and [110], the perpendicular cross section cut the neck at the center. This generates only closed orbits and the magnetoresistance saturates. However, a slightly tilted field results in an open orbit connected to the next Fermi sphere, and brings about large and unsaturated magnetoresistance. A more tilted field results in a closed orbit again, and another open orbit appears before and after the field orientation passes the next neck. The resulting peaks appearing in the angular dependence (Fig. 3.19c) prove the existence of the connected necks, from which we can discuss the size of the necks.

Similar angular dependence has been observed in organic conductors. An organic conductor θ-(BEDT-TTF)$_2$I$_3$ shows angular dependence as shown in Fig. 3.20a, when the field direction is tilted from the direction perpendicular to the conducting layer ($\theta = 0°$). The peak positions do not depend on the field strength. This compound has a typical two-dimensional cylindrical Fermi surface (Fig. 2.48), but considering the interlayer transfer, the cylinder is warping as shown in Fig. 3. 20b. When the field direction is exactly perpendicular to the conducting layer, the cross section takes a maximum at $k_c = 0$ and a minimum at $k_c = \pi/c$. By tilting the

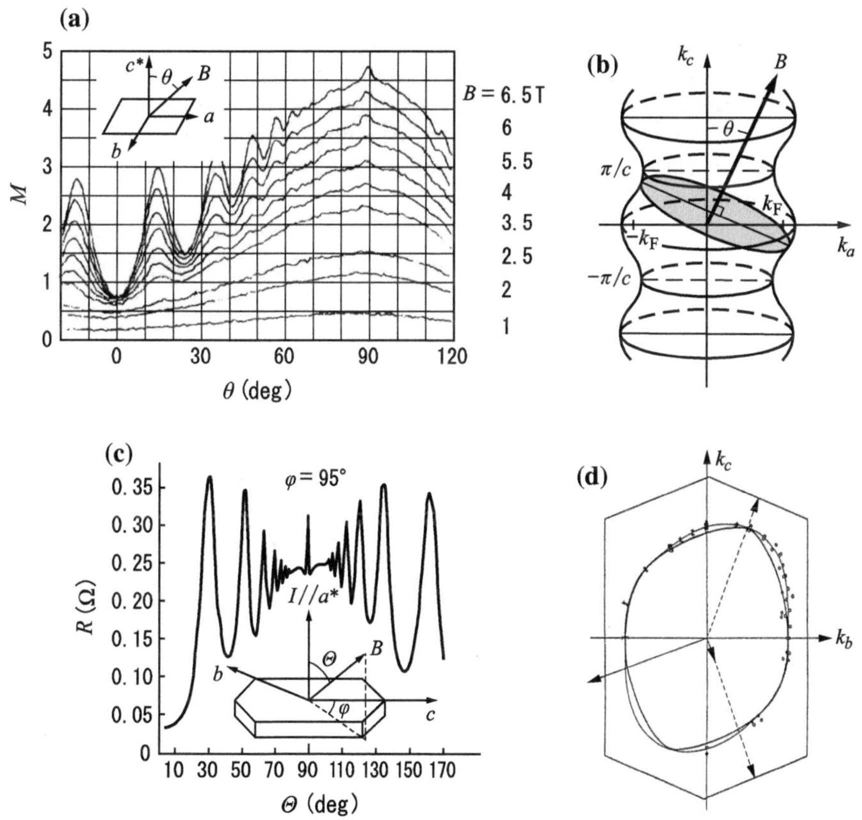


**Fig. 3.20 a** Kajita oscillation and **b** Fermi surface in θ-(BEDT-TTF)$_2$I$_3$ [16, 17] **c** Kajita oscillation and **d** Fermi surface in β-(BEDT-TTF)$_2$IBr$_2$ [18]

field direction, and when the cross section passes the left end at $k_c = \pi/c$ and the right end at $k_c = 0$, the cross section becomes approximately unchanged depending on the position of the cross section. This situation results in the magnetoresistance peak around $\theta = 15°$ in Fig. 3.20a. The magnetoresistance decreases again with increasing the tilt angle. Accordingly, the oscillation appears with the periodicity

$$\tan\theta = \frac{n\frac{\pi}{c}}{k_F} \tag{3.65}$$

which leads to

$$ck_F \tan\theta = n\pi. \tag{3.66}$$

This angular dependence is called **Kajita oscillation**. From this oscillation, we can estimate $k_F$ from the peak positions. Figure 3.20c shows the Kajita oscillation observed in β-(BEDT-TTF)$_2$IBr$_2$. The cylindrical Fermi surface has been verified from the Shubnikov-de Haas oscillation, but tilting the magnetic field to various orientations, we can plot $k_F$ as shown in Fig. 3.20d.

Although organic conductors are composed of complicated molecules, we may consider only one molecular orbital to construct the energy band in the molecular crystal. The resulting Fermi surface is very simple compared with inorganic conductors. The Fermi surface has been extensively studied by means of the Shubnikov–de Haas oscillation and the magnetoresistance, which is called **Fermiology** of organic conductors.

The Shubnikov–de Haas oscillation and the Kajita oscillation are observed only in two-dimensional conductors, but the following method is applicable to

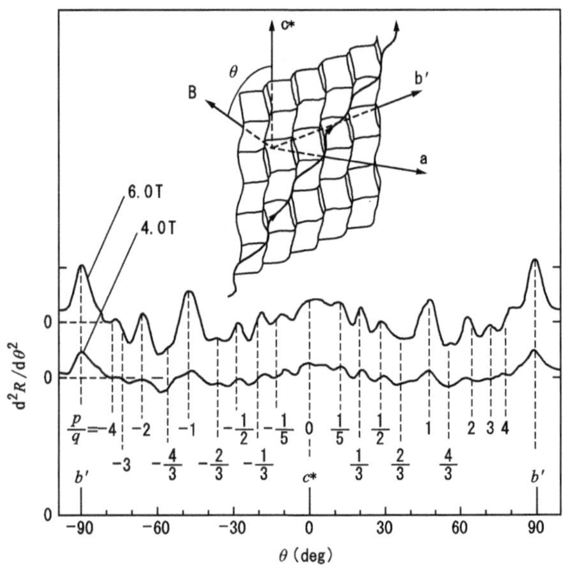


**Fig. 3.21** **a** Fermi surface, and the Lebed oscillation in (TMTSF)$_2X$ [19]

one-dimensional conductors. Suppose the one-dimensional Fermi surface (Fig. 3.21a) warping owing to the transfers in the other directions. When the magnetic field is rotated in the directions perpendicular to the one-dimensional chain, the electron orbit is a warping line. When the warping line is rational times of the lattice constants, $c$ and $b$

$$\frac{c}{b}\tan\theta = \frac{n}{m} \tag{3.67}$$

the electron orbit is periodical, and the magnetoresistance makes peaks. Here, $n$ and $m$ are integers. Figure 3.21b shows such magnetoresistance observed in $(TMTSF)_2PF_6$. Since the anomalies are small, the second derivative is plotted. Such a phenomenon is called **Lebed oscillation**. The peaks of the Lebed oscillation are determined only by the lattice constants, and not useful to estimate $k_F$. However, the Lebed oscillation is used as an evidence of the one-dimensional Fermi surface.

## 3.6 Transport Equation and Thermoelectric Power

Electrical and heat transport phenomena are treated in a unified form using the **transport equation** or the **Boltzmann equation** [20]. We can estimate conductivity and heat transport by assuming the Fermi surface is distorted under an electric field. The electron number is given by $N = 2 \times$ (Volume of the Fermi surface)/$(2\pi/L)^3$, so the electron density $n$ is

$$n = \frac{N}{V} = \frac{2}{8\pi^3}\int \mathrm{d}S\mathrm{d}k = \frac{1}{4\pi^3\hbar}\int \frac{\mathrm{d}S}{v_k}\mathrm{d}E \tag{3.68}$$

where the factor two comes from the upward and downward spins. d$S$ represents integration along the Fermi surface, and d$k$ designates integration perpendicular to the Fermi surface. The final form is derived from the definition of the velocity (Eq. 3.4). We suppose the Fermi distribution $f$ deviates from the equilibrium under an electric field $\varepsilon$.

$$\frac{\mathrm{d}f}{\mathrm{d}t} = \frac{\partial f}{\partial \mathbf{p}}\frac{\partial \mathbf{p}}{\partial t} = \frac{\partial f}{\partial E}\frac{\partial E}{\partial \mathbf{p}}\frac{\partial \mathbf{p}}{\partial t} = \frac{\partial f}{\partial E}ve\boldsymbol{\varepsilon}$$

We also assume the scattering occurs proportionally to the deviation from the Fermi distribution $\Delta f = f - f_0$.

$$\frac{\mathrm{d}f}{\mathrm{d}t} = -\frac{\Delta f}{\tau}$$

Here, $\tau$ is the relaxation time (**relaxation time approximation**). Then, we obtain

$$\Delta f = \left( -\frac{\partial f_0}{\partial E} \right) v e \varepsilon \tau.$$

Accordingly, the current is represented by

$$\mathbf{J} = e \iiint \mathbf{v} \Delta f \mathrm{d}k = e \int \mathbf{v} \Delta f \frac{1}{4\pi^3 \hbar} \frac{\mathrm{d}S}{v_k} \mathrm{d}E = \frac{e^2 \tau \varepsilon}{4\pi^3 \hbar} \int \mathbf{v}^2 \left( -\frac{\partial f_0}{\partial E} \right) \frac{\mathrm{d}S}{v_k} \mathrm{d}E.$$

When we approximate $\left( -\frac{\partial f_0}{\partial E} \right)$ as a delta function at the Fermi level, the integral at the Fermi level is written as

$$K_0 = \frac{\tau}{4\pi^3 \hbar} \int \mathbf{v}^2 \left( -\frac{\partial f_0}{\partial E} \right) \frac{\mathrm{d}S}{v_k} \mathrm{d}E = \frac{\tau}{4\pi^3 \hbar} \int \mathbf{v}^2 \frac{\mathrm{d}S}{v_k}.$$

Then, the current is $J = e^2 K_0 \varepsilon$, and the conductivity is $\sigma = e^2 K_0$. Similarly, when $v^2$ in $K_0$ is replaced by $v_x v_y$, we obtain $\sigma_{xy}$.

When temperature gradient $\nabla T$ is applied, the electric force $e\varepsilon$ is substituted by the increase of the potential energy due to the temperature difference.

$$e\varepsilon \rightarrow \frac{E(k) - \mu}{T} (-\nabla T)$$

Then instead of $K_0$, we obtain

$$K_1 = \frac{e^2 \tau}{4\pi^3 \hbar} \int \mathbf{v}^2 (E((k) - \mu) \left( -\frac{\partial f_0}{\partial E} \right) \frac{\mathrm{d}S}{v_k} \mathrm{d}E.$$

This leads to $J = \frac{e}{T} K_1 (-\nabla T)$.

When both the electric field and the temperature gradient are applied, the resulting current is represented by

$$J = e^2 K_0 \varepsilon + \frac{e}{T} K_1 (-\nabla T). \quad (3.69)$$

The heat flow $U$ is calculated by increasing the order of $K_\mathrm{n}$.

$$U = e K_1 \varepsilon + \frac{1}{T} K_2 (-\nabla T) \quad (3.70)$$

Here, the general form of the integral is

$$K_n = \frac{e^2 \tau}{4\pi^3 \hbar} \int \mathbf{v}^2 (E((k) - \mu)^n \left( -\frac{\partial f_0}{\partial E} \right) \frac{\mathrm{d}S}{v_k} \mathrm{d}E.$$

Equations 3.69 and 3.70 provide generalized forms of transport equations including electrical field and temperature gradient. The coefficients are represented by the integrals $K_n$ over the whole Fermi surface. In principle, we can calculate $\boldsymbol{v}$ from the band structure, and $K_n$ is obtained. However, we cannot directly estimate $\sigma$, because $\tau$ is temperature dependent. An integral of any function $\Phi(E)$ along the Fermi surface satisfies generally [20]

$$\int \Phi(E)\left(-\frac{\partial f}{\partial E}\right)\mathrm{d}E = \Phi(E_F) + \frac{\pi^2}{6}(k_BT)^2\left[\frac{\partial^2\Phi(E)}{\partial E^2}\right]_{E=E_F} + \cdots$$

so using $\Phi(E) = (E_F - \mu)K_0(E)$, we obtain

$$K_1 = \frac{\pi^2}{3}(k_BT)^2\left[\frac{\partial K_0(E)}{\partial E}\right]_{E=E_F}. \tag{3.71}$$

Similarly, we obtain from $\Phi(E)=(E_F - \mu)^2K_0(E)$

$$K_2 = \frac{\pi^2}{3}(k_BT)^2K_0(E_F). \tag{3.72}$$

Putting $\varepsilon = 0$ in Eq. 3.70, the thermal conductivity $\kappa$ is obtained.

$$\kappa = \frac{U}{-\nabla T} = \frac{K_2}{T}.$$

From Eq. 3.69 and $\sigma = e^2K_0$, we obtain

$$\kappa = \frac{\pi^2k_B^2K_0T}{3} = \frac{\pi^2k_B^2}{3e^2}\sigma T$$

which represents the Wiedemann–Franz law (Eq. 3.26).

Suppose two metals A and B are connected as shown in Fig. 3.22, and the two contacts are kept at different temperatures $T_1$ and $T_2$. Then, we observe an electromotive force $V = S\,\Delta T$ proportional to the temperature difference $\Delta T = T_1 - T_2$. This phenomenon is called **thermoelectric power** or **Seebeck effect**, and $S$ is called **Seebeck coefficient**. To extract the thermoelectric power from the transport equation, we put $J = 0$ in Eq. 3.69 to obtain

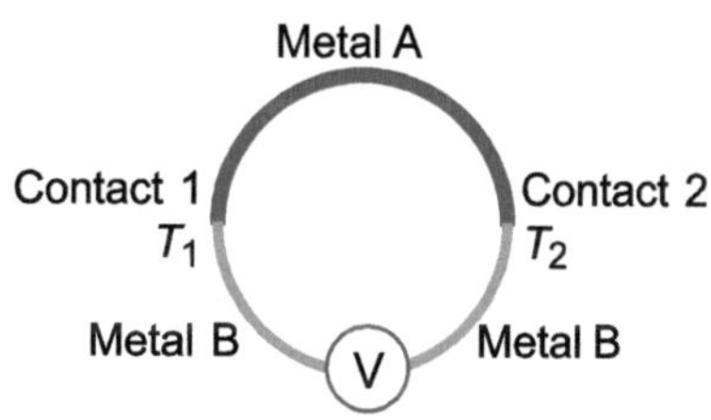


**Fig. 3.22** Thermoelectric power

$$\varepsilon = \frac{1}{eT}\frac{K_1}{K_0}\nabla T. \tag{3.73}$$

The electromotive force is obtained when we integrate $\varepsilon$ along the closed circuit in Fig. 3.22 starting from the voltmeter

$$V = \int_0^1 \varepsilon_B dx + \int_1^2 \varepsilon_A dx + \int_2^0 \varepsilon_B dx = \int_{T_1}^{T_2} (S_A - S_B)\mathrm{d}T. \tag{3.74}$$

Thus, the thermoelectric power is given by $S_A - S_B$ at $T_1$ and $T_2$. A thermocouple is an application of the thermoelectric power, where $T_1$ of the contact 1 is measured by keeping $T_2$ as the reference temperature, for example at 0 °C.

$K_1$ in Eq. 3.71 is put in Eq. 3.73, and using $\sigma = e^2 K_0$ we obtain

$$S = \frac{\pi^2 k_B^2 T}{3e}\left[\frac{\partial \ln \sigma(E)}{\partial E}\right]_{E=E_F}. \tag{3.75}$$

Here $\sigma(E)$ represents the conductivity when $E_F$ is located at $E$. From

$$\sigma = \frac{ne^2\tau}{m}$$

(Eq. 3.16), the energy dependence mainly comes from $n(E)$, and other factors such as $\tau$ and $m$ disappear because the numerator and the denominator cancel to each other in the logarithmic derivative.

$$S = \frac{\pi^2 k_B^2 T}{3e}\left[\frac{\partial \ln N(E)}{\partial E}\right]_{E=E_F} \tag{3.76}$$

For the three-dimensional free electrons, this relation is further reduced using Eq. 2.26.

$$S = \frac{\pi^2 k_B^2}{6eE_F} T \tag{3.77}$$

In general, the thermoelectric power of a metal is proportional to $T$. The coefficient is inversely proportional to $E_F$ (see Fig. 3.26). Accordingly, an energy band with a large bandwidth affords small thermoelectric power. Since Eq. 3.77 is proportional to $e$, the thermoelectric power depends on the carrier polarity; $S > 0$ for holes, and $S < 0$ for electrons. Then, we can determine the carrier polarity by measuring the thermoelectric power.

For a one-dimensional energy band, Eq. 2.21 affords

$$D = \frac{\partial N}{\partial E} = \frac{L}{\pi}\frac{\partial k}{\partial E}$$

by considering spins. The square bracket in Eq. 3.76 gives

$$\frac{1}{D}\frac{\partial D}{\partial E} = \frac{1}{D}\frac{\partial}{\partial E}\left(\frac{\partial N}{\partial E}\right) = \frac{1}{\frac{L}{\pi}\frac{\partial k}{\partial E}}\frac{\partial k}{\partial E}\frac{\partial}{\partial k}\left(\frac{L}{\pi}\frac{\partial k}{\partial E}\right) = \frac{\partial}{\partial k}\left(\frac{1}{\frac{\partial E}{\partial k}}\right) = -\frac{\frac{\partial^2 E}{\partial k^2}}{\left(\frac{\partial E}{\partial k}\right)^2}.$$

In the final step, we used a derivative of a fractional function. Then, Eq. 3.76 affords

$$S = -\frac{\pi^2 k_B^2 T}{3e}\left[\left(\frac{\partial^2 E}{\partial k^2}\right) / \left(\frac{\partial E}{\partial k}\right)^2\right]_{E=E_F}. \tag{3.78}$$

The thermoelectric power is proportional to the curvature of the band, and is negative for electron-like downward curvature, and positive for hole-like upward curvature. For a one-dimensional tight-binding band, Eq. 2.76 in Example 2.6 gives

$$S = \frac{\pi^2 k_B^2 T}{6e\beta}\frac{\cos k_F a}{1-\cos^2 k_F a} = \frac{\pi^2 k_B^2 T}{6e\beta}\frac{\cos(\pi\rho/2)}{1-\cos^2(\pi\rho/2)} \tag{3.79}$$

where $\rho$ is the charge-transfer degree, which is $\rho = 1$ for a half-filled band [21]. The thermoelectric power is proportional to $T$, and the slope is inversely proportional to $\beta$. The final cosine factor is uniquely determined from ρ, so we can estimate $\beta$ and the bandwidth from the slope of the thermoelectric power.

*Example 3.6* Calculate the room-temperature thermoelectric power of a 1/4-filled one-dimensional energy band with the bandwidth of 1 eV.

The cosine factor at $\rho = 1/2$ is

$$\frac{\cos(\pi\rho/2)}{1-\cos^2(\pi\rho/2)} = \frac{\sqrt{2}/2}{1-1/2} = \sqrt{2} = 1.414.$$

We put $4\beta = 1$ eV $= 1.602 \times 10^{-19}$ J in Eq. 3.79 to give

$$\frac{\pi^2 k_B^2}{6e\beta} = 4.8857 \times 10^{-8}\ \mathrm{V/K^2}$$

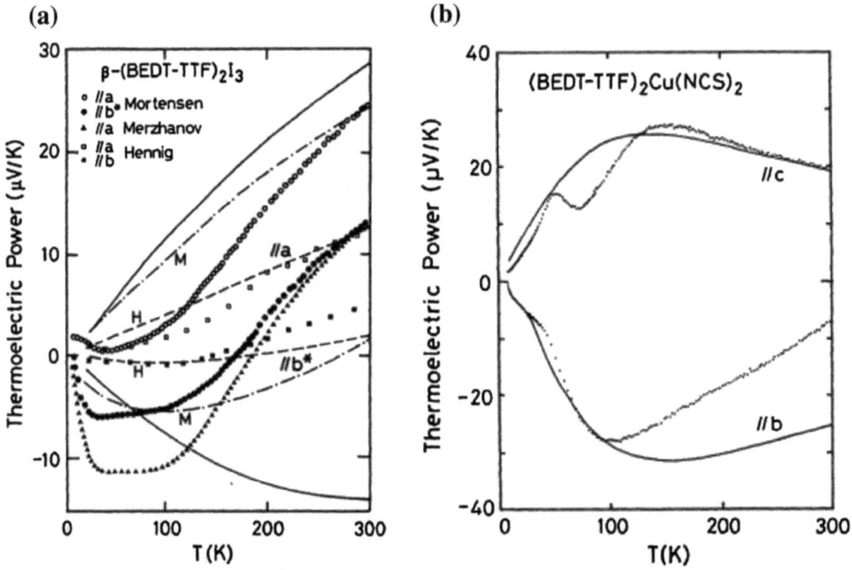


**Fig. 3.23** Thermoelectric power of **a** β-(BEDT-TTF)$_2$I$_3$ and **b** κ-(BEDT-TTF)$_2$I$_3$ [22]

then

$$\left(4.8857 \times 10^{-8}\,\mathrm{V/K^2}\right) \times (298\,\mathrm{K}) \times 1.414 = 20.6\mu\,\mathrm{V/K}$$

is the magnitude of the thermoelectric power at room temperature.

In principle, we can calculate thermoelectric power of a more complicated energy band by evaluating $K_0$ and $K_1$ using Eq. 3.73 [22]. In Fig. 3.23, the calculated thermoelectric power is compared with experimental results for β-(BEDT-TTF)$_2$I$_3$ and κ-(BEDT-TTF)$_2$Cu(NCS)$_2$ [19]. The thermoelectric power in the β-phase is essentially positive because the holes on the donors are responsible for the conduction. However, the low-temperature part along $b$ becomes negative because the Fermi surface is close to the Y point in Fig. 3.16a, and this part is regarded as electron-like. For the κ-phase, the thermoelectric power is positive along $c$, but negative along $b$. This is because the rugby ball around the Z point is hole-like (Fig. 3.16a), whereas the open Fermi surface near the Y point is electron-like. In this way, we can quantitatively evaluate $K_0$ and $K_1$ from the energy band, and the thermoelectric power is calculated. These organic conductors do not show simple $T$-linear thermoelectric power on account of the highly correlated nature. In particular, the thermoelectric power of the κ-phase exceeds 20 μV/K at 100 K, which is several times larger than the expectation in Example 3.6. This implies the effective density of states around the width of $k_BT$ near $E_F$ is several times larger than the value calculated from the energy band. This is because the state density around $E_F$ is enhanced due to the electronic correlation (Fig. 5.19c). The superconducting transition temperature is also determined by the state density around $E_F$ (Sect. 6.6),

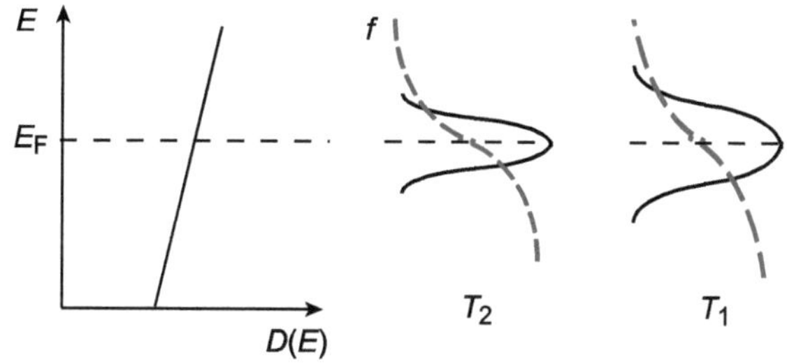


**Fig. 3.24** Origin of the thermoelectric power

so the large thermoelectric power of the κ-phase is related to the comparatively high $T_c$ = 10.4 K. As listed in Table 3.1, it is a general tendency that high $T_c$ organic superconductors have large state density at $E_F$, and large effective mass [23].

In Eq. 3.73, τ chancels each other in the denominator and numerator. The temperature dependence comes from $(-\partial f/\partial E)$ in $K_0$ and $K_1$, in which the slope of the energy band is integrated with the width of $k_BT$ around $E_F$ (Fig. 3.24). When the state density changes gradually, the parts above and below $E_F$ contribute differently. This is the origin of the thermoelectric power. The formula of the thermoelectric power does not include τ, and the thermoelectric power is measured without the actual current flow. Accordingly, thermoelectric power is insensitive to sample imperfections and domain walls in comparison with conductivity.

Thermoelectric power of a semiconductor is obtained from $\sigma \propto \exp(E_a/T)$ as

$$S = \frac{1}{e}\frac{E_a}{T} \tag{3.80}$$

where $E_a$ is the energy of the carriers measured from $E_F$. In a semiconductor, thermoelectric power increases with decreasing the temperature, in inversely proportion to $T$. From the coefficient, we can estimate the activation energy $E_a$.

From thermodynamics, internal energy $U$ is given by

$$dU = -pdV + TdS' + \mu dn$$

where the entropy is written as $S'$ to distinguish from the thermoelectric power $S$. When $U$ and $V$ are constant, we obtain

$$\frac{\mu}{T} = -\frac{\partial S'}{\partial n}.$$

$E_a$ in Eq. 3.80 is the carrier energy measured from $E_F$, so replaced by $\mu$ to afford

$$S = -\frac{1}{e}\frac{\partial S'}{\partial n}.$$

This equation means thermoelectric power is equal to the entropy per carrier in eV unit. Since entropy is represented by $S' = k_B \ln W$, we obtain

$$S = -\frac{k_B}{e}\frac{\partial \ln W}{\partial n}.$$

When the electrons are localized on each site (Sect. 5.8), we can neglect the transfers, and $W$ is estimated from the statistical number of the occupied states. For instance, the combinations of $N$ spinless electrons placed in $N_0$ sites are (Fig. 3.25a)

$$W = \frac{N_0!}{N!(N_0 - N)!}.$$

Using Stirling's equation and $\rho = N/N_0^2$, this gives [24]

$$S = -\frac{k_B}{e}\ln\frac{1-\rho}{\rho}. \tag{3.81}$$

When the electrons have spins, $W$ is replaced by

$$W = \frac{2^N N_0!}{N!(N_0 - N)!}$$

which gives the thermoelectric power

$$S = -\frac{k_B}{e}\ln\frac{2(1-\rho)}{\rho}. \tag{3.82}$$

Accordingly, for these strongly correlated systems, thermoelectric power is the entropy coming from the combinations of the site occupancy, so it is quite reasonable to write the Seebeck coefficient as $S$.

*Example 3.7* Calculate thermoelectric power of a 1/4-filled band ($\rho = 1/2$) from Eq. 3.82.

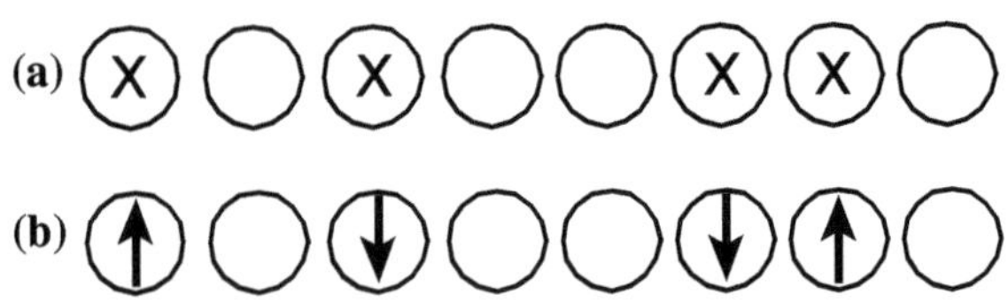


**Fig. 3.25** Examples of site occupancies in a 1/4-filled localized band **a** without and **b** with spins

Substituting $\rho = 1/2$ in Eq. 3.82, the logarithmic part is ln 2, and

$$S = -\frac{(1.38 \times 10^{-23}\,\mathrm{J/K})}{(1.60 \times 10^{-19}\,\mathrm{C})} \ln 2 = -60\,\mu\mathrm{V/K}.$$

Most organic conductors have a 2:1 composition and a 1/4-filled band. Accordingly, in strongly correlated conductors exhibiting almost temperature-independent or weakly decreasing (metallic) conductivity, we observe essentially temperature-independent thermoelectric power with the magnitude of 60 μV/K.

Variations of the thermoelectric power are summarized in Fig. 3.26.

Putting $\nabla T = 0$ in Eqs. 3.69 and 3.70, we obtain

$$U = eK_1\varepsilon,\ J = e^2K_0\varepsilon$$

which leads to

$$U = \frac{K_1}{eK_0}J = \Pi J. \tag{3.83}$$

Similarly to a thermocouple, two metals are connected, and instead of the voltmeter, current $J$ is applied. The heat flow accompanied by the current is $\Pi_A J$ in A, and $\Pi_B J$ in B. Then, heat generation or absorption with an amount of $(\Pi_A - \Pi_B)$

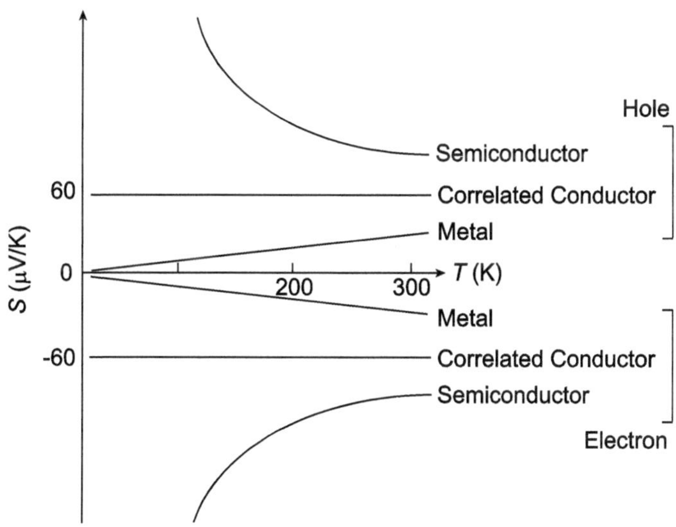


**Fig. 3.26** Temperature dependence of thermoelectric power for metals, semiconductors, and correlated conductors

$J$ occurs at the contacts. This is called **Peltier effect**, and $\Pi$ is called **Peltier coefficient**. The Peltier effect is used in a small-scale freezer. Comparing Eqs. 3.73 and 3.83, the Peltier and Seebeck coefficients satisfy $\Pi = ST$; this is a kind of Onsager's reciprocal theorem.

Electric power generation between high and low heat sources has been investigated on the basis of the Seebeck effect. The efficiency of such a thermoelectric device is represented by the **figure of merit**.

$$Z = \frac{S^2\sigma}{\kappa} \tag{3.84}$$

Thermoelectric figure of merit is large when the electrical conductivity $\sigma$ is large, the thermoelectric power $S$ is large, and the thermal conductivity $\kappa$ is small. A dimensionless quantity $ZT$ larger than one is desirable for the practical use. Organic conductors are not advantageous due to the small state density and small thermoelectric power, but advantageous due to the small $\kappa$. Polymers such as polythiophene (PEDOT, Figure 8.19e) and some small-molecule conductors like the $\tau$-phase (Fig. 7.44) have been investigated owing to the comparatively large $ZT$ values.

*Example 3.8* Calculate the $ZT$ value for $\sigma = 100$ S/cm, $S = 100$ μV/K, and $\kappa = 0.2$ W/mK

$\sigma = 100$ S/cm is converted to $10^4$ S/m. The numerator of Eq. 3.84 is called **power factor** $S^2\sigma = (100\ \mu\text{V/K})^2 \times (10^4\ \text{S/m}) = 100\ \mu\text{W/mK}^2$

which leads to

$$Z = \left(100\,\mu\text{W/mK}^2\right)/(0.2\,\text{W/mK}) = 0.0005\,\text{K}^{-1}$$
$$ZT = \left(0.0005\,\text{K}^{-1}\right) \times (300\,\text{K}) = 0.15.$$

A similar value has been realized in an optimally doped PEDOT:PSS [25]. However, $\sigma = 100$ S/cm is a comparatively large value among organic conductors, and $S = 100$ μV/K is a value near the semiconducting region, so it is not easy to attain these values at the same time. As a typical example, for κ-(BEDT-TTF)$_2$Cu(NCS)$_2$, $S = 20$ μV/K from Fig. 3.21, and $\sigma = 10$ S/cm from Fig. 5.34b. When $\kappa$ is approximated to be the same, these values afford $S^2\sigma = 0.4$ μW/m K$^2$ and $ZT = 6 \times 10^{-4}$. Several polymers with $ZT = 10^{-3}$–$10^{-2}$ are investigated as potential thermoelectric materials [26].

Fig. 3.27 Quantum Hall effect

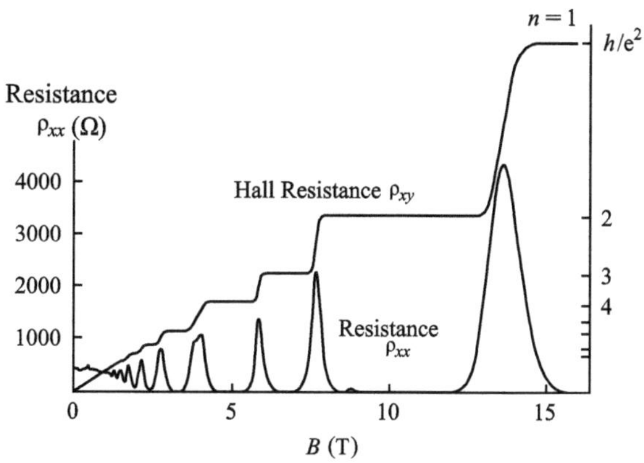


## 3.7 Quantum Hall Effect

In the Shubnikov–de Haas oscillation shown in Fig. 3.16b, the observed magnetoresistance is several times larger than the original resistance. In an ideal two-dimensional electron system such as a GaAs field-effect transistor, the lower resistance reaches to zero (Fig. 3.27) when the magnetic field is so strong that only a few Landau levels are occupied (Fig. 3.15f). In such a case, the Hall conductivity $\sigma_{xy}$ is integer times of $e^2/h$. This phenomenon is called **quantum Hall effect** [27].

From Eq. 3.50, the state density of the Laudau level is given by $N = neB/h$ per area. This is put in $N$ in the Hall effect (Eq. 3.21) to give

$$\sigma_{xy} = \frac{J_x}{E_y} = -\frac{Ne}{B} = -\frac{neB}{h}\frac{e}{B} = -\frac{e^2}{h}n.$$

This implies $\sigma_{xy}$ is quantized to integer times of $e^2/h$.

As shown in Fig. 3.27, the plateau of the quantum Hall effect satisfies $\rho_{xx} = 0$ and $\rho_{xy} = h/ne^2$, so the resistivity matrix is

$$\begin{pmatrix} \rho_{xx} & \rho_{xy} \\ \rho_{yx} & \rho_{yy} \end{pmatrix} = \begin{pmatrix} 0 & h/ne^2 \\ -h/ne^2 & 0 \end{pmatrix}.$$

The inverse matrix affords the conductivity matrix.

$$\begin{pmatrix} \sigma_{xx} & \sigma_{xy} \\ \sigma_{yx} & \sigma_{yy} \end{pmatrix} = \begin{pmatrix} 0 & -ne^2/h \\ ne^2/h & 0 \end{pmatrix}$$

These matrices mean $\rho_{xx} = 0$ and $\sigma_{xx} = 0$ at the same time. In a two-dimensional system, the carriers are localized due to the Anderson localization (Sect. 5.10), and consequently $\sigma_{xx} = 0$ is required in the quantum Hall effect. It is the most intriguing point of the quantum Hall effect that zero resistance is realized in a non superconducting state together with $\rho_{xx} = 0$.

The quantum Hall effect has been observed in graphene, a monolayer of graphite. In graphene, the valence and conduction bands contact to each other at a point, and the Fermi surface shrinks to a point. This leads to an additional phase called Berry's phase, which shifts the phase of the quantum Hall effect by $\pi$; the plateaus appear at half integers such as 1/2, 3/2, and 5/2 instead of integers. A system with a point Fermi surface is called the zero-gap semiconductor, which has been found in an organic conductor $\alpha$-$(BEDT\text{-}TTF)_2I_3$ as well (Sect. 7.7).

*Example 3.9* Calculate the resistance corresponding to $h/e^2$.

$$\frac{h}{e^2} = \frac{(6.6262 \times 10^{-34}\,\mathrm{Js})}{(1.6022 \times 10^{-19}\,\mathrm{C})^2} = 25.813\,\mathrm{k\Omega}$$

The unit of current is C/s, and the unit of voltage is W/(C/s) = J/C. Then, the unit of resistance is (J/C)/(C/s) = $\mathrm{Js/C^2}$ from $R = V/I$.

From this, we can measure $h/e^2$ exactly. The quantum Hall effect is used as a standard of resistance.

## References

1. K. Murata, M. Ishibashi, Y. Honda, M. Tokumoto, N. Kinoshita, H. Anzai, J. Phys. Soc. Jpn. **58**, 3469 (1989)
2. K. Murata, M. Ishibashi, Y. Honda, N.A. Fortune, M. Tokumoto, N. Kinoshita, H. Anzai, Solid State Commun. **76**, 377 (1990)
3. G. Mihaly, I. Kezsmarki, F. Zamborszky, L. Forro, Phys. Rev. Lett. **84**, 2670 (2000)
4. D.W. van Krevelen, *Properties of Polymers* (Elsevier, Amsterdam, 1972)
5. C.S. Jacobsen, D.B. Tanner, K. Bechgaard, Phys. Rev. Lett. **46**, 1142 (1981)
6. A.A. Abrikosov, *Fundamentals of the Theory of Metals* (North-Holland, Amsterdam, 1988)
7. T. Mori, A. Kobayashi, Y. Sasaki, H. Kobayashi, G. Saito, H. Inokuchi, Chem. Lett. **13**, 957 (1984)
8. W. Kang, G. Montambaux, J.R. Cooper, D. Jérome, P. Batail, C. Lenoir, Phys. Rev. Lett. **62**, 2559 (1989)
9. T. Mori, H. Mori, S. Tanaka, Bull. Chem. Soc. Jpn. **72**, 179 (1999)
10. N. Harrison, J. Caulfield, J. Singleton, P.H.P. Reinders, F. Herlach, W. Kurmoo, P. Day, J. Phys. Cond. Matter. **8**, 5415 (1996)
11. J. Wosnitza, *Fermi Surfaces of Low-Dimensional Organic Metals and Superconductors* (Springer, Berlin, 1996)
12. D. Schoenberg, *Magnetic Oscillations in Metals* (Cambridge University Press, Cambridge, 1984)
13. T. Kawamoto, T. Mori, C. Terakura, T. Terashima, S. Uji, K. Takimiya, Y. Aso, T. Otsubo, Phys. Rev. B **67**, 020508R (2003)
14. T. Ishiguro, K. Yamaji, G. Saito, *Organic Superconductors*, 2nd edn. (Springer, Berlin, 1998)

15. C. Kittel, *Introduction to Solid State Physics* (Wiley, New York, 1953)
16. K. Kajita, Y. Nishio, T. Takahashi, W. Sasaki, R. Kato, H. Kobayashi, A. Kobayashi, Solid State Commun. **70**, 1189 (1989)
17. K. Yamaji, J. Phys. Soc. Jpn. **58**, 1520 (1989)
18. M.V. Kartsovnik, V.N. Laukhin, S.I. Pesotskii, I.F. Schegolev, V.M. Yakovenko, J. Phys. I France **2**, 89 (1992)
19. T. Osada, A. Kawasumi, S. Kagoshima, N. Miura, G. Sito, Phys. Rev. Lett. **66**, 1525 (1991)
20. J.M. Ziman, *Principles of the Theory of Solids* (Cambridge University Press, Cambridge, 1972)
21. P.M. Chaikin, R.L. Greene, S. Etemad, E. Engler, Phys. Rev. B **13**, 1627 (1976)
22. T. Mori, H. Inokuchi, J. Phys. Soc. Jpn. **57**, 3674 (1988)
23. T. Kawamoto, T. Mori, Phys. Rev. B **74**, 212502 (2006)
24. P.M. Chaikin, G. Beni, Phys. Rev. B **13**, 647 (1976)
25. O. Bubnova, Z.U. Khan, A. Malti, S. Braun, M. Fahlman, M. Berggren, X. Crispin, Nature Mater. **10**, 429 (2011)
26. N. Dubey, M. Leclerc, J. Poly. Sci. B **49**, 467 (2011)
27. R.E. Prange, S.M. Grivin (eds.), *Quantum Hall Effect* (Springer, Berlin, 1989)

# Chapter 4
# Magnetism

**Abstract** Localized electrons make magnets. This chapter describes fundamental aspects of magnetism, placing a particular emphasis on the mean-filed theory of magnetic order.

**Keywords** Ligand-field theory · High spin · Low spin · Pascal's diamagnetism · Paramagnetism · Curie law · Spin Hamiltonian · Heisenberg model · Ising model · Ferromagnetism · Antiferromagnetism · Molecular-field approximation · Effective (internal) field · Weiss temperature · Néel temperature · Frustration · πd systems · Molecular magnetism

## 4.1 Molecular Orbitals of Transition-Metal Compounds

In the preceding chapters, we have investigated how transport properties in metals and semiconductors are discussed in view of the energy bands, and how the Fermi surface is examined experimentally. This chapter describes fundamental aspects of magnetism as the first step to deal with phenomena beyond the energy band theory. In the next chapter, the origin of magnetism will be revealed by considering the electron–electron Coulomb interaction.

Many transition-metal compounds show intriguing magnetic properties. There are several paramagnetic compounds composed of only main-group elements like NO and $O_2$ as well as stable organic radicals. However, a rich variety of magnetic materials are found in transition-metal compounds. Molecular orbitals of such transition-metal compounds have been investigated using the ligand-field theory.

As an example of transition-metal compounds, Fig. 4.1 describes molecular orbitals of $[Fe(H_2O)_6]^{2+}$ existing in an aqueous solution of $Fe^{2+}$. In an aqueous solution, six oxygen atoms of water coordinate to $Fe^{2+}$ in an octahedral manner. The Fe–O bond is a polar bond with highly ionic nature, then the Fe–O bond is approximately composed of $Fe^{2+}$ and $O^{2-}$. In such a case, the 2p levels of O are located below the 3d levels of Fe (Sect. 1.5.1). The 2p levels are fully occupied.



T. Mori, *Electronic Properties of Organic Conductors*,
DOI 10.1007/978-4-431-55264-2_4

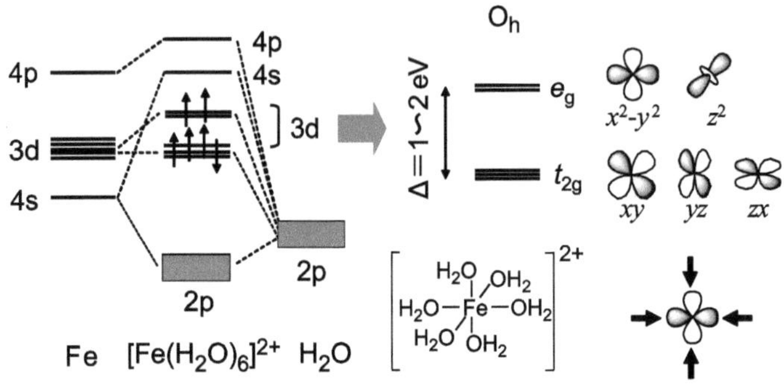


**Fig. 4.1** Molecular orbitals of a transition-metal ion $[Fe(H_2O)_6]^{2+}$ with octahedral coordination

A neutral Fe atom has an electron configuration of $4s^23d^6$. In the compound, the 4s orbital makes a stronger chemical bond with the ligand orbitals than the 3d orbitals because the 4s orbital is more extended. Since the metal orbitals act mainly as antibonding orbitals in the polar bond, the 4s level is lifted above the 3d levels (Fig. 4.1). The 4p levels are located above the 3d levels even in the atoms. Accordingly, all valence electrons enter the d levels in transition-metal compounds. A neutral Fe has $3d^8$ electrons, but the 2+ ion has $3d^6$ electrons. In general, transition-metal compounds are categorized as $d^n$ using the number of the whole valence electrons $n$. As described in Chap. 2, an elemental metal has one s electron; for instance, Fe has an electron configuration of $4s^13d^7$. The electron configuration of $Fe^0$ is $4s^23d^6$ in an atom, $4s^13d^7$ in the elemental metal, and $3d^8$ in the compounds. The number of s electrons decreases successively from two to one and zero.

When chemical bonds are formed, the originally degenerated five d orbitals split to several different levels. Here, the manner of splitting is determined from the symmetry. Suppose octahedral coordination, and six oxygen atoms are located at $\pm x$, $\pm y$, and $\pm z$ positions. The $d_{x^2-y^2}$ and $d_{z^2}$ orbitals make σ-like bonds with oxygen 2p orbitals, whereas the $d_{xy}$, $d_{yz}$, and $d_{zx}$ orbitals are orthogonal to the ligand σ-orbitals (Fig. 4.1). Then, the energy levels of the 'nonbonding' $d_{xy}$, $d_{yz}$, and $d_{zx}$ orbitals do not change, but the energy levels of the 'antibonding' $d_{x^2-y^2}$ and $d_{z^2}$ orbitals go up. At the same time, the corresponding 'bonding' 2p levels (the gray box in Fig. 4.1) go down. Using the symbol of the octahedral point group, the former three are called $t_{2g}$ orbitals, while the latter two are called $e_g$ orbitals. Consequently, the five *d* orbitals split to the lower three and the upper two. Such splitting is called **ligand-field splitting**. The magnitude of the splitting is usually $\Delta = 1–2$ eV. This is the origin of characteristic colors of transition-metal compounds.

Electron configurations of octahedrally coordinated metal complexes are shown in Fig. 4.2. From $d^1$ to $d^3$, the electrons enter the $t_{2g}$ levels. The spins are parallel according to the Hund rule. Strictly speaking, a spin is a vector, but the total $S_z$ component is simply given by multiplying the number of electrons to 1/2. Then, the spin of the $d^3$ state is $S = 3/2$. In the $d^4$ state, the fourth electron enters either the $e_g$ level with a parallel spin or the $t_{2g}$ orbital with an antiparallel spin. The former is

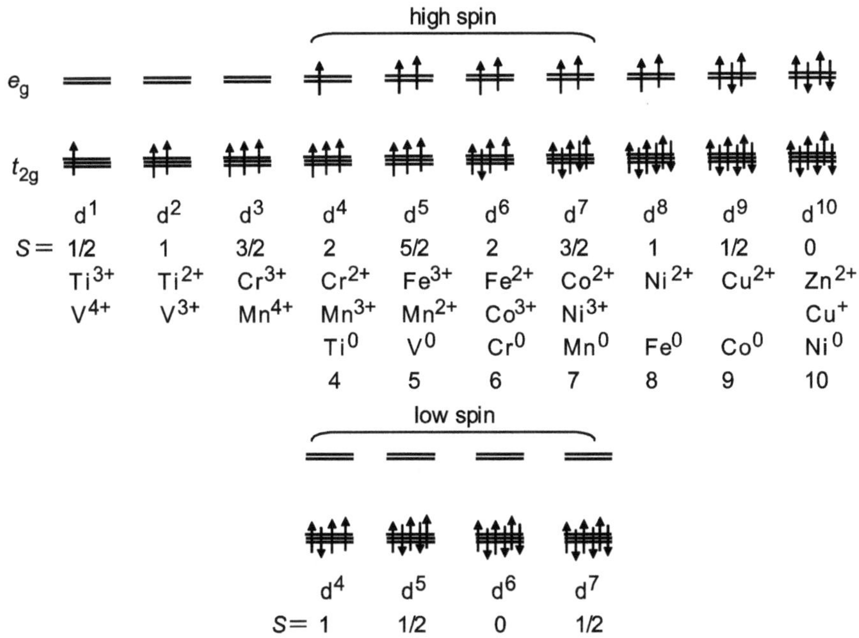


**Fig. 4.2** Electron configuration of octahedrally coordinated metal complexes

more usual because the Hund rule is ordinarily more important than the ligand-field splitting. Such an $S = 2$ state is called **high-spin state**. When the ligand-field splitting is larger than the exchange energy, the fourth electron enters the $t_{2g}$ level, and the $S = 1$ **low-spin state** is realized. The difference between the high-spin and low-spin states appears from $d^4$ to $d^7$, and the difference disappears above $d^8$. The representative metal ions are listed in Fig. 4.2. The valence electron number of a neutral metal is evident like $Fe^0$: $d^8$, and the *d* electron number of the ion is obtained by subtracting the charge like $Fe^{2+}$: $d^6$. Then, we know the spin as $S = 2$.

In a tetrahedrally coordinated complex, the $d_{xy}$, $d_{yz}$, and $d_{zx}$ orbitals make stronger bonds with the ligands, and the $d_{x^2-y^2}$ and $d_{z^2}$ orbitals are basically non bonding. Then, the splitting is upside down, and the three upper $t_2$ and two lower *e* levels are generated (Fig. 4.3). The resulting electron configurations are shown in Fig. 4.3. In a tetrahedral complex, the ligand field is weak because the orbitals are not exactly directed to the ligands. Therefore, the low-spin state does not appear. In the high-spin states, the *S* value is the same in octahedral and tetrahedral complexes, and also the same as a complex without any ligand-field splitting. For instance, a tetrahedral anion $FeCl_4^-$ has a 1− total charge, but from the $Cl^-$ charge, the formal charge of Fe is 3+. The valence electron number is $8 - 3 = 5$, and the $d^5$ state leads to $S = 5/2$ in the high-spin state. When the metal complex has a lower symmetry, the degeneracy of the d levels is further removed. In such a case, the splitting manner is examined using the index table of the point group.

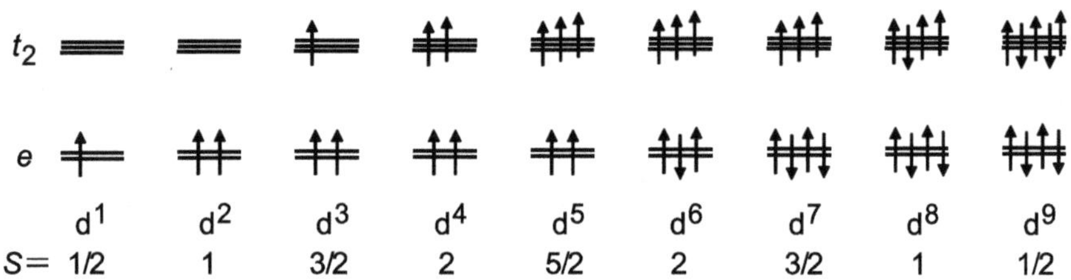


**Fig. 4.3** Electron configuration of tetrahedrally coordinated metal complexes

Many metal complexes have an infinitely connected crystal structure instead of a discrete molecular unit, but the above ligand-field model is also applied to such a solid. For instance, hematite $Fe_2O_3$ has infinitely connected chemical bonds in the solid, but the oxygen atoms octahedrally coordinate to the $Fe^{3+}$ atom with a $d^5$ state. Then, the spin state is $S = 5/2$. In transition-metal compounds, the coordination number and the valence electron number determine almost everything on the color and the magnetism. This does not depend on whether the metal atom is included in a discrete molecule or an infinitely connected solid. The same model applies to metal ions dissolved in a solution. The molecular orbital theory of transition metals is called **ligand-field theory** or **crystal field theory**.

When electrons are placed in the energy levels in Figs. 4.2 and 4.3, the ground state and the excited states split to several multiplets The splitting depends on the ligand-field splitting $\Delta$. The excitation energies to such multiplets are plotted as a function of $\Delta$ in the **Tanabe-Sugano diagram** [1]. By comparing this diagram with the actual absorption spectra, we can determine $\Delta$ and identify the respective excitations.

The magnitude of $\Delta$ systematically changes depending on the ligands in the order of $I^- < Br^- < Cl^- < F^- < OH^- < H_2O < NCS^- \sim NH_3 <$ ethylenediamine (en, $H_2NCH_2CH_2NH_2$) < phenanthroline (phen, Fig. 4.5) < $CN^-$. Ethylenediamine and phenanthroline are frequently used ligands, where the nitrogen atoms coordinate to the central metal atom. This order is known as **spectrochemical series**. The magnitude of $\Delta$ increases with strengthening the chemical bond between the metal atom and the ligand atom. When the electronegativity of the ligand atom increases, the polarization of the metal–ligand bond increases (Fig. 4.1). Then, the chemical bond weakens, and $\Delta$ decreases. Thus, $\Delta$ decreases when the ligand atom goes to the right in the periodic table (Fig. 4.4). Since the metal atomic orbital is comparatively small, an extended ligand orbital weakens the chemical bond. Then, $\Delta$ decreases when the ligand atom goes down in the periodic table. An atom with an

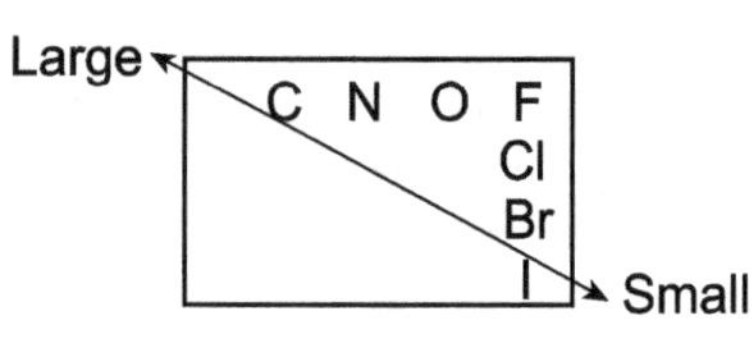


**Fig. 4.4** Ligand-field splitting $\Delta$ depending on the ligand atoms

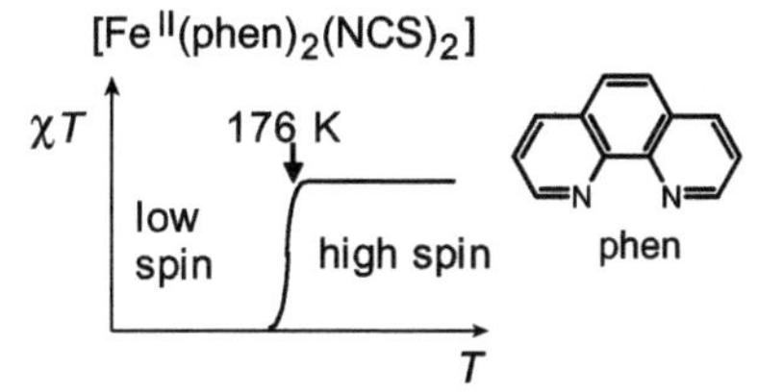


**Fig. 4.5** Temperature-dependent magnetic susceptibility in $[Fe^{II}(phen)_2(NCS)_2]$. phen: phenanthroline

extended orbital is called soft, whereas an atom with a small orbital is called hard. An anionic ligand gives smaller Δ than a neutral ligand. Therefore, halogen atoms afford small Δ, and oxygen follows, whereas nitrogen and carbon afford comparatively large Δ. Low-spin states in octahedral complexes are limited in nitrogen and carbon ligands, and are observed in comparatively exceptional cases.

A transition from a high-spin state to a low-spin state is sometimes observed. As shown in Fig. 4.5, $[Fe^{II}(phen)_2(NCS)_2]$ undergoes a transition from a paramagnetic state to a diamagnetic state at 176 K. The whole complex is neutral, and NCS is 1−, so Fe is 2+ and $d^6$. Since phen is a bidentate, the Fe atom is coordinated by six nitrogen atoms octahedrally. Then, the transition is from an $S = 2$ high-spin state to an $S = 0$ low-spin state. Such a transition is called **spin crossover transition**. In a spin crossover transition, the magnetism changes abruptly, and sometimes a paramagnetic state changes to a nonmagnetic state. At the same time, the color changes largely. In some cases, the paramagnetic state is resumed by irradiating light to the low-temperature nonmagnetic state.

*Example 4.1* Calculate *S* in the following transition-metal complexes and ions.

| | | | |
|---|---|---|---|
| (1) $MnO_2$ | (2) $KMnO_4$ | (3) $MnCl_4{}^{2-}$ | (4) $TiO_2$ |
| (5) $CuSO_4$ | (6) $CoCl_4{}^{2-}$ | (7) Cu in $La_2CuO_4$ | (8) Ti in $BaTiO_3$ |

All are in a high-spin state because the ligand is O or Cl.

(1) $7 - 4 = 3 \rightarrow d^3 \rightarrow S = 3/2$
(2) $7 - 7 = 0 \rightarrow d^0 \rightarrow S = 0$
(3) $7 - 2 = 5 \rightarrow d^5 \rightarrow S = 5/2$
(4) $4 - 4 = 0 \rightarrow d^0 \rightarrow S = 0$
(5) $11 - 2 = 9 \rightarrow d^9 \rightarrow S = 1/2$
(6) $9 - 2 = 7 \rightarrow d^7 \rightarrow S = 3/2$
(7) $2 \times 4 - 3 \times 2 = +2 \rightarrow 11 - 2 = 9 \rightarrow d^9 \rightarrow S = 1/2$
(8) $2 \times 4 - 2 \times 2 = +4 \rightarrow 4 - 4 = 0 \rightarrow d^0 \rightarrow S = 0$

To calculate the $S$ value in a high-spin state, it is not necessary to consider the ligand-field splitting. Then, $S$ is simply given by the electron (or hole) number × 1/2. Accordingly, the maximum appears at $d^5$, where $S = 5/2$. Magnetic interaction increases roughly in proportion to $S^2$ (Eq. 4.19). Therefore, such ions as $Mn^{2+}$ and $Fe^{3+}$ located near the center of the periodic table are most likely to form magnetic compounds.

## 4.2 Diamagnetism

When an external magnetic field $H$ is applied to a material, the material is magnetized by $M$. The resulting magnetic field in the material is represented by magnetic flux density $B = H + 4\pi M$ (Fig. 4.6a). The ratio $\chi = M/H$ is called magnetic susceptibility. $M$ and $\chi$ are either positive or negative. Ordinary compounds with no unpaired spins have negative $\chi$, and $B$ is smaller than $H$. Such a material is called **diamagnetic**. Most organic compounds are diamagnetic. Compounds with unpaired spins show positive $\chi$, and $B$ is larger than $H$. Such a material is called **paramagnetic**. Transition-metal compounds with $S \neq 0$ and organic radicals are paramagnetic. A magnetic field line starts from the N pole, and ends at the S pole. In the material, however, the field line runs from an S pole to an N pole to make a closed line. Accordingly, a diamagnet has such poles as Fig. 4.6a. Since the materials's N pole is repelled by the outer N pole, a diamagnet is repelled in a magnetic field. In contrast, a paramagnet is attracted to a magnetic field.

A diamagnet without a spin shows negative susceptibility because an electron orbit is distorted in a magnetic field so as to generate current that cancel the magnetic field. When the distortion of the atomic orbital generates a net flow of charge $Ze$ (Fig. 4.6b), the rotating current is given by (charge) × (revolution), that is,

$$I = (-Ze)\left(\frac{1}{2\pi}\frac{eB}{m}\right)\frac{1}{2} \tag{4.1}$$

where $\omega_c = eB/m$ is the cyclotron frequency. The final 1/2 appears because the current coming from such a Larmor motion is half the free charge current. The magnetic moment is afforded by $I$ × (orbit area), which leads to

$$\mu = -\frac{Ze^2B}{4m}\langle x^2 + y^2\rangle = -\frac{Ze^2B}{6m}\langle r^2\rangle \tag{4.2}$$

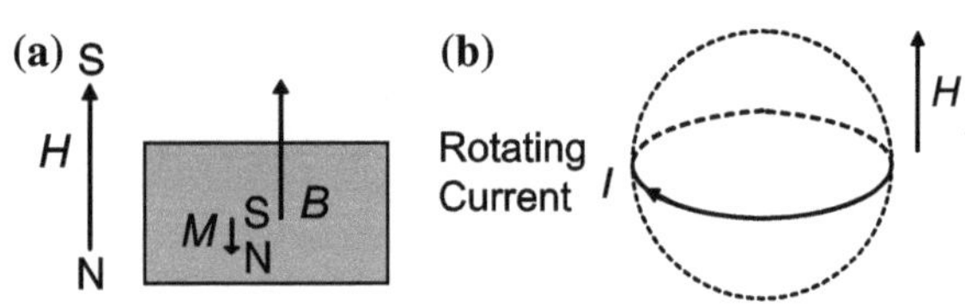


**Fig. 4.6** **a** $H$, $M$, and $B$ in a diamagnet. **b** Rotating current in a magnetic field

**Table 4.1** Pascal diamagnetic susceptibility in organic compounds ($10^{-6}$ emu/K)

| | | | | | |
|---|---|---|---|---|---|
| H | −2.93 | C | −6.00 | N (aliphatic) | −5.57 |
| N (aromatic) | −4.61 | N (monoamide) | −1.54 | N (diamide) | −2.11 |
| O (ether) | −4.61 | O (carbonyl) | +1.73 | O (carboxyl) | −3.36 |
| F | −6.3 | Cl | −20.1 | Br | −30.6 |
| I | −44.6 | S | −15.0 | Se | −23.0 |

where we used $\langle x^2 + y^2 \rangle = \frac{2}{3}\langle r^2 \rangle$ by assuming the electron distribution is spherical. This gives the susceptibility

$$\chi = \frac{N\mu}{B} = -\frac{NZe^2}{6m}\langle r^2 \rangle. \tag{4.3}$$

The resulting diamagnetism is determined by the atom species. Then, we can estimate the magnitude of the diamagnetism from the chemical composition. Such diamagnetism is called **Pascal's diamagnetism**. The values of Pascal's diamagnetism are listed in Table 4.1 [2]. Here, emu is a unit of magnetic susceptibility in the cgs electromagnetic unit. However, since $M$ and $H$ have the same unit and $\chi = M/H$, emu is dimensionless. Because many scientific papers dealing with magnetism still use the cgs unit, we will use the cgs unit in this chapter. Because $\langle r^2 \rangle$ in Eq. 4.3 represents the size of the orbital, Pascal's magnetism is large for heavy atoms. From Table 4.1, we can calculate methane's susceptibility as $-17.72 \times 10^{-6}$ emu/mol and ethanol's susceptibility as $-22.33 \times 10^{-6}$ emn/mol; these values are in good agreement with the actual observations. Even after summed over a large molecule, $\chi$ is in the order of $10^{-5}$, and $M$ is a very small fraction. In aromatic compounds, the ring current rotating around the benzene ring shields the magnetic field. It is known that a benzene ring makes a large diamagnetic contribution, $-15.1 \times 10^{-6}$ emu/mol. Diamagnetism coming from the ring current only contributes to the perpendicular direction, so a crystal of aromatic compounds has an extraordinarily large diamagnetism only in the perpendicular direction. Such a crystal shows large anisotropy in the susceptibility. By measuring the susceptibility along the different crystal axes, we can predict the orientation of the aromatic ring without knowing the crystal structure. In particular, graphite shows as large diamagnetism as $-4 \times 10^{-4}$ emu/mol in the interlayer direction. Even paramagnetic compounds have diamagnetic contributions. Then, we evaluate Pascal's diamagnetism from the chemical composition, and subtract the value from the observed susceptibility to estimate the true paramagnetic contribution.

Here, we shall investigate the difference between a magnet and a dielectric [3]. Suppose a dielectric inserted in a capacitor (Fig. 4.7). When voltage $V$ is applied to a capacitor $C$, charge $Q = CV$ is stored (Fig. 4.7a). The electric field in the capacitor is $E_0 = V/d$, where $d$ is the spacing between the electrodes. The capacitance is given by $C = \varepsilon_0 S/d$, where $S$ is the area of the electrode. When a dielectric is inserted (Fig. 4.7b), dipole $E' = -4\pi P$ (in the cgs unit) appears in the dielectric, and the

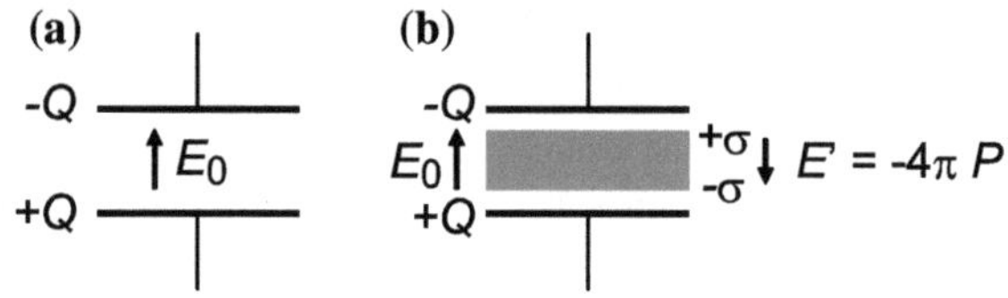


**Fig. 4.7** Electric charge accumulated in a capacitor **a** without and **b** with a dielectric material

actual electric field is reduced to $E = E_0 + E' = E_0 - 4\pi P$. When $E_0$ is substituted by the electric flux density $D$, this relation leads to $D = E + 4\pi P$. Here, $\varepsilon = D/E$ is the relative dielectric constant. By inserting the dielectric, the actual $E$ is reduced to $1/\varepsilon$ times of the value $D$ in vacuum. Since $\varepsilon$ is 11.9 for Si, 4.6 for $SiO_2$, and 11.6 for $Al_2O_3$, most electric field is screened by the electric charge $\sigma$ appearing on the dielectric surface. For instance, the actual $E$ in $Al_2O_3$ is 1/11.6 of $D$. At the same time, $C$ increases $\varepsilon$ times, and $V$ is reduced by $1/\varepsilon$ times for a given $Q$.

The difference between a dielectric and a magnet is depicted in Fig. 4.8. Magnetization $M$ corresponds to polarization $P$, where $P$ is always positive but $M$ is either positive or negative. $P$ always cancels $D$, whereas $M$ cancels $H$ when $M$ is negative. In Fig. 4.8b, $H$ and $M$ are oriented in the same direction, implying a paramagnet with $M > 0$. A diamagnet is depicted in Fig. 4.6b. The external field is reduced when $P > 0$ and $M < 0$, so the sign of $P$ and $M$ are oppositely defined. The form of the equations indicates $E$ corresponds to $H$, and $D$ corresponds to $B$, either in the cgs unit or in the MKS unit. However, the external field $D$ in vacuum corresponds to $H$, and the actual field $E$ in the material corresponds to $B$. When $\varepsilon = 12$, most electric field in the material is canceled to 1/12. In contrast, magnetic susceptibility in a diamagnet is at most in the order of $10^{-4}$. Then, the very small reduction of the magnetic field in a diamagnetic material is another serious difference. The exceptions are ferromagnets and superconductors. In particular, a superconductor shows perfect diamagnetism $M = -H/4\pi$ so as to entirely expel the magnetic field from the material, to result in $B = 0$. This is summarized as follows.

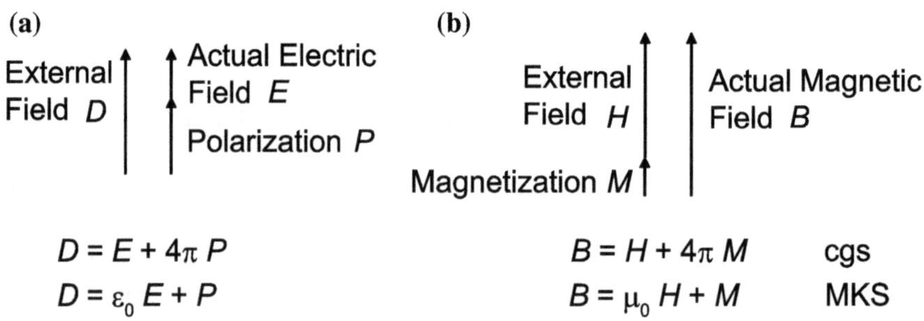


**Fig. 4.8** Difference between a dielectric and a magnet. **a** Dielectric materials, **b** magnetic materials

| | Weak | Strong |
|---|---|---|
| $M > 0$ | Paramagnetism | Ferromagnetism |
| $M < 0$ | Diamagnetism | Superconductor |

Among dielectric materials, a metal works like a superconductor. When a metal is inserted in a capacitor, the electric field does not go into the metal to result in $E = 0$ and $D = 4\pi P$. A metal works as a material with infinite $\varepsilon$. In the usual MKS unit of electromagnetism, $B$ is defined as $B = \mu_0 H + M$ (Fig. 4.8), but in materials science, $B$ is sometimes defined as $B = \mu_0\ (H + M)$. We will use this definition for describing superconductivity (Chap. 6).

The above electromagnetic equations are derived from the controlling parameters. In a capacitor, $E = D - 4\pi P$ is controlled by an applied voltage. In a coil, $H = B - 4\pi M$ is controlled by an applied current. In contrast, in Fig. 4.7, $Q$ is given instead of $V$.

When magnetization of a material is increased from zero to $M$ in a magnetic $H$, the mechanical work is given by $H\mathrm{d}M$. Then, the thermodynamical internal energy is $\mathrm{d}U = T\mathrm{d}S + H\mathrm{d}M$, where $S$ is the entropy. This equation has the same form as the equation of the volume $V$ and pressure $p$: $\mathrm{d}U = T\mathrm{d}S - p\mathrm{d}V$. Here, $M$ is an extensive variable corresponding to $V$, and $H$ is an intensive variable corresponding to $p$. Since a paramagnetic material with $M > 0$ is attracted, the sign of $H\mathrm{d}M$ is positive. The corresponding Helmfoltz free energy is defined by $F = U - HM$, and we obtain $\mathrm{d}F = T\mathrm{d}S - M\mathrm{d}H$. This quantity is integrated from zero to $H$, and the energy per volume is $\chi H^2/2$.

## 4.3 Paramagnetism

Next, we shall investigate paramagnetic materials with unpaired spins. Paramagnetic organic materials are rare; the exceptions are radicals. As described in Sect. 4.1, many transition-metal compounds are paramagnetic. Some compounds composed of main-group elements are paramagnetic; examples are $O_2$ and NO. When a magnetic field is applied to a compound with an unpaired spin, energy difference is induced depending on the parallel and antiparallel spins (Fig. 4.9a). This is called **Zeeman splitting**. For simplicity, we suppose $S = 1/2$. Then, two energy levels appear depending on the parallel and antiparallel spins (Fig. 4.9a), where the energy difference is $2\mu_B H$. Here, $\mu_B$ is called **Bohr magneton**, which is the magnetization of one electron. Suppose the Boltzmann distribution of the parallel spins $N_\uparrow$ and the antiparallel spins $N_\downarrow$.

$$\frac{N_\uparrow}{N} = \frac{\mathrm{e}^{\mu_B H/k_B T}}{\mathrm{e}^{\mu_B H/k_B T} + \mathrm{e}^{-\mu_B H/k_B T}} \quad \frac{N_\downarrow}{N} = \frac{\mathrm{e}^{-\mu_B H/k_B T}}{\mathrm{e}^{\mu_B H/k_B T} + \mathrm{e}^{-\mu_B H/k_B T}} \tag{4.4}$$

The denominator represents the total electron number, $N = N_\uparrow + N_\downarrow$. The magnetization of the whole system is obtained by multiplying $\mu_B$ to the difference of the parallel and antiparallel spins.

$$M = \mu_B(N_\uparrow - N_\downarrow) = N\mu_B \tanh\frac{\mu_B H}{k_B T} \approx N\mu_B\left(\frac{\mu_B H}{k_B T}\right) \tag{4.5}$$

Since the Zeeman splitting $2\mu_B H$ is as small as 1 K under a magnetic field of 1 T (Example 4.2), $\frac{\mu_B H}{k_B T} \ll 1$ is fulfilled at ordinary temperatures. Then, in the above equation we have approximated $\tanh x \sim x$ because of $x \ll 1$. This leads to the susceptibility

$$\chi = \frac{M}{H} = \frac{N\mu_B^2}{k_B T} = \frac{C}{T} \tag{4.6}$$

where $C$ is called **Curie constant**. For a spin other than $S = 1/2$, the Curie constant is

$$C = \frac{NS(S+1)g^2\mu_B^2}{3k_B}. \tag{4.7}$$

Here, $g$ is called $g$-value, that is, 2.0023 for a free electron. The free-electron $g$-value is slightly deviated from $g = 2$ owing to the relativistic quantum mechanics, but we can usually approximate $g = 2$. The $g$-value of an unpaired electron is more deviated from $g = 2$ in actual molecules and solids owing to the orbital angular momentum. When $S = 1/2$ is put in Eq. 4.7, $S = 1/2$ cancels with $g = 2$, and we obtain Eq. 4.6.

Equation 4.6 tells us with decreasing the temperature the susceptibility of a paramagnetic material increases in inversely proportion to the temperature (Fig. 4.9b). Spins in a paramagnetic material is randomly oriented (Fig. 4.9c), but at low temperatures the spins are more easily oriented in a magnetic field. This is the reason the susceptibility increases at low temperatures. Here, $\chi T = C =$ constant is maintained, and the Curie constant is determined only from the magnitude of $S$. Then, we can determine $S$ by measuring the magnitude of the susceptibility. In particular, the Curie constant for $S = 1/2$ is 0.375 emu K/mol (Example 4.3). With increasing the magnetic field, the approximation $x \ll 1$ does not hold. Then, we have to consider a tanh function in Eq. 4.5, and $M$ saturates as shown in Fig. 4.9d.

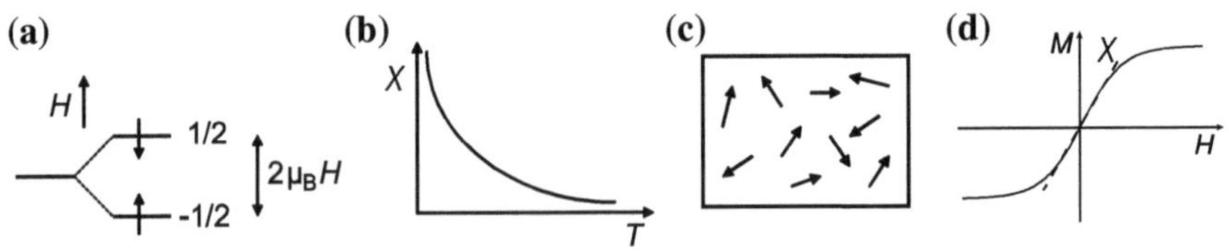


**Fig. 4.9** **a** Zeeman splitting, **b** temperature dependence of the susceptibility, **c** spin distribution, and **d** magnetization curve in a paramagnetic material

At large $H$, $M$ becomes gradually flat. This is the state all spins are aligned parallel to the magnetic field. The resulting situation is the same as a ferromagnetic material (Fig. 4.11a). Such a situation is realized only when $\mu_B H$ is comparable to $k_B T$. Accordingly, $M$ becomes flat around 1 $T$ at 1 K. In contrast, $M$ does not become flat around room temperature even under a very strong magnetic field. The gradient of the tanh function around $H = 0$ is the susceptibility.

*Example 4.2* Evaluate the Zeeman splitting at $H = 1$ T in a unit of temperature and eV.

Since one Bohr magneton is $\mu_B = 9.274 \times 10^{-24}$ J/T, 1 T corresponds to $2\mu_B H = 18.548 \times 10^{-24}$ J. This is divided by the Boltzmann constant to give $(18.548 \times 10^{-24}$ J$)/(1.3806 \times 10^{-23}$ J/K$) = 1.343$ K. Alternatively, divided by 1 eV $= 1.602 \times 10^{-19}$ J to give $1.1578 \times 10^{-4}$ eV.

*Example 4.3* Calculate the Curie constant $C = 0.375$ emu K/mol for $g = 2$ and $S = 1/2$.

Substitution to Eq. 4.6 gives in the cgs unit

$$C = \frac{N\mu_B^2}{k_B} = \frac{(6.022 \times 10^{23}\,\text{mol}^{-1}) \times (9.2741 \times 10^{-21}\,\text{erg/G})^2}{(1.3806 \times 10^{-16}\,\text{erg/K})}$$
$$= 0.375\,\text{emu K/mol}.$$

When $\chi T$ is plotted with respect to $T$, the result does not depend on the temperature, and shows this constant value (Fig. 4.17b). From the approximate relation $\frac{N\mu_B^2}{3k_B} = 0.12505$ emu K/mol $\sim$ 1/8 emu K/mol, $C$ for $S = 1/2$ is $C = 4S(S+1)\frac{N\mu_B^2}{3k_B}$ $\sim (1 \times 3)/8 = 0.375$ emu K/mol. Therefore, $C$ for $S = 5/2$ is $C \sim (5 \times 7)/8 = 4.375$ emu K/mol. The exact value is 4.377 emu K/mol. If $g$ is larger than 2, the $\chi T$ value is larger than this value.

## 4.4 Pauli Paramagnetism

A metallic electron is regarded as a kind of unpaired electron, and exhibits paramagnetism. However, paramagnetism of a metallic electron is different from the ordinary Curie paramagnetism owing to the Fermi distribution. The density of states of three-dimensional free electrons is represented by $D(E) \propto \sqrt{E}$ (Eq. 2.26). Under a magnetic field, the up (parallel) spin has slightly lower energy than the down (antiparallel) spin due to the Zeeman splitting $2\mu_B H$. This leads to a slight difference of the up and down spin numbers (Fig. 4.10). Here, the right and left sides represent

Fig. 4.10 Density of states of a metal under a magnetic field

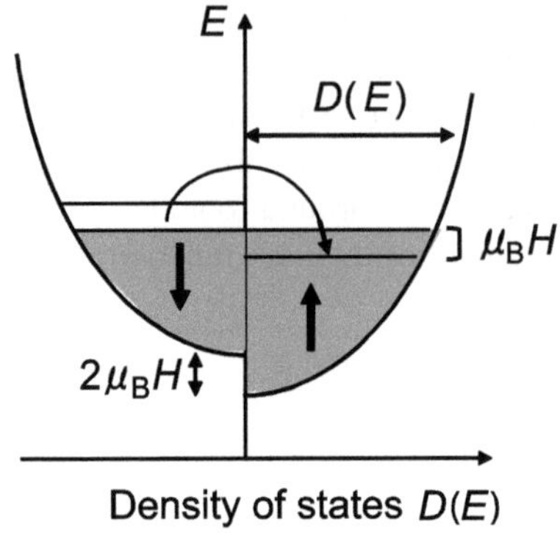


the up and down spin density of states separately. The respective energies are given by

$$E_\uparrow = \frac{h^2k^2}{2m} - \mu_B H$$

$$E_\downarrow = \frac{h^2k^2}{2m} + \mu_B H. \tag{4.8}$$

Since the up and down spin numbers are different, the metal has the magnetization, $M = \mu_B(N_\uparrow - N_\downarrow)$. The difference of $N_\uparrow$ and $N_\downarrow$ is given by the density of stated $D(E)$ multiplied by the energy difference $2\mu_B H$. Then, $M$ is

$$M = \mu_B D(E) \times 2\mu_B H \tag{4.9}$$

which leads to the susceptibility

$$\chi = \frac{M}{H} = 2\mu_B^2 D(E_F). \tag{4.10}$$

The susceptibility is proportional to the density of states $D(E)$, and independent of the temperature. Such paramagnetism is called **Pauli paramagnetism**. The above derivation does not depend on the $k$ dependence or the energy dependence of $D(E)$, so that Eq. 4.10 is generally valid for metals other than three-dimensional metals. The Pauli paramagnetism of a conducting metal is considerably smaller than the Curie paramagnetism observed in non conducting magnets. Since electron heat capacity coefficient $\gamma$ of a metal is also proportional to $D(E)$ (Eq. 2.47), the ratio of $\chi$ and $\gamma$ is

$$\frac{\chi}{\gamma} = \frac{6\mu_B^2}{\pi^2 k_B^2} \tag{4.11}$$

which is a constant known as **Wilson ratio**. Pauli susceptibility and electron specific heat coefficient depend on the effective mass and bandwidth, but the ratio is constant. For example, lanthanide and actinide compounds called heavy Fermion systems have effective masses exceeding one thousand, but roughly speaking, $\chi$ and $\gamma$ increase proportionally.

*Example 4.4* Calculate the susceptibility (emu/mol) of Pauli paramagnetism for a band with the density of states, $D(E_F)$ = 1 states/eV.

Substitution to Eq. 4.10 gives

$$\frac{2\left(6.022\times10^{23}\,\mathrm{mol}^{-1}\right)\times\left(9.274\times10^{-21}\,\mathrm{erg/G}\right)^2}{(1\,\mathrm{eV})\times\left(1.602\times10^{-12}\,\mathrm{erg/eV}\right)}=6.466\ \mathrm{x}10^{-5}\,\mathrm{emu/mol}.$$

If the density of states is constant everywhere in the band, this corresponds to the bandwidth of 1 eV. This is a typical value of the Pauli paramagnetism in organic conductors. Since the Curie paramagnetism of an $S = 1/2$ spin affords $1.25\times10^{-3}$ emu/mol at 300 K (Example 4.3), the Pauli paramagnetism is considerably smaller than the Curie paramagnetism. The Pauli paramagnetism is rather comparable to Pascal's diamagnetism (Sect. 4.2). Susceptibility of a metal is given by the sum of the Pauli paramagnetism and Pascal's diamagnetism, so it may be positive or negative. Both the Pauli paramagnetism and Pascal's diamagnetism do not depend on the temperature.

The above density of states is defined as $D(E_F)$ = 1 states/eV when two electrons with up and down spins are occupied. However, the state when only the up spin states are fully occupied is sometimes defined as $D(E_F)$ = 1 spin states/eV, and the doubly occupied state is $D(E_F)$ = 2 spin states/eV. Then, Eq. 4.10 is replaced by $\chi=\mu_B^2D'(E_F)=(3.233\times10^{-5}$ emu/mol) $\times$ ($D'(E_F)$/spin states/eV). The bandwidth estimated from the Pauli paramagnetism reflects the density of states very close to $E_F$. This is similar to the bandwidth obtained from the thermoelectric power (Example 3.6 and Fig. 3.23). Experimental susceptibility is sometimes considerably enhanced by the correlation effect (Sect. 5.7).

## 4.5 Ferromagnetism

In a paramagnet, the spin orientation is random even though the spins are regularly placed on a crystal lattice (Fig. 4.9c). When all spins are orientated in the same direction (Fig. 4.11a), the resulting state is **ferromagnetic**. When the spins are alternately aligned in an antiparallel manner, the resulting state is **antiferromagnetic**. These spin alignments are called **magnetic order**. An overwhelmingly large number of magnetically ordered states are antiferromagnetic, but we first investigate ferromagnetism.

Magnetic order occurs when the up spin state has lower energy than the down spin state owing to the interaction with the adjacent spins. Putting this energy difference as $J$, the total energy is obtained from the **spin Hamiltonian**

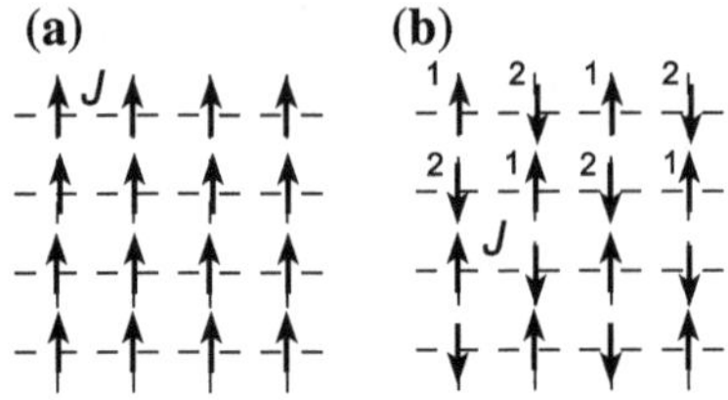


**Fig. 4.11** **a** Ferromagnetic, and **b** antiferromagnetic spin alignments

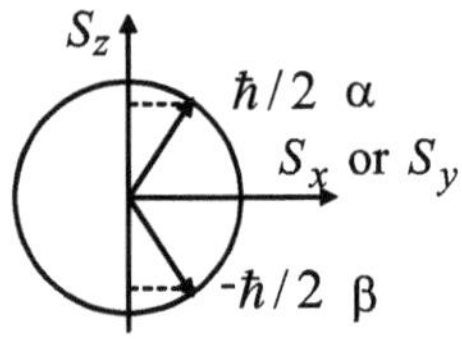


**Fig. 4.12** Vector spin

$$\widehat{H} = -\sum_{i,j} 2J_{ij}S_iS_j - g\mu_B H \sum_i S_i \tag{4.12}$$

where the second term represents the Zeeman energy under a magnetic field $H$. Spin is a vector like ($S_x$, $S_y$, $S_z$) (Fig. 4.12), and Eq. 4.12 should be represented by an inner product of two $S$ vectors. Such a spin Hamiltonian is called **Heisenberg model**. However, in a low-symmetry compound, the spin is always oriented in a definite direction due to the crystal field. In such a case, only an element $S_z$ is important, and Eq. 4.12 is represented by the scaler products. This is called **Ising model**. The following discussion is represented by the Ising model for simplicity. For $S = 1/2$, the first term takes two values depending on $S_i$ (on the site $i$) and $S_j$ (on the site $j$).

$$S_i = 1/2 \quad \text{and} \quad S_j = 1/2 \quad \text{leads to} \quad -J/2$$
$$S_i = -1/2 \quad \text{and} \quad S_j = 1/2 \quad \text{leads to} \quad J/2$$

Accordingly, the energy difference is $J$ depending on $S_i = \pm 1/2$. The more stable state is realized when

$S_i$ and $S_j$ have the same sign (ferromagnetism) for $J > 0$
$S_i$ and $S_j$ have the opposite signs (antiferromagnetism) for $J < 0$.

$J > 0$ is called ferromagnetic interaction, and $J < 0$ is called antiferromagnetic interaction. In old textbooks, however, the spin Hamiltonian is sometimes defined as

$$\widehat{H} = -\sum_{i,j} J_{ij}\overline{S}_i\overline{S}_j \tag{4.13}$$

without multiplying two. In such a case, the $J$ value is doubled.

It is a rather difficult task to solve the spin Hamiltonian exactly, and the exact solutions are known only for a few low-dimensional models. However, the **molecular-field approximation** or the **mean-field approximation** is an effective

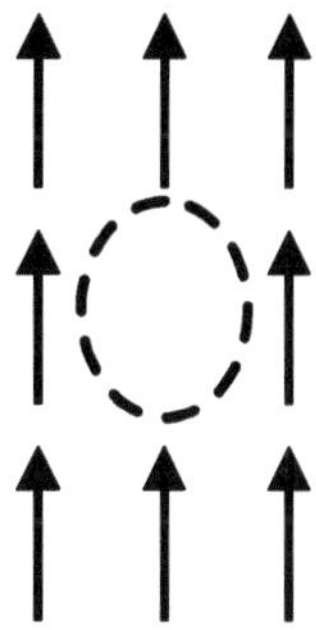

**Fig. 4.13** Molecular field approximation

tool, in which the surrounding spins are replaced by the average $\langle S \rangle$. We shall focus on the *i*-site (Fig. 4.13), and the sum of the surrounding *j*-sites is assumed to be calculated independently.

$$\hat{H} = \sum_i S_i \left( -\sum_j 2J_{ij}S_j - g\mu_B H \right) = -g\mu_B (H_{\text{eff}} + H) \sum_i S_i \tag{4.14}$$

Here $H_{\text{eff}}$ is the sum of the surrounding *j*-sites represented by the unit of magnetic field

$$H_{\text{eff}} = \frac{1}{g\mu_B} \sum_j 2J_{ij} \langle S \rangle \tag{4.15}$$

where $S_j$ is approximated by the bulk average $\langle S \rangle$. $S_i$ and $S_j$ are to be taken for all combinations in the solid, but usually only the nearest neighbor combinations are taken into account. Sometimes the next nearest neighbor sites are considered. For $J > 0$, and when the surrounding spins are oriented upwards, the central spin prefers the upward orientation. In the above equations, this effect is substituted by the **effective** or **internal field** $H_{\text{eff}}$. The response of the *i*-site spin is represented by the Curie law, namely by Eq. 4.6, but when $H$ is substituted by $H_{\text{eff}} + H$, it leads to $M = \chi_0 (H_{\text{eff}} + H)$. When the average spin $\langle S \rangle$ is oriented in a direction, it leads to the magnetization $M = Ng\mu_B \langle S \rangle$. This relation is put in Eq. 4.15 to give

$$H_{\text{eff}} = \frac{\sum_j 2J_{ij}}{N(g\mu_B)^2} M = aM \tag{4.16}$$

so the effective field is proportional to $M$. The coefficient $a$ is called the **molecular-field coefficient**. These relations give

$$M = \chi_0 (aM + H) \quad \text{which leads to} \quad (1 - \chi_0 a)M = \chi_0 H. \tag{4.17}$$

By using $\chi_0 = C/T$, this is reduced to

$$\chi = \frac{M}{H} = \frac{\chi_0}{1 - \chi_0 a} = \frac{C}{T - \theta} \tag{4.18}$$

where

$$\theta = aC = \frac{1}{k_B} \sum_j 2J_{ij} = \frac{2zJ}{k_B} \tag{4.19}$$

is **Weiss temperature**. The sign of the Weiss temperature is the same as the sign of $J$ that is positive for ferromagnetism and negative for antiferromagnetism. When the site $i$ is surrounded by $z$ equivalent neighboring sites interacting with $J$, $\theta$ is given by the last form.

In a ferromagnet with $\theta > 0$, the susceptibility from Eq. 4.18 diverges to infinity at a finite temperature (Fig. 4.14a). The diverging susceptibility means $M \neq 0$ even at $H = 0$. Below this temperature, the spins are oriented in a direction, and show finite spontaneous magnetization $M \neq 0$ even without applying a magnetic field. Above this temperature, the spins are randomly oriented similarly to the ordinary paramagnet. The transition temperature from paramagnetism to ferromagnetism is called the **Curie temperature** $T_c$, which is close to $\theta$ but not exactly identical to $\theta$. $T_c$ is the transition temperature of the material, whereas $\theta$ is the magnitude of the ferromagnetic interaction. $T_c$ is 1043 K for iron and 627 K for nickel. These metals are ferromagnetic at room temperature, but become paramagnetic above these temperatures.

If all spins in a material are oriented in one direction, the material is a magnet. However, the ordinary iron is not a magnet. This is because spins are oriented in one direction in an approximately 1 μm domain, but the magnetization is randomly oriented in different domains, and cancels each other in the whole material (Fig. 4.14b). When a magnetic field is applied to a ferromagnet, the initial magnetization curve (Fig. 4.14c) is similar to a paramagnet (Fig. 4.9d), and the magnetization of each domain is gradually ordered parallel to the magnetic field. At a high-field flat region, the magnetization is aligned in a direction. With lowering the

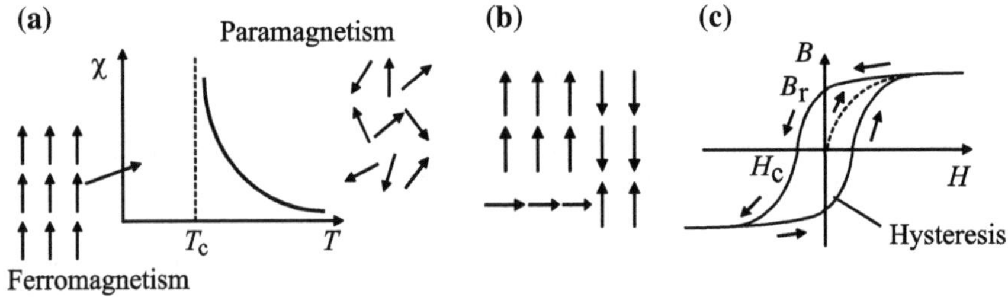


**Fig. 4.14** **a** Temperature dependence of the susceptibility, **b** magnetic domains, and **b** magnetization curve in a ferromagnet

magnetic field, the magnetization does not drop to zero even at $H = 0$ because the domains keep somewhat aligned magnetization. $B$ at $H = 0$ is called **residual magnetization** $B_r$. When an iron needle or a screw driver is repeatedly rubbed with a magnet, these materials are magnetized and attract other iron materials. This is because the magnetization orientation is aligned by the magnetic field. For $H < 0$, the magnetization follows a similar curve to reach the negative-side saturation. The field $H_c$ crossing at $B = 0$ is called **coercive field**. The residual magnetization scanned from the positive side and the negative side is oriented in the opposite direction. Such a **hysteresis** of the magnetization curve is characteristic of ferromagnets.

## 4.6 Antiferromagnetism

Next, we shall investigate the molecular-field approximation of an antiferromagnet [4, 5]. Since the total averaged spin $\langle S\rangle$ is zero, we consider two sublattices; the up and down spin sites are labeled as sublattices 1 and 2, respectively (Fig. 4.11b). The average spin of the respective **sublattice** is represented by $\langle S_1\rangle$ and $\langle S_2\rangle$, and the magnetization is $M_1$ and $M_2$. We assume $J < 0$ between the nearest neighbor 1 and 2 sites, and the molecular-field coefficient $a$ is defined as

$$a = \frac{4zJ}{N(g\mu_B)^2}. \tag{4.20}$$

The factor four appears because the number of the sublattice sites is $N/2$. Similarly to Eq. 4.17, we obtain

$$\begin{aligned} M_1 &= \chi_0(aM_2 + H) \\ M_2 &= \chi_0(aM_1 + H). \end{aligned} \tag{4.21}$$

Here, since the number of the sites is half, we have to use $\chi_0 = C/2T$. By adding the above two equations, we obtain

$$M = M_1 + M_2 = \chi_0(a(M_1 + M_2) + 2H) = \chi_0(aM + 2H)$$

which leads to

$$\chi = \frac{M}{H} = \frac{2\chi_0}{1 - \chi_0 a} = \frac{C}{T - \theta}. \tag{4.22}$$

The resulting form is the same as Eq. 4.18. From $J < 0$, the Weiss temperature $\theta$ is negative (Eq. 4.20). Then, $\chi$ does not diverge at a finite temperature (Fig. 4.15b). Zero of the coefficient determinant of $M_1$ and $M_2$ (Eq. 4.21) at $H = 0$ affords

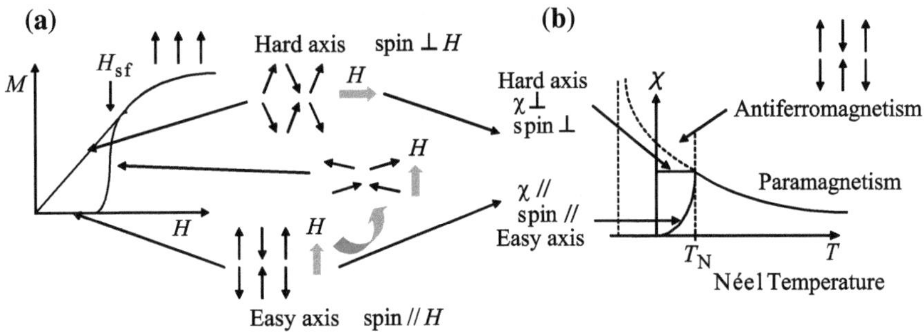


**Fig. 4.15** **a** Magnetization curve, and **b** temperature dependence of the susceptibility in an antiferromagnet. **a** Magnrtization field dependence below $T_N$, **b** magnetization susceptibility temperature dependence

$$\begin{vmatrix} -2T/C & a \\ a & -2T/C \end{vmatrix} = 0$$

which gives the transition temperature, usually called **Néel temperature**, as $T_N = |Ca/2| = |\theta|$. The transition temperature is expected to be the same as the absolute value of the Weiss temperature.

Susceptibility in the antiferromagnetic phase largely depends on the orientation of the magnetic field. The spin direction is called **easy axis**, and other two directions are called **hard axes**. When the magnetic field is applied parallel to the easy axis, the susceptibility is nearly zero except for a slight thermal excitation (Fig. 4.15b). When the magnetic field $H_\perp$ is parallel to the hard axis, the internal field coming from the tilted $M_\perp$ balances with $H_\perp$ to give

$$H_\perp + aM_\perp = 0 \,\text{leads to}\, \chi_\perp = M_\perp / H_\perp = 1/|a|$$

so the susceptibility is constant below the Néel temperature.

Magnetization curve in an antiferromagnetic phase is depicted in Fig. 4.15a. Magnetization curve along the hard axis is the same as paramagnetism (Fig. 4.9d) due to the perpendicular magnetization. Magnetization along the easy axis is essentially zero because the spins are already aligned. However, when the magnetic field is increased, the spin abruptly tilts to the hard axis, and follows a paramagnetic curve at a higher field. This is called **spin-flop transition**. The energy difference $K$ between the spin // easy axis state and the spin // hard axis state is

$$\frac{1}{2}\chi_\perp H^2 - \frac{1}{2}\chi_{//} H^2 = K \,\text{which leads to}\, H = \sqrt{\frac{2K}{\chi_\perp - \chi_{//}}}. \tag{4.23}$$

By putting $\chi_{//} = 0$ and $\chi_\perp = 1/|a|$, the spin-flop field is obtained as $H_{sf} = \sqrt{2K|a|}$. $K$ is called **magnetic anisotropy**, which comes from the lowered symmetry of the spin site.

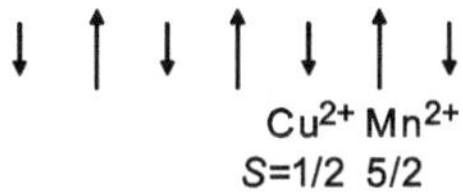


**Fig. 4.16** Ferrimagnetism

Suppose two kinds of metals with different $S$ are aligned alternately (Fig. 4.16). When an antiferromagnetic order is established, the longer spins are aligned in the same direction, and the whole material has magnetization to act as a magnet. Such a case is called **ferrimagnetism**. **Ferrite** $Fe_3O_4$ is a magnet known from the ancient times, in which $Fe^{3+}$ and $Fe^{2+}$ are aligned alternately. Other types of ferrites with the composition of $MFe_2O_4$ have been known, in which $Fe^{2+}$ is replaced by such divalent ions as $M = Mn^{2+}$, $Co^{2+}$, and $Ni^{2+}$. Since the magnetic interaction is usually $J < 0$, ferrimagnetism is an important strategy to obtain molecular magnets. For example, a molecular magnet composed of an alternating alignment of $Cu^{2+}$ and $Mn^{2+}$ has been realized (Fig. 4.27). A ferrimagnet shows susceptibility and magnetization that are basically similar to a ferromagnet.

From Eqs. 4.6, 4.18, and 4.22, a polt of inverse susceptibility $\chi^{-1}$ versus temperature gives a straight line (Fig. 4.17a). The extrapolated line passes the origin in the Curie paramagnetism, crosses the horizontal axis in the positive region in a ferromagnet and in the negative region in an antiferromagnet. The crossing point affords the Weiss temperature $\theta$. Another frequently used plot is a plot of $\chi T$ versus $T$, which affords a flat line in the Curie paramagnetism (Fig. 4.17b). Here, the magnitude of $\chi T$ provides the $S$ value according to Eq. 4.7. The actual $\chi T$ tends to be larger than the ideal value estimated in Example 4.3 because the $g$-value is usually slightly larger than two. For ferromagnetic interaction, the line goes up at low temperatures, and for antiferromagnetic interaction, the line goes down at low temperatures (Fig. 4.17b). When we obtain experimental $\chi$ data, $\theta$ is evaluated from the $\chi^{-1}$ versus $T$ plot, and the $g$-value is estimated from the $\chi T$ versus $T$ plot.

We sometimes observe $\chi T$ once increases and drops at lower temperatures (Fig. 4.17c). This is an evidence that the in-plane ferromagnetism coexists with the interplanar antiferromagnetism (Fig. 4.17d). If the absolute magnitude of the in-plane $J > 0$ is larger than the interplanar $J < 0$, the in-plane ferromagnetism develops at higher temperatures to enhance $M$ and $\chi$. However, the low-temperature development

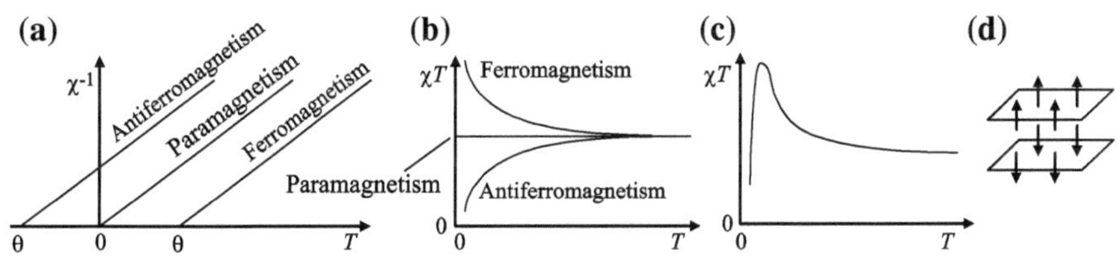


**Fig. 4.17** **a** $\chi^{-1}$, and **b** $\chi T$ plots. **c** $\chi T$ plots of a ferrimagnet. **d** In-plane ferromagnetism coexisting with interplanar antiferromagnetism

of the interplanar antiferromagnetic order diminishes $M$ and $\chi$. The in-plane ferromagnetism is sometimes substituted by ferrimagnetism. A ferromagnetic chain instead of plane leads to a similar result.

## 4.7 Low-Dimensional Magnetism

The molecular-field model is a good approximation for a three-dimensional magnet, but a low-dimensional magnet needs a more exact treatment. In particular, it has been known that a one-dimensional system does not undergo a phase transition. This is not only for a magnetic transition but also for a phase transition in general. For simplicity, we shall prove the absence of a one-dimensional ferromagnetic transition [6]. Suppose a one-dimensional ferromagnet consisting of $L$ sites with $n$ domain walls (Fig. 4.18), the statistical weight is

$$W = \frac{L!}{(L-n)!n!}. \tag{4.24}$$

Using Stirling's approximation, this leads to

$$\begin{aligned} \ln W &= L \ln L - (L-n)\ln(L-n) - n\ln n \\ &= L \ln \frac{L}{L-n} + n \ln \frac{L-n}{n} \approx -n \ln \frac{n}{L}. \end{aligned} \tag{4.25}$$

The final step is based on the approximation $L \gg n$. The free energy is given by

$$G = G_0 - TS + nJ = G_0 + k_B T n \ln \frac{n}{L} + nJ. \tag{4.26}$$

The last term comes from the energy loss due to the domain walls. The minimum $G$ is obtained by putting

$$\frac{\partial G}{\partial n} = k_B T \ln \frac{n}{L} + J \tag{4.27}$$

to be zero. When $n/L$ is small, $\ln(n/L)$ is a large negative value, and this equation is negative around $n \sim 0$. Accordingly, $G$ decreases when $n$ increases. This means that $G$ takes a minimum at a finite $n \neq 0$. $n$ becomes zero only at $T = 0$. At a finite temperature, a few domain walls are incorporated in a one-dimensional ferromagnet, and the magnetic order is not established. After my lecture, when the students

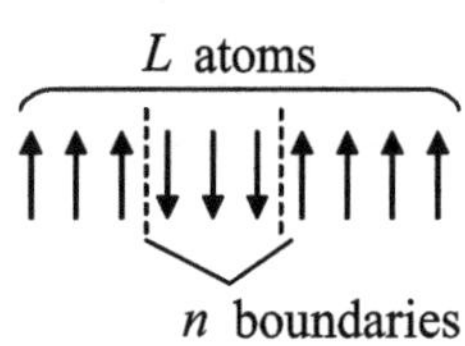


**Fig. 4.18** One-dimensional ferromagnet

submit the quiz papers on my desk, they put them upside ordered. However, when a careless student puts it upside down, the next student follows the wrong orientation. It is difficult to avoid such a disorder entirely in one dimension. In two and three-dimensional systems, the domain walls consist of lines or surfaces, and a macroscopic number of misfits are necessary to make the disorder. By contrast, in one dimension the global order is lost only by a single careless student. Since the spins in a small domain are aligned in the same orientation, this is called **short-range order**, in contrast to the **long-range order**. Then, a one-dimensional system is considered to have large **fluctuation**.

Figure 4.19 shows susceptibility of a low-dimensional antiferromagnet. A three-dimensional antiferromagnet follows the Curie–Weiss law at high temperatures, and undergoes an antiferromagnetic transition at the Néel temperature. By contrast, this low-dimensional antiferromagnet does not show such a clear transition, and the susceptibility gradually drops at low temperatures. This is due to the short-range antiferromagnetic order. The high-temperature susceptibility approaches to the Curie law. The usual low-dimensional antiferromagnet has a small three-dimensional interaction, by which a true three-dimensional long-range order occurs at a very low temperature. This particular material exhibits a three-dimensional antiferromagnetic transition at 15 K.

Temperature-dependent susceptibility of a low-dimensional magnet is theoretically calculated by solving the spin Hamiltonian (Eq. 4.12) for a low-dimensional lattice. The exact solution is, however, difficult except for a few simple models, and instead the results of numerical calculation have been obtained. Temperature-dependent susceptibility of several low-dimensional magnets is shown in Fig. 4.20 [7]. The results are given in a form of polynomials and easily compared with experimental results.

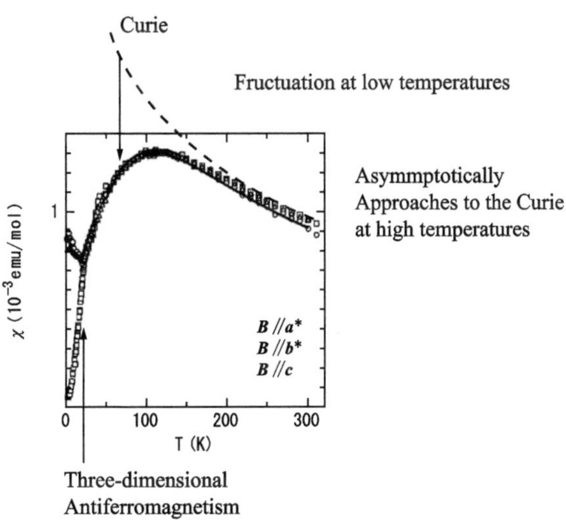


**Fig. 4.19** Temperature dependence of the susceptibility in an organic antiferromagnet, β′-(BEDT-TTF)$_2$ICl$_2$ [11]

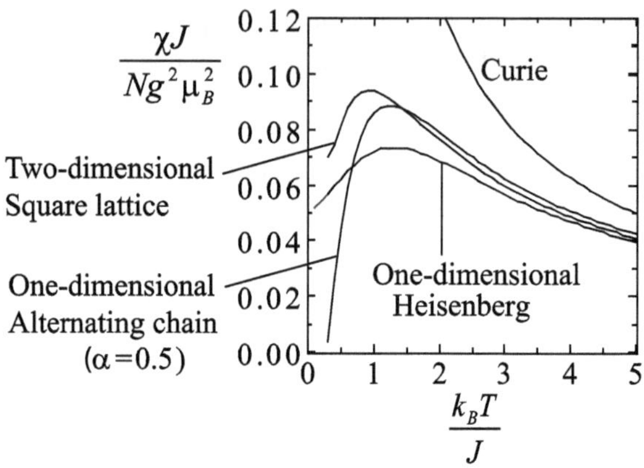


**Fig. 4.20** Susceptibility of low-dimensional magnets. One-dimensional Heisenberg antiferromagnet: W. E. Estes, D. P. Gaval, W. E. Hatfield, and D. J. Hodgson, *Inorg. Chem.*, **17**, 1415 (1978). One-dimensional alternating Heisenberg antiferromagnet: J. M. Hall, W. E. Marsh, R. R. Weller, and W. E. Hatfield, *Inorg. Chem.*, **20**, 1033 (1981). Two-dimensional square antiferromagnet: M. E. Lines, *J. Phys. Chem. Solids*, **31**, 101 (1970); J. Wang, *Phys. Rev.* B, **44**, 2396 (1991). Two-dimensional triangular antiferromagnet: N. Elstner, R. R. P. Singh, and A. P. Young, *Phys. Rev. Lett.*, **71**, 1629 (1993);M. Tamura and R. Kato, *J. Phys. Cond. Matter.*, **14**, L729 (2002). One-dimensional ferrimagnet: R. Georges, J. J. Brorrás-Almenar, E. Coronado, J. Curély, and M. Drillon, *Magnetism: Molecules to Materials*, ed. by J. S. Miller and M. Drillon, Wiley (2001), p1

## 4.8 Magnetic Frustration

In an antiferromagnetic square lattice (Fig. 4.11b), suppose an oblique interaction $J'$ (Fig. 4.21a) in addition to $J$ along the ordinary square lattice. Since $J'$ works between sites 1 and 1 (and between sites 2 and 2), an antiferromagnetic $J'$ connects parallel spins and increases the energy. Accordingly, $J'$ is **frustrated**. The molecular-field coefficient is written as $a' = \frac{4J'}{N(g\mu_B)^2}$ in addition to $a = \frac{4J}{N(g\mu_B)^2}$, and Eqs. 4.21 for the internal field are replaced by

$$\begin{aligned} M_1 &= \chi_0(aM_2 - a'M_1 + H) \\ M_2 &= \chi_0(aM_1 - a'M_2 + H). \end{aligned} \tag{4.28}$$

In general, such simultaneous equations are obtained for more complicated systems with more than two $M_1$, $M_2$, …, and the solution gives the transition temperature. By adding the two equations, we obtain

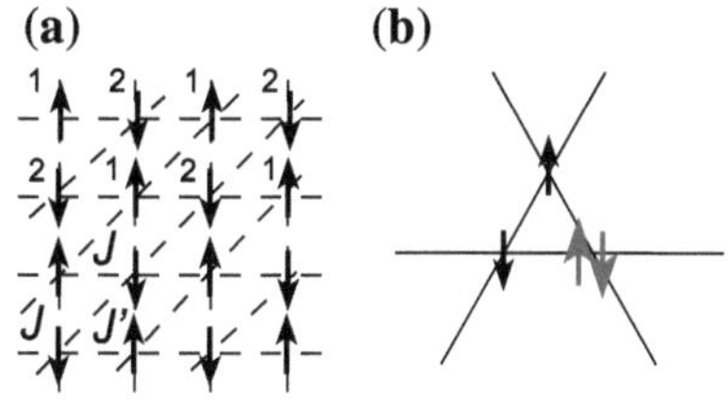

Fig. 4.21 **a** A square lattice with a frustrated interaction $J'$, and **b** a triangular lattice

$$M = M_1 + M_2 = \chi_0((a - a')(M_1 + M_2) + 2H) = \chi_0((a - a')M + 2H)$$

which leads to

$$\theta = (a - a')C = \frac{2(zJ - z'J')}{k_B}.$$

Then, in the presence of the frustrated interactions, $a$ is replaced by $a - a'$, and $zJ$ is replaced by $zJ - z'J'$. In the mean-field approximation, the Weiss temperature is given by the sum of $J$ considering the coordination number in a unit cell (Eq. 4.20). In the case with frustrated interactions, this is substituted by $J - J'$.

In a triangular lattice with three equivalent $J$ (Fig. 4.21b), after antiparallel spins are placed on sites 1 and 2, we could not determine the stable spin orientation on site 3. Consequently, an ideal triangular lattice does not undergo an antiferromagnetic transition, and instead becomes a **spin liquid** which does not show antiferromagnetism down to low temperatures. Many organic conductors have an approximate triangular lattice, and when $J$ becomes equivalent, a considerable number of organic conductors exhibit such a spin liquid state.

## 4.9 πd System

Organic conductors including magnetic anions, for example, κ-$(BETS)_2FeBr_4$, exhibits antiferromagnetism of the magnetic anions with keeping the metallic conductivity of the organic molecules (Sect. 7.8.1). Such an organic conductor is called **πd system** [8]. Some inorganic conductors in which metallic conduction and magnetism coexist are called sd systems because the conduction is mediated by the s electrons, and the magnetism is maintained by the d electrons. The Kondo effect observed when itinerant and magnetic electrons are coexisting is described in Sect. 5.9, but here we will deal with the antiferromagnetism of a πd system in analogy with the ordinary antiferromagnetism.

Fig. 4.22 Interactions in a πd system

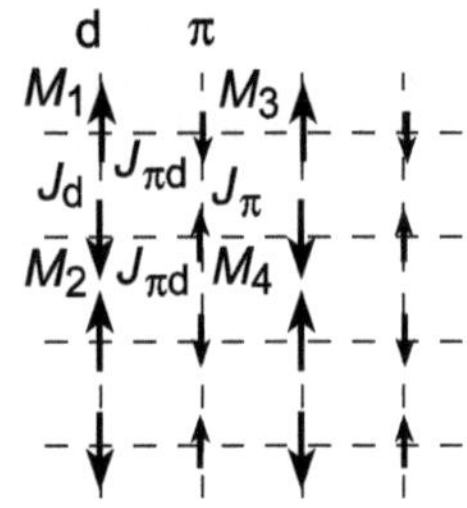


Similarly to an antiferromagnet (Fig. 4.21a), a unit cell contains d spins, $M_1$ and $M_2$, and π spins, $M_3$ and $M_4$ (Fig. 4.22). In the actual organic πd system like κ-$(BETS)_2FeBr_4$, the d spin is $S$ = 5/2, and the π spin is $S$ = 1/2. The π electron shows metallic conductivity, but the difference of the up and down spin numbers gives rise to small magnetism, $M_3$ and $M_4$, that are much smaller than $\langle S \rangle = 1/2$ There are direct interaction $J_d$ between the d spins, interaction $J_\pi$ between the π spins, and πd-interaction between the π and d spins. In analogy with Eq. 4.28, we obtain

$$\begin{aligned} -1/\chi_d M_1 + a_d M_2 + a_\pi d M_3 + H &= 0 \\ a_d M_1 - 1/\chi_d M_2 + a_\pi d M_4 + H &= 0 \\ a_{\pi d} M_1 - 1/\chi_\pi M_3 + a_\pi M_4 + H &= 0 \\ a_{\pi d} M_2 + a_\pi M_3 - 1/\chi_\pi M_4 + H &= 0. \end{aligned} \tag{4.29}$$

The molecular-field coefficients, $a_d$ and $a_{\pi d}$, respectively, correspond to $J_d$ and $J_{\pi d}$ (Eq. 4.20). The localized d electron follows the Curie law, $\chi_d = C_d/T$ (Eq. 4.6), whereas the metallic π electron shows the Pauli paramagnetism $\chi_\pi = 2\mu_B^2 D(E_F)$ (Eq. 4.10). The second equation is subtracted from the first equation, and the fourth equation is subtracted from the third equation.

$$\begin{aligned} (a_d + 1/\chi_d)(M_1 - M_2) &= a_\pi d(M_3 - M_4) \\ (a_\pi + 1/\chi_\pi)(M_3 - M_4) &= a_\pi d(M_1 - M_2) \end{aligned} \tag{4.30}$$

In the coefficient determinant

$$\begin{vmatrix} a_d + 1/\chi_d & -a_\pi d \\ -a_\pi d & a_\pi + 1/\chi_\pi \end{vmatrix} = 0$$

$\chi_d = C_d/T$ includes $T$, while $\chi_\pi$ does not involve $T$. Then, the solution for $T$ affords the transition temperature $T_N$.

$$\frac{T_N}{C_d} = -a_d + \frac{\chi_\pi a_\pi d^2}{1 + \chi_\pi a_\pi} \tag{4.31}$$

Here, the staggered susceptibility

$$\chi_q = \frac{\chi_\pi}{1 + \chi_\pi a_\pi}$$

makes maximum in the direction of the spin-density wave vector $\boldsymbol{q} = (\pi/a, \pi/b)$ of the π system. Then, Eq. 4.31 leads to

$$\frac{T_N}{C_d} = -a_d + \chi_q a_{\pi d}^2. \tag{4.32}$$

The first term represents the direct interaction between the d spins, and $T_N = C_d|a_d|$ is the Néel temperature of the ordinary d spin network. The second term represents indirect interaction mediated by the π electron. Because $a_d < 0$, both the first and second terms are positive. Because $a_\pi < 0$, the denominator of Eq. 4.31 becomes zero when $\chi_\pi$ is as large as $\chi_\pi = -1/a_\pi$, and $\chi_q$ diverges. In this extreme case, the π electrons make an antiferromagnetic order spontaneously. This should be more exactly called spin-density wave (SDW) state (Fig. 7.17f). An SDW state is a kind of antiferromagnetic state, in which the up and down spin densities of metallic electrons become alternately large. An elemental chromium shows the SDW state with keeping the metallic conduction at low temperatures. Organic conductors usually undergo metal–insulator transitions when SDW is established, but elemental iron and nickel show ferromagnetism with keeping the metallic conduction even at room temperature. Coexistence of metallic conduction and magnetic order is very common. Ordinary metals show $|\chi_\pi a_\pi| \ll 1$, but $\chi_\pi$ is proportional to the density of states, and inversely proportional to the bandwidth. Then, the second term of Eq. 4.31 is usually approximated as $\chi_\pi a_{\pi d}^2 \sim J_{\pi d}^2/E_F$. Since $E_F$ is much larger than $J_{\pi d}$, the indirect interaction is considerably smaller than $J_{\pi d}$. Even when $J_{\pi d}$ is comparatively large, the antiferromagnetic transition temperature is mainly determined by the direct interaction. Such a phenomenon as the metallic electron mediates the magnetic interaction has been known as the **RKKY** (Ruderman–Kittel–Kasuya–Yoshida) **interaction** [4]. In the RKKY interaction, we suppose a magnetic spin in a metallic background; for example, Fe impurity in Au. Then, the spin polarization induced on the metallic electron propagates as

$$M(r) \propto \frac{\cos(2k_F r)}{r^3} - \frac{\sin(2k_F r)}{r^4}. \tag{4.33}$$

This equation means the spin polarization becomes alternately positive and negative depending on the distance from the Fe atom $r$ with the periodicity of $2k_F$. In Fig. 4.22, a π site has one spin, corresponding to a half filled state, and the up and down spins form a two-site periodicity. In a one-dimensional band, Fig. 2.5 or Eq. 2.12 implies $2k_F = \pi/a$. When $r = 2a$ is put in Eq. 4.33, we obtain $2k_F r = (\pi/a)(2a) = 2\pi$, and $2a$ certainly corresponds to $2k_F$ as depicted in Fig. 4.22. If the filling is smaller than half filled, the periodicity is larger than this, but in any filling, one electron corresponds to one spin and two electrons constitute the periodicity of one up spin and one down spin. κ-$(BETS)_2FeBr_4$ has one spin per two BETS molecules, so $M_3$ and $M_4$ in Fig. 4.22 are located on BETS dimers.

When the energy band is $E(k)$, the susceptibility $\chi(q)$ of a π system is generally represented as

$$\chi(q) = \frac{g^2 \mu_B^2}{2V} \sum_{k<k_F} \frac{f(k)(1 - f(k+q))}{E(k+q) - E(k)}.$$

When periodical perturbation $V_q$ is applied to the π system, this relation is derived from the resulting density waves $\rho_q = -\chi(q)\, V_q$ using the perturbation theory. Here, $f(k)$ is the Fermi distribution function. The numerator is one only when $f(k)$ is occupied and $f(k + q)$ is unoccupied, while it is zero for other cases. Accordingly, suppose the excitation from an occupied state $E(k)$ to an unoccupied state $E(k + q)$, and the inverse of the energy difference is integrated over the $k$ space to afford $\chi(q)$. Since $E(k + q) - E(k) = 0$ makes the denominator zero, the integral diverges. In a one-dimensional conductor, the translation $q = 2k_F$ makes the Fermi surface overlap to another piece of the Fermi surface (Fig. 7.19a). Then, the Fermi surface nests, and $\chi(q)$ becomes very large at $q = 2k_F$ to result in the corresponding SDW. In general, $\chi(q)$ is calculated from the band structure $E(k)$.

When the antiferromagnetism is established, finite magnetizations $M_1 = -M_2 = M_d$ and $M_3 = -M_4 = M_\pi$ are put in Eq. 4.29 to give the internal fields

$$H_d = \left(\frac{T}{C_d} + a_d\right) M_d - a_\pi d M_\pi$$

$$H_\pi = -a_\pi d M_d + \left(\frac{1}{\chi_q}\right) M_\pi.$$

These relations are put in $E = \sum M_i H_i$ to calculate the magnetic energy per site.

$$E = \left[\frac{T}{C_d} + a_d - \chi_q a_\pi d^2\right] M_d^2 + \frac{1}{\chi_q} \left[M_\pi - \chi_q a_\pi d M_d\right]^2$$

In order to discuss finite temperatures, the free energy is obtained by calculating the canonical ensemble, and the result includes higher terms of $M_d$ and $M_\pi$.

$$F = \left[\frac{T}{C_d} + a_d - \chi_q a_\pi d^2\right] M_d^2 + \frac{1}{\chi_q} \left[M_\pi - \chi_q a_\pi d M_d\right]^2 + b_d M_d^2 + b_\pi M_\pi^4 + \cdots$$

This is a kind of the Laudau expansion (Eq. 6.14.). The coefficient of the $M_d^2$ term changes sign at

$$\frac{T_N}{C_d} = -a_d + \chi_q a_\pi d^2$$

which is the same result as Eq. 4.32. For $T > T_N$, this coefficient is positive, and $M_d = 0$ is the energy minimum. For $T < T_N$, $M_d \neq 0$ affords the energy minimum, implying an antiferromagnetic order. In order to minimizing the second term, finite spin $M_\pi = \chi_q a_{\pi d} M_d$ is induced on the π sites.

From the molecular orbital calculation, the direct interaction of the first term in Eq. 4.32 is estimated to be 2.10 K in κ-$(BETS)_2FeBr_4$, and the second term is 0.62 K [8]. Then, most of the Néel temperature 2.5 K comes from the direct interaction. However, the direct and indirect interactions work in different directions, and both are necessary for the antiferromagnetic order. The spin induced on the π system ($M_\pi$) is as small as several % of $S = 1/2$. Like this, the πd system is understood similarly to the ordinary antiferromagnet by assuming the susceptibility of the π system is the temperature-independent Pauli paramagnetism.

## 4.10 Molecular Magnetism

A variety of attempts have been made to realize organic ferromagnets [9, 10]. Figure 4.23 depicts organic radicals that are stable at room temperature. Galvinoxyl radical shows ferromagnetic interaction at room temperature, but becomes non magnetic below 85 K due to the dimerization. p-NPNN and TEMPOL show ferromagnetism around 0.5 K. Several related ferromagnetic derivatives have been known. In order to achieve ferromagnetism, molecular arrangement is important. According to the Kanamori–Goodenough role (Sect. 5.5), the adjacent molecular orbitals have to be orthogonal. For example, p-NPNN makes several polymorphisms, but only the β-phase shows ferromagnetism.

In order to investigate molecular magnetism, suppose benzoquinodimethanes where two $CH_2$ groups are attached to a benzene ring (Fig. 4.24). For ortho- and para-isomers, we can put double bonds so as to eliminate unpaired electrons, and the ground state is singlet ($S = 0$). For the meta-isomer, we cannot put double bonds so as to erase unpaired electrons. The resulting biradical has parallel spins

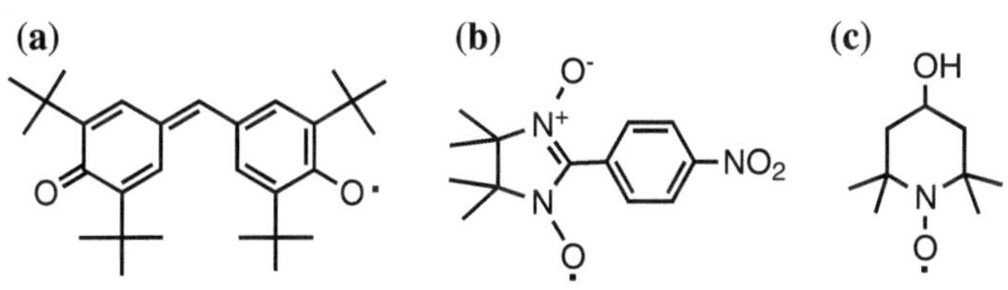


**Fig. 4.23** Molecular magnets **a** Galvinoxyl radical, **b** *p-NPNN* p-nitrophenyl nitronyl nitroxide, **c** *TEMPOL* 2,2,6,6-tetramethylpiperidinol-1-oxyl

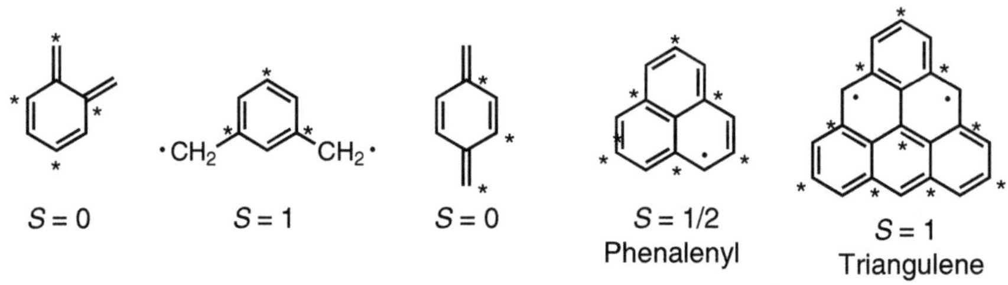


**Fig. 4.24** Benzoquinodimethanes and several magnetic aromatic hydrocarbones

according to the Hund rule to afford the triplet ($S$ = 1) state. When we write asterisks alternately on the carbon atoms, the ortho- and para-isomers have the same number of asterisked and non-asterisked carbons. Such a molecule is called **alternating hydrocarbon**. In the meta-isomer, it is impossible to write the same number of asterisked and non-asterisked carbons. Such a case is called non-alternating hydrocarbon. An alternating hydrocarbon has the $S = 0$ state, whereas a non-alternating hydrocarbon has unpaired electrons whose number is equal to the difference between the asterisked and non-asterisked carbons. The last two molecules in Fig. 4.24 are called phenalenyl and triangulene radicals. Although these molecules are hydrocarbones composed of only benzene rings, it is impossible to put double bonds alternately. These compounds are paramagnetic materials with unpaired spins.

There is a long history of research on polycarbenes (Fig. 4.25) [10]. These compounds are produced by irradiating light to the corresponding $N_2$ compounds at low temperatures. The bridging carbon has two radical spins, which generate an $S$ = 1 spin according to the Hund rule. Ferromagnetic interaction works between these carbenes because these spins are located in the meta positions similarly to Fig. 4.24. Then, all carbene spins are parallel to each other, and very large $S$ is generated.

The energy bands of polycarbene consist of π and π* bands. The energy band of polyacetylene has an energy gap between the π and π* bands (Example 2.9), and polyacetylene is non magnetic. In non-alternating hydrocarbons, the SOMO (singly occupied molecular orbital) levels carry unpaired electrons.

In a conjugated π system, each carbon atom has one electron, and when the whole π and π* bands are regarded as a single band, that band is half filled. The spin on a carbon atom has antiferromagnetic interaction with the spin on the adjacent carbon atom due to the transfer integral $t$ (Sect. 5.5), and the alternating (antiferromagnetic) spin arrangement is preferred. In other words, considering the conjugated π system as a metal, the RKKY interaction (Sect. 4.9) aligns the spins alternately. Carbon atoms other than the carbene atoms do not have spins, but the induced spins (small spins in Fig. 4.25) follow this alternate arrangement. This is the reason that the meta spins are parallel to each other. From the same logic, ortho- and para-spins are antiparallel to each other.

**Fig. 4.25** Magnetism of polycarbene

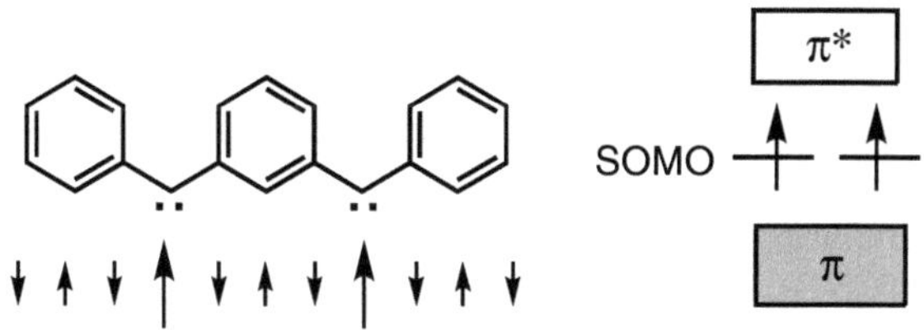


**Fig. 4.26** Magnetism of an allyl radical

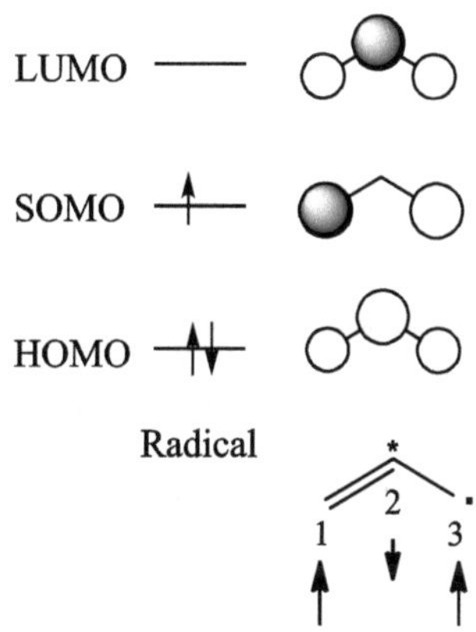


Suppose the magnetism of an allyl radical (Example 1.10), where an unpaired electron exists on the SOMO (Fig. 4.26). Since only carbons 1 and 3 have the SOMO population, these carbons have spins of the same orientation (up spins). In such a case, the HOMO population also depends on the spin orientation, and carbon 2 has larger population of down spin. Note that an allyl radical is a non-alternating hydrocarbon. The alternating spin polarization is evident from the molecular orbitals.

The Curie temperatures of purely organic ferromagnets are below the liquid helium temperature. Molecular ferromagnets are achieved more easily using metal spins. The material shown in Fig. 4.27 consists of alternate arrangement of manganese and cupper, where $Mn^{2+}$ has $S = 5/2$ spin and $Cu^{2+}$ has $S = 1/2$ spin. Antiferromagnetic interaction between Mn and Cu makes all Mn spins aligned in the same orientation to make a ferrimagnet. Here, the parallel Mn spin alignment

**Fig. 4.27** Alternate arrangement of manganese and cupper

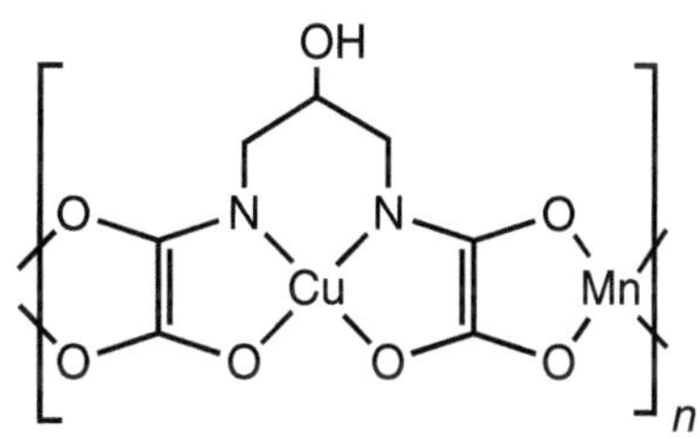

between the different chains is maintained by the interchain interaction. This compound shows a ferrimagnetic transition at 4.2 K [10].

A charge-transfer complex composed of decamethylferrocene (Cp*: pentamethylcyclopentadienyl) and tetracyanoethylene (TCNE, Fig. 7.5) [FeCp*$_2$] [TCNE] has a mixed stack structure, but shows ferromagnetism at 4.8 K. V $(TCNQ)_x$ shows ferromagnetism even at room temperature, though the structure is not certain (TCNQ in Fig. 7.5). Similarly to metal TCNQ complexes (Sect. 7.3), it is expected that the CN part of TCNQ coordinates to $V$, and the charge-transfer complex consisting of $V^{n+}$ and $TCNQ^-$ realizes the ferrimagnetism.

## References


1. P. Atkins, T. Overton, J. Rourke, M. Weller, F. Armstrong, *Inorganic Chemistry* (Oxford, 2006), Chapter 19
2. A. Weiss, H. Witte, *Magnetochemie* (Verlag Chemie, 1973)
3. E.M. Purcell, *Berkeley Physics Course, Electricity and Magnetism* (McGraw-Hill, New York, 1963), Chapters 10 and 11
4. A.H. Morrish, *The Physical Principles of Magnetism* (R. E. Krieger Pub., 1965)
5. S. Brundell, *Magnetism in Condensed Matter* (Oxford, 2001)
6. L.D. Landau, E.M. Lifshitz, *Statistical Physics* (Pergamon, 1958), Chapter 15
7. L.J. Jongh, A.R. Miedema, *Experiments on Simple Magnetic Model Systems* (Taylor, 1974)
8. T. Mori, M. Katsuhara, J. Phys. Soc. Jpn. **71**, 826 (2002)
9. O. Kahn, *Molecular Magnetism* (Wiley, 1993), p. 187
10. J.S. Miller, M. Drillon (eds.), *Magnetism: Molecules to Materials I-V* (Wiley, New York, 2001–2005)
11. N. Yoneyama, A. Miyazaki, T. Enoki, G. Saito, Bull. Chem. Soc. Jpn. **72**, 639 (1999)

# Chapter 5
# Electron Correlation

**Abstract** The energy band theory predicts that all partially filled conductors are metallic, but many organic conductors are actually semiconducting due to the Coulomb interaction. In this chapter, Coulomb interaction between electrons is considered in the molecular orbital theory, and successively in the solid states. In particular, the Hubbard model accounts for the origin of the magnetic interactions as well as the Mott insulating state. The discussion is extended to charge-order systems with intermolecular Coulomb repulsion and the Kondo effect, where metallic and localized electrons are coexisting. Electron correlation is important to explain various phenomena that are not understood within the framework of the energy band theory.

**Keywords** Electron correlation · Coulomb interaction · Configuration interaction · Metal–insulator transition · Mott insulator · Hubbard model · Stoner model · Charge order · Kondo effect · Anderson localization

## 5.1 Molecular Orbital Theory Considering the Coulomb Interaction

In Chaps. 1–3, we have discussed molecular orbitals ignoring the Coulomb repulsion between electrons. Here, we first investigate how the molecular orbital theory changes including the averaged Coulomb interaction between electrons. The results show we can use the preceding results with some modifications. The electron motion is, however, not entirely independent of other electrons' motion, so the averaged treatment is sometimes insufficient. In general, such a phenomenon is called **electron correlation**. Magnetism described in the last chapter is also closely related to electron correlation. Then, we first discuss the influence of electron correlation in the molecular orbital theory. Second we examine the Hubbard model as a tool to investigate electron correlation in solids. From this, the origin of magnetic interaction is elucidated. After investigating the meaning of the Hubbard model, we describe the Kondo effect, where localized and itinerant electrons



T. Mori, *Electronic Properties of Organic Conductors*,
DOI 10.1007/978-4-431-55264-2_5

coexist. Finally, we deal with localization in a disordered system. Accordingly, this chapter deals with various phenomena beyond the energy band theory.

The Hamiltonian including the Coulomb repulsion between electrons is

$$H = \sum_i \left[ -\frac{\hbar^2}{2m}\nabla_i^2 - \frac{1}{4\pi\varepsilon_0}\sum_n \frac{Z_n e^2}{r_{ni}} \right] + \frac{1}{4\pi\varepsilon_0}\sum_{i \neq j} \frac{e^2}{r_{ij}}. \tag{5.1}$$

The Σ for $i$ is a summation for electrons, where the first term is electron's kinetic energy, and the second term is attraction from the nuclei $n$ with the nuclear charge = the atomic number, $Z_n$. The last term represents the Coulomb repulsion between electrons $i$ and $j$. Previously, we neglected this term.

A wave function of more than two electrons is represented by the product like $\psi_1(1)\psi_2(2)$ (Sect. 1.5). Here, $\psi_1$ and $\psi_2$ designate different orbitals, and the numbers in the parentheses denote electron's numbering. Since we cannot distinguish the same kind of particles in quantum mechanics, the exact wave function has to fulfill the following condition; when we exchange electron's numbering, the wave function does not change except for the sign.

$$\Psi(1,2) = \pm\Psi(2,1) \tag{5.2}$$

Here, ± should be taken positive for Bose particles like photons and phonons with integer spins, and negative for Fermi particles like electrons, protons, and neutrons with half integer spins. Now an electron wave function has to satisfy the negative sign, and is asymmetric with respect to the electron numbering. Consequently, the exact electron wave function is constructed by adding $-\psi_1(2)\psi_2(1)$ to $\psi_1(1)\psi_2(2)$.

$$\Psi(1,2) = \frac{1}{\sqrt{2}}(\psi_1(1)\psi_2(2) - \psi_1(2)\psi_2(1)) = \frac{1}{\sqrt{2}}\begin{vmatrix} \psi_1(1) & \psi_1(2) \\ \psi_2(1) & \psi_2(2) \end{vmatrix} \tag{5.3}$$

When we exchange the electron numbering, the result, $\psi_1(2)\psi_2(1) - \psi_1(1)\psi_2(2)$, has certainly the opposite sign to the original form. For more than two electrons, this is generally represented by a determinant.

$$\begin{aligned} \Psi(1,2,\ldots) &= \frac{1}{\sqrt{N!}}\begin{vmatrix} \psi_1(1) & \psi_1(2) & \cdots & \psi_1(N) \\ \psi_2(1) & \psi_2(2) & \cdots & \psi_2(N) \\ \vdots & \vdots & \ddots & \vdots \\ \psi_N(1) & \psi_N(2) & \cdots & \psi_N(N) \end{vmatrix} \\ &= \frac{1}{\sqrt{N!}}\sum P^{(-1)}\psi_1(1)\psi_2(2)\ldots\psi_N(N) \end{aligned} \tag{5.4}$$

This is called **Slater determinant.** $P^{(-1)}$ adds a negative sign when a couple of electron numbers are exchanged once, and Σ implies a sum of all possible exchanges. When we exchange electrons 1 and 2, the columns 1 and 2 are exchanged, and the sign of the determinant is changed. When the wave functions $\psi_1$ and $\psi_2$ are the same, the first and second lines are the same, and the determinant becomes zero. Therefore, two electrons cannot enter the same state according to the Pauli exclusion principle. Here, $\psi_1$ and $\psi_2$ include the spin part.

The Slater determinant $\Psi$ is put in

$$E = \int \Psi^* H \Psi \mathrm{d}\tau \tag{5.5}$$

to give

$$E = \sum_k H_k + \sum_{k,l} (J_{kl} - K_{kl}). \tag{5.6}$$

Here, the one-electron energy

$$H_k = \int \psi_k^* \left[ -\frac{\hbar^2}{2m} \nabla^2 - \frac{1}{4\pi\varepsilon_0} \sum_n \frac{Z_n e^2}{r_n} \right] \psi_k \mathrm{d}\tau \tag{5.7}$$

comes from the sum for an electron $i$ in the Hamiltonian (Eq. 5.1). Other integrals

$$J_{kl} = \int \psi_k^*(1)\psi_l^*(2)\left(\frac{1}{4\pi\varepsilon_0}\frac{e^2}{r_{12}}\right)\psi_k(1)\psi_l(2)\mathrm{d}\tau = \int \frac{e^2 \rho_k(1)\rho_l(2)}{4\pi\varepsilon_0 r_{12}} \mathrm{d}\tau \tag{5.8}$$

$$K_{kl} = \int \psi_k^*(1)\psi_l^*(2)\left(\frac{1}{4\pi\varepsilon_0}\frac{e^2}{r_{12}}\right)\psi_k(2)\psi_l(1)\mathrm{d}\tau \tag{5.9}$$

originate from the summation related to two electrons $i$ and $j$ in the Hamiltonian (Eq. 5.1). These integrals, $H_k$, $J_{kl}$, and $K_{kl}$, are called **molecular integrals**. The Slater determinant contains a product of all electrons, but products other than $i$ and $j$ afford one after the integration. Then, we finally obtain an integral in which the last Coulomb repulsion term of Eq. 5.1 is sandwiched by $\psi_1(1)\psi_2(2)$. Since $\rho_k(1) = \psi_k(1)^*\psi_k(1)$ is the electron distribution of the molecular orbital $k$, $J_{kl}$ is the average Coulomb repulsion between the molecular orbitals $\rho_k$ and $\rho_1$. $J_{kl}$ is called **Coulomb integral**. This is different from the Coulomb integral $\alpha$ in Sect. 1.5. Although $\alpha$ corresponds to the energy of the atomic orbital, $J_{kl}$ is the average Coulomb repulsion, which is more suitable for this name. Since the Slater determinant involves $-\psi_1(2)\psi_2(1)$, where the electrons 1 and 2 are exchanged, the **exchange integral** $K_{kl}$ appears as well. Note that electrons 1 and 2 are exchanged in the latter part of $K_{kl}$. The resulting negative sign is included in Eq. 5.6. Accordingly, it holds always $J_{kl} > 0$ and $K_{kl} > 0$.

$K_{kl}$ includes $\psi_k(1)^*\psi_l(1)$. If the spin parts of $\psi_k$ and $\psi_l$ are different, this product contains a product $\alpha(1)\beta(1)$. $\alpha$ is the spin wave function for upspin, and $\beta$ for

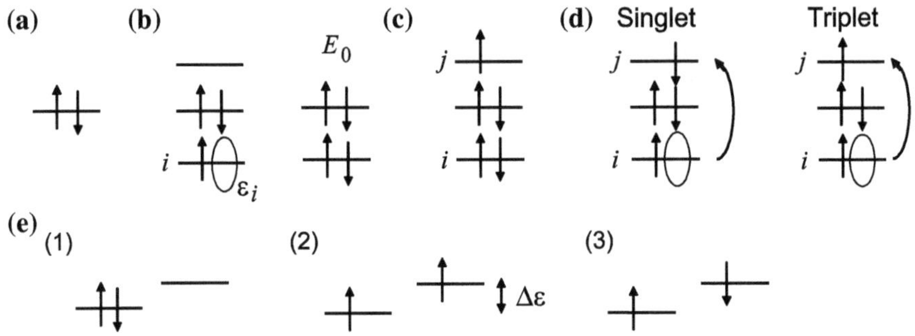


**Fig. 5.1** Various electron configurations

downspin. Since $\alpha$ and $\beta$ are associated with different states, these functions are orthogonal, then $\int \alpha(1)\beta(1)d\tau = 0$. In contrast, $J_{kl}$ includes only $\psi_k(1)^*\psi_k(1)$, which is nonzero because the integral of $\alpha(1)\alpha(1)$ is not zero. From this, the Coulomb repulsion between electrons with the same spin is $J_{kl} - K_{kl}$, whereas the Coulomb repulsion between electrons with opposite spins is $J_{kl}$. The Pauli exclusion principle is applicable to the electron position, and electrons with the same spin cannot come to the same position. Such electrons move keeping away from each other, to reduce the Coulomb repulsion by $K_{kl}$. This phenomenon is called **Fermi hole**, where electrons with the same spin cannot come to the same position. The exchange integral is correction of the Coulomb integral for the Fermi hole.

In order to investigate the meaning of the exchange integral, we shall calculate the energy of electron configurations shown in Fig. 5.1. The energy of the state in which up and down electrons enter the same level (Fig. 5.1a) is

$$E = 2H_1 + J_{11}. \tag{5.10}$$

Since the two electrons have opposite spins, $K$ does not appear. In general, Eq. 5.6 for a closed shell molecule without unpaired spins is reduced to the form

$$E = 2\sum_{i=1}^{n} H_i + \sum_{i,j=1}^{n} (2J_{ij} - K_{ij})$$

where $1 \sim n$ designates occupied molecular orbitals. The second sum includes $J_{ii} \neq 0$ only once but $K_{ii} = 0$.

When an electron is removed from the $i$ level in a molecule with the total energy $E_0$ (Fig. 5.1b), the resulting energy is obtained by subtracting the one-electron energy $H_i$ and the Coulomb repulsions between the $i$ level and the other occupied levels $k$

$$E = E_0 - \left[H_i + \sum_k (J_{ik} - K_{ik})\right]. \tag{5.11}$$

Accordingly,

$$\varepsilon_i = H_i + \sum_k (J_{ik} - K_{ik}) \tag{5.12}$$

is regarded as the one-electron energy of the $i$ level. In photoelectron spectroscopy, an electron is removed by irradiating ultraviolet light or x-ray, and $\varepsilon_i$ is directly observed as the binding energy.

Similarly, when an extra electron is added to the $j$ level (Fig. 5.1c), the energy is

$$E = E_0 + \left[ H_j + \sum_k (J_{jk} - K_{jk}) \right] = E_0 + \varepsilon_j. \tag{5.13}$$

Then, $\varepsilon_j$ is treated as the one-electron energy when an electron is removed or added. This is called **Koopmans' theorem**. By summing $\varepsilon_i$ for all occupied levels, the total energy is

$$E = \sum_i \varepsilon_i - \sum_{i,k} (J_{ik} - K_{ik})$$

where the second term emerges because the Coulomb terms are added twice in the summation of $\varepsilon_i$. By including the average Coulomb repulsions $J_{kl}$ and $K_{kl}$, $\varepsilon_i$ is regarded as the one-electron energy, and the discussion in Chaps. 1–3 is unchanged.

When an electron is excited from $i$ to $j$ (Fig. 5.1d), the energy difference is given by $\varepsilon_j - \varepsilon_i$ with correcting the Coulomb repulsion between $i$ and $j$.

$$\begin{array}{ll} \text{Singlet} & \Delta E = \varepsilon_j - \varepsilon_i - (J_{ij} - K_{ij}) \\ \text{Triplet} & \Delta E = \varepsilon_j - \varepsilon_i - J_{ij} \end{array} \tag{5.14}$$

Note that $\varepsilon_j$ includes the Coulomb repulsion with the $i$ level. This should be removed because one $i$ electron has been removed. For the singlet, the $j$ electron and the $i$ hole have the same spin, so $K$ is included, whereas for triplet, $K$ does not appear. Because of $K > 0$, the triplet state has lower energy than the singlet state by $K$. Accordingly, the energy difference between the singlet and triplet states is $K$.

Finally, suppose two energy levels with energy difference $\Delta\varepsilon$ occupied by two electrons (Fig. 5.1e). For simplicity, we assume the energy of the lower level $H_1$ to be 0. The energies of the states (1), (2), and (3) are

1. $E_1 = J_{11}$
2. $E_2 = \Delta\varepsilon + J_{12} - K_{12}$
3. $E_3 = \Delta\varepsilon + J_{12}$.

The singlet state (3) always has by $K_{12}$ higher energy than the triplet state (2). The comparison of (1) and (2) is

$$J_{11} < \Delta\varepsilon + J_{12} - K_{12}\text{: the singlet state (1) is more stable.}$$

$$J_{11} > \Delta\varepsilon + J_{12} - K_{12}\text{: the triplet state (2) is more stable.}$$

We shall assume $J_{11}$ and $J_{12}$ are approximately the same. When $\Delta\varepsilon$ is larger than $K_{12}$, the singlet state is most stable because each level is independent. This is the case of most chemical bonds, in which the up and downspins make a pair, resulting in a nonmagnetic molecule. When $\Delta\varepsilon$ is smaller than $K_{12}$, the triplet state is most stable. In the extreme case of $\Delta\varepsilon = 0$, the degenerate levels have parallel unpaired spins; this is known as **Hund's rule**. The general tendency that degenerate levels stabilize a triplet state is called **triplet instability**. In a pericyclic reaction following the Hoodward-Hoffman rule, the frontier orbitals do not cross in the [4 + 2] case due to the conservation of the orbital symmetry. The [2 + 2] case does not preserve the orbital symmetry, and the crossing levels give rise to the triplet instability due to the degeneracy. Such a reaction is regarded to proceed through the biradical state.

*Example 5.1* Figure 5.2 shows five $\pi$ states of ethylene. Calculate the energies by assuming: $H_1 = -23.6$ eV, $H_2 = -17.6$ eV, $J_{11} = 13.0$ eV, $J_{12} = 13.1$ eV, and $K_{12} = 4.2$ eV. From these values, obtain the ionization energy, the singlet excitation energy, and the triplet excitation energy [1, 2].

$$2H_1 + J_{11} = -34.2 \text{ eV}$$

| | |
|---|---|
| (a) Ground state | |
| (b) Cation | $H_1 = -23.6$ eV |
| (c) Anion | $2H_1 + H_2 + J_{11} + 2J_{12} - K_{12} = -29.8$ eV |
| (d) Singlet Excitation | $H_1 + H_2 + J_{12} = -28.1$ eV |
| (e) Triplet Excitation | $H_1 + H_2 + J_{12} - K_{12} = -32.3$ eV |
| Ionization energy | (b) − (a) = 10.6 eV |
| Singlet excitation energy | (d) − (a) = 6.1 |
| Triplet excitation energy | (e) − (a) = 1.9 eV |

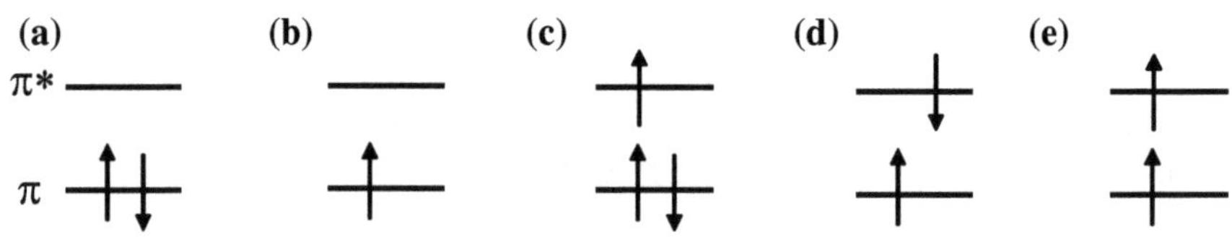


**Fig. 5.2** Electronic states of ethylene. **a** Ground state, **b** cation, **c** anion, **d** singlet, **e** triplet

In the preceding discussion, $\psi_i$ represents a molecular orbital. Here, we consider the LCAO.

$$\psi_k = \sum_i c_i^k \chi_i$$

Applying this relation in Eqs. 5.7–5.9, the respective molecular integrals are

$$H_k = \sum_r c_r^k c_r^k I_{rr} + \sum_{r \neq s} c_r^k c_s^k I_{rs}$$

$$I_{rs} = \int \chi_r^* \left[ -\frac{\hbar^2}{2m} \nabla^2 - \frac{1}{4\pi\varepsilon_0} \sum_n \frac{Z_n e^2}{r_n} \right] \chi_s \mathrm{d}\tau$$

$$J_{kl} = \sum_r \sum_s \sum_t \sum_u c_r^k c_s^k c_t^l c_u^l (rs|tu)$$

$$K_{kl} = \sum_r \sum_s \sum_t \sum_u c_r^k c_t^k c_s^l c_u^l (rt|su)$$

$$(rs|tu) = \int \chi_r(1)\chi_s(1) \left( \frac{1}{4\pi\varepsilon_0} \frac{e^2}{r_{12}} \right) \chi_t(2)\chi_u(2) \mathrm{d}\tau$$

where Eq. 5.6 is unchanged. In order to calculate $H_k$, $J_{kl}$, and $K_{kl}$, we need coefficients of the molecular orbitals, $c_r^i$. However, $c_r^i$ is obtained after the molecular orbital calculation. For the molecular orbital calculation, the molecular integrals are calculated assuming appropriate $c_r^i$. From Eq. 5.6, $\partial E/\partial c_r^i = 0$ affords the matrix elements corresponding to $\alpha$ and $\beta$. The resulting secular equation is solved to give the energy levels and $c_r^i$. The obtained $c_r^i$ is used to recalculate the molecular integrals, and the secular equation is solved again. Such calculation is repeated until $c_r^i$ no more changes. This type of itinerant calculation is called **Hartree–Fock approximation** or **self-consistent-field (SCF) method** in molecular orbital calculation. The ordinary molecular orbital calculation contains such repeating calculation (iteration).

In this calculation, the Coulomb interaction has been evaluated by assuming the distribution of the other electron unchanged depending on the motion of the focused electron. When calculating the motion of a focused electron, the position of other electrons should be influenced by the position of the focused electron due to the repulsive force between them, but the distribution of other electrons is replaced by the average values. In this respect, the Hartree–Fock method and the ordinary molecular orbital calculation are a kind of the mean-field approximation.

Since the above **atomic integrals** $I_{rs}$ and $(rs|tu)$ do not involve $c_r^i$, these integrals are calculated only once before the itinerant calculation. Although the Coulomb integral designates the repulsion between two electrons, the atomic integrals $(rs|tu)$ are related to orbitals of at most four different atoms. Since the calculation is tedious, the three- and four-centered integrals are sometimes neglected in such calculations as the complete neglect of differential overlap (CNDO) method. The atomic integrals are much easily calculated by assuming the Gaussian-type atomic orbital, $e^{-ar^2}$, instead of the Slater-type orbital, $e^{-\zeta r}$. For example, a Slater-type orbital is represented by three Gaussian-type orbitals in STO-3G.

The Coulomb integral (Eq. 5.8) is designated by an integral of the electron density $\rho(r)$, but the exchange integral is a little more complicated. In the **density functional theory**, the exchange integral is also represented by a functional of $\rho(r)$. In the $X\alpha$ method, the exchange integral of a molecule is given by $\int \rho(r)^{4/3} dr$. This approximation significantly reduces the calculation time, and enables the calculation of large molecules.

## 5.2 Electron Correlation in the Molecular Orbital Theory

Here, electron correlation is examined in the framework of the molecular orbital theory [3]. For simplicity, we consider the molecular orbital of a hydrogen molecule, where two electrons 1 and 2 enter the bonding orbital, $\chi_A + \chi_B$. The wave function of the whole molecule is

$$\Psi = (\chi_A(1) + \chi_B(1))(\chi_A(2) + \chi_B(2))(\alpha(1)\beta(2) - \beta(1)\alpha(2)). \quad (5.15)$$

Since the spin part $\alpha(1)\beta(2) - \beta(1)\alpha(2)$ is singlet, this wave function is asymmetric, and satisfies the form of the Slater determinant. Except for the spin part, the orbital part is expanded.

$$\chi_A(1)\chi_A(2) + \chi_A(1)\chi_B(2) + \chi_B(1)\chi_A(2) + \chi_B(1)\chi_B(2) \quad (5.16)$$

These terms represent the four states in Fig. 5.3. In the first term $\chi_A(1)\chi_A(2)$, the A atom has two electrons. This state is designated as $H^-H^+$, which represents a doubly occupied ionic state. The second and third terms represent the covalent bond. The final term corresponds to $H^+H^-$, where the atom B is doubly occupied. Since the four terms have the same coefficient, the molecular orbital contains 50 % of the doubly occupied states. This is obviously overestimation; owing to the mean-field approximation, the electron correlation is not considered. On account of the average electron distribution, an electron stays equivalently on A and B, but the actual electrons move keeping away from each other.

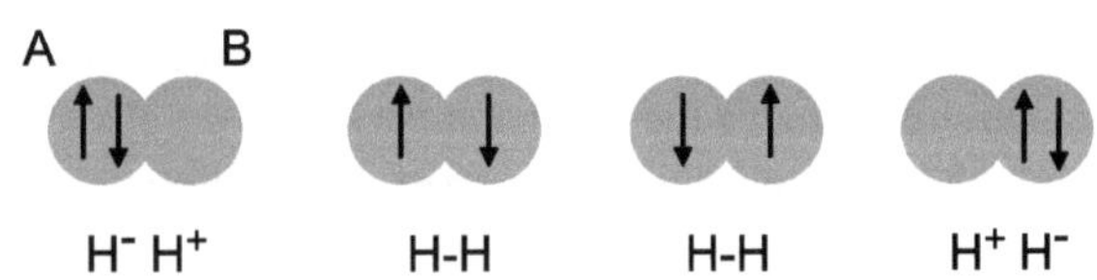


**Fig. 5.3** Ionic and covalent contributions to the molecular orbital of a hydrogen molecule

In order to eliminate the contribution of double occupancy, the first and the fourth terms are dropped in Eq. 5.16.

$$\Psi = \chi_A(1)\chi_B(2) + \chi_B(1)\chi_A(2) \tag{5.17}$$

This is the **valence-bond method**, which has been introduced by Heitler and London to explain the covalent bond of a hydrogen molecule before the molecular orbital theory. Applying this wave function in Eq. 5.5, and using Eq. 5.1, the energy is obtained,

$$E_+ = 2\alpha + \frac{J+K}{1+S^2} \tag{5.18}$$

where

$$J = \iint \chi_A(1)\chi_B(2)\frac{1}{4\pi\varepsilon_0}\left[\frac{1}{r_{12}} + \frac{1}{r_{AB}} - \frac{1}{r_{A2}} - \frac{1}{r_{B1}}\right]\chi_A(1)\chi_B(2)d\tau_1 d\tau_2 \tag{5.19}$$

$$K = \iint \chi_A(1)\chi_B(2)\frac{1}{4\pi\varepsilon_0}\left[\frac{1}{r_{12}} + \frac{1}{r_{AB}} - \frac{1}{r_{A2}} - \frac{1}{r_{B1}}\right]\chi_A(2)\chi_B(1)d\tau_1 d\tau_2. \tag{5.20}$$

$S$ is the overlap integral between $\chi_A$ and $\chi_B$. From $J < 0$ and $K < 0$, $E_+$ is lower than the atomic energy before the bond formation, $2\alpha$, and the bond formation of a hydrogen molecule is explained. These $J$ and $K$ are also called Coulomb and exchange integrals, but these are different from the Coulomb and exchange integrals in the molecular orbital theory (Eqs. 5.8 and 5.9). The above equations include not only the electron repulsion $r_{12}$ but also the nuclear attraction $r_{A2}$ and $r_{B1}$, so the above $J$ and $K$ are negative. In the molecular orbital theory, these contributions are included in $H_k$ (Eq. 5.7). The origin of chemical bonding is this nuclear attraction. $J$ and $K$ in the molecular orbital theory involve only $r_{12}$, and are positive repulsion.

The valence bond theory and the associated concept of hybrid orbitals are an excellent method to predict molecular structures. However, the valence bond theory is not convenient for discussing conjugated $\pi$-electron systems and $d$-electron systems. The energy band theory is an extension of the molecular orbital theory. Since the molecular orbital theory overestimates the contribution of doubly occupied states, the valence bond theory affords more precise bonding energy than the molecular orbital theory even in the hydrogen molecule. The valence bond theory takes account of electron correlation more exactly, whereas the molecular orbital theory entirely neglects the influence of electron correlation. Therefore, it is necessary to improve the molecular orbital theory by taking electron correlation into account.

In order to involve the effect of electron correlation in the molecular orbital theory, we consider **configuration interaction**. A hydrogen molecule has a ground state in which the bonding orbital is occupied by two electrons. To this ground state, an excited state which has two electrons in the antibonding orbital is mixed by multiplying $c$ (Fig. 5.4).

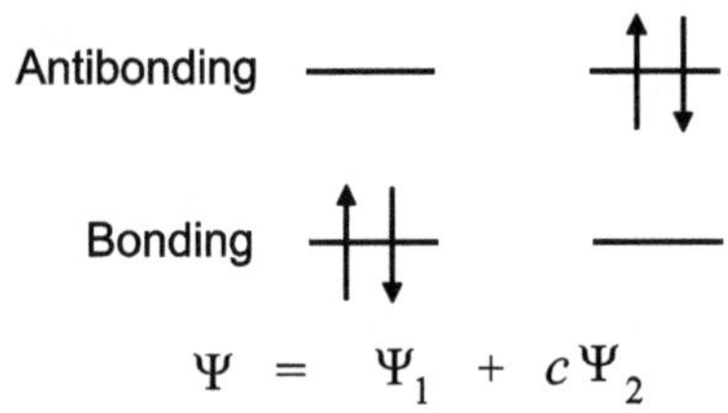


**Fig. 5.4** Configuration interaction

$$
\begin{aligned}
\Psi &= \Psi_1 + c\Psi_2 \\
&= (\chi_A(1) + \chi_B(1))(\chi_A(2) + \chi_B(2)) + c(\chi_A(1) - \chi_B(1))(\chi_A(2) - \chi_B(2)) \\
&= (1 + c)(\chi_A(1)\chi_A(2) + \chi_B(1)\chi_B(2)) + (1 - c)(\chi_A(1)\chi_B(2) + \chi_B(1)\chi_A(2))
\end{aligned} \tag{5.21}
$$

The first term represents the doubly occupied state, and the second term comes from the covalent bond. When $c$ is put $-1$ to make the first term zero, the result is equivalent to the valence bond theory. In contrast, the limit at $c = 0$ is the molecular orbital theory. Actually, $c$ is determined so as to minimize the energy according to the variational method. Then, the state with the intermediate $c$ within $-1 \leq c \leq 0$ represents a state with appropriate mixing of the double occupancy, and is better than the valance bond and molecular orbital theories. Such a configuration interaction (CI) calculation is performed in the program of molecular orbital calculation. After the ordinary molecular orbital calculation, the CI calculation is carried out by considering the excited states. Since the usual molecular orbital calculation optimizes only the occupied states, the CI calculation is indispensable when we discuss unoccupied states like LUMO and excited states.

## 5.3 Creation Annihilation Operators

In order to describe correlated systems and superconductivity, we shall introduce creation and annihilation operators, and discuss how to describe the Hamiltonian by using them [4]. Owing to the Pauli exclusion principle, an electronic state is either occupied or unoccupied. Therefore, we can describe an electron configuration as $|11100100\rangle$, where 1 and 0 represent the occupied or unoccupied states. For instance, suppose $|11110000\rangle$ is the ground state, and the energy levels are successively aligned from the bottom. Then, $|11101000\rangle$ represents the first excited state.

In order to represent such situation more elegantly, consider operators $c_i$ and $c_i^+$, which satisfy the following relations.

$$\left\{c_i, c_j^+\right\} = c_i c_j^+ + c_j^+ c_i = \delta_{ij} \tag{5.22}$$

$$\left\{c_i, c_j\right\} = c_i c_j + c_j c_i = 0 \tag{5.23}$$

$$\left\{c_i^+, c_j^+\right\} = c_i^+ c_j^+ + c_j^+ c_i^+ = 0 \tag{5.24}$$

Here { } is similar to the commutator described in Sect. 1.6, but the subtraction is replaced by the addition. This is called **anticommutator**. Only the anticommutator of the conjugated $c_i$ and $c_i^+$ concerning to the same state gives one, while all other anticommutators afford zero. If we use a commutator instead of an anticommutator, we can similarly define creation and annihilation operators of Bose particles.

Next, we define $N_i = c_i^+ c_i$ and the eigenvalue $N_i\psi = n\psi$. When we operate $N_i$ to $c_i\psi$, the result is

$$N_i(c_i\psi) = c_i^+ c_i c_i \psi = \left(1 - c_i c_i^+\right) c_i \psi = c_i\psi - c_i N_i \psi = (1-n)(c_i\psi).$$

In the second equality, we have used Eq. 5.22. Since Eq. 5.23 gives $c_i c_i \psi = 0$, the second form means the above equation should be zero. If $n = 1$, $c_i\psi$ is an eigenfunction of $N_i$ with the eigenvalue zero, or if $n \neq 1$, $c_i\psi = 0$ is required. Namely, if the original $i$ state $\psi$ is occupied and $n = 1$, $c_i$ changes the state to the $n = 0$ state. Similarly, from

$$N_i(c_i^+ \psi) = c_i^+ c_i c_i^+ \psi = c_i^+ (1 - c_i^+ c_i)\psi = c_i^+ \psi - c_i^+ N_i \psi = (1-n)(c_i^+ \psi)$$

for $n = 0$, $c_i^+ \psi$ is an eigenfunction of $N_i$ with the eigenvalue, one. For other $n$, we have to impose $c_i^+\psi = 0$. When the original $\psi$ state is unoccupied and $n = 0$, $c_i^+$ changes this state to the $n = 1$ state. These derivations demonstrate that the eigenvalues of $N_i\psi = n\psi$ are either zero or one, and designates the occupancy number of the $i$ state. $c_i^+$ changes the eigenvalue $n$ from zero to one, and $c_i$ changes $n$ from one to zero. These relations are written as $N_i|1\rangle = |1\rangle$, $N_i|0\rangle = 0$, $c_i^+|0\rangle = |1\rangle$, and $c_i|1\rangle = |0\rangle$. $N_i = c_j^+ c_i$ is called **number operator**, $c_i^+$ is the **creation operator**, and $c_i$ is the **annihilation operator**.

When all levels are unoccupied, such a state is written as $|0\rangle$ and called **vacuum state**. By using $|0\rangle$, the state in which the 1–4 levels are occupied is designated as

$$|11110000\rangle \rightarrow c_1^+ c_2^+ c_3^+ c_4^+ |0\rangle.$$

Since $c_i^+ c_j^+ = -c_j^+ c_i^+$ is derived from Eq. 4.52, the exchange of $i$ and $j$ changes the sign of this representation. For example, $c_1^+ c_3^+ c_2^+ c_4^+|0\rangle = -c_1^+ c_2^+ c_3^+ c_4^+|0\rangle$. From $c_i^+ c_i^+ = 0$, if $i$ and $j$ are the same, this representation gives zero. Then, the above formula satisfies the properties of the Slater determinant. In a metal at $T = 0$ K, the levels up to the Fermi wave vector $k_F$ are occupied. Such a state is represented as

$$|\mathrm{FS}\rangle = \prod_{k<k_F} c_k^+ |0\rangle.$$

If $k$ is continuous, this expression contains a product of an infinite number of $c_k^+$, but this is better than writing the Slater determinant with an infinite order. In the above formula, the spins are not explicitly written. If $|11110000\rangle$ means a molecular orbital $|1\uparrow, 1\downarrow, 2\uparrow, 2\downarrow, 0000\rangle$, this is written as $c_{1\uparrow}^+ c_{1\downarrow}^+ c_{2\uparrow}^+ c_{2\downarrow}^+|0\rangle$. The Fermi sea in a metal is designated as $|\mathrm{FS}\rangle = \prod_{k<k_\mathrm{F}} c_{k\uparrow}^+ c_{k\downarrow}^+|0\rangle$.

Using the above representation, we can write $\psi_i = |i\rangle = c_i^+|0\rangle$, and $\psi_1^* = \langle i| = \langle 0| c_i$. From this, an operator F is represented like $\langle j|\mathrm{F}|i\rangle = \langle 0|c_j \mathrm{F} c_i^+|0\rangle$. In order to make this representation appropriately give the matrix element $f_{ji} = \langle j|\mathrm{F}|i\rangle$, we should rewrite $\mathrm{F} \to f_{ji} c_j^+ c_i$. This substitution gives $\langle 0|c_j\ c_j^+ f_{ji} c_i\ c_i^+|0\rangle$. This is non zero only when $i = j$, because $\langle 0|c_i c_i^+|0\rangle = 1$ and $\langle 0|c_j c_j^+|0\rangle = 1$ make $f_{ji}$ survive, but otherwise this equation becomes zero. In particular, when $i$ is an eigenfunction of $f$, this leads to $\mathrm{F} \to f_{ii} c_i^+ c_i = f_{ii} N_i$. The summation over all occupied levels affords

$$\sum_i f_i N_i$$

which is a sum of the eigenvalues $f_i$ multiplied by the occupancy numbers $N_\mathrm{i}$.

In addition to the above substitution of one-electron operators, two-electron operators like Eqs. 5.8 and 5.9 are related to four states like $\langle kl|\mathrm{F}|ij\rangle$. This is similarly substituted as $\mathrm{F} \to f_{klij} c_k^+ c_l^+ c_j c_i$. For example, the Hamiltonian of the Coulomb energy part is written as

$$\sum_{k,l} J_{kl} c_k^+ c_l^+ c_l c_k.$$

## 5.4 Electron Correlation in Solids

Consider electron correlation of free electrons in a homogeneous potential. Assume an electron is located at $r = 0$, and the possibility to find another electron at $r$ is represented by the **pair distribution function** $g(r)$. When the influence of other electrons is entirely absent, $g(r)$ is constant everywhere, so we suppose $g(r) = 1$ (Fig. 5.5). In the Hartree-Hock approximation (Sect. 5.1), an electron with the same spin cannot come to the $r = 0$ position due to the Pauli exclusion principle, whereas an electron with the opposite spin is not excluded. Then, the possibility should be $g(0) = 0.5$ at $r = 0$ (Fig. 5.5); this is called **exchange hole**. In more improved approximation, it is more difficult for an electron to come to the $r = 0$ position due to the Coulomb repulsion, and $g(0)$ is less than 0.5 (Fig. 5.5), which is called **correlation hole** [3]. The molecular orbital and energy band theories do not consider the electron correlation entirely, and fundamental correction is sometimes necessary; an example is magnetic materials discussed in Chap. 4.

Energy levels of magnetic materials are partially filled by unpaired electrons. Such materials have a metal-like energy band. When each site has one electron, the

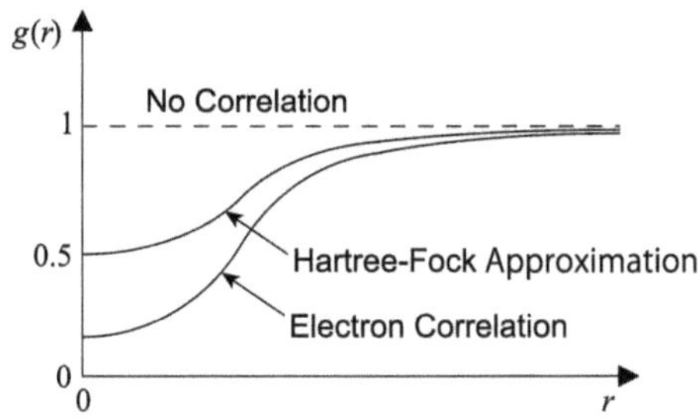


**Fig. 5.5** Pair distribution function

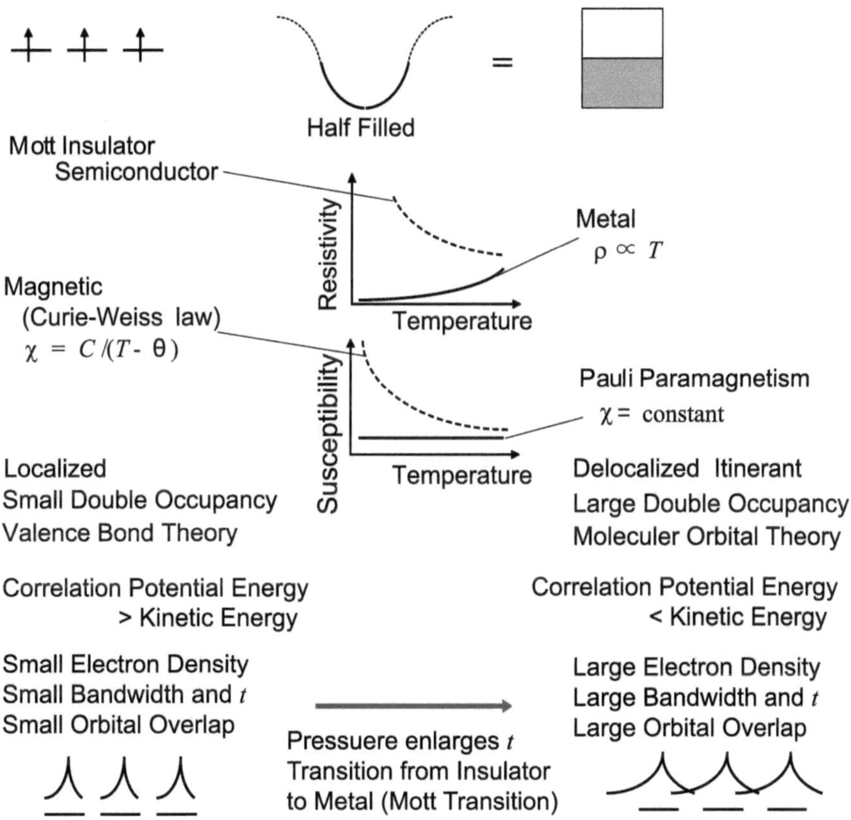


**Fig. 5.6** Comparison of the Mott insulator and a metal

energy band is half filled (Fig. 5.6). In a metal, the resistance decreases with lowering the temperature, while the resistance of a magnetic material increases because magnetic materials are insulators. The magnetic susceptibility of a metal is small and constant due to the Pauli paramagnetism, whereas the susceptibility of a magnetic material increases following the Curie–Weiss law. In a metal, electrons are delocalized, but electrons in a magnetic material are localized. A metallic conductance is realized when the orbital overlap with the adjacent orbitals is large. This is achieved when the bandwidth is large and the electron density is large. In contrast, the orbital overlap in a magnetic material is small, and the bandwidth is small. In the discussion in Sect. 5.2, the molecular orbital and energy band theories are appropriate to describe metals, whereas the valence bond theory is suitable for magnetic materials. In the former, the contribution of double occupancy is large, while it is small in the latter model. When the electron kinetic energy is larger than the potential energy coming from the electron correlation, the electron becomes metal-like, but when the potential energy is large, the electron does not move and forms a magnetic material. If partially filled electrons have local magnetic moments due to the electron correlation, the resulting paramagnetic or antiferromagnetic insulator is called **Mott insulator** [5]. When pressure is applied to a Mott insulator close to the boarder to a metal, orbital overlap increases and a transition from the Mott insulator to a metal takes place. Such a metal–insulator transition is called **Mott transition**.

We shall attempt more precise description of the Mott insulator. It is a way to add the influence of correlation to the energy band. Starting from the wave function of the Fermi sea $|\mathrm{FS}\rangle$, consider

$$\Psi_{\mathrm{G}} = \prod_i (1 - \eta n_{i\uparrow} n_{i\downarrow})|\mathrm{FS}\rangle \sim \exp\left(-\eta \sum_i n_{i\uparrow} n_{i\downarrow}\right)|\mathrm{FS}\rangle \qquad (5.25)$$

where $n_{\mathrm{i}\uparrow}$ is one when the $i$ site has an upspin electron, and zero when it is unoccupied. $n_{\mathrm{i}\downarrow}$ is a similar operator for the downspin electron. Therefore, the product $n_{\mathrm{i}\uparrow} n_{\mathrm{i}\downarrow}$ is one only when the $i$ site is doubly occupied. This wave function $\Psi_{\mathrm{G}}$ is called **Gutzwiller function**, in which doubly occupied states are subtracted from the Fermi sea. In Eq. 5.25, $\eta = 1$ means the complete subtraction of the doubly occupied states, whereas finite $0 < \eta < 1$ partly maintains the doubly occupied states. Accordingly, the Gutzwiller function is an extension of the configuration interaction of molecules to solids.

As another attempt, suppose a model in which the doubly occupied site has by $U$ higher energy (Fig. 5.7) than the singly occupied site. Since $U$ designates the Coulomb repulsion between the electrons on the same site, $U$ is called **on-site Coulomb repulsion**. Consider the Hamiltonian with the additional $U$ term.

$$H = \sum_{i \neq j} t_{ij} c_{i\sigma}^{+} c_{j\sigma} + U \sum_i n_{i\uparrow} n_{i\downarrow} \qquad (5.26)$$

**Fig. 5.7** On-site Coulomb interaction

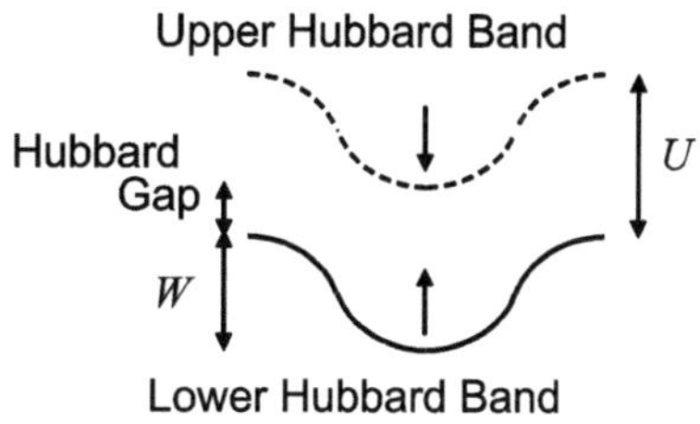


**Fig. 5.8** Hubbard gap

This is called **Hubbard model**. The first term means the electron transfer from $i$ to $j$ occurs with the probability of $t$. Accordingly, this affords the ordinary energy band. For instance, in the one-dimensional case, this term is reduced to $2t\cos ka$. Here, $\sigma$ implies $\uparrow$ or $\downarrow$. The second term adds $U$ to the doubly occupied site. Note $n_{i\uparrow}n_{i\downarrow}$ is nonzero only when the $i$ site is doubly occupied. It is not easy to solve the Hubbard model exactly. However, the first electron entering a site has zero second term, but the second electron has additional $U$. If we may roughly assume these two electrons make the energy band independent, the first electron makes an energy band $E_\uparrow = 2t\cos ka$, and the second electron makes $E_\downarrow = 2t\cos ka + U$. Then, an energy band splits to two as shown in Fig. 5.8. These bands are called the **lower** and **upper Hubbard bands**. When $U$ is larger than the bandwidth $W$, the **Hubbard gap**, $U - W$, opens between the lower and upper Hubbard bands. In the half-filled case, the lower Hubbard band is just entirely filled, and the upper Hubbard band is empty. Accordingly, the half-filled Hubbard band makes the Mott–Hubbard insulator. We will investigate a more elaborated treatment based on the mean-field theory in Sect. 5.7.2.

From Fig. 5.8, the Hubbard model affords an insulating state only in the half-filled band. As shown in Fig. 5.9, an entirely filled band gives a nonmagnetic insulator, whereas the half-filled Mott insulator affords a paramagnetic insulator. All other states that are below and above the half filling lead to a metal state. In the above discussion, the "first" and "second" electrons have opposite spins, but the spin directions are not specified. Then, the Mott insulator is paramagnetic in general, though usually an antiferromagnetic state appears at low temperatures.

**Fig. 5.9** Mott insulator and various band fillings

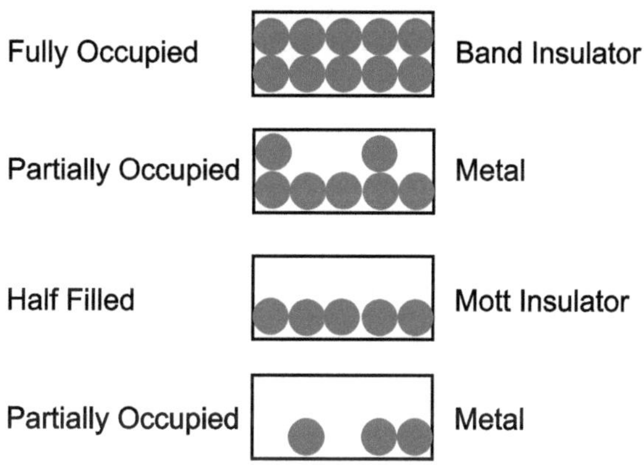

Electron correlation in various models in molecules and solids is summarized as follows.

| | | |
|---|---|---|
| Molecules | Molecular orbital theory<br>Large double occupancy<br>No electron correlation | Valence-bond theory<br>Small double occupancy<br>Strong electron correlation |
| Solids | Energy band<br>Metals | Hubbard model<br>Magnetic materials |

## 5.5 Origin of Magnetic Interaction

The origin of magnetic interaction is derived from the Hubbard model.

*Example 5.2* Solve the Hubbard model when two electrons are put in two sites (Fig. 5.10).

There are six electron configurations (Fig. 5.10). The states 2 and 4 have by $U$ higher energy than the others. This energy enters the diagonal term. If two states are connected by transferring one electron, the nondiagonal term has the transfer $t$. The triplet states 5 and 6 do not mix with other singlet states. Make the matrix of the singlet states 1–4, and diagonalize it.

Eigenvalues of the singlet states 1–4 are obtained from the following determinant, which is constructed according to the above rule.

$$\begin{vmatrix} -E & t & 0 & t \\ t & U-E & t & 0 \\ 0 & t & -E & t \\ t & 0 & t & U-E \end{vmatrix} = 0.$$

Since the determinant has the form $\begin{vmatrix} A & B \\ B & A \end{vmatrix} = 0$, we can use $|A+B||A-B| = 0$, and this is reduced to

$$\begin{vmatrix} -E & 2t \\ 2t & U-E \end{vmatrix} = 0 \quad \text{and} \quad \begin{vmatrix} -E & 0 \\ 0 & U-E \end{vmatrix} = 0.$$

This affords the solutions.

$$E = \frac{U \pm \sqrt{U^2 + 16t^2}}{2} \approx U + \frac{4t^2}{U}, -\frac{4t^2}{U} \quad \text{and } E = 0, U \tag{5.27}$$

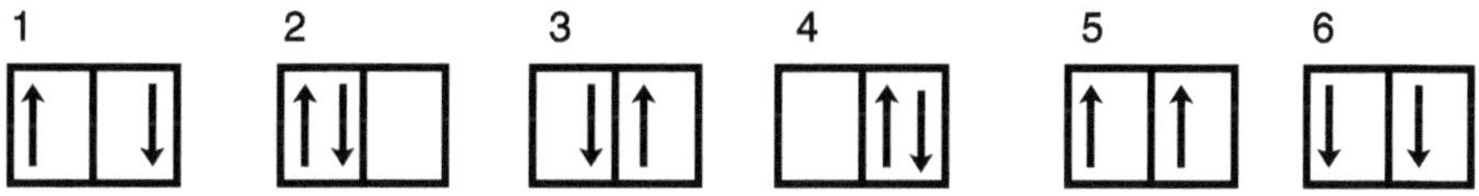


**Fig. 5.10** Electron configurations in the 2-site 2-electron Hubbard model

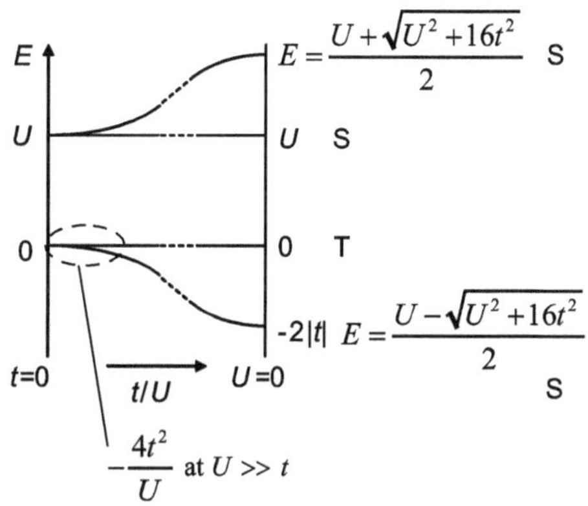


**Fig. 5.11** Solution of the two-site two-electron Hubbard model

Here we approximate $U \gg t$. The results are depicted in Fig. 5.11. The $E = 0$ solution is combined with the states 5 and 6 to make the triplet $T$. Other states represent the singlet states $S$. The left end corresponding to $t = 0$ has only two states: the singly occupied state with $E = 0$ and the doubly occupied state with $E = U$. For $t \neq 0$, the ground state is a singlet state constructed by the states 1 + 3 in Fig. 5.10. This corresponds to the ground state of the valence bond state (Eq. 5.17), namely the Heitler-London state. The singlet state has by $-4t^2/U$ lower energy than the triplet state. Accordingly, there is an antiferromagnetic interaction $J = -4t^2/U$ between these two sites owing to the transfer $t$ [6].

When one electron is put in the two sites (Fig. 5.12), the ground state energy is $-|t|$. This is equivalent to the situation where one electron enters the bonding orbital of a hydrogen molecule. If two electrons are put in these two sites, the energy is given by Eq. 5.27. The difference corresponds to $U$ of the dimer.

$$U_{\text{eff}} = \frac{U - \sqrt{U^2 + 16t^2}}{2} - (-2|t|) \rightarrow 0 + 2|t| = 2|t|(U \gg t) \tag{5.28}$$

In 2:1 composition organic conductors with a dimer structure, one electron enters each dimer. For the charge transport, the dimer has to have two electrons at a moment, but such a state has by $2|t|$ higher energy than the ground state. Here, $t$ is the intradimer transfer integral. Owing to this effective $U_{\text{eff}}$, such a conductor becomes a Mott insulator.

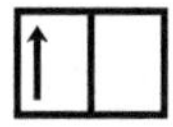

Fig. 5.12 Two-site one-electron model

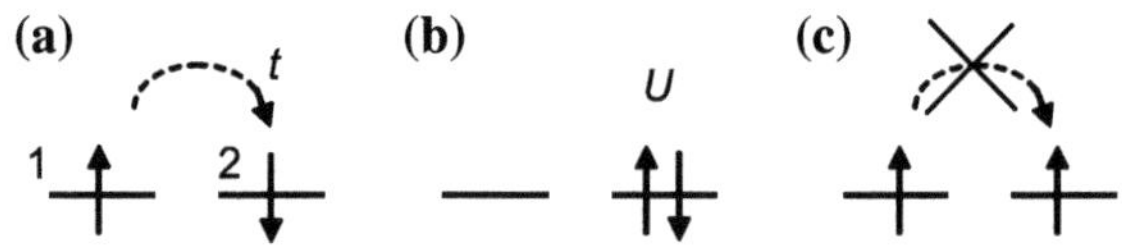


Fig. 5.13 Kinetic exchange

The above 2-site Hubbard model demonstrates that the singlet state is more stable than the triplet state by $J = -4t^2/U$. This is the origin of the antiferromagnetic interaction when there is the transfer $t$. This is alternatively explained by the perturbation theory applied to the 2-site 2-electron model (Fig. 5.13). If one electron is moved from site 1 to 2 (Fig. 5.13b), the resulting state is doubly occupied, having by $U$ higher energy. By restoring the electron to the site 1, the (b) state is mixed by the second-order perturbation theory, to lower the ground state energy by $-4t^2/U$. If the spins are parallel (Fig. 5.13c), such a process does not work due to the Pauli exclusion principle, and the energy does not change. Consequently, the singlet state has by $J$ lower energy than the triplet state. Such antiferromagnetic interaction originating from $t$ is called **kinetic exchange** [7, 8].

There is another origin of magnetic interaction. The exchange integral (Eq. 5.9)

$$K_{12} = \int \psi_1^*(1)\psi_2^*(2)\frac{e^2}{r_{12}}\psi_1(2)\psi_2(1)\mathrm{d}\tau > 0 \tag{5.29}$$

implies that the triplet state is more stable than the singlet state by $K$. This ferromagnetic interaction is called **potential exchange**. If two electrons have antiparallel spins, the antisymmetrical condition of Eq. 5.4 is already satisfied, and the two electrons may come to the same position, and enter the same orbital. Such antiferromagnetic electrons freely move to gain the kinetic energy. In contrast, electrons with parallel spins have to enter different orbitals. Since different orbitals are orthogonal, these electrons run away from each other to reduce the Coulomb repulsion and gain the potential energy. These are the meaning of the kinetic and potential exchange.

Magnetic interaction is a sum of the kinetic and potential exchange.

$$J = K - \frac{4t^2}{U} \tag{5.30}$$

Ordinary magnetic materials have the orbital overlaps between the adjacent sites, resulting in $t \neq 0$. Then, the second term (kinetic exchange) is larger than the first term (potential exchange), so $J$ is antiferromagnetic ($J < 0$). As a result, an overwhelmingly large number of magnetic materials show antiferromagnetism. In Fig. 5.1, we have shown that the energy difference $\Delta\varepsilon$ stabilizes the singlet state.

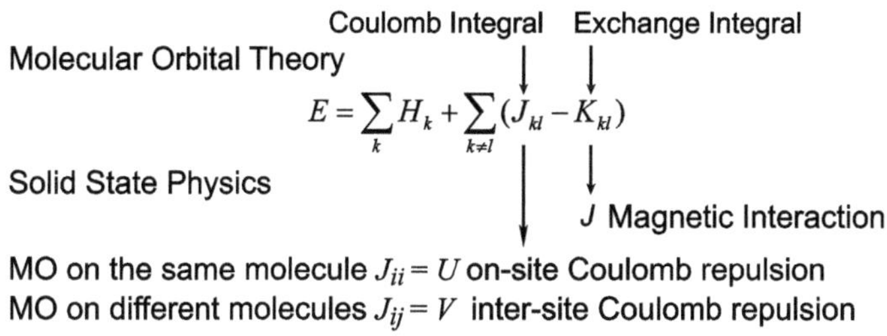


**Fig. 5.14** Symbols of the Coulomb and magnetic interactions in the molecular orbital theory and solid state physics

Here, we see $t$ instead of $\Delta\varepsilon$ stabilizes the singlet (antiferromagnetic) state. However, since $t$ makes the bonding orbital with lowered energy, these two models mean the same situation. The exception occurs when the orbitals of the adjacent sites are orthogonal. This leads to $t = 0$, and the kinetic exchange does not work. The remaining potential exchange results in ferromagnetic interaction ($J > 0$). This corresponds to the triplet instability in the exactly degenerate case in Fig. 5.1. Accordingly, ferromagnetism is achieved when the orbitals are orthogonal; this is known as **Goodenough-Kanamori rule** [9, 10].

Magnetic interaction, written as $J$ in physics (and in Chap. 4), is equivalent to the exchange interaction $K$ appearing in the molecular orbital theory (Fig. 5.14). The on-site Coulomb interaction $U$ in the Hubbard model corresponds to $J_{ii}$ in the molecular orbital theory, which is the Coulomb repulsion on the same orbital. Note $J$ designates different quantities in physics and quantum chemistry. In Sect. 5.8, the inter-site Coulomb repulsion $V$ will appear, but this should be written as $J_{ij}$ in the molecular orbital theory. This is the Coulomb repulsion between different orbitals (usually on the adjacent sites).

We can extend the above discussion for $S = 1/2$ to large $S$ [11]. For example, a $d^5$ cation $Fe^{3+}$ has five unpaired spins on the five d-orbitals (Fig. 4.2), and $S = 5/2$. Since the exchange interaction (Hund's rule) between the orbitals on the same atom is much larger than the interatomic $J$, the five electrons have always parallel spins, and $S = 5/2$ is established. In a πd-system, we have to consider the magnetic interaction between the $S = 5/2$ spin on $Fe^{3+}$ and then $S = 1/2$ spin on the organic molecule. Suppose the respective spin $s_i$ of the $S = 5/2$ spin has a magnetic interaction $J_i = -2t_i^2/U$ with the organic spin $S_p$. Then, the total spin Hamiltonian is reduced to

$$H = -\sum_{i=1}^{5} 2J_i s_i S_p = -2\sum_{i=1}^{5} \frac{2t_i^2}{U} s_i S_p.$$

Since $s_i$ on the same atom is always parallel, these are replaced by $S_q = 5/2$. In order to make the resulting form $H = -2JS_qS_p$ equivalent to the above relation, we have to apply

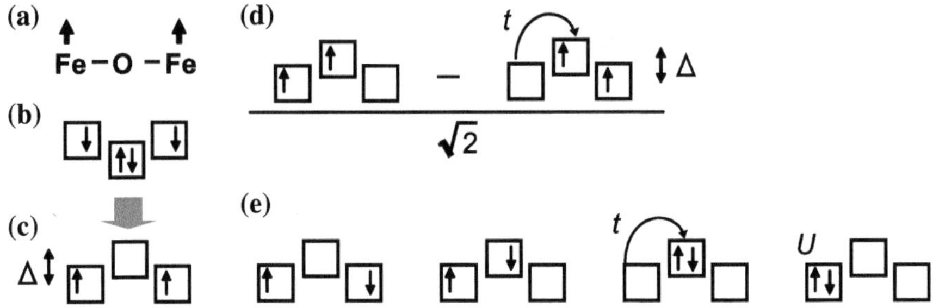


**Fig. 5.15** 3-site 2-electron Hubbard model for superexchange interaction

$$J = \frac{1}{5}\sum_{i=1}^{5} J_i = \frac{1}{5}\sum_{i=1}^{5}\frac{2t_i^2}{U}. \tag{5.31}$$

Then $J$ is represented by the average of the five $J_i$. For the interaction between the d-orbitals, there are 5 × 5 = 25 combinations, and the average

$$J = \frac{1}{25}\sum_{i=1}^{5}\sum_{j=1}^{5} J_{ij} = \frac{1}{25}\sum_{i=1}^{5}\sum_{j=1}^{5}\frac{2t_{ij}^2}{U} \tag{5.32}$$

gives the magnetic interaction between the $S = 5/2$ spins.

*Example 5.3* In transition metal compounds, magnetic interaction is usually not due to the direct overlap of metal orbitals, but mediated by the bridging ligand (for instance, oxygen) orbitals (Fig. 5.15a). Such magnetic interaction is called **superexchange interaction**. Calculate $J$ of the superexchange interaction as follows.

The oxygen 2p levels are occupied and located below the metal d-levels (Fig. 5.15b), but instead we suppose hole spins (Fig. 4.15c), and the empty ligand levels are located above the metal levels. Then, we consider a 3-site 2-electron Hubbard model, in which $U$ works on the metal level. The triplet states represented by (c) and (d) afford the matrix

$$\begin{vmatrix} 0 - E & \sqrt{2}t \\ \sqrt{2}t & \Delta - E \end{vmatrix} = 0$$

where 0 and Δ are the energies of (c) and (d). $\sqrt{2}$ appears because the transfer $t$ jumping from (c) to each term of (d) gives $2t/\sqrt{2} = \sqrt{2}t$. There are four singlet states depicted in (e). Note the second state contains two terms similarly to (d). Calculate the ground state energies of the singlet and triplet states, and the difference $J$.

The determinant of the triplet state is solved to

$$-E(\Delta - E) - 2t^2 = 0$$

Since the ground state is just below $E \sim 0$, we may approximate $E \ll \Delta$ and $\Delta - E \rightarrow \Delta$. Then, the ground state energy is approximated to $E \approx -\frac{2t^2}{\Delta}$.

Hamiltonian for the four singlet states (e) is

$$\begin{vmatrix} 0-E & \sqrt{2}t & 0 & 0 \\ \sqrt{2}t & \Delta - E & \sqrt{2}t & \sqrt{2}t \\ 0 & \sqrt{2}t & 2\Delta - E & 0 \\ 0 & \sqrt{2}t & 0 & U-E \end{vmatrix} = 0.$$

Note that $\sqrt{2}$ in the second state comes from the two terms similar to (d). This determinant is reduced to

$$-E\begin{bmatrix} \Delta - E & \sqrt{2}t & \sqrt{2}t \\ \sqrt{2}t & 2\Delta - E & 0 \\ \sqrt{2}t & 0 & U-E \end{bmatrix} - \sqrt{2}t\begin{bmatrix} \sqrt{2}t & \sqrt{2}t & \sqrt{2}t \\ 0 & 2\Delta - E & 0 \\ 0 & 0 & U-E \end{bmatrix} = 0.$$

Because of $E \ll \Delta$ and $U$ in the ground state, $E$ appearing in the matrices are approximated to zero and

$$E = -\frac{2t^2}{\Delta - \frac{2t^2}{U} - \frac{t^2}{\Delta}} \approx -\frac{2t^2}{\Delta}\left(1 + \frac{2t^2}{\Delta U} + \frac{t^2}{\Delta^2}\right).$$

The difference with the triplet state is

$$J = E_{AF} - E_F = -\frac{2t^2}{\Delta}\left(\frac{2t^2}{\Delta U} + \frac{t^2}{\Delta^2}\right) = -2b^2\left(\frac{2}{U} + \frac{1}{\Delta}\right) \tag{5.33}$$

here, we put $b = \frac{t^2}{\Delta} = \frac{t}{\Delta} \times t$. When we take the Fe–O bond as a polar bond in a molecule (Sect. 1.5.1), the coefficient of the molecular orbital on the oxygen is represented by $t/\Delta$. Then $b$ is regarded as the transfer between the “Fe–O molecule” and Fe. If the local spin is located on an independent molecule or ion such as $FeCl_4^-$, the d orbital spreads to the Cl atoms to some extent, and $b$ is the intermolecular transfer between the molecular orbitals. In the case of $U \gg \Delta$ [12], we obtain

$$J = -\frac{2b^2}{\Delta}. \tag{5.34}$$

The denominator is substituted from $U$ to $\Delta$, that is the energy difference between the d-level and the ligand level. Such superexchange is called charge-transfer-type.

By using $b$ instead of $t$, and $\Delta$ instead of $U$, superexchange interaction is represented in a similar form to the ordinary kinetic exchange interaction.

## 5.6 Conducting Metal Oxides

Energy bands of transition-metal oxides are schematically shown in Fig. 5.16. In early transition-metal oxides such as $VO_2$ and $V_2O_3$, the oxygen 2p bands are located below the lower Hubbard band (LH), and make an ordinary Mott–Hubbard insulator with $U < \Delta$. However, in later transition-metal oxides such as high-temperature superconducting cuprates, the oxygen 2p bands are located above the lower Hubbard band because of $U > \Delta$. This case is called **charge-transfer-type insulator**, which corresponds to the above Example 5.3. This difference comes from the comparatively shallow d-levels in earlier transition metals like vanadium, in contrast to the deep d-levels in later transition metals like copper. In Sect. 1.5, we investigated the energy level of atomic orbitals deepens gradually in proportion to $-Z^2$. For the carrier transport, the lowest excited state of the Mott insulator is generated by a process such as $2d^n \rightarrow d^{n-1} + d^{n+1}$, whereas that of the charge-transfer-type insulator is due to $d^n L \rightarrow d^{n+1}\underline{L}$. Here, $\underline{L}$ represents a hole state generated in the oxygen 2p band. Accordingly, excitation from 2p to UH occurs instead of the excitation from LH to UH.

One may suppose metal oxides are insulators owing to the ionic nature, but many transition-metal oxides are not only paramagnetic but also conducting [13]. $VO_2$ and $V_2O_3$ undergo metal–insulator transitions at 323 K and 160 K, respectively, which have been the subjects of intensive study for a long time. Titanium and vanadium construct nonstoichiometric oxides represented by $Ti_nO_{2n-1}$ and $V_nO_{2n-1}$, which are called **Magneli phases**. The metal–insulator transitions of $V_nO_{2n-1}$ are shown in Fig. 5.17 [15]. Pressure phase diagram of $V_2O_3$ doped with Cr and Ti is depicted in Fig. 5.18. Doping of $Cr^{3+}$ and $Ti^{3+}$ in $V^{3+}$ works as chemical pressure. The resulting transition from an insulator to a metal is recognized as a typical Mott transition.

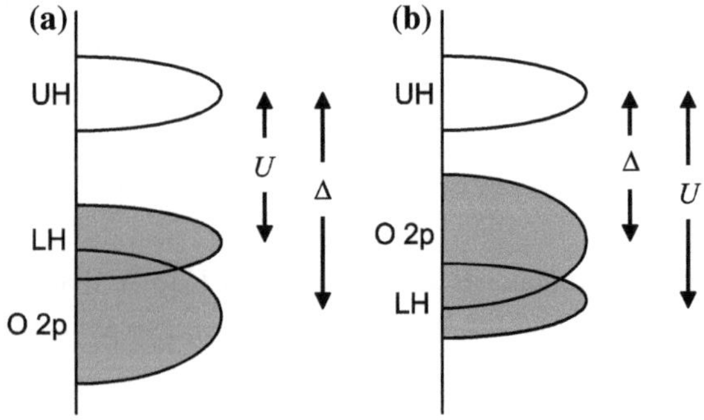


**Fig. 5.16** Energy bands in **a** the Mott–Hubbard insulator, and **b** the charge-transfer-type insulator. *UH* upper Hubbard band, and *LH* lower Hubbard band

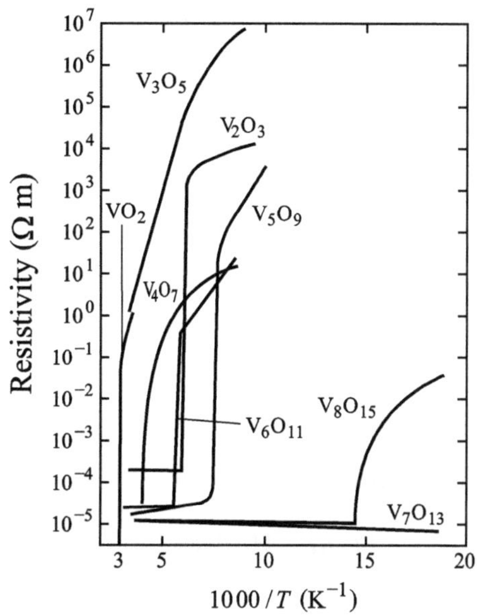


**Fig. 5.17** Metal–insulator transition in $V_nO_{2n-1}$

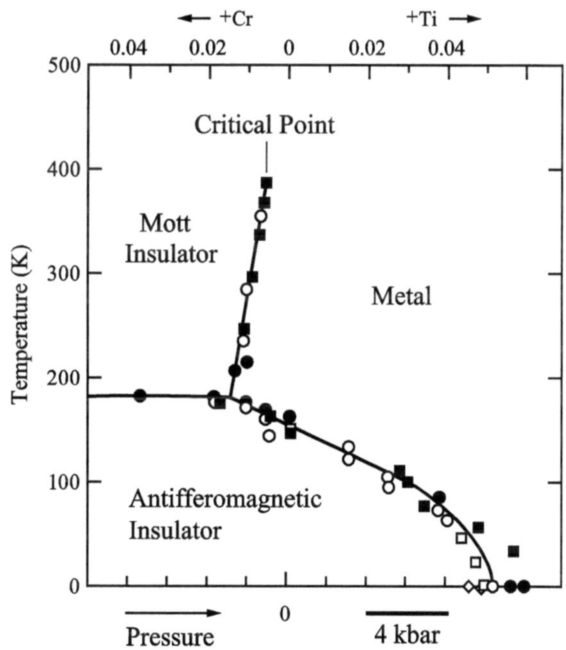


**Fig. 5.18** Pressure phase diagram of $V_2O_3$ doped with Cr and Ti [14]

Some transition-metal oxides exhibit metallic conductivity down to low temperatures [14]. $ReO_3$ shows as high conductivity as $10^5$ S/cm at room temperature, and metallic conductivity down to low temperatures. Tungsten bronzes, $Na_xWO_3$, are compounds with characteristic color and metallic luster, and show metallic conductivity down to low temperatures. $LiTi_2O_4$ and $Bi_{1-x}K_xBiO_3$ are superconductors with the transition temperatures higher than 10 K, and cuprates such as $La_{1-x}Sr_xCuO_4$ and $YBa_2Cu_3O_7$ show superconducting transitions as high as 37 K and 92 K, respectively. Among main group elements, $In_2O_3$ with several % $SnO_2$ dopant is called indium tin oxide (ITO), and widely used as a transparent conductor. For example, ITO thin films on a glass substrate are widely used in liquid crystal displays.

## 5.7 Hubbard Model

### 5.7.1 *Stoner Model*

In Sect. 5.4, we discussed the Hubbard model naively and elucidated the splitting to the lower and upper Hubbard bands. Here, we discuss a mean-field approximation known as **Stoner model** [15].

In the Hubbard model (Eq. 5.26), the energy coming from the $U$ term is reduced when the up and downspin numbers are different. In an extreme case, if all spins are up, there is no double occupancy, and the $U$ term is zero. The difference between the up- and downspins is called **spin polarization**. Similarly to the Pauli paramagnetism (Sect. 4.4), we shall investigate how the spin polarization occurs. Suppose the numbers of the up- and downspins are different by $m$ due to the energy difference $\Delta$ (Fig. 5.19a).

$$m = n_\uparrow - n_\downarrow = \int_{E_F-\Delta}^{E_F+\Delta} D(E)\mathrm{d}E = 2D\Delta \tag{5.35}$$

Here, $D = D(E_F)$ is the density of states at the Fermi level. The kinetic energy of metal electrons increases due to the spin polarization.

$$\Delta E = \int_0^{E_F+\Delta} ED\mathrm{d}E + \int_0^{E_F-\Delta} ED\mathrm{d}E - 2\int_0^{E_F} ED\mathrm{d}E = \int_{E_F-\Delta}^{E_F+\Delta} ED\mathrm{d}E = D\Delta^2 = \frac{m^2}{4D}$$
$$= \frac{1}{4}Wm^2$$

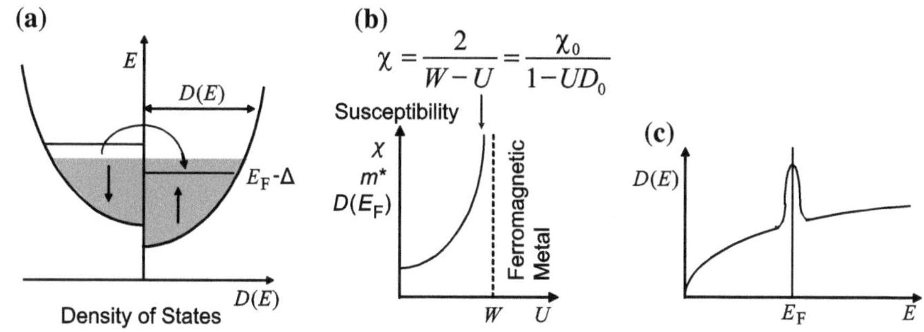


**Fig. 5.19** Stoner model

Here, $\Delta$ is replaced by $m$ using Eq. 5.35, and $1/D$ is approximated by the bandwidth $W$. Since the total electron number is $n = n_\uparrow + n_\downarrow$, the $U$ term is represented as

$$E = Un_\uparrow n_\downarrow = \frac{U}{4}(n^2 - m^2).$$

The total energy is obtained by adding these two contributions.

$$E = \frac{W-U}{4}m^2 + \frac{U}{4}n^2 \tag{5.36}$$

Since the total electron number $n$ is constant, the second term is constant. When $W > U$ (or $UD < 1$), the first term takes minimum at $m = 0$, and the most stable state does not have spin polarization. However, when $W < U$ (or $UD > 1$), the first term takes minimum at $m \neq 0$ because the coefficient is negative. Then, the most stable state has spontaneous spin polarization. This simple approximation seems $m$ to increase infinitely, but $m$ is never larger than $n$. The transition point at $U = W$ is the same as the simple intuition of the Hubbard model. The Stoner model results in a ferromagnetic metal instead of the Mott insulator.

In general, $U$ is the origin of spin polarization, but not necessarily the origin of the insulating state. Actually, iron, nickel, and cobalt show ferromagnetism with keeping the metallic conductivity. Therefore, these metals have such fractional magnetic moments as 2.2, 1.7, and 0.6 $\mu_B$ per atom. Ferromagnetism is observed in the later transition metals because the atomic orbital shrinks to enlarge $U/W$.

When a magnetic filed is applied to the above state, the energy is

$$E = \frac{W-U}{4}m^2 + \frac{U}{4}n^2 - mH. \tag{5.37}$$

This is differentiated by $m$, and the energy minimum is obtained.

$$\frac{1}{\chi} = \frac{H}{m} = \frac{W - U}{2}$$

The magnetic susceptibility is

$$\chi = \frac{2}{W - U} = \frac{\chi_0}{1 - UD_0}. \tag{5.38}$$

When $U$ approaches to $W$, the susceptibility diverges as shown in Fig. 5.19b [16]. Since the Pauli paramagnetism is related to $D(E_F)$ by $\chi = 2\mu_B^2 D(E_F)$, $D(E_F)$ and $m^*$ increase at the same time. The density of states and the effective mass generally increase not only towards a ferromagnetic metal but also towards the Mott insulator. The electron becomes heavy due to the correlation. In this occasion, the density of states is enhanced around the Fermi level as shown in Fig. 5.19c.

### 5.7.2 Hartree–Fock Approximation of the Hubbard Model

Next, we shall try to solve the Hubbard model under the Hartree–Fock approximation [17]. We will use a similar diagonalization technique to a superconductor (Sect. 6.6). Hamiltonian involving correlation is written in general

$$H = \sum_{i \neq j} t_{ij} c_{i\sigma}^{+} c_{j\sigma} + \sum_{k,k'} V_{k,k'} c_{k'\sigma'}^{+} c_{k\sigma}^{+} c_{k\sigma} c_{k'\sigma'}. \tag{5.39}$$

Here, $\sigma$ means $\uparrow$ or $\downarrow$. In the Hubbard model, the second $V_{k,k'}$ is $U$ only at the time of double occupancy. The application of the mean-field theory leads to

$$Un_\uparrow n_\downarrow = U\langle n_\uparrow \rangle n_\downarrow + Un_\uparrow \langle n_\downarrow \rangle - U\langle n_\uparrow \rangle \langle n_\downarrow \rangle. \tag{5.40}$$

In the first and second terms, one of the $n_\sigma$ product is approximated by the average $\langle n_\sigma \rangle$, but the third term is subtracted in order to remove the double counting. Similarly to the Stoner model

$$\langle n_\uparrow \rangle = \frac{n + m}{2}, \quad \langle n_\downarrow \rangle = \frac{n - m}{2}$$

is used to rewrite Eq. 5.40.

$$Un_\uparrow n_\downarrow = U\frac{n + m}{2} n_\downarrow + U\frac{n - m}{2} n_\uparrow - U\frac{n^2 - m^2}{4} \tag{5.41}$$

Since $n_\sigma$ square terms are removed by the mean-field approximation, this is solved by the diagonalization. The solution of the first term in Eq. 5.39 is represented by the band structure $E(k)$. In the $k$-space, the Hamiltonian is rewritten as

$$H=\sum_k\left[E(k)c_{k\sigma}^{+}c_{k\sigma}+E(k+Q)c_{k+Q\sigma}^{+}c_{k+Q\sigma}-\frac{Um\sigma}{2}\left(c_{k+Q\sigma}^{+}c_{k\sigma}+c_{k\sigma}^{+}c_{k+Q\sigma}\right)\right]$$
$$+\text{ (constant terms)} \tag{5.42}$$

In a one-dimensional system, the ground state of a Mott insulator is the SDW state. Then, the $E(k+Q)$ term is explicitly written because a band gap opens at $Q = 2k_{\mathrm{F}}$ due to the degeneracy of $E(k)$ and $E(k + Q)$. The periodicity of the antiferromagnetism is $Q = \pi/a$ in a one-dimensional system, but it should be replaced by a vector in a two- and three-dimensional system. The third term represents the Fourier transform of the on-site Coulomb interaction. $\sigma$ after $Um$ appears because the signs of the $m$ terms in Eq. 5.41 change depending on $\sigma = \pm 1$. The other terms in Eq. 5.41 generate constant terms. Since the term including $U$ is non diagonal, we try to diagonalize this Hamiltonian using

$$c_{k\sigma}=\cos\varphi_k a_{k\sigma}-\sin\varphi_k b_{k\sigma}$$
$$c_{k+Q\sigma}=\sin\varphi_k a_{k\sigma}+\cos\varphi_k b_{k\sigma} \tag{5.43}$$

where $a_k$ and $b_k$ represent annihilation operators $c_k$ on the antiferromagnetic A and B sublattices. These relations are put in Eq. 5.42, and the nondiagonal elements are

$$\left[-(E(k)-E(k+Q))\cos\varphi_k\sin\varphi_k+\frac{Um}{2}(\cos^2\varphi_k-\sin^2\varphi_k)\right](a_{k\sigma}^{+}b_{k\sigma}+b_{k\sigma}^{+}a_{k\sigma}).$$

The condition to make this zero is

$$\tan 2\varphi_k=\frac{Um}{E(k)-E(k+Q)} \tag{5.44}$$

which affords $\varphi_k$. This gives

$$\cos^2 2\varphi_k=\frac{1}{1+\tan^2 2\varphi_k}=\frac{(E(k)-E(k+Q))^2}{(E(k)-E(k+Q))^2+(Um)^2} \tag{5.45}$$

and

$$\cos^2\varphi_k=\frac{1+\cos 2\varphi_k}{2}=\frac{1}{2}\left(1+\frac{E(k)-E(k+Q)}{\sqrt{(E(k)-E(k+Q))^2+(Um)^2}}\right)$$

$$\sin^2 \varphi_k = 1 - \cos^2 \varphi_k = \frac{1}{2}\left(1 - \frac{E(k) - E(k+Q)}{\sqrt{(E(k) - E(k+Q))^2 + (Um)^2}}\right).$$

The diagonal term is given by

$$(E(k)\cos^2 \varphi_k + E(k+Q)\sin^2 \varphi_k + 2Um\cos \varphi_k \sin \varphi_k)a^+_{k\sigma}a_{k\sigma}$$
$$+ (E(k)\cos^2 \varphi_k + E(k+Q)\sin^2 \varphi_k - 2Um\cos \varphi_k \sin \varphi_k)b^+_{k\sigma}b_{k\sigma}.$$

The above $\cos^2\varphi_k$ is put in this equation to give

$$W(k) = \frac{E(k) + E(k+Q) \pm \sqrt{(E(k) - E(k+Q))^2 + (mU)^2}}{2}. \tag{5.46}$$

This is the solution of the Hubbard model under the Hartree–Fock approximation. If $E(k) = E(k + Q)$, this leads to $W(k) = E(k) \pm Um/2$, which gives the lower and upper Hubbard bands as shown in Fig. 5.8. The above method is called Hartree–Fock approximation because $n$ is approximated by the average $\langle n \rangle$. The concept is the same as the Hartree–Fock approximation in the molecular orbital theory, but note that the Hartree–Fock approximation of the Hubbard model looks considerably different.

### 5.7.3 *Exact and Approximate Solutions of the Hubbard Model*

The exact solution is known for the one-dimensional Hubbard model. As shown in Fig. 5.20a, the half-filled system gives an antiferromagnetic insulator for however small non zero $U$, but all other cases are paramagnetic metals. For more than two dimensions, numerical calculations tell us that only the half-filled system becomes an antiferromagnetic insulator [18]. Accordingly, the Hubbard model gives the Mott–Hubbard insulator only in the half-filled case.

The one-dimensional Hubbard model shows characteristic properties such as **spin-charge separation**. Suppose one electron is removed from a one-dimensional antiferromagnet (Fig. 5.20b). An electron on the right of the hole is successively moved to the left, and the hole moves to the right. As a result, the positive charge moves to the right (Fig. 5.20c), but the spin stays at the original position because the two successive ↓↓ do not move. This simple example demonstrates that spin and charge move independently in a one-dimensional system.

There are a large number of investigations concerning to the Hubbard model of a various dimensions and lattice systems. Figure 5.21 is a phase diagram extracted from the Hartree–Fock approximation for a three-dimensional simple cubic lattice.

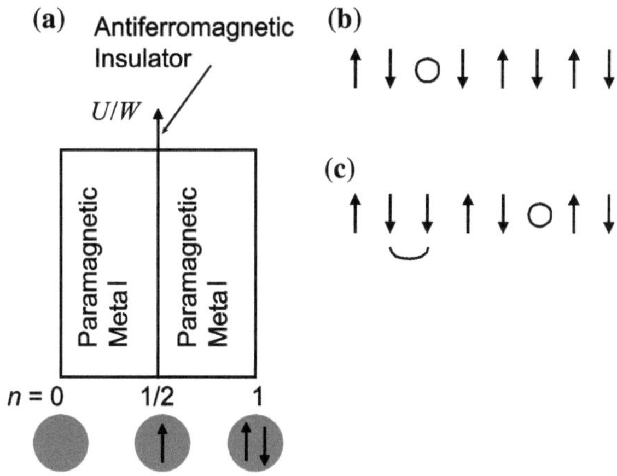


**Fig. 5.20** **a** Phase diagram of the one-dimensional Hubbard model, and **b**, **c** the spin-charge separation [19]

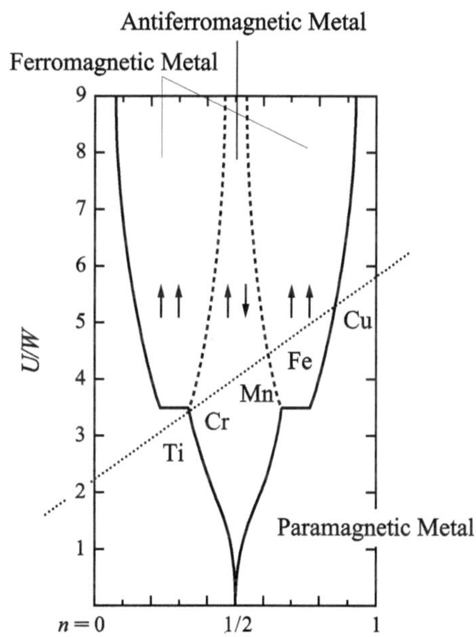


**Fig. 5.21** Phase diagram of the three-dimensional Hubbard model in the Hartree–Fock approximation [22]

The region around the half filled shows antiferromagnetism, and the region far from the half filled is paramagnetic metals. However, the large $U$ region is occupied by ferromagnetism [20]. Ferromagnetism in the large $U$ region is convincing from the Stoner model. This does not conflict with the ferromagnetism coming from the potential exchange at the $t \rightarrow 0$ limit (Sect. 5.5). This also agrees with the triplet instability at $\Delta\varepsilon = 0$ (Fig. 5.1e), and Hunt's rule between degenerate orbitals. It has been predicted by Nagaoka that the state slightly doped to the half filling in the large $U$ limit shows ferromagnetism [23]. We may say generally that ferromagnetism is related to $t \rightarrow 0$.

If the whole $d$ band is regarded as a single band, the elements located in the right end of the periodic table have small atomic orbitals and large $U$. Then, the transition metals are placed on the right-up line as shown in Fig. 5.21. Starting from paramagnetic metals, antiferromagnetic metals like chromium and manganese appear around the half filling. Then, ferromagnetic metals follow in iron, cobalt, and nickel. Finally, paramagnetic metals appear again in copper. This diagram satisfactorily reproduces the magnetism of elemental transition metals. The mathematically exact solution indicates only the half-filled system is exceptional, but this may be because we have extracted only the on-site Coulomb $U$. The above approximate solution suggests a rather realistic picture owing to the average Coulomb interaction. Then, we need to consider the Coulomb interaction other than the on-site interaction.

## 5.8 Charge Order

In the exact solutions of the Hubbard model, only the half-filled system becomes the Mott insulator. However, the Coulomb repulsion works in other cases. Suppose low density electrons distributed in a uniform potential. When the potential energy is more important than the kinetic energy, the most stable state is a lattice of electrons aligned with equal intervals. Such a state is called **Wigner lattice**.

Next, suppose the quarter-filled case in which two sites have one electron (Fig. 5.22a). In an average, each site has 1/2 electron. Then, the equal interval lattice is an alternate arrangement like 101010. When an electron moves, the electron has to come to the adjacent site of another electron at a moment. However, we suppose the Coulomb repulsion works between the adjacent sites. Such

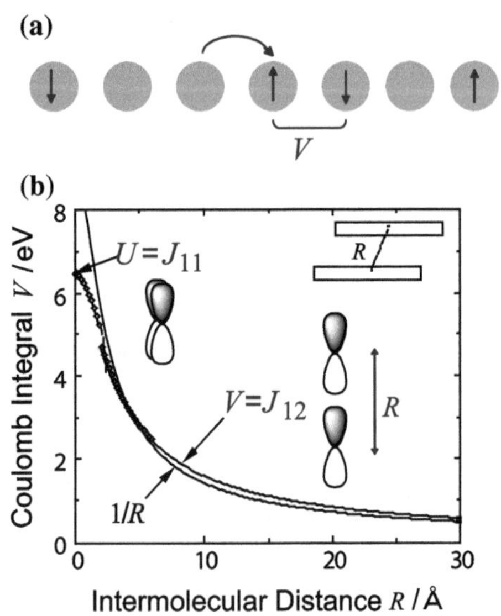


**Fig. 5.22** **a** Inter-site Coulomb interaction $V$ in the 1/4-filled band, **b** calculated Coulomb repulsion between BEDT-TTF molecules [21]

repulsion is called **inter-site Coulomb repulsion**, and designated as $V$. We consider $V$ together with $U$.

$$H = \sum_{i \neq j} t_{ij} a_i^+ a_j + U \sum_i n_{i\uparrow} n_{i\downarrow} + V \sum_{i \neq j} n_i n_j \tag{5.47}$$

This model is called **extended Hubbard model.** If $V$ is sufficiently large, electrons cannot move without breaking the alternating arrangement, and the resulting **charge-order** state is insulating.

Figure 5.22b is a plot of the Coulomb repulsion between two BEDT-TTF HOMO's (Sect. 7.7) as a function of the interplanar spacing $R$. The left end, $R = 0$, corresponds to $U = J_{11}$, which is the Coulomb integral between the same HOMO. With increasing $R$, the Coulomb integral decreases. The real spacing occurs at $V = J_{12}$. Note that $U$ corresponds to $J_{11}$, and $V$ is $J_{12}$ (Fig. 5.14). When the molecules are separated to some extent, the Coulomb repulsion between the HOMO is equivalent to the repulsion between the point charges placed on the centers. At large $R$, the simple $R$ inverse curve in Fig. 5.22b is very close to the calculation based on the molecular orbitals. Using the atomic unit (Sect. 1.5), $R$ is divided by the Bohr radius, and its inverse gives the Coulomb repulsion in Hartree unit (27.2 eV). The $R$ inverse diverges at $R \to 0$, but the spreading HOMO gives a finite $U$. The calculated $U$ is about 6 eV, but the experimentally obtained value is around 1 eV. This discrepancy comes from the screening of other electrons. The actual $V$ becomes small proportionally, which is usually several 0.1 eV.

Since the solution of the extended Hubbard model is more difficult than the Hubbard model, we shall investigate the $t \to 0$ **atomic limit**. Suppose the one-dimensional quarter-filled system (Fig. 5.23). Without any spin polarization, the mean-field electron number in the $U$ term is $\langle n_\uparrow \rangle = \langle n_\downarrow \rangle = n/2$. Then, the energy per site in the metal state with homogeneous $n = 1/2$ charge is

$$E = U \times \frac{1}{4} \times \frac{1}{4} + V \times \frac{1}{2} \times \frac{1}{2} = \frac{U}{16} + \frac{V}{4}. \tag{5.48}$$

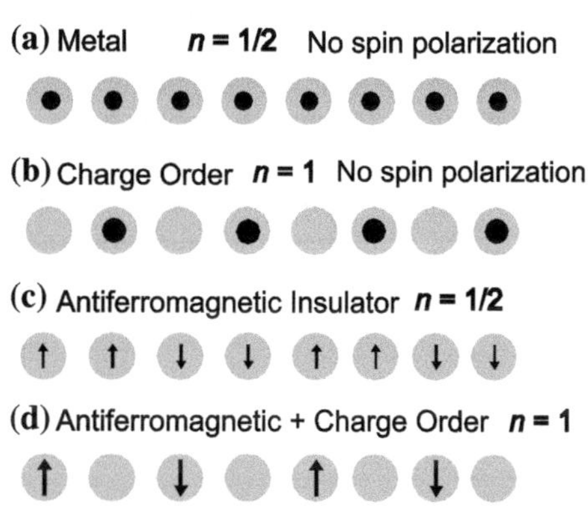


**Fig. 5.23** Various states in the one-dimensional quarter-filled system

The charge-order state (Fig. 5.23b) has energy per site.

$$E = \frac{1}{2}\left[U \times \left(\frac{1}{2} \times \frac{1}{2} + 0 \times 0\right) + 2V \times 1 \times 0\right] = \frac{U}{8} \tag{5.49}$$

Comparison of Eqs. 5.48 and 5.49 gives

$$\frac{U}{16} + \frac{V}{4} > \frac{U}{8} \text{ leads to } \ 4V > U.$$

This is the condition where the charge order is more stable than the homogeneous metal state. Then, the charge order occurs when $V$ is larger than one quarter of $U$.

We have assumed no spin polarization and the average $\langle n_\uparrow \rangle = \langle n_\downarrow \rangle = n/2$. Instead, we consider an antiferromagnetic (SDW) state which has alternately $\langle n_\uparrow \rangle = 1/2$ and $\langle n_\downarrow \rangle = 0$ or $\langle n_\uparrow \rangle = 0$ and $\langle n_\downarrow \rangle = 1/2$. The $U$ term in Eq. 5.48 disappears, and the energy per site is

$$E = U \times \frac{1}{2} \times 0 + V \times \frac{1}{2} \times \frac{1}{2} = \frac{V}{4}.$$

In comparison with Eq. 5.49, this spin-polarized state is more stable than the non spin-polarized state at large $2V > U$. Note that the antiferromagnetic Mott insulating state appears when $V$ is included even in the non half-filled system. We further consider a spin-polarized charge-order state (Fig. 5.12d).

$$E = \frac{1}{2}[U \times (1 \times 0 + 0 \times 0) + 2V \times 1 \times 0] = 0$$

This state is more stable than the states (a) and (c) as far as $V > 0$, and more stable than (b) as far as $U > 0$.

We can calculate entropy of the localized states from Eq. 3.82. According to $F = E - TS$, the free energy of the state with large entropy decreases at $T \neq 0$ K. Since a metal phase has the largest entropy, a metal state becomes stable at high temperatures above a metal–insulator transition.

*Example 5.4* Figure 5.24 illustrates charge-order states in a quarter-filled square lattice. The black balls represent occupied sites, and the white balls correspond to empty sites. Compare the stability at the atomic limit. Assume $V$ works along the bond direction and there is no spin polarization.

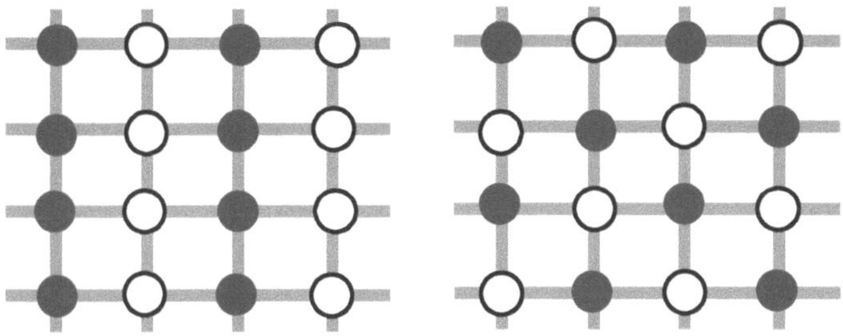


**Fig. 5.24** Charge-order patterns in the square lattice. **a** Stripe, **b** checkerboard

Homogeneous metal $E = U \times \frac{1}{4} \times \frac{1}{4} + V \times 2 \times \frac{1}{2} \times \frac{1}{2} = \frac{U}{16} + \frac{V}{2}$

Stripe $E = \frac{1}{2}\left[U \times \left(\frac{1}{2} \times \frac{1}{2} + 0 \times 0\right) + V \times (1 \times 1 + 0 \times 0 + 2 \times 1 \times 0)\right] = \frac{U}{8} + \frac{V}{2}$

Checkerboard $E = \frac{1}{2}\left[U \times \left(\frac{1}{2} \times \frac{1}{2} + 0 \times 0\right) + V \times 4 \times 1 \times 0\right] = \frac{U}{8}$

Then, the checkerboard is more stable than the stripe as far as $V > 0$. Comparison of the homogeneous metal and the checkerboard affords

$$\frac{U}{16} + \frac{V}{2} > \frac{U}{8} \text{ namely } 8V > U.$$

This is the condition that the checkerboard is stable. In the stripe, the charge is aligned linearly, so the $V$ term contains $1 \times 1$, whereas the checkerboard contains $1 \times 0$ for all $V$ terms. Therefore, the inter-site repulsion is completely avoided in the checkerboard pattern. In the stable pattern, the black balls are aligned so as to avoid contacts in the bond direction. Then, the charges are aligned along the bisector of two bond directions, namely along the diagonal direction. We can compare stability of various charge-order patterns in a similar way.

In the atomic unit, since double occupancy increases the $U$ term, $U$ works like $t$ to spread the electron distribution. At high temperatures, the electron kinetic energy increases, and the phase that is stable at large $t$ and $U$ is realized. When spin polarization is considered, the $U$ term usually becomes zero, and we have to involve $t$. If we take into account of the trick coming from the replacement of $t$ by $U$, the atomic limit provides a roughly reasonable phase diagram. In Sect. 7.7.4, we will mention that the triangular lattice leads to a little more complicated charge-order pattern [22].

For more improved approximation considering the energy band formed in the charge-order state, the mean-field Hamiltonian is calculated similarly to Eq. 5.41 using $\langle n_\downarrow \rangle$ and $\langle n_\uparrow \rangle$ according to the Hartree–Fock approximation [23]. In the Hubbard model, $U\langle n \rangle$ appears only in the diagonal terms, and the calculation is reduced to the inclusion of the self-consistent site potentials. In the extended

Hubbard model, the nondiagonal terms contain $V\langle n\rangle$, and we have to determine these terms self-consistently.

Since the Coulomb interaction is a long range force which decreases following $1/R$, it is natural to consider inter-site $V$ and the resulting charge order. The Hubbard model without considering the inter-site $V$ has been investigated extensively, but the exact mathematical treatment makes the truncation effect of the Coulomb potential more important.

*Example 5.5* Another simple model to discuss the charge order is inclusion of the **site potential**. Suppose an alternating chain similarly to Example 2.9, and the site potential on 2 is assumed to be by $\Delta$ higher than the site 1. The transfer $t$ is uniform. The secular equation is represented by

$$\begin{vmatrix} \alpha - E & 2t\cos ka \\ 2t\cos ka & \alpha + \Delta - E \end{vmatrix} = 0.$$

Calculate the energy band.

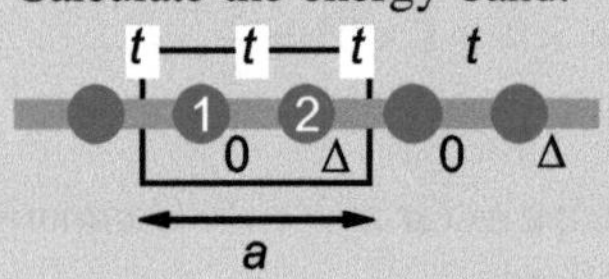


We obtain the square equation

$$(E-\alpha)^2 - \Delta(E-\alpha) - 4t^2\cos^2 ka = 0$$

which is solved to

$$E = \alpha + \frac{\Delta \pm \sqrt{\Delta^2 + 16t^2\cos^2 ka}}{2}.$$

This energy band has an energy gap at the half filling (Fig. 5.25).

To investigate the Peierls transition, the modulation of the transfer was considered (Example 2.9). Alternatively, the energy gap is obtained like this from the site potential. In Sect. 7.7, it will be demonstrated that when a BEDT-TTF complex has such charge order as $D^+D^0$, we can include the effect by assuming the HOMO level of the $D^+$ state by $\Delta = 0.2 \sim 0.3$ eV higher. The Hartree–Fock approximation of the Hubbard model contains $\sum_i U\langle n_{i\uparrow}\rangle\langle n_{i\downarrow}\rangle$ in the diagonal terms, but supposing $\langle n_{i\uparrow}\rangle$ and $\langle n_{i\downarrow}\rangle$ have been known, $\Delta$ is added to the diagonal term. However, the difference should be noted that Fig. 5.25 affords a nonmagnetic insulator with a band gap, whereas the charge order state due to $V$ gives a paramagnetic insulator.

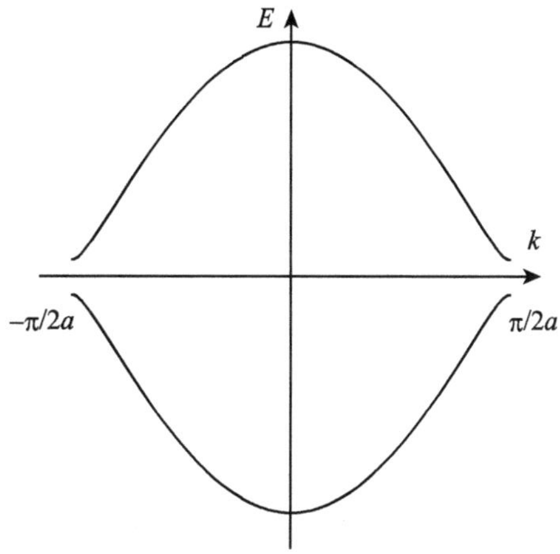


Fig. 5.25 Energy band of an alternating chain with the site potential, $\Delta/t = 0.3$

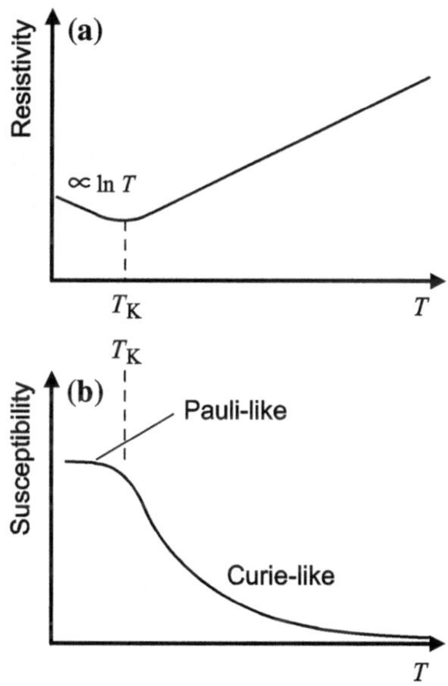


**Fig. 5.26** **a** Temperature dependence of electric resistance, and **b** susceptibility of dilute magnetic alloys

## 5.9 Kondo Effect

In Sect. 4.9, we discussed the πd-system, in which the π electrons are spin polarized due to the d spins. By contrast, in the **Kondo effect**, the d spin makes a singlet state with the metal electrons, and the spin disappears [24]. Suppose a dilute magnetic alloy; for instance, a small amount of magnetic metal such as iron is incorporated in copper. In such a case, the resistivity increases logarithmically at low temperatures (Fig. 5.26a), where a resistivity minimum appears at the **Kondo temperature** $T_K$.

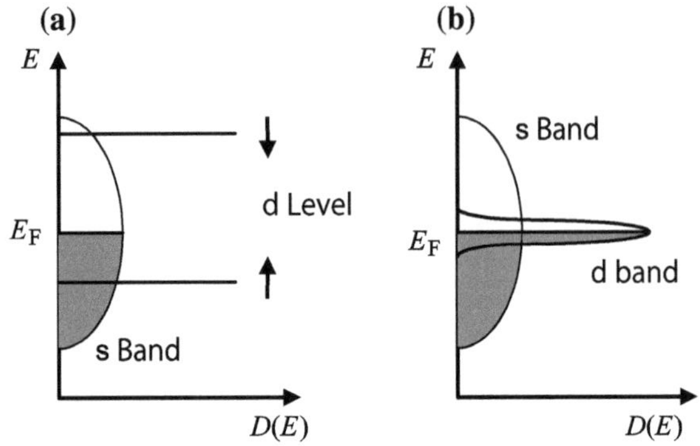


**Fig. 5.27** Energy bands in the Kondo metal. **a** $T > T_K$, **b** $T < T_K$

In **heavy Fermion systems** such as $CeCu_6$, $UBe_{13}$, and $UPt_3$, the f spins are no more dilute, but the resistivity similarly increases to reach a constant value.

Such a phenomenon is qualitatively explained as follows. The energy band is depicted in Fig. 5.27. Copper forms a partially filled wide s band. Iron exhibits magnetism, where the occupied upspin level (lower Hubbard level) is located below the Fermi level, whereas the empty downspin level (upper Hubbard level) is located above the Fermi level (Fig. 5.27a). Suppose a transfer $\Delta$ between the s and d electrons. If the temperature is higher than $k_B T_K \sim \Delta$, the d-level is localized. The s electrons show the very small Pauli paramagnetism, but most magnetic susceptibility comes from the Curie contribution of the localized d spins. At temperatures lower than $\Delta$, the d electrons are hybridized with the s electrons (Fig. 5.27b). In particular, the heavy Fermion systems with periodical f sites make a narrow band below $T_K$, and the density of states is given by the sum of the s and f bands, where the susceptibility is extraordinarily large. The Curie magnetism turns to the flat Pauli magnetism at low temperatures (Fig. 5.26b), where the effective mass is hundreds to thousands times larger than that of the ordinary free electron. Here, the *f* electron joins to the metallic electron and moves as a heavy electron.

For a little more quantitative analysis, consider a d-level located at $E_0$ below $E_F$ (Fig. 5.28a). We start from the **Anderson Hamiltonian** [24–26].

$$H = \sum_k E(k) c_k^+ c_k + E_0 \sum_\sigma n_{d\sigma} + U n_{d\uparrow} n_{d\downarrow} + V \sum_{k\sigma} (c_k^+ d_o + d_\sigma^+ c_k) \qquad (5.50)$$

The first term represents the s band, the second term is the d-level at $E_0$, the third term is the on-site Coulomb energy on the d-level, and the final term means that the s and d electrons exchange with the probability of $V$.

We may assume the density of states of the d-level represented by

$$D(E) = \frac{\Delta}{\pi} \frac{1}{(E - E_0 - nU)^2 + \Delta^2}. \qquad (5.51)$$

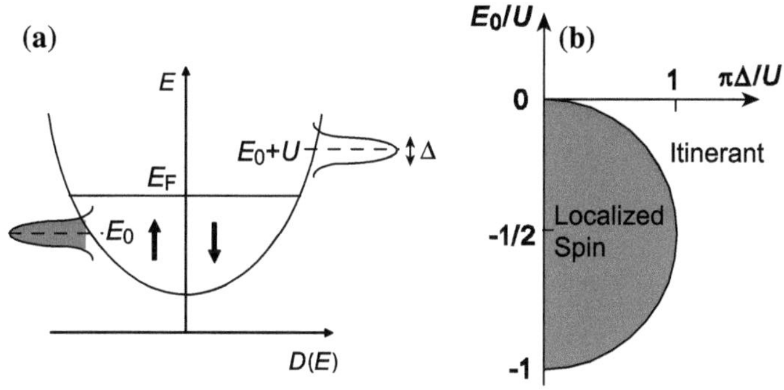


**Fig. 5.28** Mean-field approximation of the Kondo effect

The d band with the width Δ is located at $E_0 + nU$. When there is spin polarization, the d band splits to the lower Hubbard band at $E_0$ and the upper Hubbard band at $E_0 + U$ (Fig. 5.28a). From the Stoner model (Sect. 5.7.1), the spin polarization occurs at $U > W$, namely $UD(E_F) > 1$. Without spin polarization, $E_F$ comes to $E = 0$, so $D(0)$ from Eq. 5.51 is put in $UD(0) > 1$ to afford the condition of spin polarization.

$$\frac{\Delta}{\pi}\frac{U}{(E_0+nU)^2+\Delta^2} > 1$$

This relation gives a semicircle as shown in Fig. 5.28b, within which the d electron is localized. The d electron localizes most easily when $E_0$ is located by $U/2$ below $E_F$, because the spin-polarized levels appear at $E_0$ and $E_0 + U$. In this occasion, the d-level behaves as a localized spin when $\Delta < U/\pi$, but behaves as itinerant when Δ is larger than this criterion. The whole d-level is occupied at $E_0 < -U$ and empty at $E_0 > 0$, so the d electron is not magnetic out of the semicircle in Fig. 5.28b.

Figure 5.29a shows the residual resistivity when a small amount of transition metal is alloyed in copper. In titanium and iron, $E_0$ and $E_0 + U$, respectively, coincide with $E_F$, and $D(E_F)$ becomes large. As a result, the residual resistivity is large. Note that the d-level gradually deepens when going to the right in the periodic table (Sect. 1.5). Manganese, located in the middle of these elements, corresponds to $E_0 = -U/2$, and the residual resistivity is small because the manganese spin is localized as shown in Fig. 5.27a. In both ends of the periodic table, the impurity transition metal is not magnetic. Accordingly, the mean-field approximation in Fig. 5.28b explains well the results in Fig. 5.29a.

When the s and d electrons are hybridized, additional electrical resistivity $R(T)$ appears coming from the scattering of the *s* electrons by the *d* electrons. Such a contribution is calculated on the basis of the Hamiltonian in Eq. 5.50. After an elaborate calculation

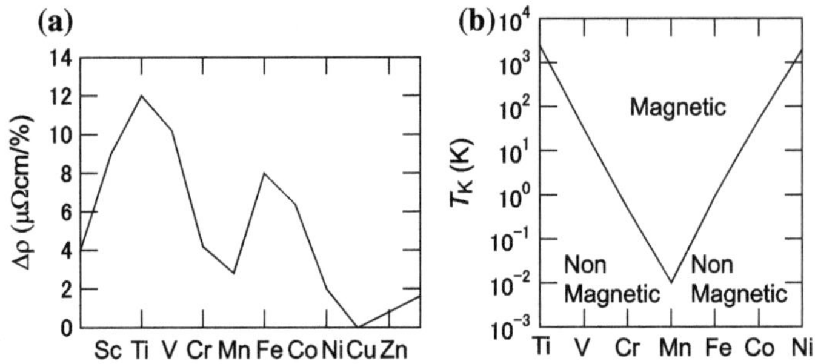


**Fig. 5.29** **a** Residual resistivity when impurity transition metal is alloyed in copper, and **b** the corresponding Kondo temperatures

$$R(T) = R_0(1 + 2JD(E_F)\ln(k_BT/\Delta)) \tag{5.52}$$

has been obtained [24]. Δ appears in the logarithm because we truncated the integral by assuming $V$ is non zero within the region of $\pm\Delta$. Here

$$J = -2V^2\left(\frac{1}{E_0} + \frac{1}{E_0 + U}\right) \tag{5.53}$$

is the antiferromagnetic interaction between the $s$ and $d$ electrons. Since $J$ is negative (antiferromagnetic), the resistivity diverges logarithmically at $T \to 0$. However, this divergence is avoided when higher order terms are included. From Eq. 5.52, the increase of the resistivity starts at

$$T_K = \Delta\exp(-1/D(E_F)|J|). \tag{5.54}$$

When transition metals are alloyed in copper or gold, the resulting $T_K$ is plotted in Fig. 5.29b. $T_K$ is as low as 0.01 K in manganese, and this metal is magnetic even below the helium temperature. Consequently, the residual resistivity in Fig. 5.29a is comparatively small. Here, we measure the residual resistivity above $T_K$. Vanadium and chromium as well as iron and cobalt are located on both sides of manganese, where $T_K$ is higher, and the residual resistivity is measured below $T_K$. Then, the residual resistivity is large. The outer most metals have $T_K$ as high as 1000 K, and the d electrons are not magnetic, so that the residual resistivity is again small.

The same discussion applies to conducting metal coordination complexes, where the conduction is mediated by ligand organic materials. The central metal atom has magnetism when the upspin level $E_0$ is located below $E_F$, and the downspin level $E_0 + U$ is above $E_F$. Hybridization of the d-level occurs when $E_0$ or $E_0 + U$ is close to $E_F$. In this occasion, $J$ in Eq. 5.53 is very large, and $T_K$ in Eq. 5.54 is above room temperature. The most interesting case is intermediate $T_K$, where the d spin is localized above $T_K$, but the magnetism disappears below $T_K$ due to the hybridization with the s electrons. $T_K$ is mostly determined by the d bandwidth Δ, but whether $E_0$ is located near $E_F$ or not is another important factor. $V$ is almost the same quantity as $t_{\pi d}$ (Sect. 4.9).

Among organic πd-systems (Sects. 4.9 and 7.8.1), iron spins in λ-$(BETS)_2FeCl_4$ and κ-$(BETS)_2$ $FeBr_4$ are localized even at helium temperatures. The Kondo effect is a phenomenon in which the d spin is erased by the antiferromagnetically coupled s spins. To erase the iron $S = 5/2$ spin, at least five $S = 1/2$ π spins are necessary. Then, this is impossible from the composition. Copper in $Cu(DMDCNQI)_2$ (Sect. 7.9) does not have a spin and is in a mixed-valent state ($Cu^{4/3+}$) even at room temperature, implying $T_K$ is above room temperature. In this case, the d-level of copper is very close to $E_F$ of DMDCNQI.

## 5.10 Anderson Localization in a Disordered System

The above discussion dealt with the problem whether an electron in a homogeneous medium is localized or itinerant. On the other hand, when phosphorus or boron is doped in silicon, an increasing amount of dopant enhances the conductivity, and a highly doped case shows almost metallic conductivity. In this case, the carrier orbital spreads to some extent, and when the overlap with the adjacent orbitals exceeds some limit, the electron becomes itinerant. Since the wave function behaves like $e^{-\zeta r}$, the transfer integral $t$ is also expected to show exponential dependence on the interatomic distance. Equation 3.11 tells us $m^*$ is inversely proportional to $t$, so from Eq. 3.16, the conductivity shows exponential dependence on the interatomic distance. Thus, the carrier transport is mediated by the **hopping conduction** where the carrier jumps between the dopant levels. By using the average dopant distance $R$, we assume the conductivity is represented as

$$\sigma \propto \exp(-2\alpha R - W/k_B T). \tag{5.55}$$

This relation considers the conductivity exponentially decreases when the dopant concentration decreases and $R$ increases. $W$ means the depth of the dopant level. It is expected that itinerant conduction is achieved when the density of states exceeds a certain limit. The density of states in a three-dimensional energy band is small near the band edge (Fig. 5.30), and we suppose the **mobility edge** $E_c$ at which the electron becomes itinerant. When the carrier concentration is small and $E_F$ is below $E_c$, the electron is localized (1), but the electron is itinerant at $E_F > E_c$ in the large dopant concentration (2) [5].

Transport is mediated by electrons thermally excited to $E_c$, whose number is given by $D(E_F)k_BT$ per volume. When the hopping is possible to the levels within the radius of $R$, the number of levels is estimated to be $(4\pi R^3/3)D(E)\mathrm{d}E$ per energy interval $\mathrm{d}E$ [27]. The average energy interval is given by the inverse $W = 3/4\pi R^3 D(E)$, which is put in Eq. 5.55 to afford

**Fig. 5.30** Mobility edge

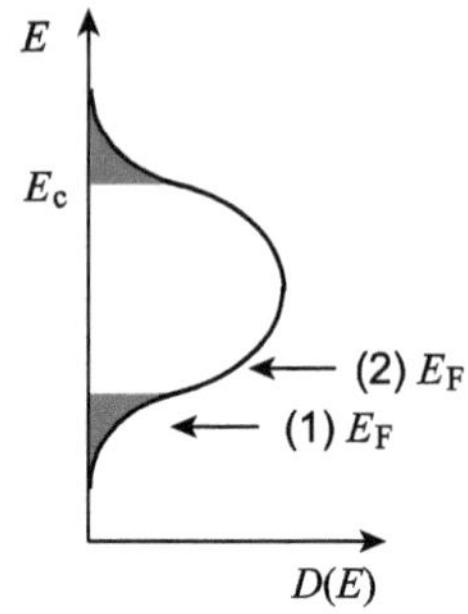


$$\sigma \propto \exp(-2\alpha R - 3/4\pi R^3 D(E) k_B T).$$

This relation has maximum when the derivative of the exponent with respective to $R$ is zero.

$$2\alpha = 9/4\pi R^4 D(E) k_B T \tag{5.56}$$

With increasing $R$, the number of hopping destinations increases, but the hopping probability to a distant site decreases rapidly in proportion to $\exp(-2\alpha R)$. Thus, $R$ given by Eq. 5.56 contributes most largely to the conduction. If $R$ is smaller than the actual dopant distance $R_D$, there are many hopping destinations, and the first term in Eq. 5.55 is unimportant. This case affords a simple activated conductivity, $\sigma \propto \exp(-W/k_B T)$. If $R$ is larger than $R_D$, the first term is Eq. 5.55 is important. In this case, the optimal $R$ given by Eq. 5.56 is put in Eq. 5.55 to afford

$$\sigma \propto \exp(-a/T^{1/4}) \quad \text{or} \quad \log\rho \propto 1/T^{1/4} \tag{5.57}$$

where the resistivity $\rho$ depends on the exponential of $1/T^{1/4}$ instead of $1/T$ [27]. The logarithmical plot of resistivity versus inverse temperature becomes more flat at low temperatures (Fig. 5.31). This is because the temperature dependent hopping distance in the first term of Eq. 5.55 determines the conductivity instead of the thermal excitation in the second term. Such a process is called **variable-range hopping** (VRH). In this case, the plot of log $\rho$ versus $1/T^{1/4}$ gives a straight line. The above derivation is based on three-dimensional hopping, but in the two-dimensional hopping a circular region instead of a sphere is used to give $\log\rho \propto 1/T^{1/3}$. The one-dimensional VRH depends on $\log\rho \propto 1/T^{1/2}$. In general, the $n$-dimensional VRH is represented as $\log\rho \propto 1/T^{1/d+1}$. In low dimension, the hopping destinations are restricted, and the saturation of the rise in log $\rho$ is moderated.

We shall mention the **percolation theory** [28]. Suppose conducting and insulating balls are put in a box, and we shall investigate the overall conductance. In Fig. 5.32, conducting and insulating balls are black and white, respectively. If the black balls are connected, a conducting **cluster** is formed. When the cluster is

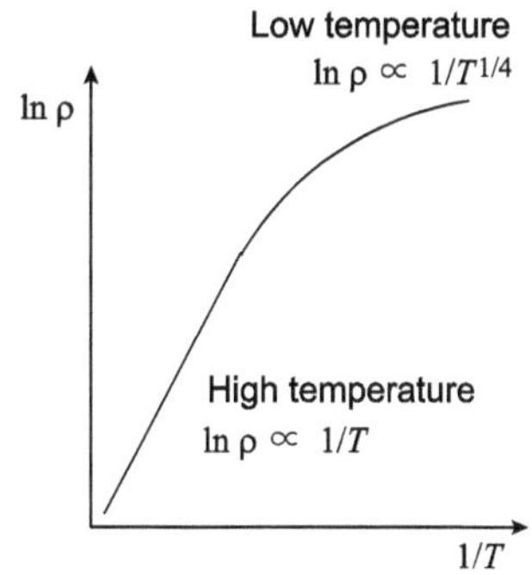


**Fig. 5.31** Temperature dependent resistivity in the variable-range hopping

connected from an end to another end, the overall material is conducting. With increasing the ratio of black balls, the conductance does not increase gradually, but the conductance is abruptly achieved as a critical ratio. This is called **percolation limit**. Computer simulation tells us the percolation limit is 0.593 for the two-dimensional square lattice, 0.50 for the two-dimensional triangular lattice, 0.311 for the three-dimensional simple cubic lattice, 0.246 for the body-centered lattice, and 0.198 for the face-centered lattice. The percolation limit decreases with increasing the dimension. Among the three-dimensional lattices, the increasing coordination number from six, eight to twelve reduces the percolation limit in the above order. In one dimension, only a single white ball makes the whole material insulating. This is obvious supposing a one-lane highway, where only a single troubled car makes the whole way jammed.

Localization due to disorder is discussed in view of Anderson's **scaling theory**. Suppose conductance $G$ of a cube with an edge $L$ (Fig. 5.33a) [29]. Conductance is inverse of resistance, and in the d dimensional case, represented by the conductivity $\sigma$ as

$$G = \sigma L^{d-2}. \tag{5.58}$$

Three-dimensional resistivity is $\rho = RS/L$, where $S = L^2$ is the cross sectional area. Then, the unit of resistivity is Ω cm. This leads to $G = \sigma L$ in the three-dimensional case, in agreement with the above relation. In the two-dimensional system, $G = \sigma$ does not depend on $L$, which is known as **sheet resistance**, Ω/square.

When the cube is enlarged to $\nu L$, the resulting conductance $G(\nu L)$ is determined only from $G(L)$ and $\nu$. This is called scaling hypothesis, which is mathematically represented as

$$\frac{G(\nu L)}{G(L)} = f(G(L), \nu).$$

The logarithm is divided by $\log \nu$, and the $\nu \to 1$ limit is represented by $\beta(G)$.

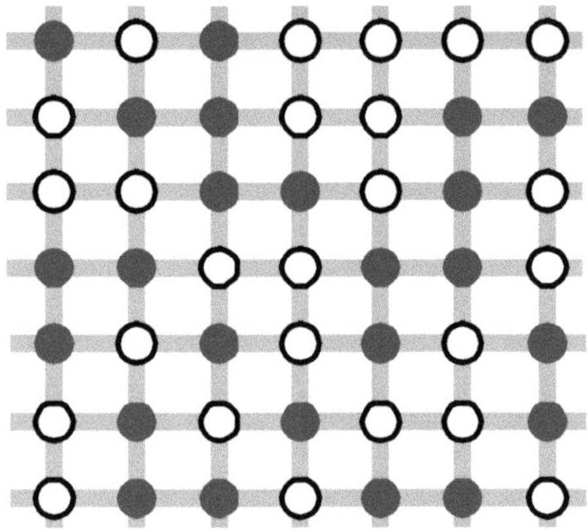

**Fig. 5.32** Two-dimensional square lattice with 50 % black balls

$$\lim_{\nu\to 1}\frac{\log(G(\nu L)/G(L))}{\log \nu}=\lim_{\nu\to 1}\frac{\log G(\nu L)-\log G(L)}{\log(\nu L)-\log L}=\frac{d\log G}{d\log L}\equiv\beta(G)$$

$\beta(G)$ is the index when $G$ is represented by a power of $L$. Conductance of a bulk metal follows Eq. 5.58, and in the large $G$ limit, $\beta(G) = d - 2$ is obvious. However, in a strongly localized system with small $G$, this macroscopically obvious relation is not applicable. When the wave function diminishes exponentially, the conductance follows

$$G = G_0 e^{-\alpha L/\xi}$$

in analogy with the hopping conduction represented by Eq. 5.55. In this case, $\beta(G)$ is

$$\beta(G) = \log(G/G_0) = -\alpha L/\xi.$$

In the limit of $G \to 0$, $\beta(G)$ diverges to infinity. The rough sketch of $\beta(G)$ is depicted in Fig. 5.33b.

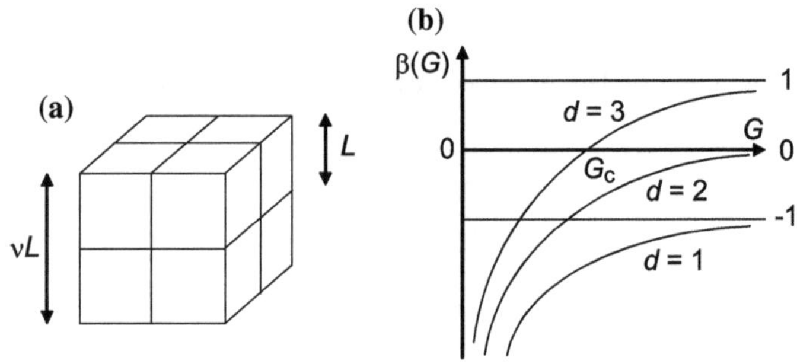


**Fig. 5.33** Scaling theory

In the one-dimensional case, $\beta(G)$ is less than $-1$ over the whole region. With increasing $L$, $G$ decreases monotonically, and a one-dimensional system is always localized. Then, a disordered one-dimensional system is not conducting. This conclusion agrees with the percolation theory. In the two-dimensional case, $\beta(G)$ is negative over the whole region, but $\beta(G)$ approaches to zero in the large $G$ limit. A disordered two-dimensional system is localized, but the localization is weak. In the three-dimensional case, $\beta(G)$ crosses zero at a certain $G_c$, where a metal–insulator transition takes place. This disorder-induced metal–insulator transition is called **Anderson transition**, and the region below $G_c$ is called **Anderson localization**. The transition from an insulator to a metal which occurs when a dopant is doped to silicon is regarded as an Anderson transition.

## References


1. R.G. Parr, B.L. Crawford Jr, J. Chem. Phys. **16**, 526 (1948)
2. R.S. Mulliken, J. Chem. Phys. **46**, 497, 675 (1949)
3. P. Flude, *Electron Correlations in Molecules and Solids* (Springer, Berlin, 1991)
4. L.D. Landau, E.M. Lifshitz, *Quantum Mechanics Non-Relativistic Theory (Chap. 9)* (Pergamon, New York, 1962)
5. N.F. Mott, *Metal-Insulator Transitions* (Taylor & Francis, London, 1990)
6. P. Ziesche, O. Gunnarsson, W. John, H. Beck, Phys. Rev. B **55**, 10270 (1997)
7. P.W. Anderson, Solid State Phys. **14**, 99 (1963)
8. P.W. Anderson, *Magnetism,* 1st edn., ed. by G.T. Rado, H. Suhl, vol. 25 (1963)
9. J. Kanamori, J. Phys. Chem. Solids **10**, 87 (1959)
10. J.B. Goodenough, Phys. Rev. **100**, 564 (1955)
11. O. Kahn, *Molecular Magnetism* (Wiley, New York, 1993), p. 187
12. J. Zaanen, G.A. Sawatzky, Can. J. Phys. **65**, 1262 (1987)
13. J.B. Goodenough, Prog. Solid State Chem. **5**, 145 (1971)
14. D.B. McWhan, A. Menth, J.P. Remeika, W.F. Brickman, T.M. Rice, Phys. Rev. B **7**, 1720 (1973)
15. E.C. Stoner, J. Phys. Radium **12**, 372 (1931)
16. W.F. Brinkman, T.M. Rice, Phys. Rev. B **2**, 4302 (1970)
17. F. Gebhard, *The Mott Metal-Insulator Transition* (Springer, Berlin, 1997)
18. J.E. Hirsch, Phys. Rev. B. **31**, 4403 (1985)
19. D.R. Penn, Phys. Rev. **142**, 350 (1966)
20. Y. Nagaoka, Phys. Rev. **147**, 392 (1966)
21. T. Mori, Bull. Chem. Soc. Jpn **73**, 2243 (2000)
22. T. Mori, J. Phys. Soc. Jpn. **72**, 1469 (2004)
23. H. Seo, J. Phys. Soc. Jpn. **69**, 805 (2000)
24. J. Kondo, Solid State Phys. **23**, 183 (1970)
25. A. Bladin, *Magnetism,* Vth edn., ed. by G.T. Rado, H. Suhl, vol. 58 (1973) p. 66
26. P.W. Anderson, Phys. Rev. B **124**, 41 (1961)
27. N.F. Mott, Phil. Mag. **19**, 835 (1969)
28. D. Stauffer, A. Aharony, *Introduction to Percolation Theory* (Taylor & Francis, London, 1992)
29. E. Abraham, P.W. Anderson, D.C. Liccardello, T.V. Ramakrishnan, Phys. Rev. Lett. **42**, 673 (1979)

# Chapter 6
# Superconductivity

**Abstract** This chapter outlines superconductivity. After various aspects of superconductivity are described, these phenomena are explained starting from the microscopic theory.

**Keywords** Superconductivity · Meissner effect · Perfect diamagnetism · London equation · Penetration depth · Coherence length · Hebel-Slichter peak · Critical field · Ginzburg-Landau expansion · Magnetic flux · Pauli limit · BCS theory · Bogoliubov transformation

## 6.1 Perfect Diamagnetism

It is well known that electric resistance becomes zero in a superconducting state. Resistance is so perfectly zero that supercurrent in a superconducting ring (permanent current) is not attenuated even after thousands of years. This conclusion may be somewhat surprising because it is only one hundred years since superconductivity was discovered. However, this conclusion has been obtained by estimating the attenuation probability of the supercurrent. "Perfectly zero" is rare in the nature, and superconductivity is an extraordinary phenomenon. We shall discuss first, the macroscopic description of superconductivity and later the microscopic origin.

Magnetic field does not enter a superconducting material. This is known as **perfect diamagnetism** or **Meissner effect**. This phenomenon is more essential than the zero resistance. As discussed in Sect. 4.2, magnetic field $H$, flux density $B$, and magnetization $M$ are related to each other by the next equation.

$$B = H + 4\pi M \,(\text{cgs}) \quad B = \mu_0(H + M)\,(\text{MKS}) \tag{6.1}$$

Magnetism is usually discussed using the cgs unit, but superconductivity is frequently described in the MKS unit. In the following, we will describe both relations. In the MKS unit, $B = \mu_0(H + M)$ is more widely used instead of $B = \mu_0 H + M$ (Fig. 4.8) for describing superconductivity, so we shall use the former notation. It is a characteristic of superconductivity that perfect diamagnetism



T. Mori, *Electronic Properties of Organic Conductors*,
DOI 10.1007/978-4-431-55264-2_6

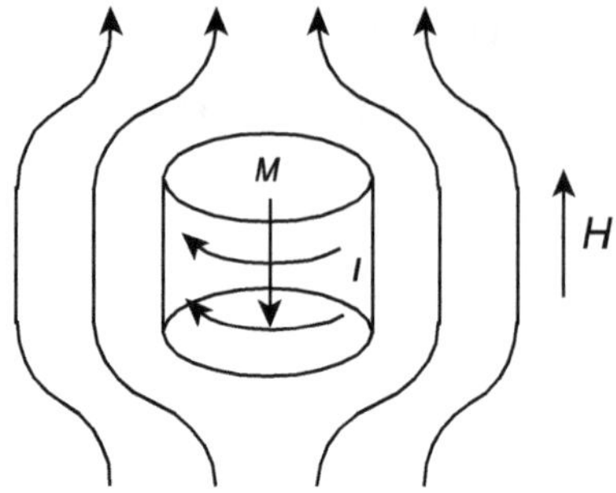


**Fig. 6.1** Perfect diamagnetism

$B = 0$ is attained due to the complete screening of the magnetic field. Thus, the magnetization is

$$M = -1/4\pi H \quad M = -H \tag{6.2}$$

and the susceptibility is

$$\chi = -1/4\pi \quad \chi = -1. \tag{6.3}$$

In order to generate such magnetization that entirely eliminates the magnetic field, a screening current flows as shown in Fig. 6.1. If the material has resistance, energy dissipation occurs in a magnetic field due to the screening current. Therefore, zero resistance is a requisite of the perfect diamagnetism. In turn, we assume only zero resistance. Suppose a material is cooled from the normal state to the superconducting state in zero magnetic field, then a magnetic field is applied. In order to cancel the change of the magnetic field, a screening current flows to result in the perfect diamagnetism. However, when the material is cooled across the transition temperature in a magnetic field, perfect diamagnetism does not appear. Therefore, perfect diamagnetism is a stronger condition than perfect conductance. There have not been known any materials that exhibit only perfect conductance but no perfect diamagnetism.

In order to describe perfect diamagnetism more precisely, we shall suppose the equation of motion of a charged particle $e^*$ with mass $m^*$ in an electric field $\varepsilon$.

$$m^* \frac{\partial v}{\partial t} = e^* \varepsilon.$$

From Eq. 3.15, the current $J = n_s e^* v$ is put in this relation.

$$\frac{\partial J}{\partial t} = \frac{n_s e^{*2}}{m^*} \varepsilon$$

$\nabla\times$ is applied from the left, and using the Maxwell equation

$$\nabla \times \varepsilon = -\frac{1}{c}\frac{\partial B}{\partial t} \quad \nabla \times \varepsilon = -\frac{\partial B}{\partial t}.$$

we obtain

$$\nabla \times \frac{\partial J}{\partial t} = \frac{n_s e^{*2}}{m^*}\nabla \times \varepsilon = \frac{n_s e^{*2}}{cm^*}\left(-\frac{\partial B}{\partial t}\right)$$

This leads to

$$\frac{\partial}{\partial t}\left(\nabla \times J + \frac{n_s e^{*2}}{cm^*}B\right) = 0.$$

In the MKS unit, we put $c \to 1$. This relation means the content in the parenthesis does not depend on the time. However, we shall assume this quantity is always zero in a superconductor.

$$\nabla \times J = -\frac{1}{c\Lambda}B \quad \nabla \times J = -\frac{1}{\Lambda}B \tag{6.4}$$

Here, $\Lambda = m^*/n_s e^{*2}$ is used. This equation is called **London equation**. To another Maxwell equation

$$\nabla \times B = \frac{4\pi}{c}J \quad \nabla \times B = \mu_0 J$$

$\nabla\times$ is applied from the left, and $\nabla \times \nabla \times B \to -\nabla^2 B$ is used. $J$ is eliminated by putting Eq. 6.4 in this equation.

$$\lambda^2 \nabla^2 B = B \tag{6.5}$$

Here

$$\lambda^2 = \frac{c^2 m^*}{4\pi n_s e^{*2}} \quad \lambda^2 = \frac{m^*}{\mu_0 n_s e^{*2}} \tag{6.6}$$

is used. This relation implies that $B$ is attenuated as

$$B = H\exp(-x/\lambda) \tag{6.7}$$

in the superconductor. Then, if the London equation (Eq. 6.4) is assumed, perfect diamagnetism is concluded. $\lambda$ is called **penetration depth**. Similarly, $\nabla\times$ is applied to Eq. 6.4, and using the Maxwell equation, we obtain

$$\lambda^2 \nabla^2 J = J.$$

This leads to the screening current

$$J = J_0 \exp(-x/\lambda)$$

that flows only at the surface of the superconductor. Here, the coordinate $x$ is taken from the superconductor surface to the inside, and $z$ is along the magnetic field. Then, the screening current flows in the $y$ direction (Fig. 6.1). $e^*$ is the charge of the Cooper pair $2e$, and $m^*$ is $2m$. $\lambda$ is in the order of 40–200 nm (Table 6.1). From

**Table 6.1** Transition temperature, coherence length, penetration depth, and superconducting gap in representative superconductors [1–3]

| Material | $T_c$/K | $\xi$/nm | $\lambda$/nm | $\kappa = \xi/\lambda$ | $2\Delta$/K | $2\Delta/k_BT_c$ |
|---|---|---|---|---|---|---|
| Al | 1.18 | 1550 | 45 | 0.03 | 4.1 | 3.5 |
| Pb | 7.2 | 87 | 39 | 0.48 | 31 | 4.3 |
| Nb | 9.25 | 39 | 52 | 1.3 | 35 | 3.8 |
| NbN | 16 | 5 | 200 | 40 | 59 | 3.7 |
| $Nb_3Sn$ | 18 | 3 | 65 | 22 | 77 | 4.3 |
| $Nb_3Ge$ | 23.2 | 3 | 90 | 30 | 77 | 4.3 |
| $PbMo_6S_8$ | 14 | 2.2 | 215 | 98 | 97 | 4.2 |
| $LiTi_2O_4$ | 11 | | | | 42 | 3.8 |
| $La_{2-x}Sr_xCuO_4$ | 37 | 2 | 200 | 100 | 151 | 4.1 |
| $YBa_2Cu_3O_7$ | 89 | // 3.4 | 36 | 11 | 348 | 4.0 |
| | | ⊥ 0.7 | 125 | 179 | | |
| $Tl_2Ba_2Ca_2Cu_3O_{10}$ | 123 | // | 173 | | 510 | 4.1 |
| | | ⊥ | 480 | | | |
| $HgBa_2Ca_2Cu_3O_8$ | 133 | // 1.5 | 130 | 87 | 557 | 4.2 |
| | | ⊥ 0.19 | 3500 | $1.8 \times 10^4$ | | |
| $MgB_2$ | 39 | // 6.5 | 100 | 15 | 140 | 3.6 |
| | | ⊥2.5 | | | | |
| $SmFeAsO_{1-x}F_x$ | 55 | | 200 | | 100 | |
| $K_3C_{60}$ | 19 | 2.6 | 240 | 92 | 57 | 3.6 |
| $Rb_3C_{60}$ | 29.6 | 2 | 247 | 124 | 87 | 3.0 |
| $(TMTSF)_2ClO_4$ | 1.2 | *a* 77 | | | 3.8 | 3.2 |
| | | *b* 36 | | | | |
| | | *c* 2 | | | | |
| κ-$(ET)_2Cu(NCS)_2$ | 10.4 | // 7 | 1400 | 200 | 30 | 2.9 |
| | | ⊥0.5 | 40,000 | $8 \times 10^4$ | | |
| κ-$(ET)_2Cu[N(NC)_2]Br$ | 11.3 | // 3.7 | 1500 | 400 | 57 | 5.0 |
| | | ⊥ 0.6 | 38,000 | $6 \times 10^4$ | | |

ET = BEDT-TTF

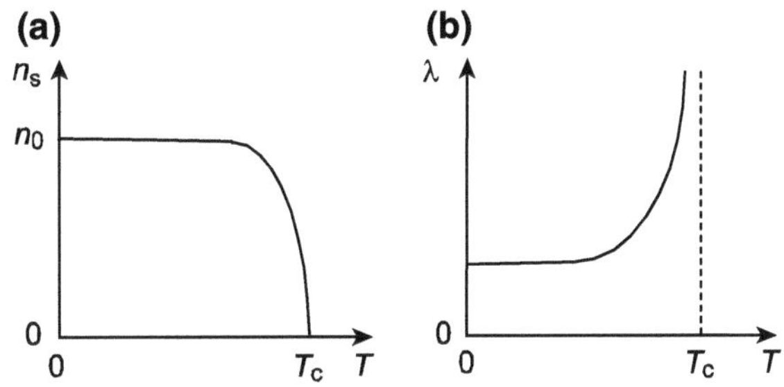


**Fig. 6.2** Temperature dependence of **a** The number of superconducting electrons, $n_s$, and **b** the penetration depth $\lambda$

Eq. 6.6, $\lambda$ decreases with increasing the number of superconducting electrons $n_s$. Later we will see $n_s$ shows the following temperature dependence.

$$n_s = n_0\left[1 - (T/T_c)^4\right] \tag{6.8}$$

Since this becomes zero at $T \rightarrow T_c$, $\lambda$ diverges to infinity at $T \rightarrow T_c$ (Fig. 6.2). Near $T_c$, the magnetic field gradually penetrates deeply into the superconductor. This is because the decreasing $n_s$ cannot screen the magnetic field.

The London equation is alternatively derived from momentum in the magnetic field.

$$\mathbf{p} = m\mathbf{v} + \frac{e^*}{c}\mathbf{A} \quad \mathbf{p} = m\mathbf{v} + e^*\mathbf{A} \tag{6.9}$$

Here, the vector potential $\mathbf{A}$ is represented by $\mathbf{H} = \nabla \times \mathbf{A}$. We assume $\mathbf{p} = 0$ in a superconductor. $\mathbf{J} = n_s e^* \mathbf{v}$ is put in the above equation.

$$\mathbf{J} = -\frac{c}{4\pi\lambda^2}\mathbf{A} \quad \mathbf{J} = -\frac{1}{\mu_0\lambda^2}\mathbf{A} \tag{6.10}$$

$\nabla\times$ is applied from the left to lead to the London equation. Accordingly, Eq. 6.10 is equivalent to the London equation.

*Example 6.1* In an organic superconductor, (BEDT-TTF)$_2X$, the lattice volume per this formula is approximately 800 Å$^3$. Then, the electron density is $n = 1.25 \times 10^{21}$ cm$^{-3}$ = $1.25 \times 10^{27}$ m$^{-3}$. Calculate the penetration depth assuming the electron mass is equal to the free electron.

$e^* = 2e$, $m^* = 2\,m$, and $n_s = n/2$ are put in Eq. 6.6.

$$\lambda^2 = \frac{c^2 m^*}{4\pi n_s e^{*2}} = \frac{c^2 m}{4\pi n e^2} = \frac{(2.998 \times 10^{10}\ \text{cm/s})^2 \times (9.1096 \times 10^{-28}\ \text{g})}{4\pi(1.25 \times 10^{21}\ \text{cm}^{-3}) \times (4.803 \times 10^{-10}\ \text{esu})^2}$$
$$= 2.26 \times 10^{-10}\ \text{cm}^2$$

Equivalently in the MKS unit

$$\lambda^2 = \frac{m^*}{\mu_0 n_s e^{*2}} = \frac{m}{\mu_0 n e^2}$$
$$= \frac{(9.1096 \times 10^{-31}\ \mathrm{kg})}{(4\pi \times 10^{-7}) \times (1.25 \times 10^{27}\ \mathrm{m^{-3}}) \times (1.6022 \times 10^{-19}\ \mathrm{C})^2}$$
$$= 2.26 \times 10^{-14}\ \mathrm{m^2}$$

which leads to $\lambda$ = 150 nm. The actually observed $\lambda$ is anisotropic and in the order of $10^3$–$10^5$ nm. This is attributed to $n_s$ smaller than $n$, and $m^*$ larger than the free-electron value.

## 6.2 Energy Gap

As shown in Fig. 6.3a, a superconductor has an energy gap $\pm\Delta$ at the Fermi energy. The density of states in the gap is piled up out of the gap and makes peaks. The gap shows such temperature dependence (Fig. 6.3b).

$$\Delta \propto \sqrt{T_c - T} \tag{6.11}$$

The formation of an energy gap is analogous to band insulators such as the Peierls insulators (Fig. 6.3c). The energy gap in the Peierls insulator is associated with the

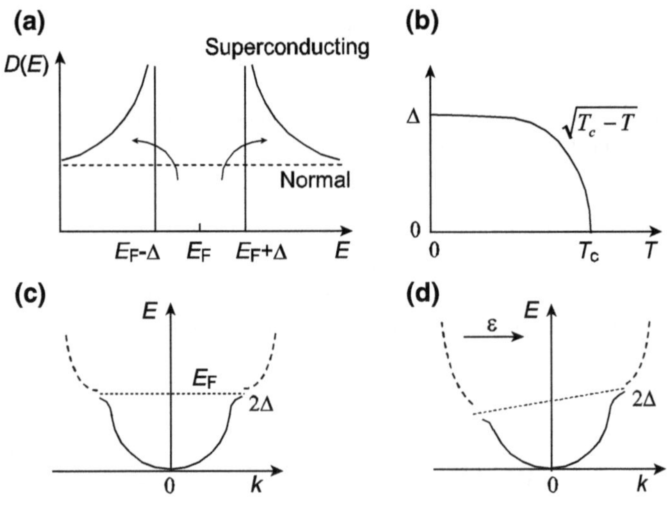


**Fig. 6.3** Superconducting gap

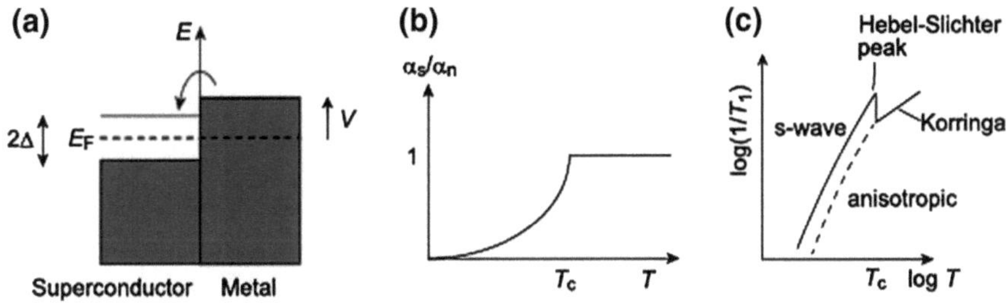


**Fig. 6.4** **a** Energy diagram observed by STS. **b** Ultrasonic attenuation, and **c** NMR relaxation time

enlarged periodicity on the lattice, and connected with the periodicity in the real space. By contrast, we will later see that the superconducting gap is related to the standing wave formed by the **Cooper pair** connecting the $k_F$ and $-k_F$ electrons in the $k$ space. When an electric field $\varepsilon$ is applied, the pair is formed from electrons at $k_F + \delta k$ and $-k_F + \delta k$ by shifting the center of gravity by $\delta k = p/\hbar$, so the gap is also shifted (Fig. 6.3d). The supercurrent flows persistently with a constant momentum **p** unless excitation occurs across this gap, so the supercurrent flows without friction. Accordingly, resistance becomes zero because the energy dissipation does not occur due to the energy gap.

The superconducting gap is directly observed by scanning tunneling spectroscopy (STS) and infrared spectroscopy (Fig. 6.4a). We will see later that the superconducting gap is related to the transition temperature like $2\Delta = 3.53k_BT_c$ (Eq. 6.37). Then in $YBa_2Cu_3O_7$ with $T_c \sim 90$ K, the superconducting gap $2\Delta$ is several 10 meV when observed in STS, and several hundred $cm^{-1}$ when observed in infrared spectroscopy. Since the superconducting gap is in the same order as the transition temperature, it is more than two orders of magnitude smaller than the bandwidth, which is usually in the order of several eV. The superconducting gap is also observed in the exponential ultrasonic attenuation (Fig. 6.4b). In ordinary metals, spin-lattice relaxation time $T_1^{-1}$ of nuclear magnetic resonance follows the **Korringa rule**, where $(T_1T)^{-1}$ is constant (Fig. 6.4c). In a superconductor, $T_1^{-1}$ once makes a peak called **Hebel-Slichter peak** just below $T_c$, and diminishes exponentially. However, in anisotropic superconductors with zero gap directions, the peak does not appear, and $T_1^{-1}$ decreases following a power law at low temperatures.

Magnetic field destroys superconductivity, and restores the normal metal state. The **critical magnetic filed** $H_c$ is approximated by a downward parabola (Fig. 6.5a).

$$H_c(T) = H_c(0)\left[1 - \left(\frac{T}{T_c}\right)^2\right] \tag{6.12}$$

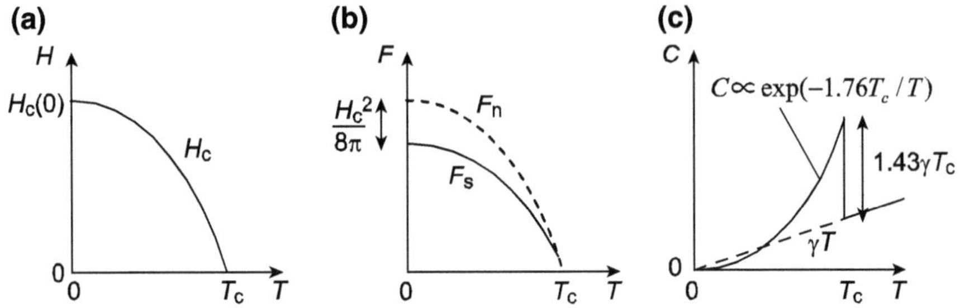


**Fig. 6.5** **a** Temperature dependence of critical magnetic field, **b** free energy, and **c** heat capacity

In a superconducting magnet, a strong magnetic field is generated by a huge current flowing in a superconducting coil. The resulting magnetic field is limited by the critical field. In a superconducting magnet in a laboratory NMR, typically about 100 A of current flows in the superconducting coil, which generates at most ten T (Tesla) of magnetic field.

Energy that is necessary to exclude magnetic fields from a material is given by $dE = MdH$ (Sect. 4.2). $M$ in the perfect diamagnetic state (Eq. 6.2) is put in this equation, and integrated from zero to $H_c$.

$$F_s - F_n = -\int_0^{H_c} \frac{H}{4\pi} dH \quad F_s - F_n = -\int_0^{H_c} \frac{H}{\mu_0} dH$$

$F_n$ designates the free energy in the **normal state**. After the integration, we see how the free energy of the superconducting state is lower than the free energy of the normal state.

$$F_s - F_n = -\frac{H_c^2}{8\pi} \quad F_s - F_n = -\frac{H_c^2}{2\mu_0}. \tag{6.13}$$

Temperature dependence of the free energy is depicted in Fig. 6.5b. The transition to the superconducting state is second order at zero magnetic field, but first order at nonzero magnetic field. On account of the second-order transition, there is no latent heat at $T_c$. However, the heat capacity jumps by $\Delta C = 1.43\gamma T_c$ as shown in Fig. 6.5c, where $\gamma$ is the coefficient of the electronic heat capacity (Sect. 4.9). Since an isotropic superconductor has an energy gap of $\Delta = 1.76 k_B T_c$, the heat capacity diminishes exponentially following $C \propto \exp(-1.76 T_c/T)$ below $T_c$. An anisotropic superconductor has finite state density even in the gap, and the heat capacity decreases following a power law, $C \propto T^m$.

## 6.3 Ginzburg-Landau Expansion

Since $\Delta$ is small around $T_c$, we can expand the free energy $F_s$ with respect to $\Delta$.

$$F_s - F_n = a|\Delta|^2 + \frac{b}{2}|\Delta|^4 + \cdots \quad (6.14)$$

$\Delta$ odd terms do not appear because $\pm\Delta$ are equivalent. This is called **Ginzburg-Landau expansion**. This is similar to the Laudau expansion of the $\pi d$-system shown in Sect. 4.9. Here, $\Delta$ and $M$ are called **order parameter**. $\Delta$ that minimizes the free energy is obtained by differentiating the above equation by $|\Delta|^2$.

$$\frac{\partial(F_s - F_n)}{\partial|\Delta|^2} = a + b|\Delta|^2 = 0 \quad (6.15)$$

Assuming such temperature dependence as $a = a'(T - T_c)$, from the shape of Eq. 6.14 (Fig. 6.6), we learn the stable state with the minimum free energy.

When $T > T_c$ $a > 0$ leads to $|\Delta|^2 = 0$

When $T < T_c$ $a < 0$ leads to $|\Delta|^2 = |a|/b$

Accordingly, $\Delta \neq 0$ is realized below $T_c$, and the nonzero $\Delta$ results in the superconducting state. The temperature dependence of $\Delta$ is

$$|\Delta| = \sqrt{\frac{a'(T_c - T)}{b}} \quad (6.16)$$

which leads to Eq. 6.11. In general, the Landau expansion is equivalent to the mean-field approximation, and such temperature dependence is universal in the mean-field theory.

In order to investigate the free energy under a magnetic field, Ginzburg and Landau have further assumed that $\Delta$ behaves like a macroscopic wave function.

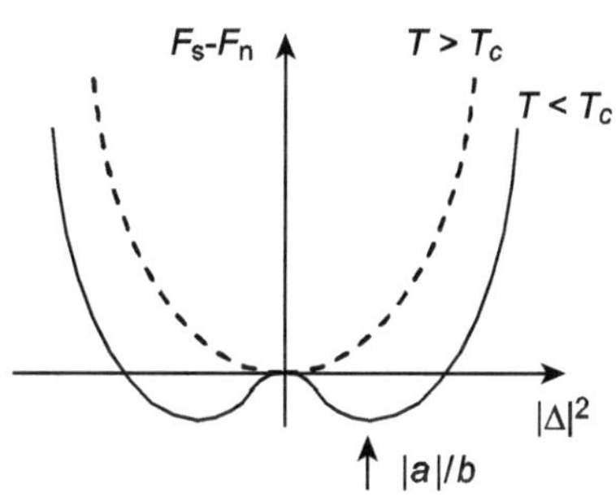


**Fig. 6.6** Free energy in the Ginzburg-Landau expansion

Since $F_s$ has a dimension of energy, Eq. 6.14 is rewritten including the macroscopic wave function $\psi = (\sqrt{2m}/\hbar^2)\Delta$.

$$F_s - F_n = \alpha|\psi|^2 + \frac{\beta}{2}|\psi|^4 + \frac{1}{2m^*}\left|\left(\frac{\hbar}{i}\nabla - \frac{e^*}{c}\mathbf{A}\right)\psi\right|^2 + \frac{H^2}{8\pi} \tag{6.17}$$

The first and second terms come from Eq. 6.14. The third term is the kinetic energy in the Schrödinger equation, where the magnetic field is involved as the vector potential **A** following Eq. 6.9. The momentum **p** is designated in the differential form according to Eq. 1.7. The last term comes from the energy of the magnetic field (Eq. 6.13). When $\psi$ is spatially constant under zero magnetic field, this equation is reduced to Eq. 6.14. If $\psi$ spatially varies under a magnetic field, the third term affords additional energy.

In order to estimate the distribution of $\psi$ and the magnetic field that minimizes the free energy, variation of Eq. 6.17 with respect to $\delta\psi^*$ and $\delta\mathbf{A}$ is calculated. The variation for $\delta\psi^*$ gives

$$\alpha\psi + \beta|\psi|^2\psi + \frac{1}{2m^*}\left(\frac{\hbar}{i}\nabla - \frac{e^*}{c}\mathbf{A}\right)^2\psi = 0 \tag{6.18}$$

and the variation for $\delta\mathbf{A}$ affords

$$\mathbf{J} = \frac{\hbar e^*}{2m^* i}(\psi^*\nabla\psi - \psi\nabla\psi^*) - \frac{e^{*2}}{cm^*}\psi^*\psi\mathbf{A} \tag{6.19}$$

Put $\mathbf{A} = 0$ in Eq. 6.18, and we perform the replacement $\psi = (|\alpha|/\beta)^{1/2}f$ and $\xi^2 = \hbar^2/2m^*|\alpha|$ in a one-dimensional system.

$$\xi^2\frac{d^2f}{dx^2} + f - f^3 = 0$$

When the edge of the superconductor is located at $x = 0$, the limiting conditions are $f = 0$ at $x = 0$ and $f = 1$ at $x = \infty$. In such a case, we easily see

$$f = \tanh\left(\frac{x}{\sqrt{2}\xi}\right)$$

is the solution of the above differential equation. At the edge of a superconductor under zero magnetic field, the superconducting gap $\Delta$ decays to zero within a distance of $\xi$ (Fig. 6.7a). The characteristic distance $\xi$ is called **coherence length**, which is typically 2–200 nm. The temperature dependence of $\xi$ is obtained from the temperature dependence of $\alpha$.

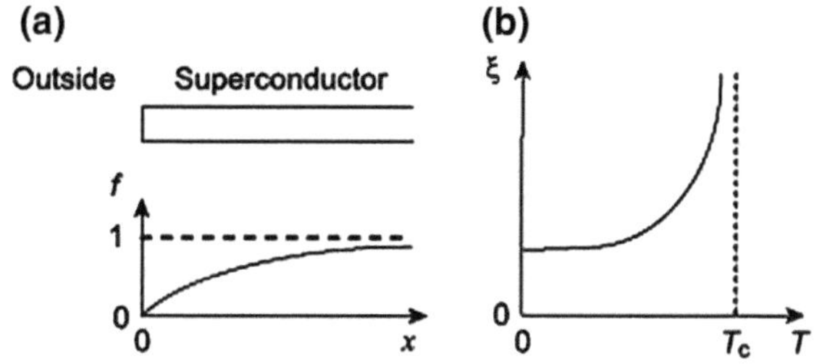


**Fig. 6.7** **a** Spatial variation of the superconducting gap, and **b** temperature dependence of the coherence length

$$\xi \propto \left(1 - \frac{T}{T_c}\right)^{-1/2}$$

As shown in Fig. 6.7b, $\xi$ diverges to infinity at $T_c$ similarly to $\lambda$. Then, if there is a zero gap region, for instance, at the superconductor's edge, the gap is not restored even at an infinitely distant place at $T_c$.

The coherence length represents the size of the Cooper pair. Since the superconducting gap is $2\Delta$, uncertainty of the momentum is

$$2\Delta \approx \delta\left(\frac{p^2}{2m}\right) \approx \frac{p_F}{m}\delta p$$

Suppose the uncertainty of $x$ to be $\xi$. Using the uncertainty principle

$$\xi = \delta x \approx \frac{\hbar}{\delta p} = \frac{\hbar p_F}{m2\Delta} = \frac{\hbar^2 k_F}{m2\Delta}.$$

This is transformed using $E_F = \hbar^2 k_F^2/2m$.

$$\xi \approx \frac{E_F}{k_F \Delta}$$

$1/k_F$ is in the order of the lattice constant, and $E_F/\Delta$ is $10^2$–$10^3$. Then, $\xi$ is expected to be $10^2$–$10^3$ times of the lattice constant, and in the order of 10–100 nm. The actual values for superconductors are listed in Table 6.1.

The first and second terms in Eq. 6.19 are respectively called **paramagnetic** and **diamagnetic currents**. Putting $\psi = \psi_0 e^{i\theta}$ in Eq. 6.19,

$$\mathbf{J} = \frac{\hbar e^*}{m^*}|\psi|^2 \nabla\theta - \frac{e^{*2}}{cm^*}|\psi|^2 \mathbf{A} \tag{6.20}$$

is obtained. Suppose a current flowing in a donut of a superconductor. When the above equation is integrated rotating around the donut, the phase θ has to be unique, so that it is integer times of $2\pi$.

$$\oint \nabla\theta \mathrm{d}l = 2\pi n$$

The integration of vector potential follows the Stokes theorem

$$\oint \mathbf{A}\mathrm{d}l = \int \nabla \times \mathbf{A}\mathrm{d}l = \int H\mathrm{d}S = \Phi$$

where $\Phi$ is the integration of the magnetic field passing through the hole of the donut. Using these relations, Eq. 6.20 is converted.

$$\frac{m^* c}{e^{*2}} \oint J\mathrm{d}l + \Phi = n\Phi_0 \quad \text{where } \Phi_0 = \frac{2\pi\hbar c}{e^*} = \frac{hc}{e^*} = \frac{hc}{2e} = 2 \times 10^{-7}\ \text{Gauss/cm}^2$$

Outside the superconductor, the current is zero, and the integration of the first term becomes zero. This leads to $\Phi = n\Phi_0$, implying the magnetic flux in the donut is quantized to integer times of $\Phi_0$. $\Phi_0$ is called **magnetic flux.** $\Phi_0$ corresponds to a magnetic field generated by a Cooper pair with charge $2e$ rotating around the donut. Since $\Delta$ works as a macroscopic wave function, its phase has to coincide with the original one after a rotation, and this condition gives rise to the quantization of the magnetic field in the unit of flux. Minimum energy $\Delta$ is necessary to destroy one Cooper pair, but if $T$ is considerably lower than $T_c$, this probability exponentially decays to a very small number. This suppresses the decay of $\Phi$, leading to the permanent current. Then, zero resistance and the very small attenuation ratio of the superconducting current are concluded.

*Example 6.2* Calculate the magnitude of a magnetic flux.

$$\Phi_0 = \frac{hc}{2e} = \frac{(6.626 \times 10^{-27}\ \text{ergs}) \times (2.998 \times 10^{10}\ \text{cm/s})}{2(4.803 \times 10^{-10}\ \text{esu})} = 2.068 \times 10^{-7}\ \text{G/cm}^2$$

Or in the MKS unit

$$\Phi_0 = \frac{h}{2e} = \frac{(6.626 \times 10^{-34}\ \text{Js})}{2(1.602 \times 10^{-19}\ \text{C})} = 2.068 \times 10^{-15}\ \text{T/m}^2.$$

In general, $e$ in the MKS unit is replaced by $e/c$ in the cgs unit. The relation $\text{T} = 10^4$ G leads to $1\ \text{T/m}^2 = 10^{-8}\ \text{G/cm}^2$. A magnetic flux is substantially smaller than the geomagnetism, 0.24–0.66 G.

## 6.4 Type-II Superconductor

A superconductor has two characteristic distances: the penetration depth $\lambda$ and the coherence length $\xi$. $\lambda$ characterizes how a magnetic field is screened, and $\xi$ defines how the superconducting gap is restored. In the absence of magnetic field, $\xi$ determines the distance in which the superconductivity grows at the edge. Since a superconductor has lower energy than the normal state by $H_c^2/8\pi$, the boundary between a superconductor and a metal has a surface energy $H_c^2(\xi - \lambda)/8\pi$ per area. Accordingly, superconductors are classified into two depending on the relative magnitude of $\lambda$ and $\xi$.

1. $\xi > \lambda$: the surface energy is positive, implying a magnetic flux increases the energy. Then, it is preferable that the whole superconducting body keeps superconductivity to expel magnetic flux, or otherwise the whole superconductivity breaks. Such a material is called **type-I superconductor**. As listed in Table 6.1, elemental metals other than niobium are type-I superconductors. $M$ and $B$ in a type-I superconductor are depicted in Fig. 6.8a. Perfect diamagnetism $M = -1/4\pi$ is observed up to $H_c$, where $M$ abruptly drops to zero, and $B$ jumps from zero to $H$.
   When a magnetic field is applied perpendicularly to a plate of a type-I superconductor, superconductivity breaks in some parts, and the plate is separated to superconducting and non superconducting regions. Such a state is called **intermediate state**. The intermediate state of a superconductor depends on the shape. In general, magnetization of a magnetic material depends on the shape through the demagnetization coefficient, and a type-I superconductor is the extreme case.
2. $\xi < \lambda$: the negative surface energy makes the **mixed state** stable, where a minimal flux unit is incorporated in the superconductor one by one. Such a material is called **type-II superconductor**. As listed in Table 6.1, all compound superconductors are type II. Around a flux, the superconducting gap $\Delta$ grows up more rapidly than $H$ (Fig. 6.9). A type-II superconductor is a "hard"

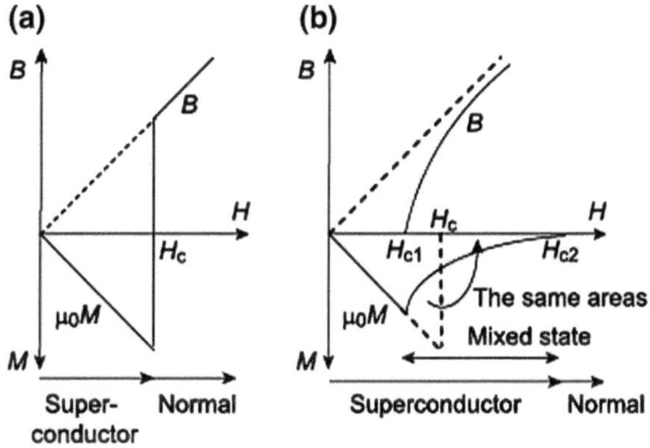


**Fig. 6.8** Magnetization $M$ and magnetic flux density $B$ inside **a** a type-I superconductor, and **b** a type-II superconductor

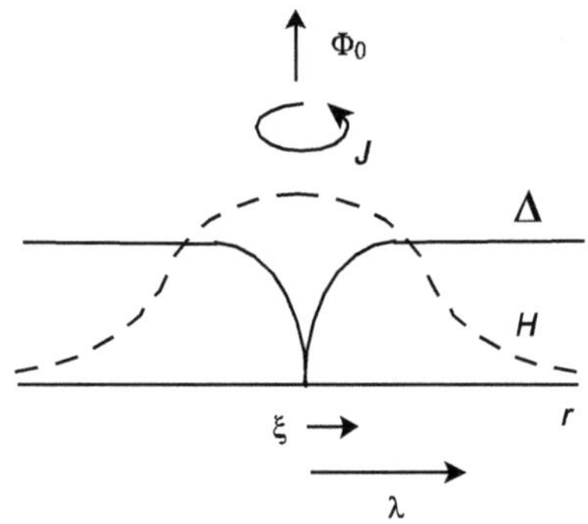


**Fig. 6.9** Magnetic field and superconducting gap around a flux

superconductor, where the superconductivity is not easily destroyed. $M$ and $B$ in a type-II superconductor are depicted in Fig. 6.8b. A flux enters the superconductor at a much lower $H_{c1}$ than $H_c$, and the superconductivity is partially maintained up to $H_{c2}$ much larger than $H_c$.

The superconductivity is broken in an area of $\pi\xi^2$ per flux. $H_{c2}$ is given by a magnetic field at which this area covers the whole material.

$$H_{c2} = \frac{\Phi_0}{2\pi\xi^2} \tag{6.21}$$

To calculate $H_{c1}$, the London equation (Eq. 6.5) is integrated around a flux.

$$\lambda^2 2\pi r \nabla \times B = \Phi_0$$

This relation is integrated again between $\xi < r < \lambda$ to afford

$$B(r) = \frac{\Phi_0}{2\pi\lambda^2} \ln\frac{\lambda}{r}.$$

$H_{c1}$ is given by the condition in which the energy of the penetrated magnetic field balances with the energy of a flux.

$$H_{c1} = \frac{\Phi_0}{4\pi\lambda^2} \ln\frac{\lambda}{\xi}$$

This relation reflects that in a flux the area of broken superconductivity is $\pi\xi^2$, whereas the area of the penetrated magnetic field is $\pi\lambda^2$. The critical fields are represented as

$$H_{c1} = H_c/\sqrt{2}\kappa, \quad H_{c2} = \sqrt{2}\kappa H_c$$

where the ratio of $\lambda$ and $\xi$, $\kappa = \lambda/\xi$, is called **Ginzburg-Landau parameter**. Since typical $\kappa$ is as large as one hundred (Table 6.1), $H_{c1}$ is considerably smaller than $H_c$, and $H_{c2}$ is much larger than $H_c$. The **mixed state** is realized in the region of $H_{c1} < H < H_{c2}$.

*Example 6.3* $H_{c2}$ in β-(BEDT-TTF)$_2$I$_3$ at 0 K is estimated to be 1.8 T in the conducting plane and 0.086 T perpendicular to the plane from the extrapolation of Fig. 7.27d. Calculate the out-of-plane and in-plane coherence lengths $\xi$. From the generalization of Eq. 6.21, $H_{c2}$ along the $a$ axis is represented by $H_{c2} = \Phi_0/2\pi\xi_b\xi_c$ using ξ along the perpendicular directions.

By assuming the in-plane $\xi_{//}$ isotropic, the perpendicular $H_{c2}^{\perp}$ is

$$H_{c2}^{\perp} = \frac{\Phi_0}{2\pi\xi_{//}^2} \text{ leading to } \xi_{//}^2 = \frac{\Phi_0}{2\pi H_{c2}^{\perp}} = \frac{(2.068 \times 10^{-15}\ \mathrm{Tm}^2)}{2\pi(0.086\ \mathrm{T})} = 3.827 \times 10^{-15}\ \mathrm{m}^2$$

which gives $\xi_{//}$ = 62 nm. The in-plane $H_{c2}^{//}$ satisfies

$$\xi_{\perp}\xi_{//} = \frac{\Phi_0}{2\pi H_{c2}^{//}} = \frac{(2.068 \times 10^{-15}\ \mathrm{Tm}^2)}{2\pi(1.8\ \mathrm{T})} = 1.829 \times 10^{-16}\ \mathrm{m}^2$$

which affords $\xi_{\perp}$ = 3.0 nm. Here, $\xi_{\perp}$ is about twice of the lattice constant, 1.5 nm, meaning the interlayer superconductivity is critically coherent. In some organic superconductors, $\xi_{\perp}$ is smaller than the interlayer spacing, where the interlayer superconductivity is incoherent, and each superconducting layer is independent. Figure 7.27d demonstrates $\xi_{//}$ is almost isotropic in the conducting plane, implying the two-dimensional superconductivity.

Under a magnetic field, parallel spins are stabilized in comparison with the singlet Cooper pair. Therefore, the ordinary singlet superconductivity does not exist exceeding the Zeemann energy. The condition that Δ balances with the Zeemann energy gives $\Delta = \sqrt{2}\mu_B H$. Using $2\Delta = 3.53k_BT_c$ (Eq. 6.37), this is converted to $H_P$ = (1.86 T/K)$T_c$. $H_P$ is called **Pauli limit** or **paramagnetic limit** (Fig. 6.10). $H_{c2}(0)$ of a superconductor with $T_c$ = 1 K never exceeds 1.86 T, whereas that of $T_c$ = 10 K is ten times larger. $H_{c2}(0)$ is usually as large as the Pauli limit. If $H_{c2}$

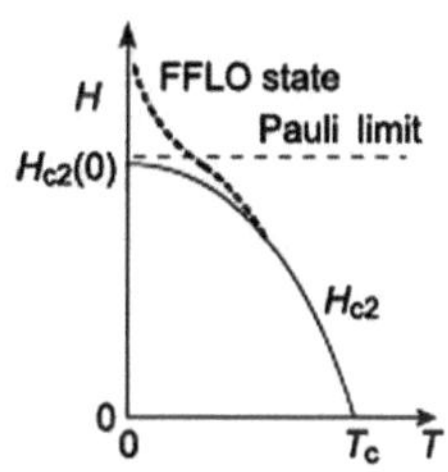


**Fig. 6.10** Pauli limit and the FFLO state

shows such a parabolic temperature dependence as Eq. 6.12, $H_{c2}(0)$ is related to the gradient of $H_{c2}$ at $T_c$ as

$$H_{c2}(0) = -0.5\ T_c \left( \frac{dH_{c2}}{dT} \right)_{T_c}.$$

However, more elaborate calculation considering the effect of the Pauli limit affords the following relation [2].

$$H_{c2}(0) = -0.69\ T_c \left( \frac{dH_{c2}}{dT} \right)_{T_c}$$

Triplet superconductivity (Sect. 6.7) is not restricted by the Pauli limit. $H_{c2}$ also exceeds the Pauli limit in the **FFLO** (Fulde-Ferrell-Larkin-Ovchinnikov) **state**, where spatially inhomogeneous superconductivity is stabilized (Fig. 6.10) [4, 5]. FFLO states have been suggested in TMTSF superconductors (Fig. 7.18b) and field-induced superconductivity in λ-$(BETS)_2FeCl_4$ (Fig. 7.41g).

*Example 6.4* Show the Pauli limit is given by $H_P$ = (1.86 T/K)$T_c$.

In the MKS unit, it is calculated as

$$H_P = \frac{1.765 k_B T_c}{\sqrt{2}\mu_B} = \frac{1.765 \times (1.38 \times 10^{-23}\ \text{J/K})}{\sqrt{2}(9.27 \times 10^{-24}\ \text{J/G})} T_c = (1.86\ \text{T/K}) T_c.$$

In a type-II superconductor with $\xi \ll \lambda$, $\nabla\psi$ in Eq. 6.19 is nonzero only in a very limited area of the radius of $\xi$. Then, the paramagnetic current in the first term of Eq. 6.19 is neglected. Thus, we obtain

$$\mathbf{J} = -\frac{e^{*2} n_s}{cm^*}\mathbf{A}.$$

This is equivalent to the London equation (Eq. 6.10). Accordingly, the London equation is valid in the $\xi \ll \lambda$ limit of a type-II superconductor. It is derived from the Ginzburg-Landau expansion by neglecting the paramagnetic current and considering only the diamagnetic current.

In the mixed state, each flux is stable in an isolated state, and repelled from each other. Then, fluxes construct a regular triangular lattice (Abrikozov's flux state). When a superconductor contains impurity and imperfection, fluxes are pinned to them, and do not move freely. In a superconducting wire, movement of fluxes is suppressed by introducing pinning centers and using fine strand wires.

When a magnetic field is applied perpendicularly to the superconducting layers of two-dimensional superconductors such as cuprate and organic superconductors with the incoherent interlayer $\xi$ shorter than the layer spacing, a flux in a layer does

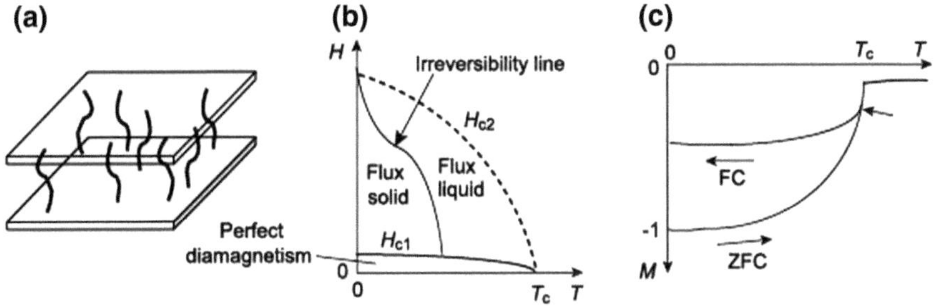


**Fig. 6.11** **a** Pancake model of a two-dimensional superconductor, **b** irreversible line, and **c** magnetization in the FC and ZFC states

not simply correspond to a flux in the next layer. Such a state is represented by the pancake model in Fig. 6.11a. In this case, the **irreversible line** in Fig. 6.11b is more important than $H_{c2}$. Outside the irreversible line is regarded as a flux liquid, composed of randomly incorporated fluxes. The low-temperature side of the irreversible line is a flux solid state. In such a case, magnetization measured under a magnetic field (in a field cooled (FC) state) is considerably smaller than the magnetization measured after cooling the sample down to the lowest temperature in zero field (in a zero-field cooled (ZFC) state). In the ZFC state, magnetic field in hollows of a superconductor is also zero, and in an ideal case, the susceptibility is very close to the perfect diamagnetism $\chi = -1/4\pi$. In the FC state, however, fluxes in hollows are maintained, and the susceptibility is considerably smaller than the perfect diamagnetism. The ratio of the ZFC susceptibility and the perfect diamagnetism is called superconducting fraction. It is close to 100 % in a bulk superconductor, but some superconductors show only less than 10 %.

## 6.5 Coulomb Repulsion in a Metal

Before investigating the microscopic origin of superconductivity, we shall discuss interaction between electrons in a metal. Coulomb repulsion between metal electrons is largely reduced owing to the screening effect of other electrons. Suppose a homogeneous metal electron. The Fourier transform of the Coulomb potential $V(r) = \frac{e^2}{4\pi\varepsilon_0 r}$ is represented as follows [6]:

$$V(k) = \int V(r) \mathrm{e}^{ikr} \frac{\mathrm{d}k}{(2\pi)^3} = \frac{e^2}{\varepsilon_o k^2} \tag{6.22}$$

For a screened electron in a metal, the Coulomb potential is replaced by the Yukawa potential $V(r) = \frac{e^2}{4\pi\varepsilon_0 r}\mathrm{e}^{-k_0 r}$, and the Fourier transform is

$$V(k) = \frac{e^2}{\varepsilon_o(k^2 + k_0^2)}. \tag{6.23}$$

This is substantially reduced, and in particular decays rapidly at large $r$. On account of this correction, the experimentally estimated Coulomb repulsion in a solid is several times smaller than the Coulomb integral evaluated on the basis of the molecular orbital calculation (Fig. 5.22b).

We have described the plasma oscillation in Sect. 3.2. The same discussion inversely holds for ions or atoms in metallic electrons. The lattice dielectric constant is obtained by replacing $m$ to $M$ and $e$ to $eZ$ in Eq. 3.30.

$$\varepsilon(\omega) = 1 - \frac{\Omega_p^2}{\omega^2} \quad \text{where} \quad \Omega_p = \frac{nZ^2e^2}{\varepsilon M} \tag{6.24}$$

Since metallic electrons are also in the lattice, the dielectric constant $\varepsilon_0$ in Eq. 6.23 is to be substituted by the relative dielectric constant of the lattice, $\varepsilon$ in Eq. 6.24. This leads to

$$V(k) = \frac{e^2}{\varepsilon_o(k^2 + k_0^2)} \frac{\omega^2}{\omega^2 - \Omega_p^2}. \tag{6.25}$$

Similarly to Fig. 3.9b, this equation is negative at $\omega < \Omega_p$. $\Omega_p$ is the representative energy of the lattice vibration, and regarded as the **Debye temperature**, which is in the order of room temperature. For slow electrons with lower energy than this energy, after an electron passes attracting the surrounding ions, the second electron goes into the comparatively positive area. As a result, the second electron is attracted to the first electron. This is the origin of the Cooper pair.

## 6.6 Microscopic Theory of Superconductivity

Similarly to the Hubbard model, we shall investigate Eq. 5.29 with attractive potential for the Cooper pair [7]. In a Cooper pair, a $k\uparrow$ electron most preferably makes a pair with a $-k\downarrow$ electron.

$$H = \sum_{i \neq j} t_{ij} c_{i\sigma}^{+} c_{j\sigma} + \sum_{k,k'} V_{k,k'} c_{k'\uparrow}^{+} c_{-k'\downarrow}^{+} c_{-k\downarrow} c_{k\uparrow}$$

In analogy with Eq. 5.41 for the Hubbard model, we apply the following mean-field approximation to the second term.

$$V c^+_{k'\uparrow} c^+_{-k'\downarrow} c_{-k\downarrow} c_{k\uparrow} = V < c^+_{k'\uparrow} c^+_{-k'\downarrow} > c_{-k\downarrow} c_{k\uparrow} + V c^+_{k'\uparrow} c^+_{-k'\downarrow} < c_{-k\downarrow} c_{k\uparrow} > \\ - V < c^+_{k'\uparrow} c^+_{-k'\downarrow} > < c_{-k\downarrow} c_{k\uparrow} > \tag{6.26}$$

This equation represents a process destroying a Cooper pair with $k\uparrow$ and $-k\downarrow$, and creating another Cooper pair with $k'\uparrow$ and $-k'\downarrow$. We assume the attractive force $V < 0$ is independent of $k$ at least for small $k$. We have used the average electron number $\langle n_{i\sigma}\rangle = \langle c^+_{i\sigma} c_{i\sigma}\rangle$ in Eq. 5.41, but the present average $\langle c^+_{k\uparrow} c^+_{-k\downarrow}\rangle$ involves two creation operators. Then, the electron number changes. The corresponding ground state has such a form

$$|\mathrm{BCS}\rangle = \prod_k \left(\cos\theta_k + \sin\theta_k c^+_{k\uparrow} c^+_{-k\downarrow}\right)|0\rangle \tag{6.27}$$

which is called BCS (Bardeen-Cooper-Schrieffer) state. $c^+_{k\uparrow} c^+_{-k\downarrow}|0\rangle$ designates a state $\psi_{k\uparrow}\psi_{-k\downarrow}$ with $k\uparrow$ and $-k\downarrow$ electrons. Therefore

$$\sin\theta_k = 1 \quad \text{for } k < k_\mathrm{F} \text{ and } \sin\theta_k = 0 \text{ for } k > k_\mathrm{F}$$

is the ordinary Fermi sea, where the states with $k < k_\mathrm{F}$ are occupied. If $\sin\theta_k$ is not a step function, $|\mathrm{BCS}\rangle$ is constructed by adding states of different particle numbers. In general, the phase $\varphi$ and the particle number $N$ satisfy an uncertainty principle $\Delta\varphi\Delta N > 1$, so a superconducting state with a standing wave and a definitely ordered phase $\varphi$ results in an uncertain particle number $N$.

We designate

$$\Delta_k = -\sum_{k'} V_{k,k'}\langle c_{-k'\downarrow} c_{k'\uparrow}\rangle, \quad \Delta^*_k = -\sum_{k'} V_{k,k'}\left\langle c^+_{k'\uparrow} c^+_{-k'\downarrow}\right\rangle \tag{6.28}$$

in Eq. 6.26, and the noninteracting diagonal term is written as $\xi_k = E(k) - \mu$. The Hamiltonian is

$$H = 2\sum_k \xi_k c^+_{k\sigma} c_{k\sigma} - \sum_k \left(\Delta^*_k c_{-k\downarrow} c_{k\uparrow} + \Delta_k c^+_{k\uparrow} c^+_{-k\downarrow} - \Delta_k\left\langle c^+_{k\uparrow} c^+_{-k\downarrow}\right\rangle\right). \tag{6.29}$$

The last term is a constant. In general, $\cos\theta_k$ and $\sin\theta_k$ in Eq. 6.27 are imaginary numbers written as $\cos\theta_k \to u_k$ and $\sin\theta_k \to v_k$. For $s$-wave superconductors, however, these quantities, together with $\Delta_k$, are taken real, so we shall regard these parameters real.

We can calculate the total energy by sandwiching the Hamiltonian with the BCS state (Eq. 6.27), but we have to determine the coefficients, $\cos\theta_k$ and $\sin\theta_k$. Similarly to the molecular orbital theory, variation of the total energy with respect to $\theta_k$ gives the ground-state energy and $\theta_k$. BCS carried out such calculation. Alternatively, in analogy with the solution of the Hubbard model (Eq. 5.43), we shall attempt to diagonalize the Hamiltonian by the **Bogoliubov transformation**

$$c_{k\uparrow}^{+} = \cos\theta_k \gamma_{k\uparrow}^{+} + \sin\theta_k \gamma_{-k\downarrow} \quad c_{k\uparrow} = \cos\theta_k \gamma_{k\uparrow} + \sin\theta_k \gamma_{-k\downarrow}^{+}$$

$$c_{-k\downarrow} = \cos\theta_k \gamma_{-k\downarrow} - \sin\theta_k \gamma_{k\uparrow}^{+} \quad c_{-k\downarrow}^{+} = \cos\theta_k \gamma_{-k\downarrow}^{+} - \sin\theta_k \gamma_{k\uparrow}. \tag{6.30}$$

Here, $\gamma_{k\uparrow}^{+}$ and $\gamma_{-k\downarrow}$ are creation and annihilation operators of the interacting quasi particles. Note that the creation and annihilation operators are "hybridized" in the BCS state; this is characteristic of superconductivity. Putting Eq. 6.30 in Eq. 6.29, the nondiagonal term is

$$(2\xi_k \sin\theta_k \cos\theta_k + \Delta_k^* \sin^2\theta_k - \Delta_k \cos^2\theta_k)(\gamma_{k\uparrow}^{+}\gamma_{-k\downarrow}^{+} + \gamma_{-k\downarrow}\gamma_{k\uparrow}).$$

Supposing $\Delta_k^* = \Delta_k$ are real, this term is zero at

$$\tan 2\theta_k = \frac{\Delta_k}{\xi(k)} \tag{6.31}$$

This gives

$$\cos^2 2\theta_k = \frac{1}{1 + \tan^2 2\theta_k} = \frac{\xi_k^2}{\xi_k^2 + \Delta_k^2}$$

leading to

$$\cos^2\theta_k = \frac{1 + \cos 2\theta_k}{2} = \frac{1}{2}\left(1 + \frac{\xi_k}{\sqrt{\xi_k^2 + \Delta_k^2}}\right)$$

$$\sin^2\theta_k = \frac{1}{2}\left(1 - \frac{\xi_k}{\sqrt{\xi_k^2 + \Delta_k^2}}\right)$$

Putting Eq. 6.30 in Eq. 6.29, the diagonal term is obtained as

$$\xi_k(\cos^2\theta_k - \sin^2\theta_k) + 2\Delta_k \cos\theta_k \sin\theta_k$$

but putting $\cos^2\theta_k$ and $\sin^2\theta_k$, the diagonal term is finally obtained.

$$E = \sum_k \sqrt{\xi_k^2 + \Delta_k^2}.$$

As shown in Fig. 6.12a, an energy gap, $2\Delta_k$, opens at $E_F$ in the original non-interacting band $\xi_k$. The coefficients, $\cos\theta_k$ and $\sin\theta_k$, which correspond to occupied and unoccupied probabilities of the Cooper pair, change not stepwise but gradually over the region of $\Delta$ (Fig. 6.12b). This reminds us that the Fermi distribution

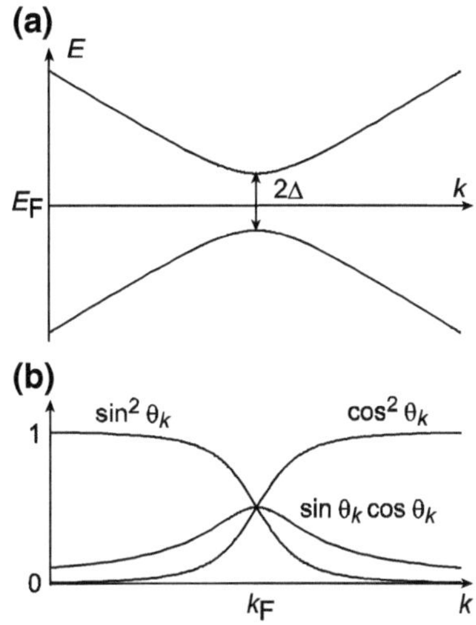


**Fig. 6.12** **a** Energy, $E(k)$, of a superconducting state, and **b** probabilities of occupied and unoccupied states in the Cooper pair

function changes gradually over the region of $k_BT$ (Fig. 2.19). However, the superconducting ground state is composed of states with different electron numbers even at 0 K as designated by Eq. 6.27.

Equation 6.30 is put in the definition of $\Delta_k$ (Eq. 6.28), and $\langle \gamma_{-k'\downarrow}\gamma_{k'\uparrow}\rangle$ and $\left\langle \gamma^+_{k'\uparrow}\gamma^+_{-k'\downarrow}\right\rangle$ are neglected. Using $\gamma_{-k'\downarrow}\gamma^+_{-k'\downarrow} = 1 - \gamma^+_{-k'\downarrow}\gamma_{-k'\downarrow}$, we obtain

$$\Delta_k = -\sum_{k'} V_{k,k'} \sin\theta_{k'} \cos\theta_{k'} \left(1 - \left\langle \gamma^+_{-k'\downarrow}\gamma_{-k'\downarrow}\right\rangle - \left\langle \gamma^+_{k'\uparrow}\gamma_{k'\uparrow}\right\rangle\right).$$

Here

$$\sin\theta_{k'}\cos\theta_{k'} = (1/2)\sin 2\theta_{k'} = \Delta_{k'}/2\sqrt{\xi^2_{k'} + \Delta^2_{k'}}.$$

Since $\left\langle \gamma^+_{-k'\downarrow}\gamma_{-k'\downarrow}\right\rangle$ and $\left\langle \gamma^+_{k'\uparrow}\gamma_{k'\uparrow}\right\rangle$ represent excited quasi particles, these averages are assumed to follow the Fermi distribution function $f(E) = 1/(\exp(E/k_BT) + 1)$. We then obtain

$$\Delta_k = -\sum_{k'} V_{k,k'} \frac{\Delta_{k'}}{2\sqrt{\xi^2_{k'} + \Delta^2_{k'}}}(1 - 2f(E)). \tag{6.32}$$

This is called the gap equation, because it determines the gap self-consistently. We usually assume that $V_{k,k'}$ is attractive for electrons whose energy is lower than the Debye frequency $\omega_D$.

$$V_{k,k'} = -V \quad |\xi_k| < \hbar\omega_D$$
$$= 0 \quad |\xi_k| > \hbar\omega_D$$

From Eq. 6.28, $\Delta_k$ follows similar relations.

$$\Delta_k = \Delta \quad |\xi_k| < \hbar\omega_D$$
$$= 0 \quad |\xi_k| > \hbar\omega_D$$

At 0 K, the Fermi distribution is $f(E) = 0$ for $E > 0$. Then, the above equation is

$$\Delta = V \sum_{k'} \frac{\Delta}{2\sqrt{\xi_{k'}^2 + \Delta^2}}.$$

Both sides are divided by $\Delta$, and the summation is converted to an integration

$$1 = V \int_0^{\hbar\omega_D} \frac{D\mathrm{d}\xi}{\sqrt{\xi^2 + \Delta^2}} = VD \sinh^{-1} \frac{\hbar\omega_D}{\Delta} \tag{6.33}$$

where $D = D(E_F)$ is the density of states at the Fermi level. In the weak coupling limit, $VD \ll 1$, this relation is reduced to

$$\Delta = 2\hbar\omega_D \exp(-1/VD). \tag{6.34}$$

This equation determines the energy gap $\Delta$ at 0 K.

At finite temperatures, the definition of the Fermi distribution function affords $1 - 2f(E) = \tanh(E/2)$ in Eqs. 6.32, and 6.33 is substituted by

$$\frac{1}{DV} = \int_0^{\hbar\omega_D} \frac{\tanh(\sqrt{\xi^2 + \Delta^2}/2k_BT)}{\sqrt{\xi^2 + \Delta^2}} \mathrm{d}\xi. \tag{6.35}$$

To estimate $T_c$, $\Delta$ is put zero. The integration is carried out by transforming $x = \xi/2k_BT_c$

$$\frac{1}{DV} = \int_0^{\hbar\omega_D/2k_BT} \frac{\tanh x}{2k_BT_cx} 2k_BT_c\mathrm{d}x = \int_0^{\hbar\omega_D/2k_BT} \frac{\tanh x}{x} \mathrm{d}x = \ln\left(\frac{2e^{\gamma}\hbar\omega_D}{\pi k_BT_c}\right)$$

where $\gamma = 0.57721$ is Euler's constant. This relation gives the transition temperature.

$$k_B T_c = 1.13\hbar\omega_D \exp(-1/VD) \tag{6.36}$$

When the attracting potential $V$ is zero, inside the exponential is $-\infty$, and $T_c$ is zero. Then, $T_c$ increases exponentially with increasing $V$. It is obvious that $T_c$ increases similarly with increasing $D$. Therefore, compounds with narrow band and large $D$ realize high $T_c$. Since the exponential is less than one, $T_c$ does not exceed $k_B T_D = \hbar\omega_D$. Comparing Eqs. 6.34 and 6.36, we obtain the relation between $\Delta$ and $T_c$.

$$\frac{2\Delta}{k_B T_c} = \frac{2\pi}{e^\gamma} = 3.53 \tag{6.37}$$

From this, $T_c$ increases in proportion to $\Delta$. $T_c$ and $\Delta$ are determined from various measurements, but as shown in Table 6.1 this relation holds satisfactorily. Integration of Eq. 6.35 at $\Delta \neq 0$ gives the temperature dependence of $\Delta$, and it is approximated by Eq. 6.11.

Accordingly, $T_c$ and $\Delta$ are derived from the microscopic Hamiltonian considering attracting force between a Cooper pair. Such mean-field approximation works very well in superconductivity. The Ginzburg-Landau equation is derived from the BCS model using the Green function technique. The Ginzburg and Landau equation gives the properties of type-II superconductors (Sect. 6.4), and the London equation is the extreme case, from which perfect diamagnetism is verified. We have started from phenomenology and discussed the microscopic model, but the former is derived from the later. Thus, various aspects of superconductivity are explained from the microscopic model.

## 6.7 Anisotropic Superconductivity

In the preceding sections, $V_{k,k'}$ and $\Delta_k$ are assumed to be independent of the $k$ direction. Such an isotropic case is called s-wave superconductivity. Anisotropic superconductivity with a variety of $k$-dependent $\Delta_k$ has been investigated. For example, $p$-wave superconductivity follows such a relation as $\Delta_k = \Delta_0 k_x \propto \Delta_0 \sin k_x a$, where $\Delta_k$ has a node at $k_x = 0$. An anisotropic gap $\Delta_k = \Delta_0 (k_x^2 - k_y^2) \propto \Delta_0(\cos k_x a - \cos k_y a)$ defines $d$-wave superconductivity, which has two nodal planes similar to the $d_{x^2-y^2}$ orbital.

The superconducting gap $\Delta_k = -\sum_{k'} V_{k,k'} \langle c_{-k'\downarrow} c_{k'\uparrow} \rangle$ in Eq. 6.28 contains wave functions of two electrons 1 and 2 in the form of $V_{k,k'}\langle\psi_1(-k')\psi_2(k')\rangle$. Here, the pair wave function $\langle\psi_1(-k')\psi_2(k')\rangle$ is a function of relative coordinates $\delta\boldsymbol{r} = \boldsymbol{r}_1 - \boldsymbol{r}_2$ of two electrons with coordinates $\boldsymbol{r}_1$ and $\boldsymbol{r}_2$ constituting the Cooper pair. In the p-wave, the pair wave function has the same form as the p orbital as a function of $\delta\boldsymbol{r}$. In the d-wave given by $\Delta_k = \Delta_0 (k_x^2 - k_y^2)$, the pair wave function has the same shape as the $d_{x^2-y^2}$ orbital.

An s-wave is an even function of $\delta \boldsymbol{r}$, but an electron wave function is to be asymmetric. Thus, the spin part is a singlet with a form of $\alpha_1\beta_2$–$\beta_1\alpha_2$ composed of up and down spins. The orbital part of a *p*-wave Cooper pair is an odd function of $\delta \boldsymbol{r}$, so the spin part is an even function consisting of a triplet pair of up spins. In a *p*-wave, the signs of $\psi(k)$ and $\psi(-k)$ are opposite, and even if $V_{k,k'}$ is a positive repulsive force, $\Delta_k$ may be positive and attractive as a whole. Since the orbital part of a *p*-wave function has a node at $\delta r = 0$, the probability in which $\boldsymbol{r}_1$ and $\boldsymbol{r}_2$ come to the same position is zero. Then, a *p*-wave is preferable when the repulsive force is important. $^3$He is a Fermi particle, and forms a pair similar to the superconducting pair when it makes a superfluid state around 1 mK. The pair is p-wave because two helium atoms do not come to the same position. The ordinary p orbital has a node along the equator, but $^3$He takes the **ABM** (Anderson-Brinkman-Morel) **state**, with nodal points at north and south poles, and the **BM** (Balian-Werthamer) **state**, where the sum of spin and orbital momenta becomes zero. The most stable B phase is the BW state, whereas the A phase appearing near the transition temperature is the ABM state. The appearance of different superfluid phases is an evidence of the anisotropic pairing. A heavy Fermion system, $UPt_3$ has several superconducting phases, and the *p*-wave superconductivity has been suggested.

Since the orbital part of a *d*-wave is symmetric, the spin part should be asymmetric and singlet. As shown in Fig. 6.13, the $d_{x^2-y^2}$ wave has opposite signs at $\boldsymbol{k} \sim (0, \pi/a)$ and $\boldsymbol{k}' \sim (\pi/a, 0)$. Then, if $V_{k,k'}$ is positive and repulsive around $\boldsymbol{k}$–$\boldsymbol{k}' \sim (\pi/a, \pi/a)$, and the contribution is larger than other parts, the overall $\Delta_k$ is attractive. Since the pair is formed between up and down spins, antiferromagnetic interaction mediates such an attractive force. Superconductivity in cuprates is suggested to have such an origin.

Owing to the node, anisotropic superconductivity has density of states within the gap $\Delta$. This leads to characteristic temperature dependence of the heat capacity and $T_1^{-1}$ in NMR. The superconductivity of $\kappa$-phase organic superconductors appears on the border of an antiferromagnetic insulating phase, and there are several experimental evidences that support *d*-wave superconductivity [8]. In cuprates, the d-wave has been established by the experiments of Josephson junctions called $\pi$-junctions. Such a result has not been reported in organic superconductors. Several organic superconductors other than the $\kappa$-phase appear on the border of charge-order or non magnetic phases. Accordingly, the pair symmetry of organic superconductors may not be unique, and conclusive results have not been obtained.

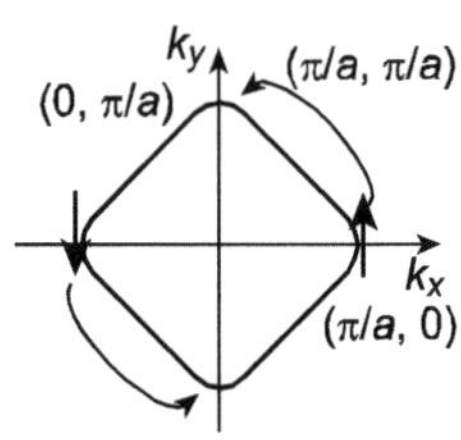


**Fig. 6.13** Scattering process generating attraction in $d_{x^2-y^2}$ wave superconductivity

## 6.8 Phase Diagram in a One-Dimensional System

The Hamiltonian including the interaction $V$ shows superconductivity and a Mott insulating state depending on the sign of $V$ (Sects. 5.7.2 and 6.6). The most stable phase has been investigated using the one-dimensional system (Fig. 6.14). The one-dimensional system is investigated because the power law of the long-range correlation is exactly calculated, though the short-range correlation is uncertain. Then, we suppose the most probable three-dimensional long-range order is related to the most slowly decaying correlation that can be exactly calculated considering the fluctuation characteristic of the one-dimensional system. In a one-dimensional band, the electron is located around either $k_F$ or $-k_F$. Two contributions are important in $V$: back scattering $g_1$, which scatters an electron around $k_F$ to -$k_F$, and forward scattering $g_2$, which scatters an electron around $k_F$ to $k_F$. When $g_1$ and $g_2$ are plotted, the border appears at $g_1 = 0$ and $g_1 = 2g_2$ (Fig. 6.14). A Mott insulator, that is usually called SDW in the present approximation, is stable in the upper right. The Peierls insulator (CDW) is stable in the lower right. The ordinary singlet superconductivity (SS) appears in the lower left, and p-wave superconductivity (TS) in the upper left.

The relation of $g_1$ and $g_2$ to $U$ and $V$ is not simple, but in a half-filled band, we may suppose $g_1 = U - 2V$ and $g_2 = U + 2V$. The resulting $U$ and $V$ are depicted in Fig. 6.14. Roughly speaking, at $U < 0$ and $V > 0$, the energy is lowered when two electrons enter the same site, and the adjacent sites are preferably empty. This leads

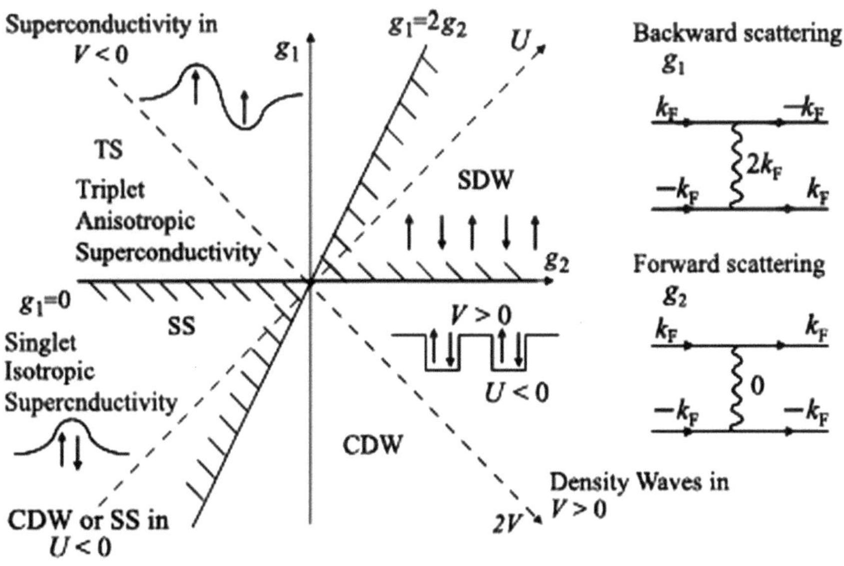


**Fig. 6.14** Phase diagram in a one-dimensional system

to a 2-0-2-0 arrangement and generates a nonmagnetic CDW state. At $U > 0$ and $V > 0$, the electrons keep away from each other, and an SDW state is stable.

The $s$-wave superconductivity appears at $U < 0$ and $V < 0$, where the interaction is generally attractive. The $p$-wave superconductivity is preferable at $U > 0$ and $V < 0$. The $p$-wave is associated with the present one-dimensional model, while the $d$-wave appears in a two-dimensional system. Anisotropic $p$-wave and $d$-wave superconductivity is stable at $U > 0$. Anisotropic superconductivity is preferable in the vicinity of the SDW or antiferromagnetic phase in the upper half of the phase diagram, because the low-energy spin fluctuation is important and works like phonons in s-wave superconductors.

## References

1. C.P. Poole, Jr., Handbook of Superconductivity, Academic Press, London (2000)
2. T. Ishiguro, K. Yamaji, G. Saito, *Organic Superconductors* (Springer, Germany, 1998)
3. N.R. Werthamer, E. Helfand, P.C. Hohenberg, Phys. Rev. **147**, 295 (1966)
4. P. Flude, R.A. Ferrell, Phys. Rev. **135**, A550 (1964)
5. A.I. Larkin, Y.N. Ovchinnikov, Sov. Phys. JETP **20**, 762 (1965)
6. http://hooktail.sub.jp/fourieralysis/coulombFourier/
7. P.G. de Gennes, Superconductivity of Metals and Alloys, Addison-Wesley, Reading (1966)
8. K. Ichimura, K. Nomura, J. Phys. Soc. Jpn. **7**(5), 051012 (2006)

# Chapter 7
# Charge-Transfer Complexes

**Abstract** Highly conducting organic materials are achieved by charge transfer or chemical doping. In this chapter, charge transfer in organic electron donors and acceptors are analyzed based on the redox potentials and energy levels. Then, following the historical sequence, representative organic metals and superconductors are described.

**Keywords** Charge transfer · Electron donor · Electron acceptor · Redox potential · Radical cation salt · Radical anion salt · Charge-density wave · Spin-density wave · Nesting · Commensurate · Incommensurate · Field-induced superconductivity

## 7.1 Requirements for Organic Conductors

This chapter describes organic conductors. First, we shall investigate requirements for organic conductors. High conductivity of graphite comes from the extended π-conjugated system. In polyacetylene, the π-electron is delocalized over the whole polymer, and as we have seen from the molecular orbital theory in Chap. 1 and the energy band theory in Chap. 2, a π-electron can move over the polymer chain. If the terminal states are not taken into account, we can write two resonance structures. This gives the first requirement.

(1) <u>A large π-conjugated system has delocalized π-electrons</u>.

However, pure polyacetylene is a semiconductor, and not highly conducting. This is ascribed to an energy gap coming from the bond alternation, which is regarded as a kind of the Peierls transition (Sect. 2.6). When polyacetylene is exposed to potassium vapor, potassium is incorporated in the form of $K^+$, and the electron is transferred to the LUMO (Fig. 7.1b). Then, the conduction (π*) band (Fig. 7.1c) is electron doped. The negative charge on the carbanion can move over the conjugated π-system (Sect. 1.5.3), and the doped polyacetylene exhibits metallic conductivity.



T. Mori, *Electronic Properties of Organic Conductors*,
DOI 10.1007/978-4-431-55264-2_7

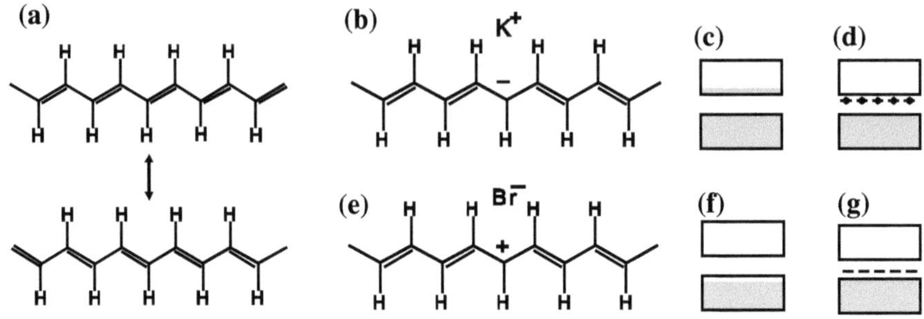


**Fig. 7.1** Structure and energy band in polyacetylene

The doped electron is bound to a distorted part, and makes a **polaron level** similar to the donor level in inorganic semiconductors (Fig. 7.1d). Donor and acceptor levels of inorganic semiconductors are typically located by 0.05 eV lower than the band edge. Since this energy is not much different from room temperature, many carriers are generated in the energy band even at room temperature. By contrast, the polaron level in an organic molecule is sometimes by 0.5 eV lower than the band edge. The polaron is localized not because it is bound to the counterion but mainly because it largely distorts the surrounding molecular structure. This is reflected by the name that represents large polarization of the surroundings.

When polyacetylene is exposed to bromine vapor, an electron is transferred from polyacetylene to bromine, and an electron is taken away from the HOMO. This forms a hole-doped valence ($\pi$) band (Fig. 7.1f). The positive charge on the carbocation can move over the conjugated $\pi$-system. This gives the second strategy to organic conductors.

(2) Carrier doping.

Graphite shows as high in-plane conductivity as $10^5$ S/cm and out-of-plane conductivity of 10 S/cm. However, graphite does not show superconductivity. When K or Br is doped, the K or Br atoms are incorporated between the graphene layers, and graphite interlayer compounds are formed. Among them, $C_8K$ shows superconductivity around 0.1 K [1]. Fullerene ($C_{60}$) itself is not conducting, but alkali-doped fullerene exhibits superconductivity at 19 K in $K_3C_{60}$ and at 38 K in $Cs_3C_{60}$ [1].

Graphite is a good electric conductor, but benzene is an insulator. Around 1950, it was found that the intermediate aromatic hydrocarbons consisting of several benzene rings, as well as large $\pi$-electron systems such as phthalocyanine, show slight conductivity, and are named **organic semiconductors** [2–7]. Perylene (Fig. 7.2b), which is an aromatic hydrocarbon composed of five benzene rings, is an example. Perylene is regarded as a small part of a graphene sheet with the edges terminated by hydrogens (Fig. 7.2a). When perylene is exposed to bromine vapor, some perylene molecules become cationic to show as high conductivity as 0.1 S/cm [8]. Then, perylene is hole doped. Since electron is transferred from perylene to

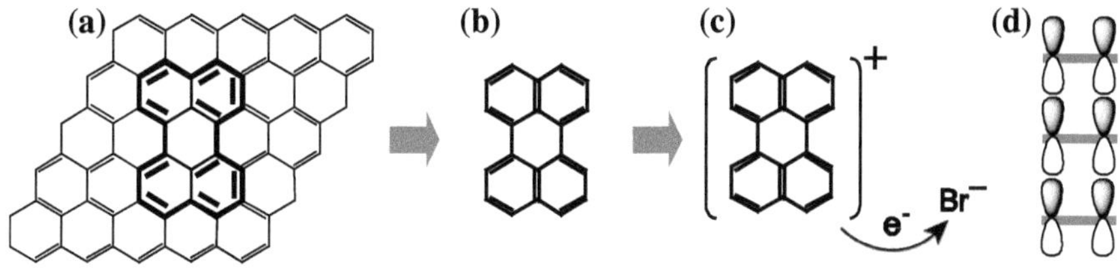


**Fig. 7.2** Conduction in perylene

bromine, such a doped small-molecule organic conductor is called **charge-transfer complex**.

Small-molecule organic conductors exhibit electric conductivity because electron moves from molecule to molecule. Such a π-conjugated system is a planar molecule, tending to make a stacking structure (Fig. 7.2d). The π-orbital incorporating the carrier extends above and below the planar molecule to make overlap with the π-orbitals of the adjacent molecules in the stack. This is the origin of the intermolecular charge transport. Even in polymers such as polyacetylene, the conductivity is mainly determined by the interchain charge transport, where the interchain π-overlap is important. This gives the third requirement to organic conductors.

(3) Intermolecular overlap of the π-orbitals.

Composition of perylene halide has been reported to be perylene:$Br_2$ = 1:2.2 and perylene:$I_2$ = 1:2.5 [9].

HOMO levels of acenes and oligothiophenes are shown in Fig. 7.3. The left scale and the numbers in the V unit represent redox potentials $E_{redox}$. The right scale and the numbers in the eV unit represent the ionization energy $I_g$ measured by photoelectron spectroscopy in solids [10]. The HOMO level goes up with increasing the size of the molecule. This tendency is also predicted from the HOMO levels of the cyclic $4N + 2$ π-electron systems (Fig. 2.7). It may be difficult for two-ring systems like naphthalene and dithiophene to carry positive charges, whereas the four-ring systems have the HOMO levels above 5.5 eV and can transport holes. Photoconductivity of anthracene is also well known. This affords the fourth strategy to electric conduction.

(4) Easily oxidized or reduced molecules.

Such a property is achieved by a large π-system, so we go back to the first condition.

As shown in Fig. 7.4, **phenes** with zigzag ring arrangement have larger $I_p$ (deeper HOMO levels) than acenes with straight ring arrangement. This principle is applied for designing organic semiconductors (Sect. 8.10). In a straight edge of graphene, an edge parallel to the C=C bond like picene is called armchair edge, while an edge perpendicular to the C=C bond like acenes is called zigzag edge [1]. An **armchair edge** has a large HOMO–LUMO gap, and the frontier orbital is

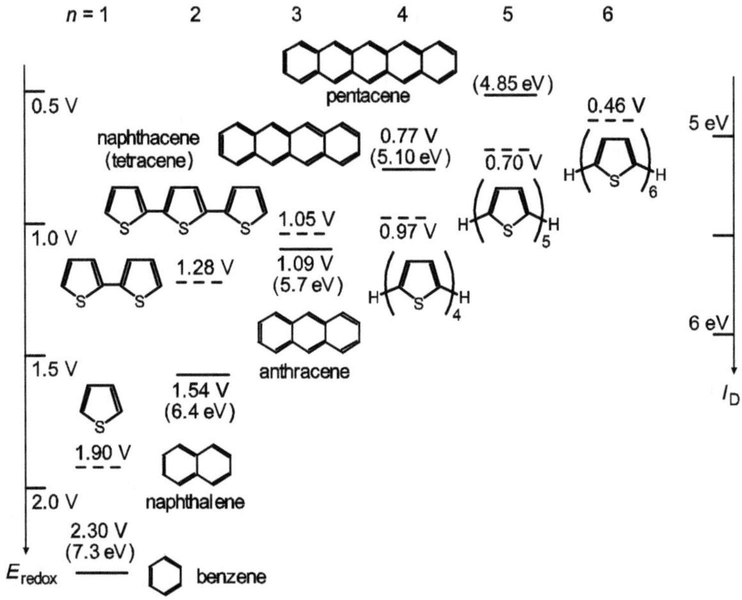


**Fig. 7.3** HOMO levels of acenes and oligothiophenes, estimated from the redox potential (*left scale* in V) and from the ionization energy (*right scale* in eV) [10, 11]

delocalized over the whole molecular sheet. A **zigzag edge** has a small HOMO–LUMO gap, and the SOMO is localized on the edge. Then, such a zigzag edge has magnetism. Phenalenyl radical (Fig. 4.24) is an example of magnetism observed in a minimal molecule unit with a zigzag edge.

## 7.2 Charge-Transfer Complex

Next, we shall investigate redox properties of molecules. Molecules capable of carrying positive charges are called **electron donors** because such molecules can easily give electrons to other molecules. An electron donor is a molecule with a comparatively high HOMO level and small $I_p$. An electron donor is easily oxidized to a cation $D^+$. By contrast, molecules capable of carrying negative charges are called **electron acceptors**, which have relatively deep LUMO levels and large electron affinity $E_A$. An electron acceptor is easily reduced to an anion $A^-$.

Redox potentials $E_{redox}$ of representative organic donors and acceptors are shown in Fig. 7.5 [12]. The values versus standard calomer electrode (SCE) is by 0.241 V smaller than the thermodynamical values versus standard hydrogen electrode (SHE) [13]. Ag/AgCl electrode is 0.222 V versus SHE, and is approximately the same as the value versus SCE. $Ag/AgNO_3$ electrode is also frequently used, which corresponds to 0.30–0.35 V versus SCE [14, 15]. The potential of reference

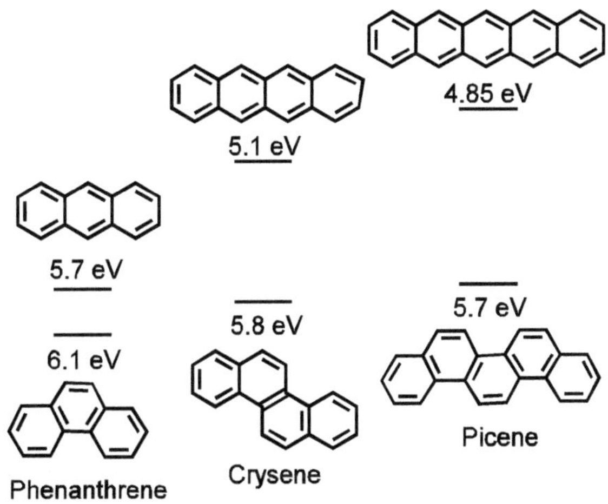


**Fig. 7.4** Ionization energy (eV) of acenes and phenes [10]

electrodes depends on the concentration and the temperature, and shows long-term shift, so usually standard compounds such as ferrocene (0.38 V vs. SCE) and TTF (0.34 V vs. SCE) are measured under the same conditions. In some papers, the values versus ferrocene is reported [16].

$I_p$ obtained from photoelectron spectroscopy in solids is approximately related to the first oxidation potential as $I_p = E_{redox}$ (vs. SCE) + 4.44 eV [17, 18]. Alternatively, $E_{redox}$ versus ferrocene is assumed to be $I_p$ = 4.8 eV [19]. $I_p$ corresponds to the HOMO level[1]. The LUMO level is similarly obtained from the first reduction potential. LUMO is an unoccupied state and is not directly observed by photoelectron spectroscopy. However, the HOMO–LUMO gap is estimated from the absorption edge of the visible spectrum (Sect. 1.2). The LUMO level is obtained by adding the HOMO–LUMO gap to the HOMO level. $I_p$ for solids is by 1.0–1.7 eV smaller than $I_p$ for gas-phase molecules because of the polarization of the surrounding molecules or the dielectric constant of the solid. Note that molecular orbital calculation sometimes affords the latter values [10, 20–22].

In Fig. 7.5, the HOMO levels of weak donors are located at the left bottom, and when we go up, we find gradually strong donors. Anthracene and perylene are weak donors and located at the left bottom. Strong donors like tetrathiafulvalene (TTF) are located around 0.5–0.3 V. Materials with negative redox potentials, located above 0 V, work as cations like ammonium ions rather than donors.

The LUMO levels of weak acceptors are located at the right top, and when we go down, we find gradually strong acceptors. Perylene and tetracene are weak acceptors, and receive electrons to the LUMO from small-work-function metals like

[1]Such a relation as $I_g = E_{ox}$ (vs. SCE) + 5.75 eV has been proposed between the gas-phase ionization energy $I_g$ and the redox potential [20-22]. However, it has been also suggested that a slope is not one ($I_g = 1.5E_{ox}$ (vs. SCE) + 5.8 V) [10]. This has been ascribed to solvation. Owing to the polarization, solid-state $I_p$ is by 1.0–1.7 eV smaller than gas-phase $I_g$ [10]. Accordingly, the difference between $I_p$ and $E_{ox}$ is approximated to 4.44 eV.

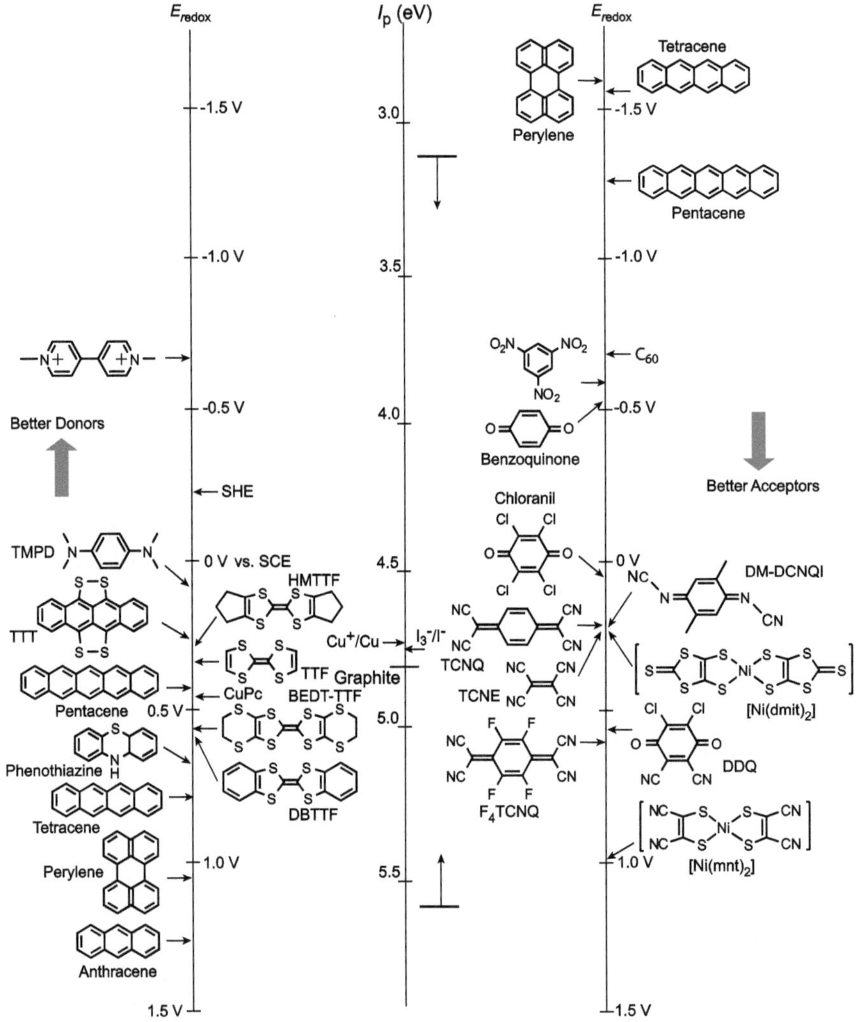


**Fig. 7.5** Representative organic donors (*left*) and acceptors (*right*). The redox potentials indicate the HOMO levels for donors and the LUMO levels for acceptors. The central scale represents the energy levels, where the arrows at 3.15 and 5.6 eV mean the stability region coming from the redox reaction of water (Sect. 8.10)

alkali metals (Fig. 8.13). Weak acceptors are located between −1.0 and 0 V. Benzoquinone (Fig. 7.6a) is a weak acceptor because by receiving two electrons, the aromatic ring is resumed and the dianionic form is stabilized. Usually, the dianion receives two protons to form hydroquinone. The reaction of benzoquinone to hydroquinone is a typical organic reduction reaction.

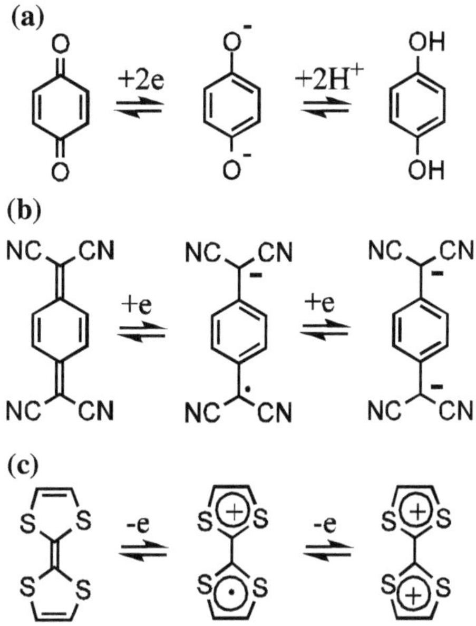


**Fig. 7.6** Resonance structures of **a** benzoquinone, **b** TCNQ, and **c** TTF

Like trinitrobenzene, compounds with many electron-withdrawing groups are acceptors. In organic chemistry, groups that stabilize carbanions are called **electron-withdrawing** groups [23]. Typical electron-withdrawing groups are

$$-NO_2,\ -CN,\ -SO_3H,\ -COOH,\ -COOCH_3,\ -CHO,\ \text{and halogen.}$$

When these groups are attached, the electrophilic aromatic substitution occurs at the meta position (Fig. 7.7a). These groups contain nitrogen, oxygen, and halogen, that are more electronegative than carbon. Then, these groups destabilize the intermediate carbocation. Since the carbocation is not generated at the meta position, the meta substitution occurs preferably.

By contrast, groups that stabilize carbocanions are called **electron-donating groups**. Examples are

$$-NH_3,\ -OH,\ -OCH_3,\ \text{and}\ -CH_3.$$

Since nitrogen and oxygen are more electronegative than carbon, these atoms are originally electron withdrawing. In the course of the electrophilic aromatic substitution, however, these groups stabilize the intermediate carbocation due to the resonance structures (Fig. 7.7b). Then, these groups result in the electron-donating effect. In such a case, the substitution occurs preferably at the ortho and para positions.

Compounds with many electron-withdrawing groups such as nitro and cyano groups are electron acceptors. Tetracyanoethylene (TCNE) and tetracyanoquinodimethane (TCNQ) are located around 0–0.3 V in Fig. 7.5, and are strong acceptors. Tetramethylphenylenediamine (TMPD) with electron-donating amino

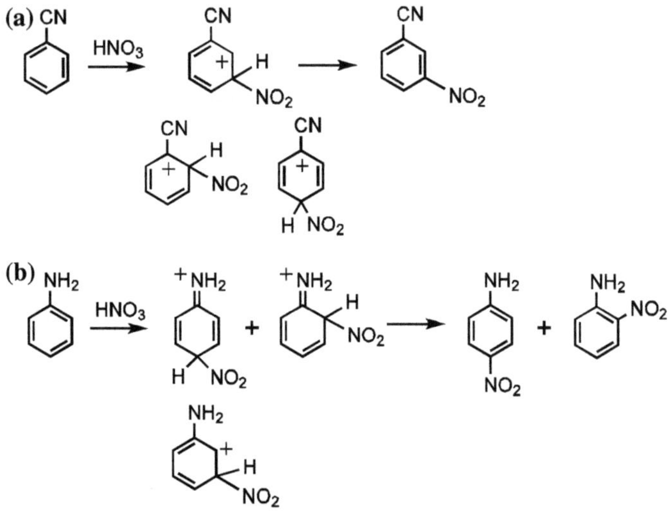


**Fig. 7.7** Electrophilic aromatic substitution (nitration) to an aromatic ring with **a** an electron-withdrawing (cyano) group and **b** electron-donating (ammino) group

groups is a strong electron donor. This gives the first strategy to make good electron acceptors and donors.

(1) An electron acceptor is formed by attaching many electron-withdrawing groups, and an electron donor is formed by attaching many electron-donating groups.

As shown in Fig. 7.6b, when electrons are incorporated in TCNQ, the central benzene ring is resumed from the quinoid structure. Then, similarly to benzoquinone, the anionic form is stabilized by the aromaticity. TCNQ accepts up to two electrons, and is reduced to the dianionic form, but even the monoanion is stabilized by delocalization. Accordingly, TCNQ is a good electron acceptor.

As shown in Fig. 7.6c, the five-membered ring of TTF has seven π-electrons. Each carbon atom gives up one π-electron, whereas among the six valence electrons of a sulfur atom, two are used to make two σ-bonds, two make a lone pair, and the remaining two enter the π-orbital. Then, three carbon and two sulfur atoms make a $3 \times 1 + 2 \times 2 = 7\pi$-system. A 7π-system is less stable than a 6π-system (Sect. 2.1), and a 7π-system gives up an electron to make a 6π-cation. Since TTF has two rings, TTF is oxidized to 2+. Even the monocation is stabilized by the delocalization, and TTF is a good electron donor. A 6π-system is restored similarly to benzoquinone and TCNQ. This is the second strategy to make good electron acceptors and donors.

(2) Use a structure whose ionic form has six π electrons.

Chloranil and dicyanoquinonediimine (DCNQI, Fig. 7.5) are other examples of acceptors. In organic chemistry, carbocation is more popular than carbanion. This

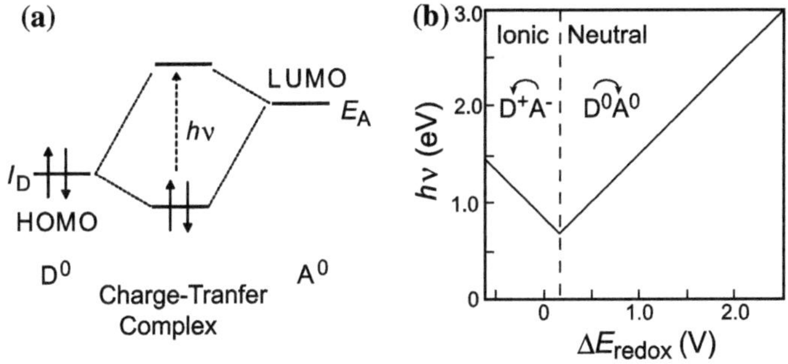


**Fig. 7.8** **a** Energy levels and **b** charge-transfer absorption in charge-transfer complexes [11]

comes from the slight electronically positive nature of carbon. As a result, good acceptors are rarer than donors. Several metal complexes like [Ni(dmit)$_2$] and [Ni(mnt)$_2$] (Fig. 7.5, where dmit is 1,3-dithiole-2-thione-4,5-dithiolato and mnt is maleonitriledithiolato) are representative electron acceptors. A neutral form of [Ni(dmit)$_2$] has been known, but the redox potential of [Ni(mnt)$_2$] corresponding to 0/1− is as large as 1.0 V, so [Ni(mnt)$_2$] usually exists in the anionic form.

A donor has a comparatively high HOMO level, and an acceptor has a comparatively low LUMO level. When the HOMO level of a donor is as high as the LUMO level of an acceptor, charge transfer occurs from the donor to the acceptor, and a $D^+A^-$ state is formed (Fig. 7.8a). Such a $D^+A^-$ state is called **ionic charge-transfer complex**. $D^{2+}A^{2-}$ is usually not formed because of the on-site Coulomb repulsion $U$, or because the second redox potential is higher than the first redox potential. The on-site Coulomb repulsion $U$ is sometimes estimated by the difference of the first (0/1+) and second (1+/2+) redox potentials.

Even when acceptor's LUMO is located considerably above donor's HOMO, the ionic state $D^+A^-$ is quantum chemically hybridized with the neutral state $D^0A^0$

$$\Psi = \Psi(D^0A^0) + c\Psi(D^+A^-) \tag{7.1}$$

to stabilize the partial charge-transfer state (Fig. 7.8a), where $c$ is a coefficient less than one. Such a state is called **neutral charge-transfer complex**. This is a somewhat conflicting notation, but it implies such a state as $D^{\delta+}A^{\delta-}$, where the charge transfer $\delta$ due to the slight hybridization of the $D^+A^-$ state is typically less than 0.1.

In a charge-transfer complex, a characteristic electronic transition appears from the lower HOMO-originating level to the upper LUMO-originating level (Fig. 7.8a). This is called **charge-transfer absorption**. When a donor and an acceptor are mixed, a new charge-transfer absorption appears. Note that neither components show such a transition. For example, when a practically colorless solution of pyrene is mixed with a pale color solution of polynitro compounds, acid anhydrids, or quinone, the charge-transfer compounds show characteristic orange or violet color. The transition energy $h\nu$ is

$$h\nu = I_D - E_A - E(D^+A^-) \tag{7.2}$$

where $I_D - E_A$ is the difference between the HOMO and LUMO levels, and is obtained from the difference between the redox potentials of the donor and the acceptor, $\Delta E_{redox}$. Here $E(D^+A^-)$ is the Madelung energy, which is a sum of the Coulomb energies of the ionic crystal. As shown in Fig. 7.8b, a plot of $h\nu$ versus $\Delta E_{redox}$ makes a 45° line, but the line does not pass the origin due to the Madelung energy. The charge-transfer transition gives a broad absorption band usually in the visible-to-near infrared region.

## 7.3 Basic Charge-Transfer Complexes

A highly conducting pyrene bromide found in 1954 is not very stable because pyrene is a weak donor. After that, strong donors and acceptors have been gradually developed as shown in Table 7.1. In 1960, TCNQ was developed by the Du Pont group [24]. Du Pont was famous for Teflon, so they prepared various organic compounds substituted with CN instead of halogen, among which they found TCNQ. TCNQ forms such 1:1 complexes as Na(TCNQ) and K(TCNQ) with alkali metals. These complexes have one-dimensional stacks of TCNQ (Fig. 7.9), between which alkali metals are incorporated [25]. In Fig. 7.9, the TCNQ molecules are stacked perpendicular to the view graph. Alkali metal TCNQ is a stable compound, but is a Peierls insulator with the room-temperature conductivity of $10^{-2}$–$10^{-4}$ S/cm, where the TCNQ column is distorted with a twofold periodicity.

**Table 7.1** History of organic charge-transfer complexes

| | | |
|---|---|---|
| 1954 | Perylene bromide: first highly conducting charge-transfer complex | [8] |
| 1960 | TCNQ | [24] |
| 1964 | Model of high-temperature superconductors by Little | [45] |
| 1970 | TTF | [35] |
| 1973 | (TTF)(TCNQ) | [39] |
| 1980 | $(TMTSF)_2PF_6$: first organic superconductor | [51] |
| 1981 | $(TMTSF)_2ClO_4$: first ambient-pressure organic superconductor | [57] |
| 1982 | $(BEDT\text{-}TTF)_2ClO_4(TCE)_{0.5}$: two-dimensional organic metal | [65] |
| 1984 | β-$(BEDT\text{-}TTF)_2I_3$: 8 K superconductor under pressure | [70, 71] |
| 1986 | $(TTF)[Ni(dmit)_2]_2$: anionic superconductor | [137, 138] |
| 1987 | κ-$(BEDT\text{-}TTF)_2Cu(NCS)_2$: ambient-pressure 10 K superconductor | [106] |
| 2001 | λ-$(BETS)_2FeCl_4$: field-induced superconductivity | [121] |
| 2003 | β′-$(BEDT\text{-}TTF)_2ICl_2$: 14.2 K superconductivity under 82 kbar | [75] |

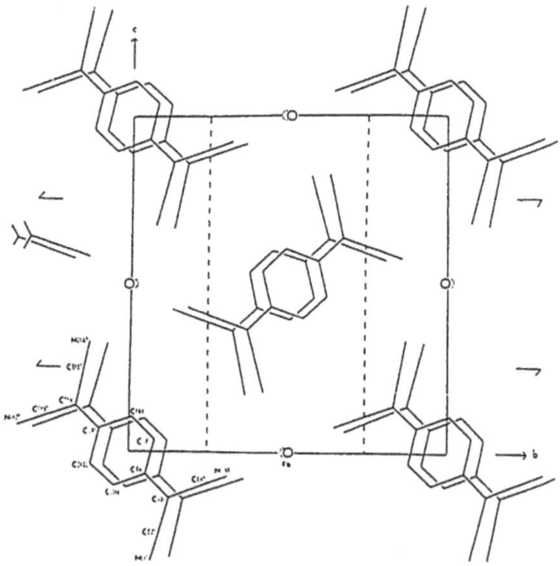

**Fig. 7.9** Crystal structure of Na(TCNQ) [25]

Cu and Ag form similar 1:1 complexes, but the structure has polymorphism [26]. Cu(TCNQ) exhibits nonlinear conductivity, in which the resistivity drops by about four orders above a certain threshold voltage; such a property is interested due to the potential application to memory devices [27].

Structures of charge-transfer complexes composed of organic donor and acceptor are depicted in Fig. 7.10 [28, 29]. Many complexes make a **mixed column**, where the donor and the acceptor are stacked alternately (Fig. 7.10a). (TMPD)(TCNQ) is an ionic charge-transfer complex consisting of $D^+A^-$, which also makes a mixed stack (Fig. 7.10b) [30]. Alternate arrangement of positive and negative ions is most favorable on account of the Coulomb attraction similarly to such an ionic crystal as NaCl. Neutral charge-transfer complexes like (Perylene)(TCNE) make a mixed stack as well [31]. In some partially charge-transferred complexes, the donors make a column, and the acceptors make an independent column, which is called **segregated column**. In a segregated column, the hole moves only in the donor column, and the electron moves only in the acceptor column, so that high conductivity is achieved. In a mixed stack, the carriers have to hop between the donor and acceptor levels in the stacking direction with the largest $\pi$-overlap, so the conductivity is low. In general, neutral and ionic charge-transfer complexes form mixed stacks, and the conductivity is low. High conductivity is attained in a segregated column with partial charge transfer. Such a situation is realized when the donor level is close to the acceptor level (Fig. 7.5) [32]. The combination of TTF and TCNQ is an example.

(NMP)(TCNQ) forms a segregated column as shown in Fig. 7.10e (NMP: *N*-methylphenadinium) [33]. This complex has a partial charge-transfer state like $NMP^{2/3+}$ $TCNQ^{2/3-}$. The conductivity slightly increases from room temperature to

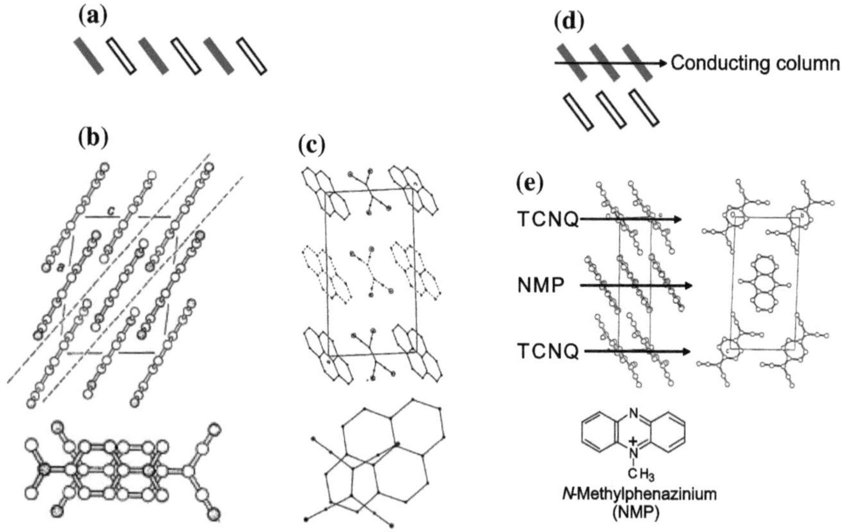


**Fig. 7.10** Mixed and segregated columns [28–33]. **a** Mixed column. **b** (TMPD)(TCNQ) [30]. **c** (Pyrene)(TCNE) [31]. **d** Segregated column. **e** (NMP)(TCNE) [33]

200 K (Fig. 7.11), and this is regarded as the first organic metal [34]. Do not confuse an organic metal with an organometallic compound; the latter is a compound that has a metal–carbon bond.

TTF was developed in 1970 [35]. TTF forms complexes with Cl, Br, I, and SCN, which are represented as (TTF)$I_{0.7}$. Charge-transfer complexes are formed not only in the combination of organic donor and acceptor but also in the combination of an organic molecule and an inorganic ion. The combination of an organic donor and an inorganic anion is called **radical cation salt**, and the combination of an organic acceptor and an inorganic cation is called **radical anion salt**. (TTF)$I_{0.7}$ is an

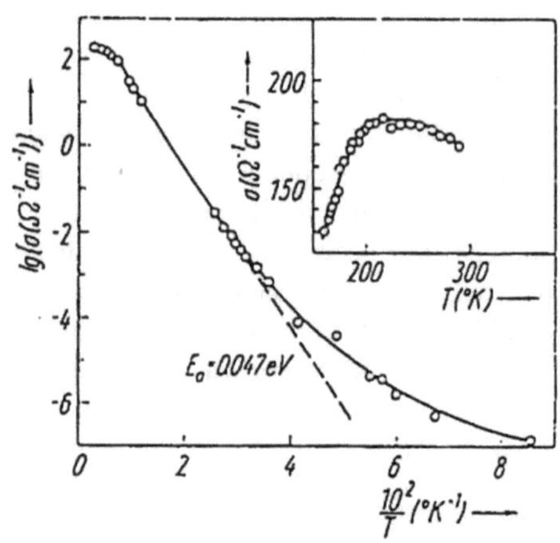


**Fig. 7.11** Temperature dependence of the conductivity in (NMP)(TCNQ) [34]

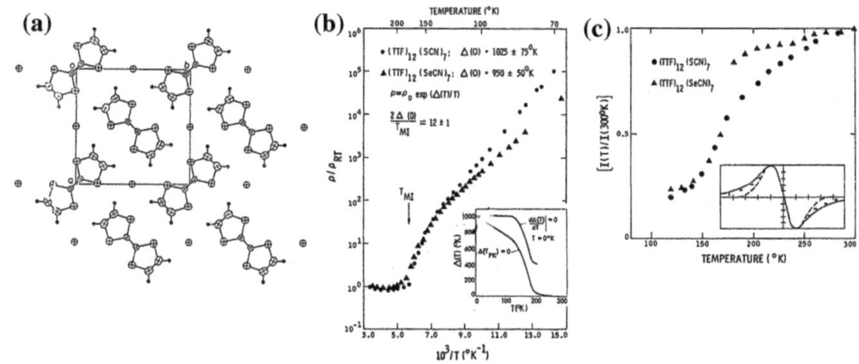


**Fig. 7.12** **a** Crystal structure [37], **b** conductivity, and **c** magnetic susceptibility of (TTF)$X_x$ [38]

example of a radical cation salt, and K(TCNQ) is a radical anion salt. TTF and halogen make 1:1 and 2:1 salts as well, but (TTF)$X_{0.7}$ is the most conducting [36]. In the crystal structure (Fig. 7.12a), TTF is stacked perpendicular to the view graph, and the halogen atoms are incorporated in between the TTF columns [37]. TTF and halogen atoms are stacked with independent periodicities, and this gives rise to the fractional composition. Depending on the halogen, the ratio changes in the range of 0.56–0.76, but the ratio is almost constant for a given halogen. This material shows as high room-temperature conductivity as 300 S/cm and almost flat temperature dependence down to about 200 K (Fig. 7.12b) [38]. This material, however, turns to an insulator below this temperature. The magnetic susceptibility drops below this temperature (Fig. 7.12c). The susceptibility is regarded as the Pauli paramagnetism above 200 K, and a nonmagnetic band insulator below 200 K. This material has distortion with the periodicity of halogen, and the low-temperature phase is considered to be a Peierls insulator. This is a typical **one-dimensional conductor** that exhibits high conductivity only in the π-stacking direction. In such a material, the Peierls transition is unavoidable. The one-dimensional conductors were extensively studied together with inorganic one-dimensional conductors such as KCP and $NbSe_3$ (Fig. 2.44).

## 7.4 (TTF)(TCNQ)

After the development of TTF in 1970, it was combined with TCNQ, and (TTF)(TCNQ) was reported in 1973 [39]. (TTF)(TCNQ) has a segregated column structure, in which TTF and TCNQ make independent columns (Fig. 7.13a) [40]. Accordingly, it shows partial charge transfer represented as ($TTF^{0.59+}$) ($TCNQ^{0.59-}$). Since TTF and TCNQ make one-dimensional energy bands independently, the energy bands cross each other at the position where the charge-transfer degree becomes 0.59 (Fig. 7.13b) [41]. This complex exhibits as high room-temperature

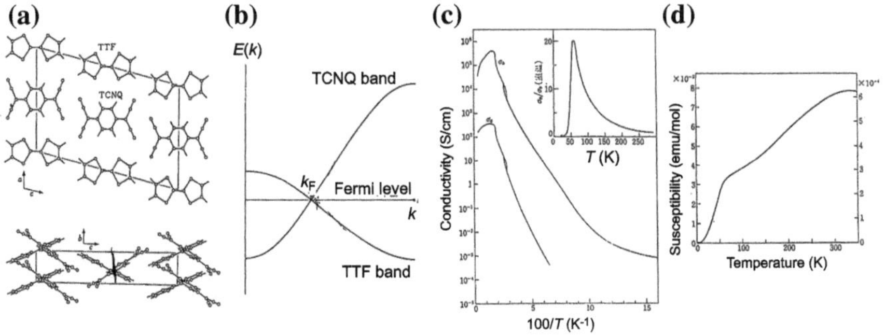


**Fig. 7.13** **a** Crystal structure [40], **b** energy band [41], **c** resistivity, and **d** magnetic susceptibility of (TTF)(TCNQ) [44]

conductivity as 600 S/cm, and the resistivity shows a metallic drop (Fig. 7.13c). However, the Peierls transition takes place at 53 K, and an insulating phase appears below this temperature. This compound largely extended the metallic region from 200 K to low temperatures. The huge enhancement of the conductivity just above the transition temperature was first attributed to superconductivity fluctuation [42], while Bardeen suggested **Flöhlich mode** superconductivity, where charge is transported by sliding **charge-density waves** (CDW) [43]. The susceptibility drops zero below the transition (Fig. 7.13d), which indicates the Peierls transition [44]. In general, the Peierls transition is equivalent to the formation of CDW.

Before these findings, Little suggested a possibility of high-temperature superconductivity in 1964. According to the BCS theory, $T_c$ is given by Eq. 5.43.

$$k_B T_c = 1.13\hbar\omega_D \exp(-1/VD)$$

Here, the lattice vibration $\hbar\omega_D$ mediating the attraction of Cooper pairs is several hundred K, so the resulting $T_c$ is several ten K. However, the polarization energy of organic molecules is several eV. Using the polarization energy instead of the lattice vibration, Little suggested that $T_c$ would increase to several hundred K [45]. His suggestion considerably enhanced the interest in organic conductors. Later, the enhancement of the conductivity in (TTF)(TCNQ) was proved to be not so large, and it was revealed that the sliding CDW does not induce superconductivity because the CDW is **pinned** by impurity or by the CDW on the adjacent columns. However, these findings accelerated the study of CDW in the 1970s.

(TTF)(TCNQ) was extensively studied as a typical one-dimensional conductor [41]. The CDW state below the Peierls transition has periodical modulation of charge density, where the TTF and TCNQ columns are distorted. Such a periodical modulation is observed by X-ray diffraction. Since the column is along the $b$-axis, extra diffractions appear at $0.295b^*$ and $0.59b^*$ in between the ordinary Bragg spots (Fig. 7.14a). These two different diffractions, respectively, correspond to $2k_F$ and $4k_F$ periodicity [46]. Since TCNQ has 0.59− charge, 0.295 of the energy band is occupied (Fig. 7.14b). The filling number becomes half because up and down spin

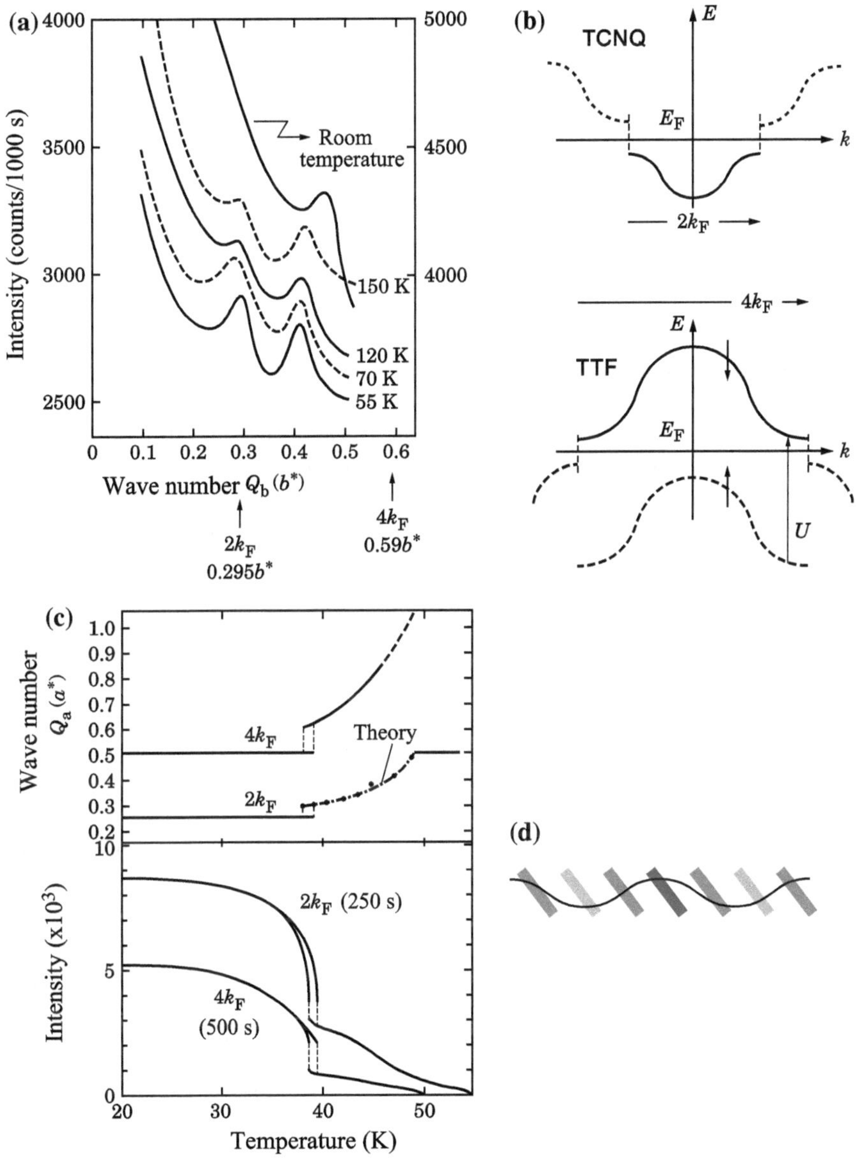


**Fig. 7.14** $2k_F$ CDW and $4k_F$ CDW in (TTF)(TCNQ) [41]. **a** $k$-dependence of the X-ray diffraction [46]. **b** Energy band with CDW modulation. **c** Temperature dependence of the X-ray diffraction [46]. **d** Charge modulation in the TCNQ $2k_F$ CDW

electrons fill the whole energy band; namely two electrons enter an energy band. The TCNQ column has charge rich and poor positions with the inverse periodicity, namely per 3.39 molecules. This extra periodicity opens a gap at $k_F$ similarly to Fig. 2.35. As shown in Fig. 7.14b, the energy gap pushes down and stabilizes the electrons around $k_F$. This stabilization gives rise to the CDW. Since TTF has large $U$, the holes enter only in the upper Hubbard band, and the gap opens due to the $4k_F$ modulation (Fig. 7.14b). Here, the holes enter top of the TTF band, and the lower Hubbard band is empty for holes, or fully occupied by electrons. The charge-transfer degree is exactly estimated from the position of the X-ray extra diffraction. The $4k_F$ CDW was first found in (TTF)(TCNQ), which comes from the combination of the Hubbard model and the Peierls transition.

As shown in Fig. 7.14a, the $2k_F$ and $4k_F$ diffractions appear from much higher temperatures than the transition temperature, 53 K. Here, the phases of the TCNQ CDW between different columns are entirely random. The periodicity is established only in a single column. Such a one-dimensional periodicity gives rise to X-ray **diffuse scattering** composed of a sheet extending perpendicular to the $b^*$ axis. Below the transition temperature, the three-dimensional order occurs, where the charge density in the adjacent columns aligns alternately, and **superlattice spots** appear in the X-ray diffraction. The CDW fluctuation or the short-range order appears much above 53 K on account of the one-dimensional nature similarly to one-dimensional magnets (Sect. 4.7). The typical Pauli paramagnetism is temperature independent, but the susceptibility in Fig. 7.13d shows a considerable decrease between room temperature and the transition temperature. This is attributed to the short-range order. Below the transition temperature, the long-range order appears in the X-ray spots. As shown in Fig. 7.14c, $2k_F$ changes below the transition temperature, but becomes constant below 38 K. This is because the charge-transfer degree decreases from 0.59 to finally 0.50. Accordingly, the low-temperature state is $(TTF^{1/2+})(TCNQ^{1/2-})$. The resulting charge density of the TCNQ column has a four-molecule periodicity (Fig. 7.14d). The Peierls transition of (TTF)(TCNQ) is characteristic of one-dimensional conductors, but the additional effect of $U$ and the temperature dependence of the charge-transfer degree considerably complicate the properties.

Conductivity of (TTF)(TCNQ) increases under pressure, but once drops above 15 kbar as shown in Fig. 7.15a [47]. $2k_F$ increases under pressure, but becomes a constant value of $b^*/3$ above 15 kbar (Fig. 7.15b), implying that the high-pressure charge state is $(TTF^{2/3+})(TCNQ^{2/3-})$ [48]. Noninteger ratio of charge-transfer degree and CDW periodicity like 0.59 is called **incommensurate**, whereas integer ratios such as 1/2 and 2/3 are called **commensurate**. Usually commensurate CDW is more stable, and the charge transfer is **locked** to a commensurate value. The extra conductivity below 15 kbar is ascribed to the sliding incommensurate CDW. The charge-transfer degree of (TTF)(TCNQ) is 0.59 under ambient conditions, whereas it decreases to 1/2 at low temperatures, and increases to 2/3 under pressures. Anyway, it goes to a commensurate value.

CDW is not superconducting because CDW is pinned by impurities and periodicity of the original crystal. In some CDW materials such as $NbSe_3$, however, sliding CDW occurs above a certain threshold voltage at which the pinning is released, and the electric resistance largely deviates from ordinary Ohm's law to give rise to **nonlinear conductivity** [49]. When the current carried by sliding CDW is $J_{CDW}$, the CDW passes over a certain impurity with a velocity of $v = J_{CDW}/n|e|$.

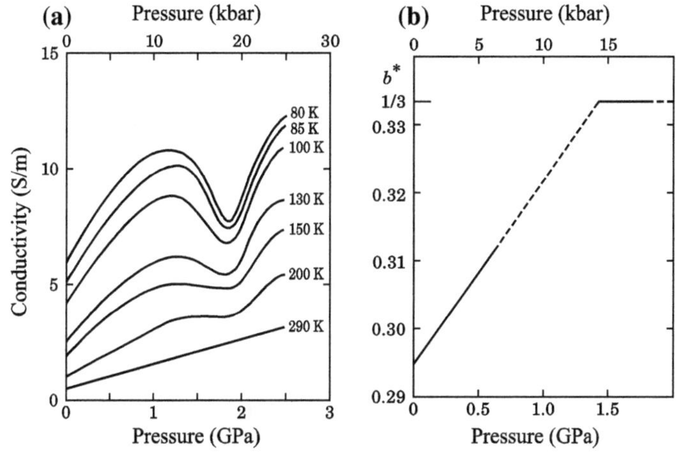


**Fig. 7.15** **a** Pressure dependence of conductivity [41, 47] and (b) $2k_F$ in (TTF)(TCNQ) [48]

If the wavelength of the CDW is $\lambda$, AC voltage with the frequency of $f = v/\lambda = J_{CDW}/n|e|\lambda$ is generated. In $NbSe_3$, such an oscillating voltage, which is called **narrow band noise**, appears around 1 MHz. The frequency of the narrow band noise increases proportionally to $J_{CDW}$. Since the pinning force is characterized by this frequency, microwave AC conductivity above this frequency is higher than the DC conductivity. Accordingly, various extraordinary conducting properties are derived from the CDW motion.

In the latter 1970s, when physicists explored unconventional sliding conduction, Bechgaard designed materials aiming at reducing the Peierls transition temperature [50]. When TTF was replaced by the selenium-substituted TSF (Fig. 7.16), the extended selenium orbitals enhanced the transfer integrals and bandwidth, and reduced the Peierls transition temperature of (TSF)(TCNQ) down to 29 K. The transition temperature of (HMTTF)(TCNQ) was 50 K, whereas (HMTSF)(TCNQ) underwent the transition at 24 K, and the resistance increase was very small. Successively, superconductivity was found in a radical cation salt of TMTSF.

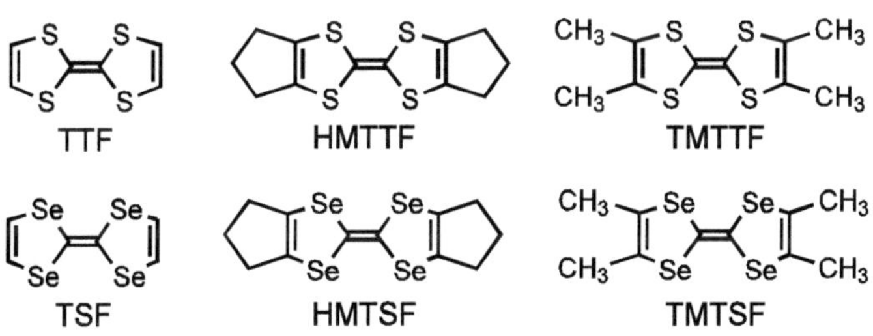


**Fig. 7.16** TTF: derivatives. TSF: tetraselenafulvalene, HMTTF: hexamethylenetetrathiafulavalene, HMTSF: hexamethylenetetraselenafulvalene, TMTTF: tetramethyltetrathiafulavalene, TMTSF: tetramethyltetraselenafulvalene

## 7.5 TMTSF

In 1980, superconductivity was found in $(TMTSF)_2PF_6$ at 0.9 K under the pressure of 12 kbar [51, 52]. In this compound, TMTSF is stacked along the crystallographic *a*-axis (Figs. 7.17a and 2.45a), but sandwiched by the adjacent columns along the *b*-axis [53]. Then, the *ab* plane forms the conducting layer, and the $PF_6^-$ anions are incorporated between the layers. Such an alternate stacking of the conducting and anion layers is generally observed in radical cation salts. In the conducting layer, the intrachain transfers are ten times larger than the interchain transfers, so the Fermi surface is warping planes (Fig. 2.45b). This intrachain/interchain ratio is much smaller than one hundred in (TTF)(TCNQ), so $(TMTSF)_2PF_6$ is regarded as a quasi-one-dimensional conductor. In addition to the $PF_6$ salt, $AsF_6$, $SbF_6$, $TaF_6$, $ClO_4$, and $ReO_4$ form isostructural superconducting salts.

Since $PF_6^-$ has 1− charge, two TMTSF molecules carry one hole. $(TMTSF)_2PF_6$ does not undergo a metal–insulator transition down to fairly low temperatures, but turns to an insulator at 12 K (Fig. 7.17b) [54]. The susceptibility is temperature independent and Pauli-like down to 12 K (Fig. 7.17c), but shows antiferromagnetic anisotropy below 12 K (Fig. 4.15b) [55]. The metal–insulator transition is suppressed under pressure, and a superconducting phase appears around 1 K (Fig. 7.17d) [56]. The insulating phase below 12 K is attributed to a **spin-density-wave** (SDW) phase. As shown in Fig. 7.17f, an SDW phase has modulated up and down spin densities (Sects. 4.9 and 5.8). The SDW phase of $(TMTSF)_2PF_6$ has a four-molecule

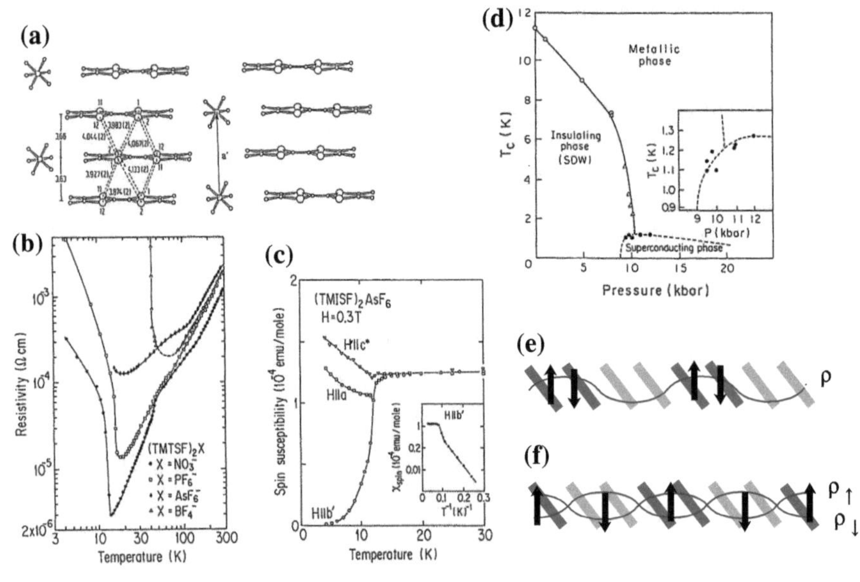


**Fig. 7.17** $(TMTSF)_2X$. **a** Crystal structure [53], **b** conductivity [54], **c** susceptibility [55], **d** phase diagram [56], **e** CDW, and **f** SDW

periodicity. The SDW phase shows similar properties to antiferromagnetism. In the SDW phase, $\rho_\uparrow - \rho_\downarrow$ is modulated, but the total charge density $\rho = \rho_\uparrow + \rho_\downarrow$ is constant. By contrast, a CDW phase has modulated total charge density $\rho$ (Fig. 7.17e). The periodicities of SDW and CDW are the same and $2k_F$. In a quarter-filled band, both show a four-molecule periodicity. Antiferromagnetic metals are common in inorganic materials, but SDW phases in organic conductors are insulating.

$(TMTSF)_2ClO_4$ is the first ambient-pressure superconductor, which exhibits superconductivity at 1.4 K under ambient pressure [57]. Anions in the TMTSF salts are located on inversion centers, while $ClO_4$ does not have an inversion center. As a result, the $ClO_4$ anions are alternately aligned along the *b*-axis below 24 K. This anion order is considered to suppress the SDW formation. Later, superconductivity has been found in the sulfur analog, TMTTF (Fig. 7.16), under a fairly high pressure of 20–60 kbar. Combining the phase diagrams of the TMTTF and TMTSF salts, Fig. 7.18a is the **generic phase diagram** [58]. $(TMTTF)_2PF_6$, located at the left end, gives rise to charge order at a fairly high temperature, and turns to a nonmagnetic (spin-Peierls) insulating state at low temperatures. However, even this compound shows superconductivity at 1.8 K under a pressure of as high as 54 kbar. Since the superconducting phase borders the SDW phase, anisotropic superconductivity has been suggested (Fig. 6.14). The Hebel–Slichter peak (Fig. 6.4c) does not appear in the NMR [59]. The critical field is not flat but continues to increase at low temperatures (Fig. 7.18b), and a spatially inhomogeneous FFLO state has been suggested (Fig. 6.10) [60].

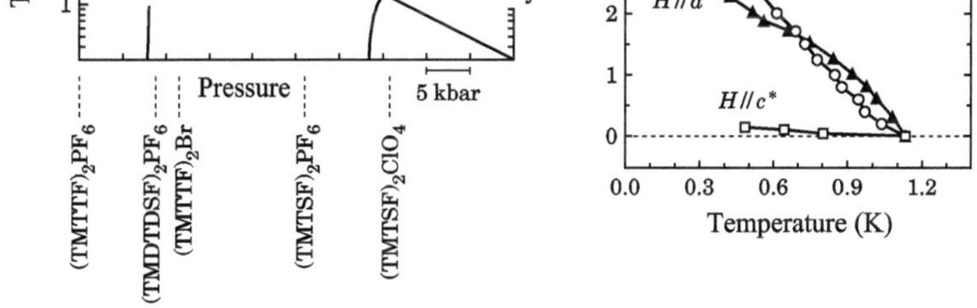


**Fig. 7.18** **a** Generic phase diagram and **b** upper critical field in $(TMTSF)_2X$

## 7.6 Metal–Insulator Transitions in Organic Conductors

Many organic conductors undergo metal–insulator transitions at low temperatures. There are various origins of the insulating states: the Peierls transition (CDW), SDW, Mott insulator, and charge order. We shall investigate the Peierls transition in more detail. Then, we shall discuss how to distinguish the origins of the insulating states.

In the Peierls transition, an energy gap opens (Fig. 2.35) when the Fermi surface at $-k_F$ is translated onto the Fermi surface at $k_F$ by the CDW periodicity $(k_a, k_b) = (\pi/a, 0)$ (Fig. 7.19a). In an ideal one-dimensional conductor, the Fermi surface consists of a pair of perfectly flat planes, which entirely overlap to each other by the $2k_F$ translation. Overlap of the Fermi surface is called **nesting**. With considerable interchain interaction, the Fermi surface warps as shown in Fig. 7.19b. Then, the straight translation ($2k_F = \pi/a$) does not lead to the complete overlap, and generates electron and hole pockets (Fig. 7.19c). This is called **imperfect nesting**. Even such warping Fermi surface **nests perfectly** by an oblique nesting vector $(\pi/a, \pi/b)$ as shown in Fig. 7.19d. When the interchain transfer runs only in one direction, such perfect nesting is possible. However, $(TMTSF)_2PF_6$ has transfers not only along the $b$-axis but also in the *p1* and *p2* directions (Fig. 2.45a), and the perfect nesting is destroyed. This is the reason that $(TMTSF)_2PF_6$ does not undergo the Peierls transition down to low temperatures.

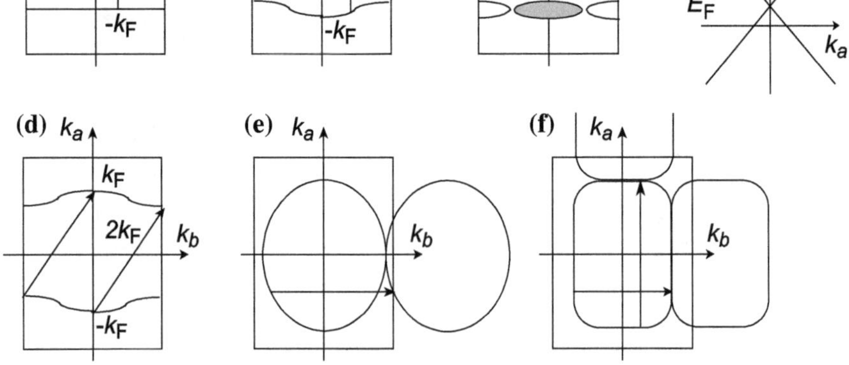


**Fig. 7.19** Nesting in the one-dimensional Fermi surface. **a** One-dimensional Fermi surface. **b** Quasi one-dimensional Fermi surface. c The resulting semimetal. **d** Nesting in the quasi one-dimensional Fermi surface. **e** Two-dimensional cylindrical Fermi surface. **f** Two-dimensional Fermi surface: nesting in the straight parts.

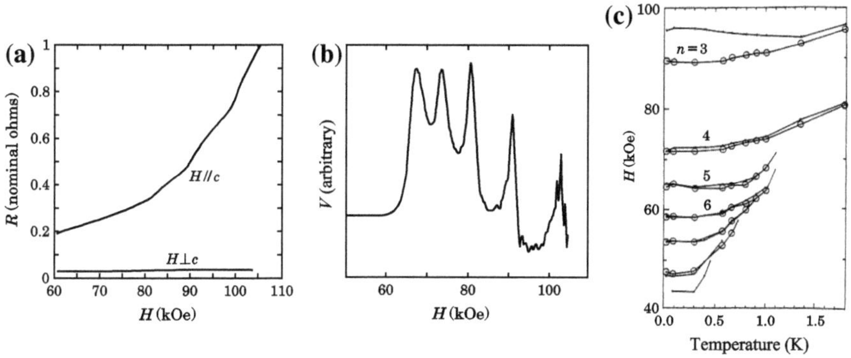


**Fig. 7.20** Field-induced SDW in $(TMTSF)_2PF_6$. **a** Resistance, **b** differential resistance: output voltage of 150 Hz harmonics when 50 Hz AC is applied [61], and **c** phase diagram of the subphases [52]

The cylindrical Fermi surface of a two-dimensional metal does not overlap except for a point or a line along the $k_c$ axis (Fig. 7.19e). Accordingly, an extra periodicity does not generate a gap in a two-dimensional metal, which therefore does not undergo the Peierls transition. Some two-dimensional metals have square and triangular Fermi surface, and such Fermi surface nests at the straight part. For instance, the Fermi surface in Fig. 7.19f mostly disappears by two kinds of nesting vectors along the $k_a$ and $k_b$ axes. Such two-dimensional CDW has been known in layered chalcogenides like $MoS_2$ and $NbS_2$.

When a magnetic field is applied perpendicular to the layer of the high-pressure metallic phase of $(TMTSF)_2PF_6$ above the SDW phase, anomalies appear in the conductivity (Fig. 7.20a), and plateaus appear in the Hall coefficient. This is a similar phenomenon to quantum Hall effect (Fig. 3.27), and called **field-induced SDW** [52, 61]. Under the magnetic field, the SDW is restored, but small Fermi pockets (Fig. 7.19c) appear owing to the imperfect nesting. The nesting vector is oblique (Fig. 7.19d), while the third-direction interchain interactions *p1* and *p2* give rise to the imperfect nesting, and small pockets are generated. This is quantized to the Laudau levels to lead to the anomalies of the resistivity and the Hall coefficient. Since the Fermi surface is very small, even a comparatively weak magnetic field attains the quantum limit around $n \sim 1$. The optimal nesting vector changes depending on the magnetic field, and many subphases appear as shown in Fig. 7. 20c. Zero resistance like the quantum Hall effect has not been observed, but the resistance is expected to go to zero at very low temperatures.

We can experimentally distinguish various insulating states according to Fig. 7.21. The Peierls transition due to CDW is characterized by the nonmagnetic insulating state, so that the magnetic susceptibility, measured by a SQUID magnetometer and electron spin resonance (ESR), drops to zero below the transition temperature $T_{MI}$. The temperature dependence below $T_{MI}$ follows the singlet–triplet excitation.

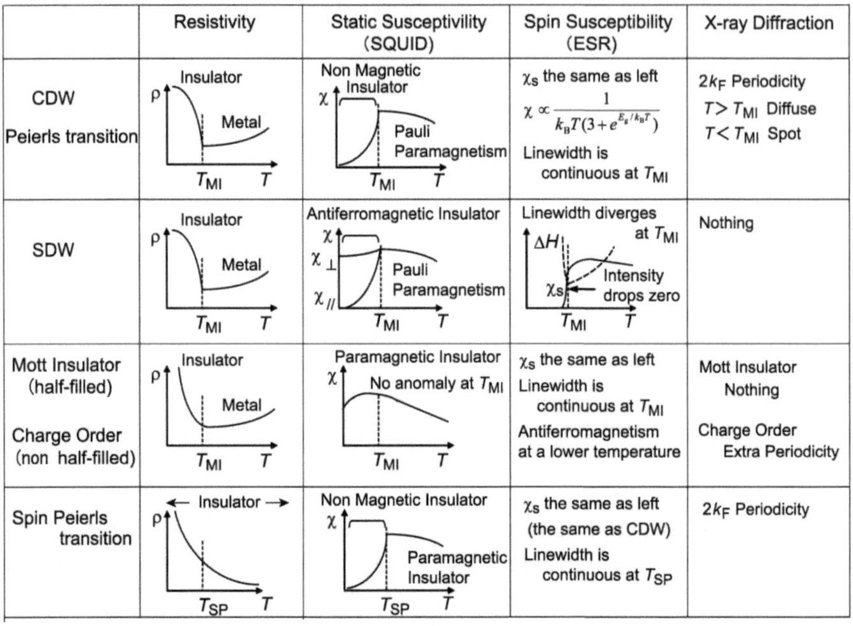


**Fig. 7.21** Experimental determination of various insulating states

$$\chi \propto \frac{1}{k_B T(3 + \exp(E_g/k_B T))} \tag{7.4}$$

The susceptibility drops rapidly below $T_{MI}$, but does not abruptly becomes zero. The susceptibility decays exponentially below $T_{MI}$, and the ESR linewidth is usually continuous at $T_{MI}$. The high-temperature susceptibility is temperature independent and Pauli-like. However, when $U$ is large, the susceptibility slightly increases at low temperatures, and looks like a low-dimensional magnet (Fig. 4.20). Observation of $2k_F$ periodicity in X-ray diffraction is the ultimate evidence of CDW.

An SDW state is a kind of an antiferromagnetic state, so the anisotropic susceptibility is the unambiguous evidence. The ESR linewidth rapidly broadens below $T_{MI}$, and sometimes the signal disappears, as far as the high-temperature linewidth is not too large.

The Mott insulator phase and the charge-order phase are characterized by the paramagnetic insulating state. However drastic is the transition of the conductivity, the susceptibility is entirely continuous at $T_{MI}$. The Mott insulator does not have any extra periodicity, whereas the superlattice periodicity as well as the splitting of the vibrational (Raman) spectrum is the final evidence of the charge order.

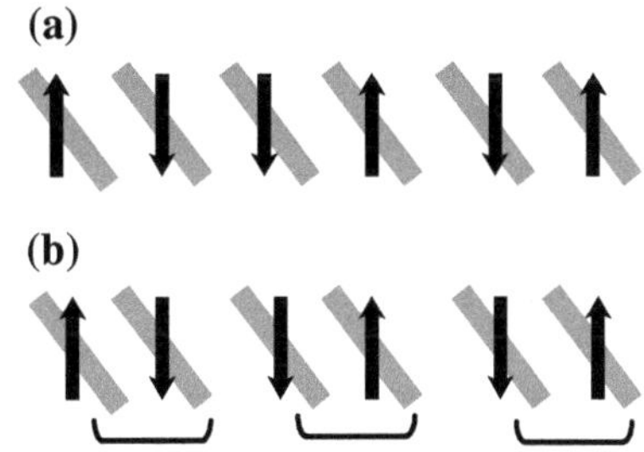


**Fig. 7.22** Spin-Peierls transition. **a** Paramagnetic insulator. **b** Singlet insulator

The **spin-Peierls transition** appeared in the generic phase diagram of the TMTSF salts (Fig. 7.18a). The Mott insulator usually turns to an antiferromagnetic state at low temperatures, in which the spin orientation is aligned alternately but the lattice periodicity is unchanged. By contrast, in the spin-Peierls transition, the spins are paired by the lattice distortion to lead to a nonmagnetic insulating state. Figure 7.22 represents a half-filled example, in which the twofold periodicity appears below the transition, but the periodicity is generally $2k_F$. The spin-Peierls transition is characteristic of a highly one-dimensional spin system, and so named because the mathematical treatment is analogous to the Peierls transition. The spin-Peierls transition is a transition from a paramagnetic insulator to a nonmagnetic insulator. Further information is obtained on the origins of metal–insulator transitions from solid-state NMR and optical measurements.

Various insulating states are related to the structures as shown in Fig. 7.23. More than 90 % of organic conductors have 2:1 composition, where a donor has 1/2+ charge, and the energy band is quarter filled. If the structure is uniform, and composed of a single transfer integral, charge order is the only possibility of the insulating state (Fig. 7.23a). The θ-phase BEDT-TTF salts are the two-dimensional examples. By contrast, a dimer structure gives rise to a single carrier per dimer, leading to an effectively half-filled band and resulting in the Mott insulating state. In this case, the

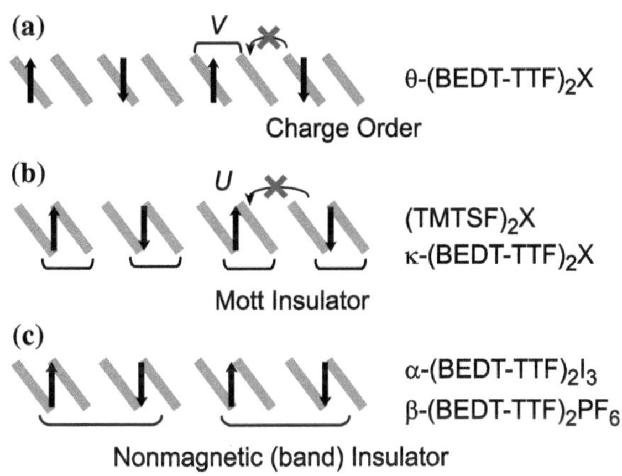


**Fig. 7.23** Insulating states of $D_2X$-type quarter-filled conductors. **a** Uniform. **b** Dimer. **c** Quatermer

effective $U$ is not the molecular $U$, but $2t$ of the dimer $t$, as verified in Example 5.2. This category includes representative organic superconductors such as $(TMTSF)_2X$ and κ-$(BEDT\text{-}TTF)_2X$. Here, the superconductivity appears near the Mott insulating state. In the tetramers (Fig. 7.23c), the corresponding insulating state is nonmagnetic. Whether the insulating state is paramagnetic or nonmagnetic is uniquely determined by the symmetry of the original structure. (TTF)(TCNQ) has such charges as $(TTF^{1/2+})(TCNQ^{1/2-})$ at low temperatures, and it is quarter filled. Although the room-temperature structure is uniform like Fig. 7.23a, the TCNQ column undergoes a $2k_F$ transition, and the low-temperature state is tetramerized like Fig. 7.23c. The TTF column has a $4k_F$ distortion, which is dimerized like Fig. 7.23b.

Figure 7.23 is schematically depicted for one-dimensional conductors, while real one-dimensional conductors are susceptible to the Peierls transition, and a non-magnetic insulating state is always most likely. The above grouping is mainly valid for the following two-dimensional BEDT-TTF conductors [62–64]. It is sometimes ambiguous whether a two-dimensional molecular arrangement is dimeric or tetrameric, but it is uniquely determined from the crystal symmetry [64].

## 7.7 BEDT-TTF

In order to realize organic superconductivity, organic conductors metallic down to low temperatures are explored. Along this line, two-dimensional conductors are investigated to suppress the Peierls transition. BEDT-TTF (see Fig. 7.40) was found in the course of such a strategy. In 1982, β″-$(BEDT\text{-}TTF)_2ClO_4$(1,1,2-trichloroethane)$_{0.5}$ was found to be metallic down to low temperatures (Fig. 7.24) [65]. This salt is not superconducting, but this compound is not composed of a strict columnar structure (Fig. 7.24a) [66]. We define an angle $\phi$ between the molecular plane and the intermolecular vector (Fig. 7.24c) [67]. The ordinary stacking structure corresponds to $\phi = 90°$, and the horizontal alignment is $\phi = 0°$. Molecules in this $ClO_4$ salt are aligned in the $\phi = 60°, 30°$, and $0°$ directions (Fig. 7.24a). A BEDT-TTF

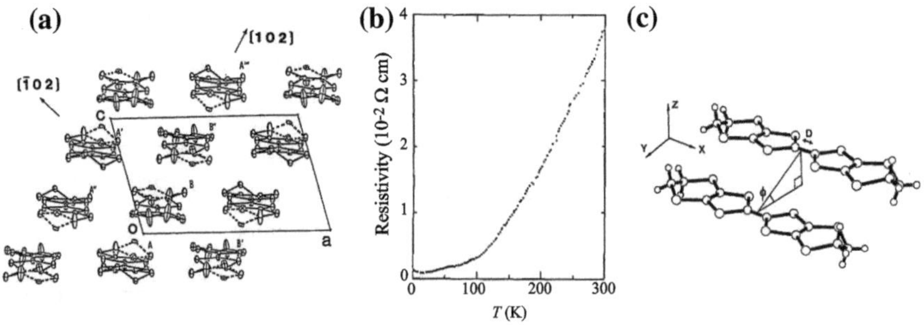


**Fig. 7.24** **a** Crystal structure [66] and **b** temperature-dependent conductivity in β″-$(BEDT\text{-}TTF)_2ClO_4$(1,1,2-trichloroethane)$_{0.5}$ [65]

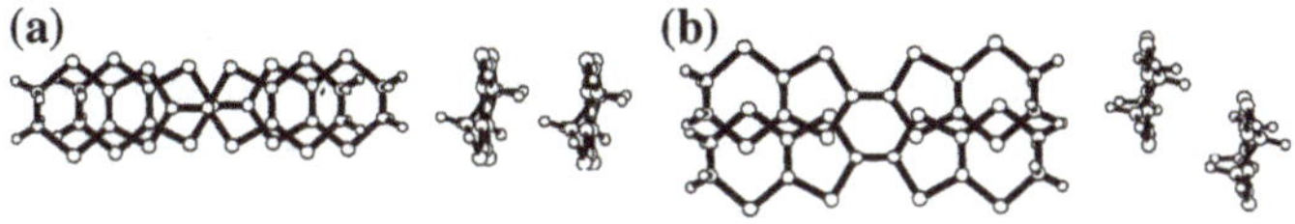


**Fig. 7.25** **a** Ring-over-bond and **b** ring-over-atom structures

molecule consists of a ladder-like arrangement of eight sulfur atoms, where the S-S distances are about 3.2 Å both along the molecular long and short axes. In the stack, the molecules are usually slipped along the molecular long axis ($D$ in Fig. 7.24c). The structure in which a sulfur atom is located in the middle of S-S of the adjacent molecule is most stable. This structure (Fig. 7.25a) corresponds to $D = 1.6$ Å, and called ring-over-bond (RB) structure because a five-membered ring of TTF is on the top of a C=C bond of another molecule. By contrast, the $\phi = 60°$ overlap (Fig. 7.25b) has $D = 0$ Å, but the molecules are slipped along the molecular short axis ($y$ in Fig. 7.24c). Since a five-membered ring is located on the top of a sulfur atom, this arrangement will be called ring-over-atom (RA) structure.

Several hundreds of BEDT-TTF salts with various counter anions have been known, among which some have more than two structures with the same composition. The variety of crystal structures are governed by some principles, where most adjacent molecular arrangements are constructed from the combinations of the RB and RA structures. The resulting BEDT-TTF structure types are depicted in Fig. 7.26 [62–64, 68]. Structure types of BEDT-TTF salts are conventionally designated as Greek characters such as α and β. This has been initiated following the crystallographers' transition to distinguish different structures with the same composition, but gradually the structure types are indicated. The repetition of the RB structure forms the ordinary stacking structure, and is called the β structure in

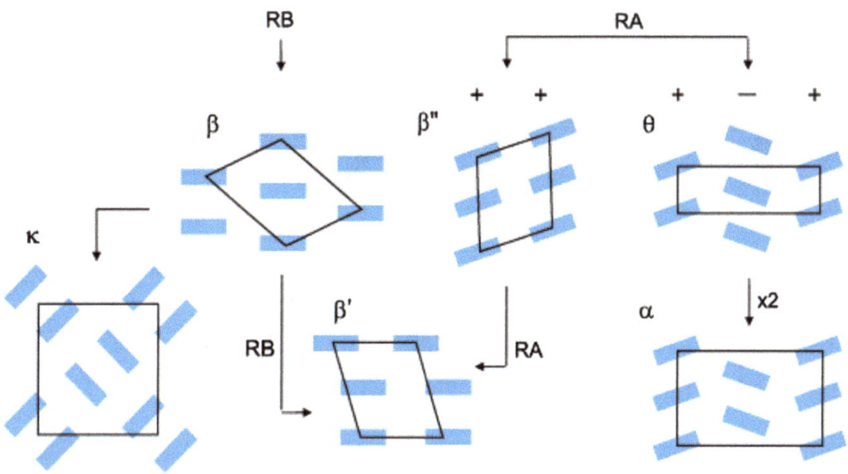


**Fig. 7.26** Representative structure types of BEDT-TTF salts [62–64, 68]

BEDT-TTF. However, BEDT-TTF never forms a uniform column, and has more than twofold (dimerized) periodicity. The repetition of the RA structure makes the above β″ structure. Alternate arrangement of the RB and RA structures forms the so-called β′ structure. The β″ structure is regarded as a pseudo-columnar structure in the $\phi = 60°$ direction, where the molecules are tilted from the columnar direction. Accordingly, a pseudo-column has two kinds: right-up and left-up. The right-up and left-up are arranged alternately in the θ-structure. This structure is analogous to the herringbone structure, which is the most popular structure in organic semiconductors such as pentacene and oligothiphene. The θ-structure dimerized along the stacking direction is called α-structure.

### 7.7.1 β-Structure

This structure corresponds to the conventional stacking structure, but is accompanied by a strong dimerization (Fig. 7.27a). In β-(BEDT-TTF)$_2$I$_3$, the dimerization is demonstrated by the largely different molecular-long-axis slips, $D = 1.5$ Å for p1 and 3.8 Å for p2 [69]. The strong dimerization balances the intrachain and

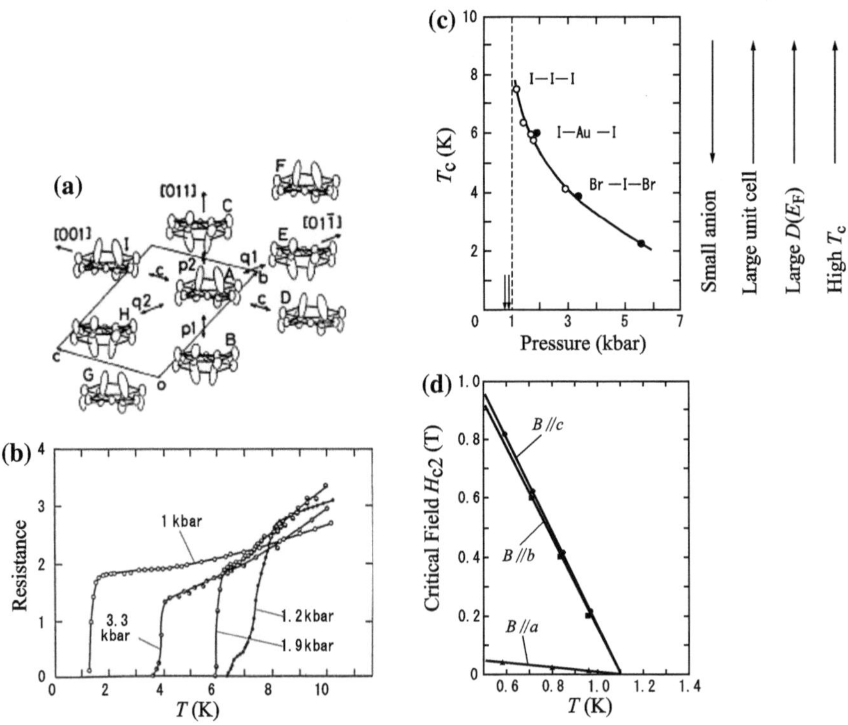


**Fig. 7.27** β-(BEDT-TTF)$_2$I$_3$. **a** Crystal structure. The transfer integrals are p1: 245, p2: 84, q1: 127, q2: 68, and c: 50 meV [69]. **b** Conductivity around the superconducting transition [71]. **c** Pressure dependence of the superconducting $T_c$ in β-(BEDT-TTF)$_2$X [70–72]. **d** Anisotropy of the critical fields in β-(BEDT-TTF)$_2$X [73].

interchain interactions to lead to the two-dimensional conduction. Here, the smaller intrachain interaction p2 is in the same order as the interchain interactions, and determines the anisotropy. The resulting cylindrical Fermi surface (Fig. 3.16) has been proved by the observations of the Shubnikov–de Haas oscillation and the Kajita oscillation (Fig. 3.20). The outer ethylene groups of BEDT-TTF are deviated from the molecular plane so that BEDT-TTF does not form a simple stacking structure. This is the origin of the two-dimensional conduction.

This salt undergoes a superconducting transition at 1.5 K [71], but $T_c$ increases to 8 K under a slight pressure of 1.2 kbar (Fig. 7.27b) [70, 71]. Higher pressure gradually reduces $T_c$ (Fig. 7.27c) [72]. At ambient pressure, incommensurate lattice distortion due to the donor ethylene groups interacting with the anions occurs around 100 K. However, this is suppressed by a slight pressure to raise $T_c$ to 8 K. When $I_3^-$ is replaced by $AuI_2^-$ and $IBr_2^-$, $T_c$ is 4.9 and 2.7 K, respectively (Fig. 7.27c) [72]. This is because the salts other than $I_3^-$ do not have the lattice modulation. $T_c$ of these salts decreases under pressure. According to the BCS theory, $T_c$ is given by Eq. 6.36.

$$k_B T_c = 1.13\hbar\omega_D \exp(-1/VD)$$

The lattice volume decreases under pressure, and the transfer increases to reduce $D$, and reduce $T_c$. In general, superconducting $T_c$ decreases under pressure. Since $I_3^-$ is larger than $AuI_2^-$ and $IBr_2^-$, the lattice volume decreases in this order, and $T_c$ decreases (Fig. 7.27c). Such a tendency is called chemical pressure.

Figure 7.27d shows the temperature dependence of the critical field in three magnetic field directions [73]. The in-plane (*//b* and *//c*) critical fields are almost the same to prove the two-dimensional conductance (Example 6.3). It is difficult to measure conductivity exactly in different directions because we have to make contacts to different crystals. By contrast, the anisotropy of the critical field is more exactly estimated by rotating the crystal with respect to the magnetic field orientation for a single set of contacts.

### 7.7.2 *β'-Structure*

For smaller anions such as $AuCl_2^-$ and $ICl_2^-$, the β-structure is no more maintained, and the β′-structure is realized (Fig. 7.28). This is an alternate stack of the RB and RA structures and shows very strong dimerization [74]. The β′-phase is a Mott insulator with the conductivity of as small as 0.1 S/cm at room temperature. The susceptibility shows low-dimensional paramagnetism (Fig. 4.19), followed by antiferromagnetism around 15 K. Following the principle that superconductivity borders to a Mott insulator phase, properties of β′-(BEDT-TTF)$_2$ICl$_2$ under very high pressure have been investigated, and superconductivity has been found at 14. 2 K under 83 kbar [75]. This is the highest $T_c$ among the BEDT-TTF conductors.

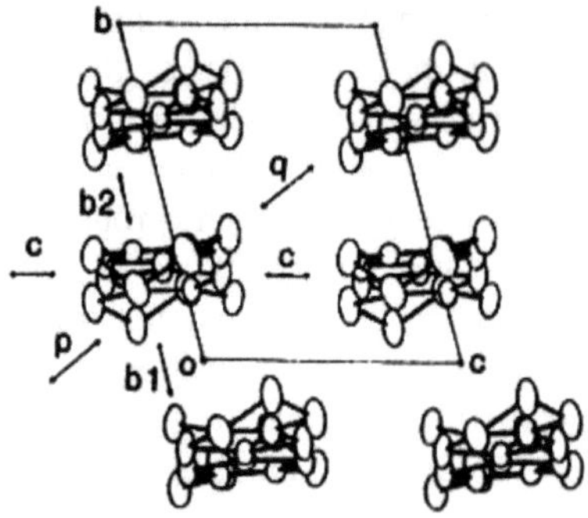


**Fig. 7.28** Crystal structure of the β′-phase. The transfers integrals are b1: 264, b2: 100, p: 65, q: −20, and c: −23 meV for the $AuCl_2$ salt [74]

(BEDT-TTF)(TCNQ) forms a mixed stack complex and two segregated complexes [76–78]. In a segregated complex, the BEDT-TTF molecules have the β′-structure, and TCNQ molecules are stacked perpendicular to the BEDT-TTF molecules (in the *c* direction in Fig. 7.28). This complex is a semiconductor at room temperature but becomes a metal above 330 K. Another segregated phase has β″-type BEDT-TTF arrangement, and the conductivity is metallic down to low temperatures.

### 7.7.3 *β″-Structure*

As exemplified by β″-(BEDT-TTF)$_2$ClO$_4$(1,1,2-trichloroethane)$_{0.5}$, the β″-structure is universally observed in the salts of small tetrahedral anions such as $ClO_4^-$, $ReO_4^-$, and $BF_4^-$. There are many variations in which the unit cell contains 2, 3, 4, and 6 donors [62–64]. A primitive example is β″-(BEDT-TTF)$_2$AuBr$_2$, involving two donors in a unit cell (Fig. 7.29) [79]. This salt is metallic down to helium temperatures but not superconducting. On account of the comparatively large

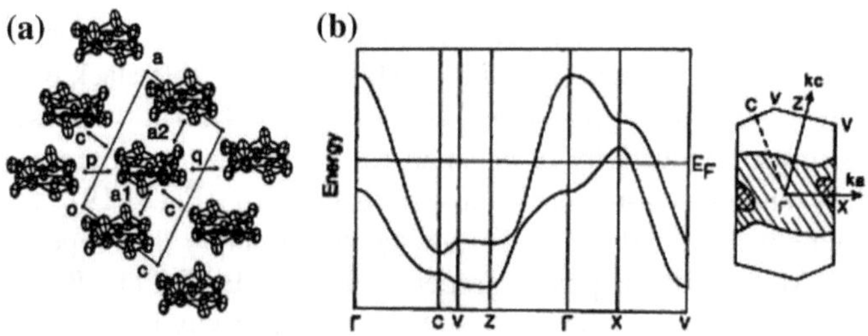


**Fig. 7.29** β″-(BEDT-TTF)$_2$AuBr$_2$. **a** Crystal structure, and **b** band structure. The transfer integrals are a1: 15, a2: 65, c: 150, p: 87, and q: 96 meV [79]

**Fig. 7.30** Crystal structure of β″-$(BEDT\text{-}TTF)_3(ClO_4)_2$. The transfer integrals are q1: −160, q2: 212, p1: 50, p2: 66, r1: −30, and r2: 20 meV [80]

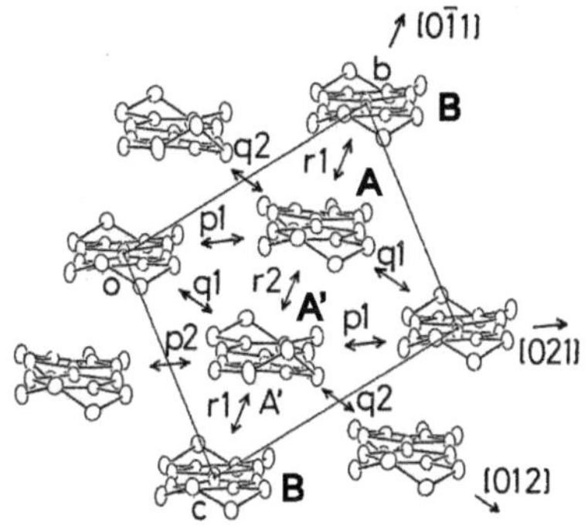


ϕ = 30° (*c*) interaction, this salt has an open Fermi surface. A small pocket coexists, with which the Shubnikov–de Haas oscillation is in agreement, though more than two oscillations have been observed.

β″-$(BEDT\text{-}TTF)_3(ClO_4)_2$ contains three donor molecules in a unit cell (Fig. 7.30) [80], and there are many isostructural members that involve $BF_4^-$ and $HSO_4^-$ [62]. These salts have extraordinary average charge, 2/3+. These materials are metals with the room-temperature conductivity of 50 S/cm, but undergo metal–insulator transitions at 170, 150, and 130 K, respectively. The crystal consists of Molecule A on a general position and Molecule B on an inversion center, and the low-temperature insulating state is ascribed to a charge-order state that is approximately $A^+A^+B^0$ [81].

Other β″-phases containing four or six donor molecules in a unit cell usually undergo metal–insulator transitions [62]. In the β″-phases, the nonplanarity of the BEDT-TTF ethylene units is sterically released by the RA structure, namely by the short-axis sliding. Accordingly, dimerization is weak in the β″-structure. As a result, charge order is the origin of the insulating state instead of the Mott insulating state [81]. Semimetallic Fermi surface consisting of small electron and hole pockets is characteristic of the β″-structure except for the primitive β″-$(BEDT\text{-}TTF)_2AuBr_2$ containing only two donor molecules in a unit cell. Several superconductors have been found in the β″-family [82]; examples are β″-$(BEDT\text{-}TTF)_4(H_3O)Fe(C_2O_4)_3PhCN$ (8 K at ambient) [83], β″-$(BEDT\text{-}TTF)_3Cl_2(H_2O)_2$ (2 K under 16 kbar) [84], and β″-$(BEDT\text{-}TTF)_4Pt(CN)_4H_2O$ (5 K under 6.5 kbar) [85]. In the first material, the Fe spin is paramagnetic down to low temperatures, and superconductivity and paramagnetism coexist in the πd-system. The second is an example of superconductors with a composition other than 2:1 [86]. Superconductivity in the latter two compounds emerges when nonmagnetic insulating states, which are probably due to charge order, are suppressed by pressure. Although the superconducting phases in the representative TMTSF and the κ-phase organic superconductors appear on the border of the antiferromagnetic insulating phases, these superconducting phases appear by suppressing the charge-order and nonmagnetic insulator phases [87]. The superconducting properties are not very well investigated in these compounds.

### 7.7.4 *θ-Structure*

In the structures so far described, all molecular planes are parallel. However, the θ-phase is composed of two kinds of differently tilted columns (Fig. 7.31a). This structure is very close to the herringbone structure. This structure has the highest symmetry among the BEDT-TTF salts, and the energy band is represented by a simple formula (Example 2.11). The cylindrical Fermi surface has been verified by the Shubnikov–de Haas oscillation and the Kajita oscillation (Fig. 3.20a). The first Kajita oscillation was found in the θ-phase.

This family shows a variety of conducting properties from a metallic superconductor, θ-$(BEDT\text{-}TTF)_2I_3$ to room-temperature insulators. A series of salts represented by θ-$(BEDT\text{-}TTF)_2MM'(SCN)_4$ contain monovalent metals $M$ = Rb and Cs, and divalent metals $M'$ = Zn and Co, so the $MM'(SCN)_4$ unit is 1- and BEDT-TTF is quarter filled [88]. The Rb and Cs salts undergo metal–insulator transitions at 190 and 20 K, respectively (Fig. 7.31b). The transition temperature $T_{MI}$ is related to the dihedral angle θ between the molecules in the adjacent columns (Fig. 7.31c); the smaller is θ, the lower is $T_{MI}$. This is known as the **universal phase diagram** of the θ-phase [74]. With reducing θ, the transfer integral increases, and the bandwidth increases (Fig. 7.31d). Then, $T_{MI}$ decreases. This is because the π-orbital overlap increases with decreasing θ (the inset in Fig. 7.31d).

Conductivity under pressure is measured using a **clamp cell** (Fig. 7.32). A sample with contact wires is dipped in an oil within a Teflon container, and pressurized under a hydraulic system. By fastening the screw, the piston is clamped at the same position, and the applied pressure is maintained without the hydraulic system. Cooling the clamp cell, the measurement is carried out at low temperatures under the pressure. An oil applies **hydrostatic pressure**, but using epoxy resin instead of oil, **uniaxial strain** is applied.

Figures 7.31e, f show the resistivity of θ-$(BEDT\text{-}TTF)_2CsZn(SCN)_4$ under uniaxial strain [89]. The *c*-axis strain enlarges the dihedral angle, and raises $T_{MI}$ (Fig. 7.31e). In the universal phase diagram (Fig. 7.31c), the material moves to the left. The *a*-axis strain reduces the dihedral angle, and lowers $T_{MI}$ (Fig. 7.31f). The material moves to the right in the universal diagram (Fig. 7.31c). Thus, the universal phase diagram is verified by the uniaxial strain. Hydrostatic pressure increases the dihedral angle and raises $T_{MI}$. When the θ-phase is cooled, the *c*-axis shows an ordinary thermal shrink, but the *a*-axis expands. Then, the cooling enlarges the dihedral angle and stabilizes the insulating phase. Hydrostatic pressure seems to have the same effect.

According to the universal phase diagram, the θ-phase undergoes a metal–insulator transition. However, on account of the uniform structure represented by Fig. 7.23a, the insulating state has to be a charge-order state. Owing to the two-dimensional structure, a variety of charge-order patterns are possible as shown in Fig. 7.33. This figure shows the most stable patterns depending on $U$, $V_p$, and $V_c$, which are determined at the atomic limit similarly to Example 5.4 [90]. Charge order is preferable when $V$ is large in comparison with $U$. When $V_c < V_p$, charge

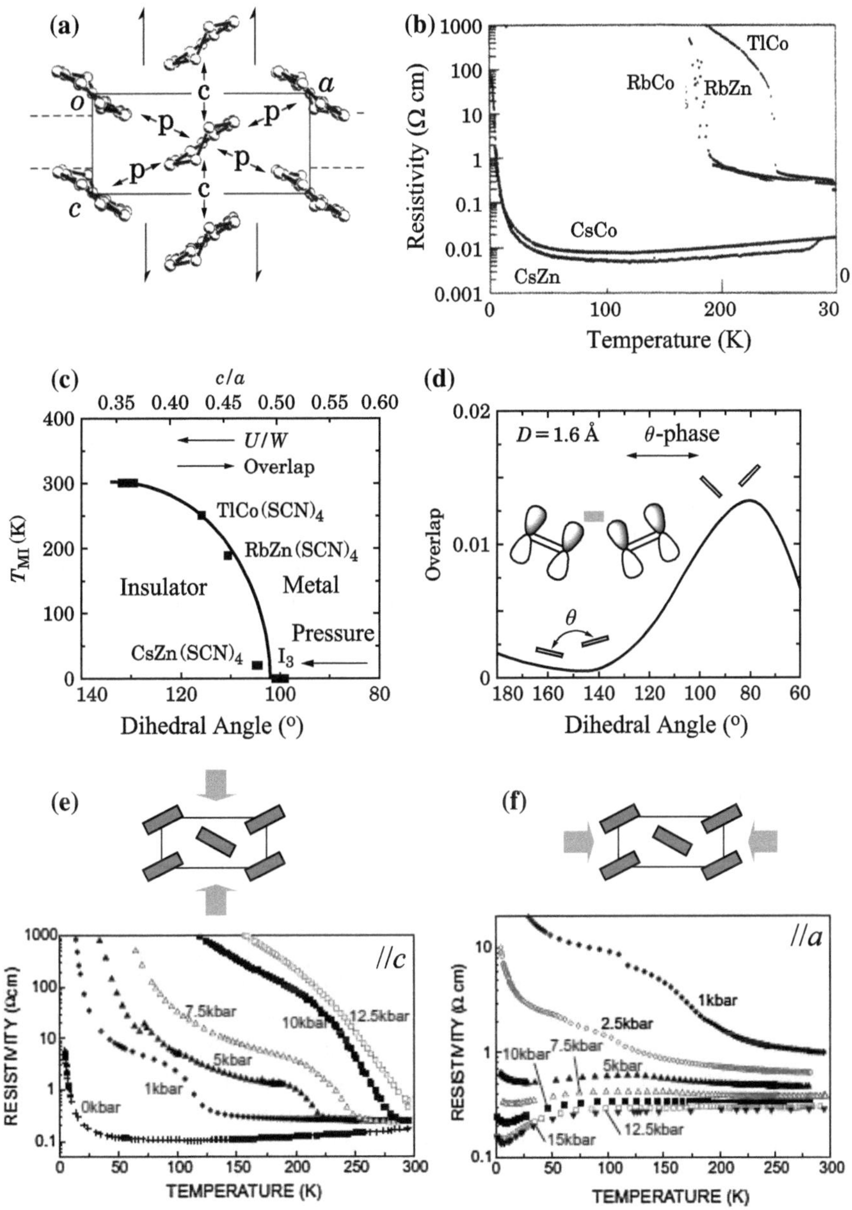


**Fig. 7.31** θ-Structure **a** Crystal structure [88]. **b** Temperature dependence of the resistivity in θ-(BEDT-TTF)$_2$*MM'*(SCN)$_4$ [88]. **c** Universal phase diagram of the θ-phase [63]. **d** Dihedral angle dependence of the transfer integral (*p* in (**a**)) [63]. Temperature dependence of the resistivity under the uniaxial pressure along the (**e**) *c*, and **f** *a* axes in θ-(BEDT-TTF)$_2$ CsZn(SCN)$_4$ [89]

**Fig. 7.32** A clamp cell

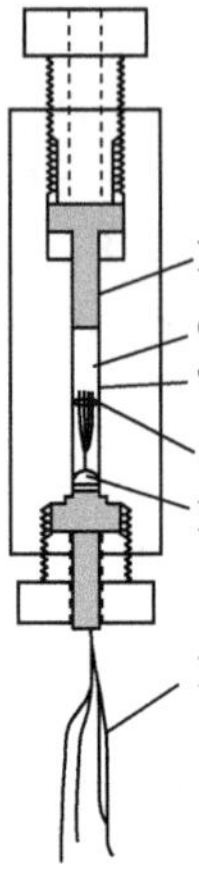


alignment along $p$ is avoided rather than charge alignment along $c$, and the charge rich sites are aligned along the vertical axis. This leads to the vertical stripe. By contrast, when $V_c > V_p$, the horizontal stripe is realized in order to avoid the vertical alignment of the charge-rich sites. The horizontal stripe has a zigzag alignment, but a straight alignment along $p$, which is called diagonal stripe, is similarly stable. Thus, the charge-order patterns are determined by the magnitude of $V_c$ and $V_p$. The Coulomb repulsion has been estimated from the distance between the point charges located at the molecular centers (Fig. 5.22). Since the interplanar distance ($c$) is shorter than the diagonal ($p$) distance, the Coulomb repulsions are $V_c > V_p$. Accordingly, the horizontal stripe is realized.

However, around $V_c = V_p$, a threefold (3-fold) charge order becomes stable (Fig. 7.33), where the charge-rich sites are discretely placed [90]. The θ-phase with equivalent $c$ and $p$ falls into a triangular lattice, but the triangular lattice does not have something like the checkerboard pattern in the square lattice (Example 5.4), where the $V$ repulsion is avoided in all bond directions. This reminds us the frustrated triangular antiferromagnet (Fig. 4.21b). In the checkerboard pattern, the charge-rich sites are aligned in the bisector of two bond directions (namely in the 45° direction). In the threefold pattern, the charge-rich sites are aligned in the bisector direction. However, the charges are separated to 1:2 as $D^{+}D^{1/4+}D^{1/4+}$, and the charge poor sites have fractional charges. Consequently, the threefold state is fairly conducting. The θ-phase Rb salt has the horizontal stripe below 190 K (Fig. 7.31b), but the state just above 190 K has such a non-stripe charge order. The low-temperature state of the Cs salt below 20 K is not a simple horizontal stripe, but a glassy mixture of the horizontal stripe and the non-stripe state. Such charge-order patterns have been investigated by X-ray diffraction. At the same time, NMR lineshape splits, and the C=C stretching of the Raman spectra splits. The dielectric constant shows an anomalous increase, and nonlinear conductivity appears in the charge-order phase [91, 92].

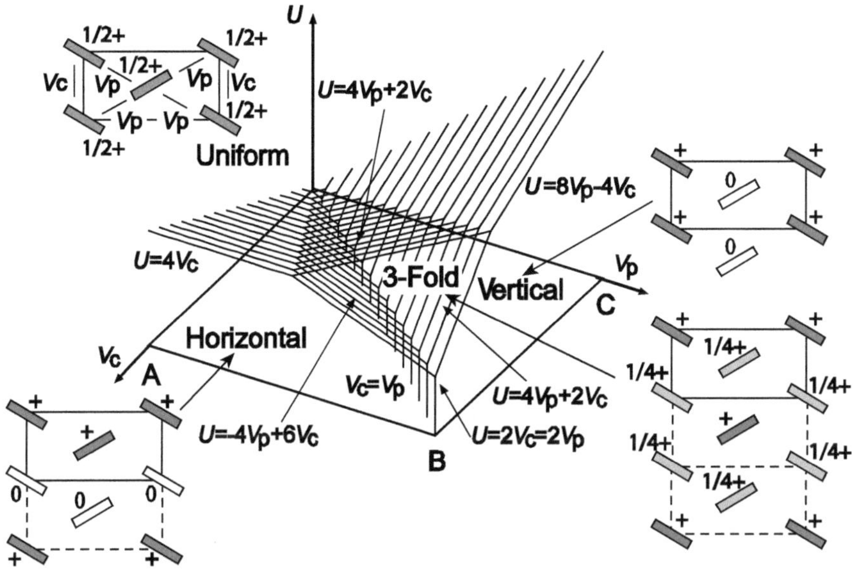


**Fig. 7.33** Charge-order patterns in the θ-phase [90]

## 7.7.5 α-Structure

α-$(BEDT\text{-}TTF)_2I_3$ was named as the α-phase because it was found prior to β-$(BEDT\text{-}TTF)_2I_3$. Both phases are easily obtained by the electrochemical crystallization; the α-phase appears predominantly but sometimes coexists with the β-phase. The β-phase is a metal down to low temperatures, and becomes a superconductor, whereas the α-phase undergoes a metal–insulator transition at 135 K (Fig. 7.34b). In the crystal structure, the vertical direction of the θ-phase is doubled, and the symmetry drops to a triclinic system (Fig. 7.34a). The crystallographically independent Molecule A is located on a general position, but Molecules B and C are located on inversion centers [68]. A unit cell contains four molecules, and the energy band consists of semimetallic small electron and hole pockets [93, 94]. Similarly to the θ-phase, the insulating phase below 135 K has a horizontal stripe, but the unit cell does not change because the original lattice is doubled in the vertical direction. Owing to the original low symmetry, the insulating state is nonmagnetic (Fig. 7.34c) [95]. On account of the charge order, Molecules A and A′ become nonequivalent, and the inversion center is lost at low temperatures. Accordingly, second harmonic generation (SHG) is observed in the insulating phase [96]. Even above 135 K, Molecule B is charge poor (close to neutral), and this non-stripe charge order results in the almost flat temperature dependence of the resistivity [97, 98]. The metal–insulator transition is suppressed under pressure, and this material becomes a metal down to low temperatures under the pressure of 10 kbar (Fig. 7.34b) [93, 94]. The Hall coefficient increases by several orders at the same time (Fig. 7.34d). This

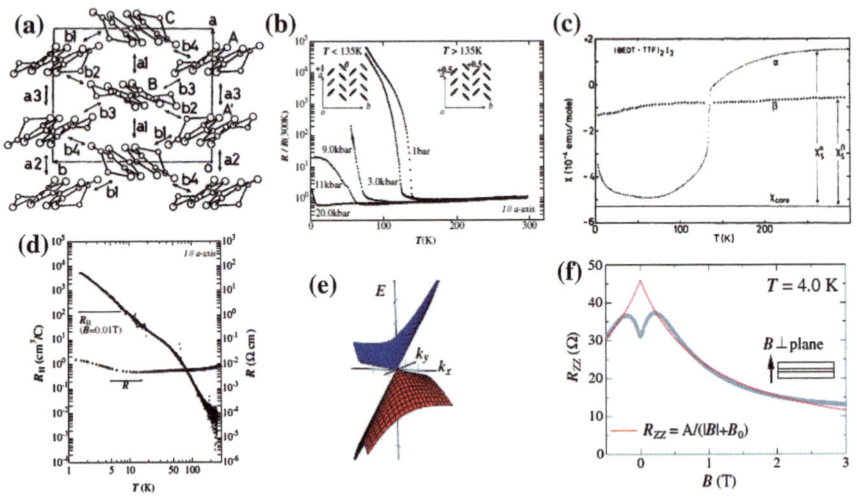


**Fig. 7.34** α-(BEDT-TTF)$_2$I$_3$. **a** Crystal structure. The transfer integrals are a1: −30, a2: −49, a3: 18, b1: 123, b2: 142, b3: 62, and b4: 23 meV [68]. **b** Temperature dependence of the resistivity under pressure [93, 94]. **c** Temperature dependence of the magnetic susceptibility [95]. **d** Temperature dependence of the Hall coefficient [93, 94]. **e** Energy band and the zero-gap state around the Fermi energy [99]. **f** Magnetoresistance [93, 94, 100, 101]

phenomenon is attributed to the **zero-gap state**, where two energy bands have a contact point and the Fermi energy comes to this point (Fig. 7.34e) [99]. This is analogous to the point Fermi surface in the single-layer graphene. The energy of the zero-gap state is proportional to $k$ instead of the ordinary $k^2$, to follow the Dirac equation instead of the Schrödinger equation. Then, this is called massless **Dirac particle**. In the quantum Hall effect of graphene, the phase of the plateau shifts by half because **Berry's phase** shifts due to the connection at the Dirac point. In the α-phase, remarkable negative magnetoresistance has been observed under the perpendicular magnetic field (Fig. 7.34f), which has been ascribed to the $n$ = 0 Landau level (zero mode) located at the Fermi level [100, 101].

There are another series of α-salts represented as α-(BEDT-TTF)$_2$$M$Hg(SCN)$_4$ ($M$ = K, Rb, and NH$_4$) [102]. The molecular packing is the same as Fig. 7.34a, but the lattice angle is very close to 90°. The resulting band structure is considerably different (Fig. 7.35a), and consists of an open Fermi surface and a hole pocket. This Fermi surface is derived from a large elliptical Fermi surface, whose area is the same as the first Brillouin zone. Splitting the large Fermi surface at the ZV boundary, the open part and a hole pocket are generated. Accordingly, the open belt and the hole pocket have the same area. This salt shows metallic conductivity down to low temperatures, but a shoulder appears in the resistivity around 10 K (Fig. 7.35b) [103]. This has been attributed to some kind of density waves, which partially erase the Fermi surface. Magnetoresistance for a perpendicular field makes

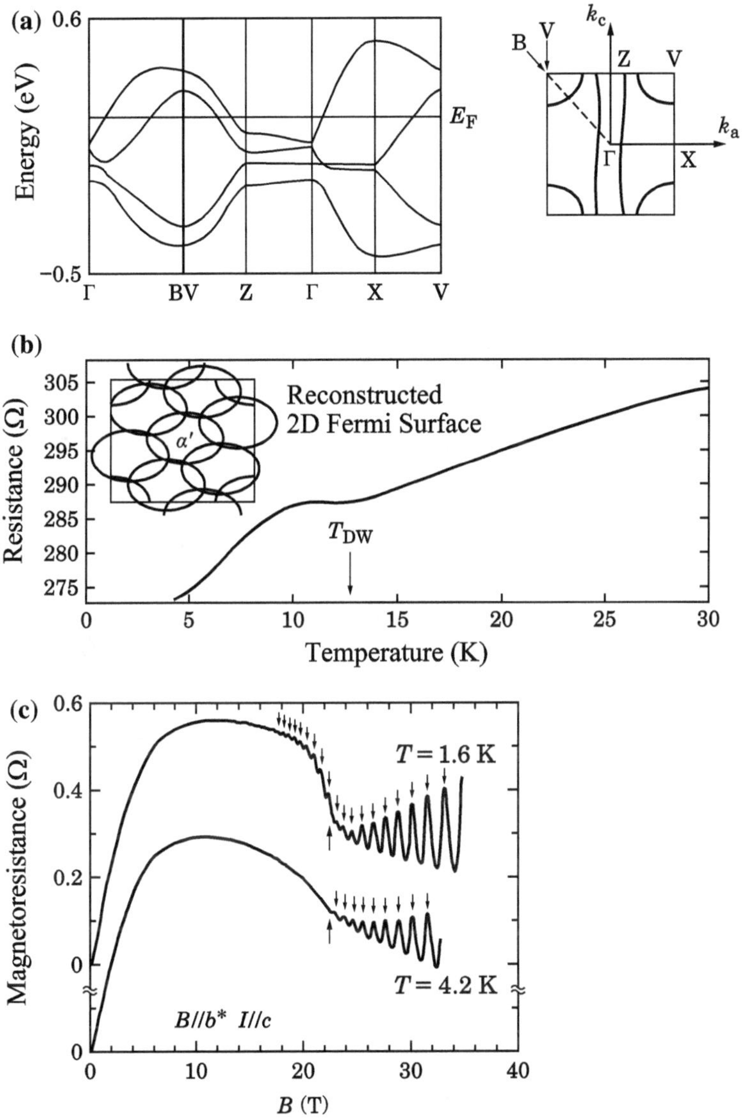


**Fig. 7.35** α-$(BEDT\text{-}TTF)_2KHg(SCN)_4$. **a** Band structure [102], **b** resistance at low temperatures [103], and **c** magnetoresistance [104]

a peak between 10 and 20 T (Fig. 7.35c), and decreases to give rise to the Shubnikov–de Haas oscillation [104]. The resistivity drops around 20 T, which is attributable to the disappearance of the density waves. The oscillation is related to the closed pocket.

### 7.7.6 κ-Structure

The κ-phase is composed of alternately 90°-rotated dimers (Fig. 7.36a), which construct an ultimate two-dimensional structure. First, κ-$(BEDT\text{-}TTF)_2I_3$ was found [105], but κ-$(BEDT\text{-}TTF)_2Cu(NCS)_2$ was discovered to be an ambient-pressure superconductor with $T_c = 10.4$ K in 1987 [106]. Then, the $Cu[N(CN)_2]Br$ salt with $T_c = 11.6$ K and the $Cu[N(CN)_2]Cl$ salt with $T_c = 12.8$ K at 0.3 kbar were found successively [107–109]. Cu is 1+, and the whole anion has 1− charge. So a donor has 1/2+ charge, and the energy band is quarter filled. The Fermi surface is cylindrical (Fig. 3.17); this is the first material in which the Shubnikov–de Haas oscillation was observed in organic conductors.

The resistivity of the $Cu(NCS)_2$ salt once increases from room temperature to 100 K, and becomes metallic below this temperature (Fig. 7.36b). The semiconducting behavior above 100 K indicates the Mott insulating phase (Fig. 7.36c) [110].

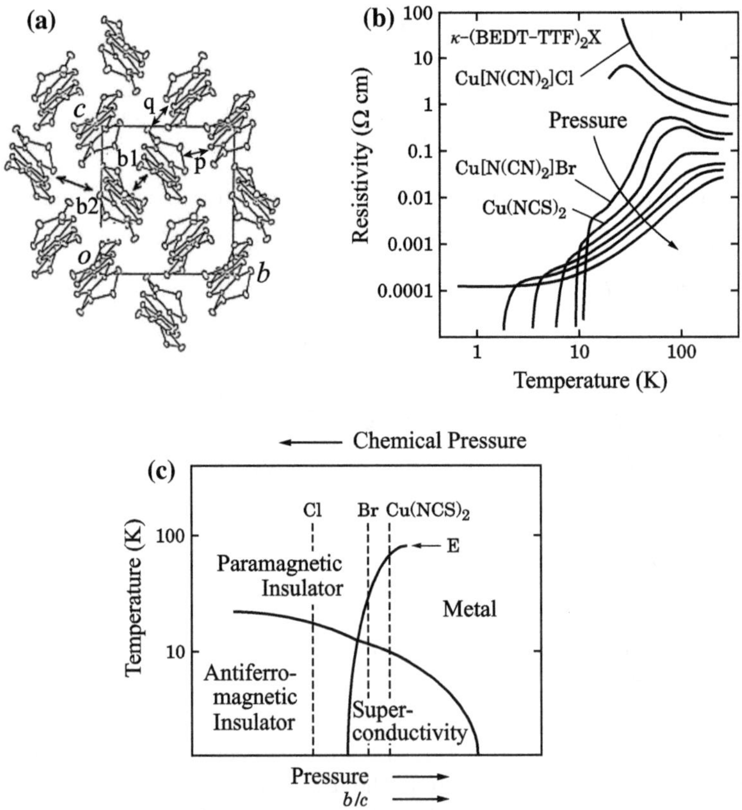


**Fig. 7.36** κ-Structure **a** Crystal structure. The transfer integrals for κ-$(BEDT\text{-}TTF)_2Cu(NCS)_2$ are b1: −257, b2: −105, p: −114, p′: −100, q: 17, and q′: 29 meV [63], **b** Temperature dependence of the resistivity, and the **c** universal phase diagram [110]. Br and Cl designate the $Cu[N(CN)_2]Br$ and $Cu[N(CN)_2]Cl$ salts, respectively

This is an example of the dimer Mott insulator, where the dimer has a hole (Fig. 7.23b). The corresponding $U$ is not the molecular $U$, but $2t$ of the intradimer transfer $t_{b1}$ (Example 5.2). κ-(BEDT-TTF)$_2$Cu[N(CN)$_2$]Br shows similar temperature dependence of the resistivity, but because this salt is located at the left of the Cu (NCS)$_2$ salt (Fig. 7.36c), $T_c$ is slightly higher. κ-(BEDT-TTF)$_2$Cu[N(CN)$_2$]Cl is located in the Mott insulator regime, and exhibits insulating behavior over the whole temperature range. Then, an antiferromagnetic phase appears below 10 K. Under a small pressure of 0.3 kbar, this salt moves to the right in the phase diagram (Fig. 7.36c), and becomes superconducting. The resulting $T_c$ is highest among the κ-phases, because they are on the border of the Mott insulating phase. Further pressure moves this salt to the right in the phase diagram, and reduces $T_c$ (Fig. 7.36b). Thus, the superconductivity of the κ-phase is well understood by the universal phase diagram (Fig. 7.36c) [110]. The chemical pressure works in the opposite direction to the physical pressure, probably because it reduces the axis ratio $c/a$ due to the polymeric anion structure [63]. The superconductivity bordering to an antiferromagnetic insulator phase resembles the cupper oxide superconductors, and d-wave superconductivity mediated by the antiferromagnetic fluctuation has been suggested. This has been discussed in view of the absence of the Hebel–Slichter peak in NMR (Fig. 6.4c), as well as the thermal conductivity, penetration depth, and specific heat. However, the superconductivity symmetry is controversial because some people prefer isotropic superconductivity [87]. The π-junction measurement has not been reported. The phase boundary between the Mott insulating and metal phases is analogous to the gas and liquid phases of ordinary matter. There is a critical (end) point $E$ in Fig. 7.36c, above which the phase boundary disappears and a supercritical state appears. The Mott insulating phase is regarded as a gas of the doubly occupied states, and the metal phase is the liquid [87]. Some evidence of the FFLO state at the low-temperature high-field end of κ-(BEDT-TTF)$_2$Cu(NCS)$_2$ has been proposed [87].

κ-(BEDT-TTF)$_2$Cu$_2$(CN)$_3$ is located in the Mott insulator region, and superconducting under a slight pressure. This compound, however, does not show antiferromagnetism down to low temperatures. This is attributed to the triangular nature of the interdimer network, in which a spin liquid state is realized (Sect. 4.8) [111]. The spin liquid state is suggested from various experiments such as specific heat [87]. The dielectric constant makes a peak around 20 K due to the unbalanced charge in the dimer, namely due to the occurrence of a kind of charge order [112]. This is schematically depicted in the ferroelectric domains in Fig. 7.37 [113]. The Mott insulator phase followed by an antiferromagnetic phase has been so far believed to be essential in these dimer Mott systems (Fig. 7.23b), but such charge order due to the unbalanced dimer charge is in general possible. This view may modify the model in Fig. 7.23, which states one of the Peierls transition, the Mott insulating state, and the charge order is realized. A large variety of two-dimensional structures are characteristic of the BEDT-TTF salts. However, the charge patterns sometimes have longer periodicity than the crystal lattice, and are more complicated. This is an origin of a variety of intriguing physical properties.

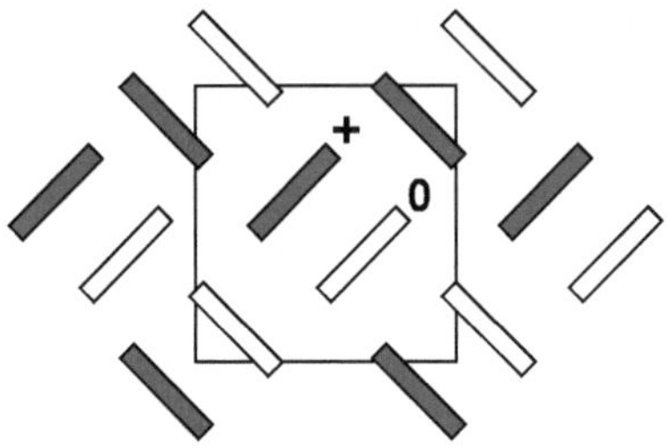


**Fig. 7.37** Charge order in κ-(BEDT-TTF)$_2$Cu$_2$(CN)$_3$

### 7.7.7 δ-Structure

This structure type has not been much interested on account of the insulating properties, but involves many salts with various anions. β-(BEDT-TTF)$_2$PF$_6$ is the representative member (Fig. 7.38), where β means the second phase among the salts of the same composition, but is not related to the β-structure. In general, this structure type is called δ-phase after δ-(BEDT-TTF)$_2$AuBr$_2$ [114]. In the stack along the *a*-axis in Fig. 7.38a, the molecular long axes are not parallel to each other in the interaction a1. The δ-phase is characterized by this nonparallel stacking. On account of the small a2, this phase shows a strong dimerization, but the transverse *c* interaction is uniform. This leads to one dimensionality along the *c*-axis and an open Fermi surface. Along this direction, the crystal is elongated, and the conductivity is largest. This salt undergoes a metal–insulator transition at 295 K accompanied by the 2*c* periodicity. Since the susceptibility becomes nonmagnetic,

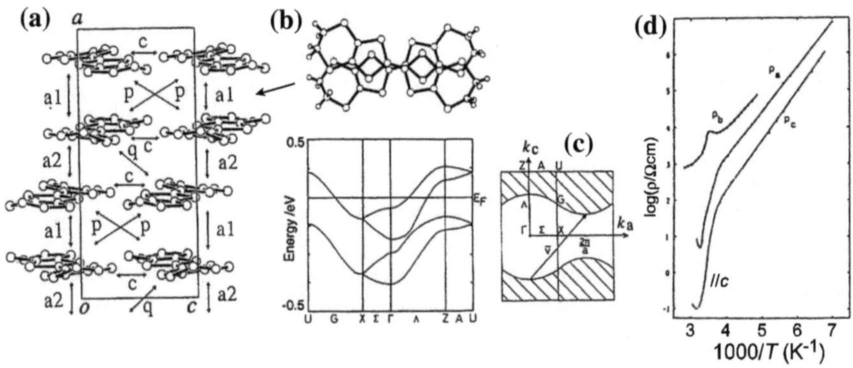


**Fig. 7.38** δ-Structure. **a** Crystal structure of β-(BEDT-TTF)$_2$PF$_6$. The transfer integrals are a1: −104, a2: 32, c: −84, p: −28, and q: −142 meV [64], **b** twisted overlap in the stack (a1), **c** energy band [64], and **d** temperature dependence of the resistivity [115]

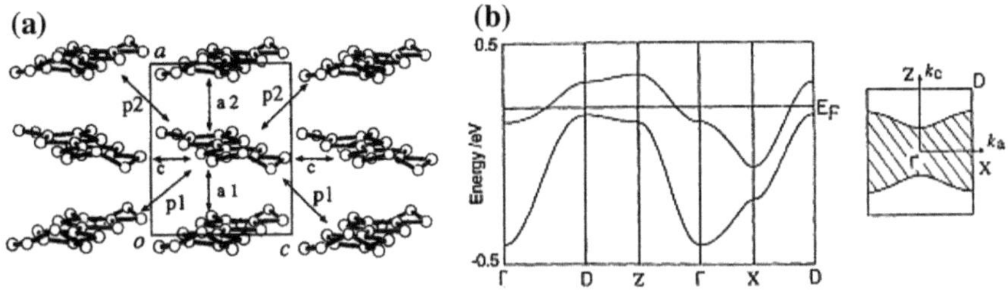


**Fig. 7.39** α′-(BEDT-TTF)$_2$AuBr$_2$. **a** Crystal structure, and **b** band structure. Transfer integrals are a1: −54, a2: −119, c: −87, p1: −19, and p2: −19 meV [64]

this is regarded as a Peierls transition along the transverse ($c$) axis. However, charge inhomogeneity has been observed, so exactly speaking, the charge order gives rise to the nonmagnetic insulating state [116]. A charge-ordered nonmagnetic insulator is not surprising because it is analogous to α-(BEDT-TTF)$_2$I$_3$.

In the δ-structure, the twisted overlap appears alternately, but all overlaps are twisted in the α′-structure (Fig. 7.39). This salt also has an open Fermi surface that is one dimensional along the transverse ($c$) axis. This salt is insulating and has a $2c$ periodicity even at room temperature. In contrast to the δ-phase, this salt is paramagnetic down to low temperatures. This difference comes from the dimeric nature of the α′-phase with considering the $2c$ periodicity, in contrast to the tetrameric nature of the δ-phase.

## 7.8 Organic Superconductivity

In addition to TMTSF and BEDT-TTF, more than one hundred organic superconductors (Table 7.2) have been found on the basis of more than 20 molecules (Fig. 7.40) [82]. About eight superconductors have been formed from DMET, which is a hybrid of TMTSF and BEDT-TTF. Many have a stacking structure like the TMTSF salt [117], but κ-(DMET)$_2$AuBr$_2$ has a BEDT-TTF-like κ-structure [118]. BEDT-TTF consists of a ladder-like array of eight sulfur atoms, but the outer six-membered ring is larger than the inner five-membered ring. An increase of this unbalance leads to a more insulating state, whereas a decrease of the unbalance stabilizes the metallic state. Since superconductivity appears at the boundary between the metal and the insulator phases, both modifications reduce the possibility to realize superconductivity. For instance, the outer sulfur atoms are replaced by oxygen atoms in BO, which form mostly simple metal salts, although a few superconductors have been known [119].

**Table 7.2** Representative organic superconductors [82]

| | | |
|---|---|---|
| $(TMTSF)_2PF_6$ | 0.9 K (12 kbar) | [51] |
| $(TMTSF)_2ClO_4$ | 1.4 K | [57] |
| β-$(BEDT\text{-}TTF)_2I_3$ | 1.5 K (1.2 kbar 8.1 K) | [69–71] |
| κ-$(BEDT\text{-}TTF)_2Cu(NCS)_2$ | 10.4 K | [106] |
| κ-$(BEDT\text{-}TTF)_2Cu[N(CN)_2]Br$ | 11.6 K | [107] |
| κ-$(BEDT\text{-}TTF)_2Cu[N(CN)_2]Cl$ | 12.8 K (0.3 kbar) | [108, 109] |
| β′-$(BEDT\text{-}TTF)_2ICl_2$ | 14.2 K (82 kbar) | [75] |
| β-$(DMET)_2Au(CN)_2$ | 0.8 K (5 kbar) | [117] |
| κ-$(DMET)_2AuBr_2$ | 1.9 K | [118] |
| β″- $(BO)_2ReO_4$ | 0.9 K | [119] |
| λ-$(BETS)_2GaCl_4$ | 5.5 K | [120] |
| λ-$(BETS)_2FeCl_4$ | 1.8 K (3 kbar)<br>Field-induced superconductivity at 17 T | [120] |
| κ-$(BETS)_2FeBr_4$ | 1.1 K | [125, 126] |
| $(DTEDT)_3Au(CN)_2$ | 4 K | [86] |
| β-$(BDA\text{-}TTP)_2SbF_6$ | 6.9 K | [129] |
| $(MDT\text{-}TSF)(AuI_2)_{0.436}$ | 4.5 K | [128] |
| $(TTF)[Ni(dmit)_2]_2$ | 1.6 K (7 kbar) | [138] |
| $(EDT\text{-}TTF)[Ni(dmit)_2]_2$ | 1.3 K | [139] |

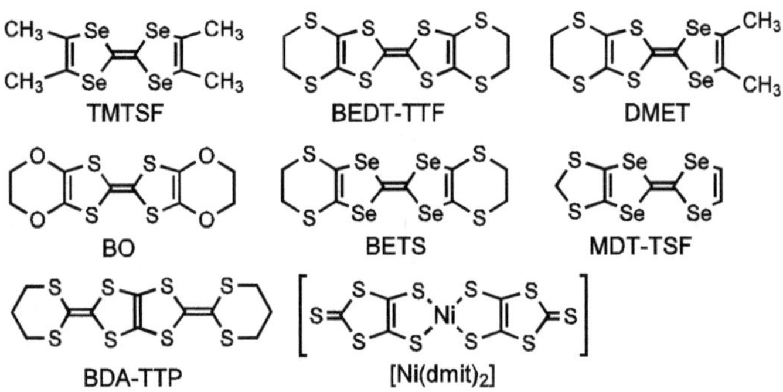


**Fig. 7.40** Representative molecules forming organic superconductors. DMET: dimethyl(ethylenedithio)diselenadithiafulvalene, BO: bis(ethylenedioxy)tetrathiafulvalene, BETS: bis(ethylenedithio)tetraselenafulvalene, MDT-TSF: methylenedithiotetraselenafulvalene, BDT-TTP: 2,5-bis (1,3-dithian-2-ylidene)- 1,3,4,6-tetrathiapentalene, dmit: 1,3-dithiole-2-thione-4,5-dithiolate

## 7.8.1 *πd-Superconductors*

BETS, where the inner sulfur atoms of BEDT-TTF are substituted by selenium atoms, gives many metallic salts [120]. However, the dimerization is strong in contrast to BO salts. The λ-phase has a tetramerized stacking structure (Fig. 7.41a, b) and a closed Fermi surface (Fig. 7.41c). The $GaCl_4^-$ salt is an ambient-pressure

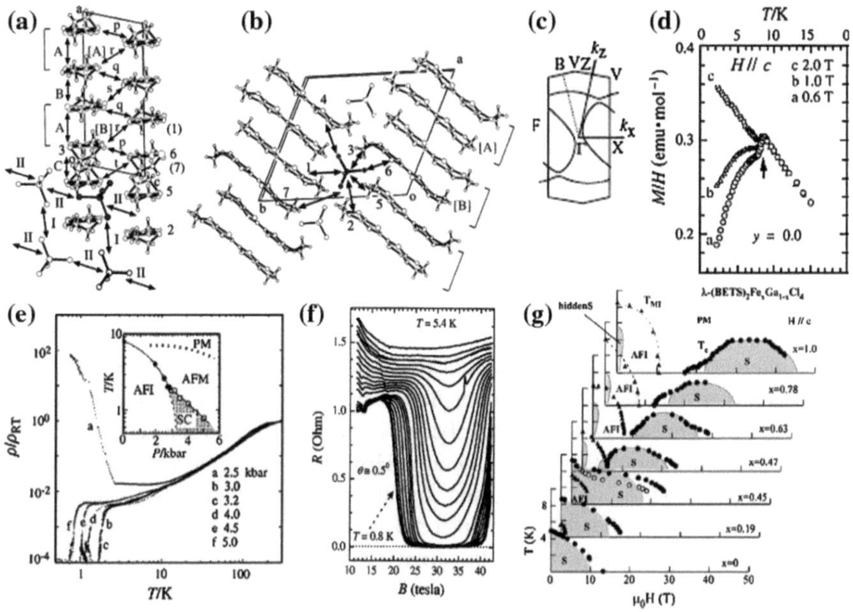


**Fig. 7.41** λ-$(BETS)_2FeCl_4$. **a**, **b** Crystal structure. The transfer integrals are, A: 335.5, B: 182.7, C: 148.3, p: −28, q: −93, r: 130, s: 171, and t: 25.6 meV. Magnetic interactions are $J_{\pi d}/k_B$ (6): −14.14 K, $J_{dd}/k_B$ I: 0.36, and II: 0.14 K [122]. **c** Fermi surface, **d** susceptibility at low temperatures, **e** resistivity [120], **f** field-induced superconductivity [121], and **g** the phase diagram in λ-$(BETS)_2Fe_xGa_{1-x}Cl_4$ [123]

superconductor, but the $FeCl_4^-$ salt undergoes a metal–insulator transition at 8 K accompanied by the antiferromagnetic order of the Fe $S = 5/2$ spins (Fig. 7.41d, e). This transition is suppressed by pressure, and superconductivity appears [121].

The $FeCl_4^-$ salt is a representative πd-system. We consider HOMO for BETS and five Fe 3d orbitals containing the unpaired spins (SOMO) for $FeCl_4^-$ [122]. The latter orbitals somewhat spread to Cl. From the overlap between the Cl orbitals and the BETS HOMO, we can estimate the πd-interaction. Equation 5.34 represents the superexchange interaction.

$$J = -\frac{2b^2}{\Delta}$$

Here, $t/\Delta$ in $b = t \times t/\Delta$ is regarded as the coefficient of the $FeCl_4^-$ SOMO on Cl, and another $t$ corresponds to the overlap between the Cl orbital and the BETS HOMO. Δ is the energy difference between the BETS HOMO and the $FeCl_4^-$ SOMO, which is about 1 eV. Thus, the five SOMOs of $FeCl_4^-$ containing five unpaired spins are located by 1 eV below the BETS HOMO. The SOMOs are the lower Hubbard bands, and the five spins are all parallel to form the $S = 5/2$ spin. The upper Hubbard bands of $FeCl_4^-$ are empty and located much above the

BETS HOMO level. According to Eq. 5.31, $J_{\pi d}$ is obtained from a square sum of the transfers of the five SOMOs and the BETS HOMO, whereas $J_{dd}$ is estimated from the 25 combinations between the SOMOs. Thus, we can estimate intermolecular $J$ from the intermolecular $t$ similarly to the superexchange interaction.

λ-$(BETS)_2FeCl_4$ contains crystallographically independent two BETS molecules (A and B in Fig. 7.41a). One of them has a particularly large πd-interaction; 6 in Fig. 7.41b is as large as $t = 55.2$ meV and $J_{\pi d}/k_B = -14.14$ K. The direct interaction between the $FeCl_4^-$ anions is $t = 19.7$ meV and $J_{dd}/k_B = -0.36$ K [122]. To estimate the Néel temperature, the Curie constant is multiplied in Eq. 4.32. The Curie constant for the $S = 5/2$ spin is proportional to $S(S + 1) \sim 35/4$ (Eq. 4.7), which is much larger than 3/4 for the $S = 1/2$ spin. Then, the contribution of the direct interaction from $J_{dd}$ is evaluated to be 3.73 K (the first term in Eq. 4.32), which is larger than the indirect interaction 2.49 K through the πd-interaction (the second term). The molecular orbital theory indicates that the antiferromagnetism essentially comes from the dd interaction. However, since the interaction in the interlayer direction is only mediated by the πd-interaction, the π-electron is important to achieve the three-dimensional magnetic order. Since the extremely large πd-interaction is only related to Molecule B, the antiferromagnetism induces the charge localization on Molecule B and gives rise to the metal–insulator transition due to the charge order.

The metal–insulator transition is suppressed under pressure, and superconductivity appears (Fig. 7.41e). The antiferromagnetism is suppressed under a magnetic field of 11 T, and the metallic phase is restored at the same time. When a stronger field above 20 T is applied exactly parallel to the conducting layer, **field-induced superconductivity** emerges (Fig. 7.41f). In general, superconductivity is destroyed by a magnetic field, but owing to the internal field from $FeCl_4^-$, the effective field on BEST is $H_{eff} = H_{ext} + a_{\pi d}M_d$. Above 11 T, the antiferromagnetism is destroyed, and all Fe spins are aligned in the direction of the magnetic field. On account of the antiferromagnetic πd-interaction ($a_{\pi d} < 0$), the internal field $a_{\pi d}M_d$ is directed in the opposite direction to $H_{ext}$. Then, the internal field cancels the external field around 32 T to result in $H_{eff} = 0$, and the field-induced superconductivity appears. This mechanism has been theoretically predicted by Jaccarino and Peter [124], and such a superconductor has been found in inorganic chalcogenides. However, the inorganic chalcogenide has a complicated composition, and the λ-phase is the first field-induced superconductor with a pure composition [121]. In an alloyed system, λ-$(BETS)_2Fe_xGa_{1-x}Cl_4$, the field of the field-induced superconductivity decreases with decreasing $x$ (Fig. 7.41g), because the effective field is given by the average. The superconducting phase has a tail in the phase diagram (Fig. 7.41g), suggesting the existence of the FFLO state. It is noteworthy that the tail exists not only on the high-field side but also in the low-field side.

κ-$(BETS)_2FeBr_4$ has a Fermi surface similar to the BEDT-TTF κ-phase, and the high-temperature resistivity is flat or slightly increasing just like the κ-phase on the border of the Mott insulator phase (Fig. 7.42a) [125, 126]. The salt shows antiferromagnetism due to the Fe spins at 2.5 K (Fig. 7.42b), but successively a superconducting phase appears at 1.1 K (inset in Fig. 7.42a). Below this temperature,

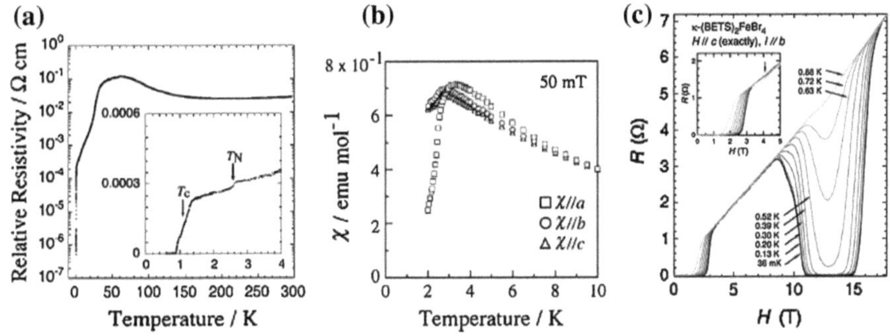


**Fig. 7.42** κ-$(BETS)_2FeBr_4$ **a** Temperature dependence of the resistivity, **b** susceptibility at low temperatures, and **c** field-induced superconductivity [125, 126]

antiferromagnetism of the Fe spins and superconductivity of the BETS π-electrons coexist. Under a magnetic field, the superconductivity once breaks, and field-induced superconductivity appears around 13 T (Fig. 7.42c). The central magnetic field at which the field-induced superconductivity occurs corresponds to $J_{\pi d}$, implying $J_{\pi d}$ in the κ-phase is less than half of the λ-phase. The calculated $J_{\pi d}/k_B = -7.62$ K is actually approximately half of the λ-phase (−14.14 K). The contributions to the Néel temperature from the direct and indirect interactions (Eq. 4.32) are 2.1 and 0.07 K, respectively. Then, the antiferromagnetism is mostly derived from the direct interaction. The induced spin on the π-system is $M_\pi = \chi_q a_{\pi d} M_d$ from Eq. 4.16, which is estimated to be several % of the completely polarized $S = 1/2$ spin [122].

In another πd-system consisting of BEDT-TTF and Fe oxalate, the anion layer shows ferromagnetism together with the metallic conduction of the π-system [127].

## 7.8.2 *Superconductors of Other Donors*

All the above superconductors have 2:1 composition with a quarter-filled band. Superconductors with other compositions are not many [86]. MDT-TSF (Fig. 7.40) forms ambient-pressure superconductors with the composition of (MDT-TSF)$(AuI_2)_{0.436}$ [128]. The MDT-TSF and $AuI_2^-$ form columns with respective independent periodicities (Fig. 7.43), so the ratio determines the composition, 0.436. The selenium substitution as well as the outer five-membered ring stabilizes the metallic states, so that this salt has a closed Fermi surface in spite of the one-dimensional columnar structure (Fig. 7.43b). The corresponding Shubnikov–de Haas oscillations have been observed. Owing to the incommensurate periodicity, the Fermi surface is reconstructed, which is discussed in detail from the oscillation frequencies. When the selenium atoms of the MDT-side TSF are replaced by sulfur, the resulting salt undergoes a metal–insulator transition accompanied by the

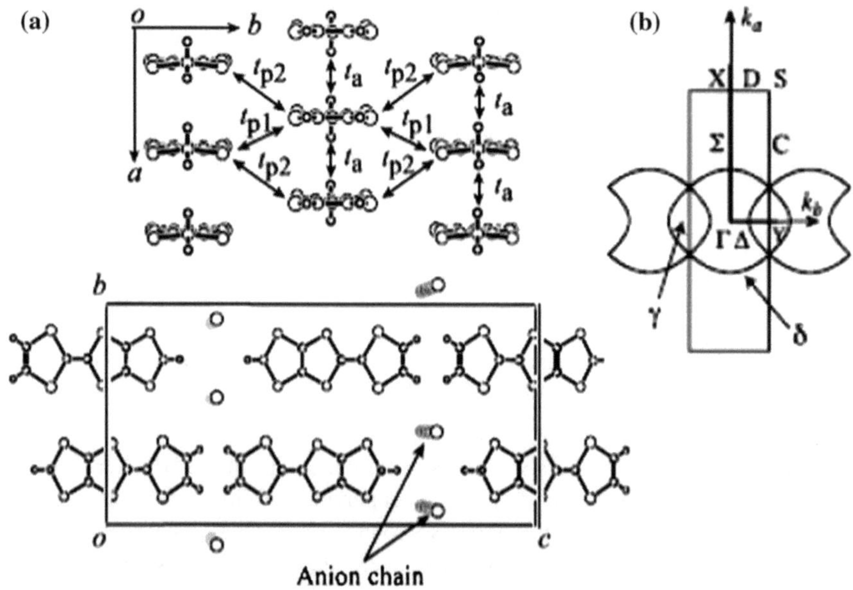


**Fig. 7.43** (MDT-TSF)$(AuI_2)_{0.436}$. **a** Crystal structure, and **b** Fermi surface. The transfer integrals are a1: 106.9, p1: −10.3, and p2: −35.4 meV [128]

antiferromagnetism. The transition is suppressed under pressure and superconductivity is restored. Then, similarly to the κ-phase, the superconducting phase borders to the incommensurate antiferromagnetic insulator.

Differently from the so far discussed superconductors, BDA-TTP (Fig. 7.40) does not have a TTF skeleton. BDA-TTP is a weaker electron donor than TTF, and strongly correlated. β-$(BDA\text{-}TTP)_2SbF_6$ forms a strongly dimerized β-phase, and exhibits a comparatively high $T_c$ (6.9 K) [129].

Recently, superconductivity as high as $T_c$ = 18 K has been reported in K-doped picene (Fig. 7.4) [130, 131]. The structure is uncertain, but the bandwidth of 0.3–0.4 eV is smaller than TTF salts, and results in a larger density of states [132]. Superconductivity has been reported in other phenes like phenanthrene.

A few comments on the nomenclature of donor molecules. All organic compounds are named according to the method of the International Union of Pure and Applied Chemistry (IUPAC), but such names as TTF and TCNQ are trivial names. Organic conductors are usually named by an abbreviation based on the trivial names, following the convention of the group who first prepared the compound. It is desirable to add the molecular structure to the abbreviation except for widely used compounds. Non-TTF molecules like BDA-TTP are named very closely to the IUPAC name. 1,3-Dithiane means a six-membered ring with two sulfur atoms at the 1 and 3 positions (Fig. 7.40), whereas 1,3-dithiole means a five-membered ring with two sulfur atoms at the 1 and 3 positions. These are correct IUPAC names. The numbering starts from a heavy atom, so a sulfur is one. There is no space after the comma between 1 and 3. However, TTP (tetrathiapentalene) is a trivial name. The IUPAC name of TTF is 2-(1,3-dithiol-2-ylidene)-1,3-dithiole, but it is difficult to imagine the molecule from such a name immediately, so we usually use the trivial name.

### 7.8.3 τ-Structure

Although the τ-phase is not a superconductor, the τ-phase has a characteristic structure (Fig. 7.44) [86]. The τ-phase is composed of donors shown in Fig. 7.44a. The upper right triangle means a methyl group extending up and a hydrogen atom extending down. The bottom right dashed triangle means a methyl group going down with a hydrogen extending up. Then, the two methyl groups are located at the *trans* position. Since four different moieties attach to the root carbon, this carbon is a chiral center. To name the absolute configuration, H is placed back of the central C, and other three moieties are rotated from the heavy group to the light group as S > ethylene > methyl. If the rotation is clockwise, the absolute configuration is named *R*, and the anticlockwise configuration is *S*. In this occasion, both the upper and lower carbon atoms have the *S* configuration, so the absolute configuration is *S*, *S*. In dimethyl BEDT-TTF, superconducting salts have been found in *S*,*S*- and *S*,*R*-molecules. The *R*,*R*-molecule is an enantiomer of the *S*,*S*-molecule, and both are optically active. The *cis*-dimethyl BEDT-TTF corresponds to the *S*,*R*-molecule and is optically inactive because there is a mirror plane. Such an isomer is called *meso* form.

Crystal structure of τ-(P-*S*,*S*-DMEDT-TTF)$_2$(AuBr$_2$)$_{1+y}$ belongs to the tetragonal system with a fourfold axis (Fig. 7.44b) [133]. The name τ comes from the tetragonal. Respectively 90° rotated molecules surround a linear anion, but additional anions are sandwiched between the layers. When anion composition is large, the anions are often incorporated in the donor layer. A considerable transfer integral exists between the 90° molecules, so the square lattice results in the Fermi surface as shown in Fig. 7.44c. The τ-phase is essentially metallic down to low temperatures, and the Shubnikov–de Haas oscillation has been observed [134]. From the oscillation, the area of the Fermi surface is 6.8–6.1 % of the Brillouin zone, leading to the composition to be $y = 0.875$–$0.898$. The resulting composition is very close to 1:1, and the donors are oxidized nearly up to 1+. Below 10 K, large negative magnetoresistance has been observed, where the resistance at 15 T becomes half of the 0 T value. The magnetoresistance at 5 T has a tetragonal symmetry with 90° periodicity (Fig. 7.44d), while the symmetry decreases to 180° periodicity at 0.5 T (Fig. 7.44e). This suggests the presence of ferromagnetic order, though the magnetic order has not been found in other experiments. The ferromagnetic fructuation has been attributed to the flat band coming from the closely 1:1 composition [135]. A flat band tends to show ferromagnetism similarly to the $U/t \rightarrow \infty$ limit of the Hubbard model (Sect. 4.13.3). Owing to the flat band, the τ-phase shows thermoelectric power as large as 150 μV/K in spite of the metallic conductivity [136].

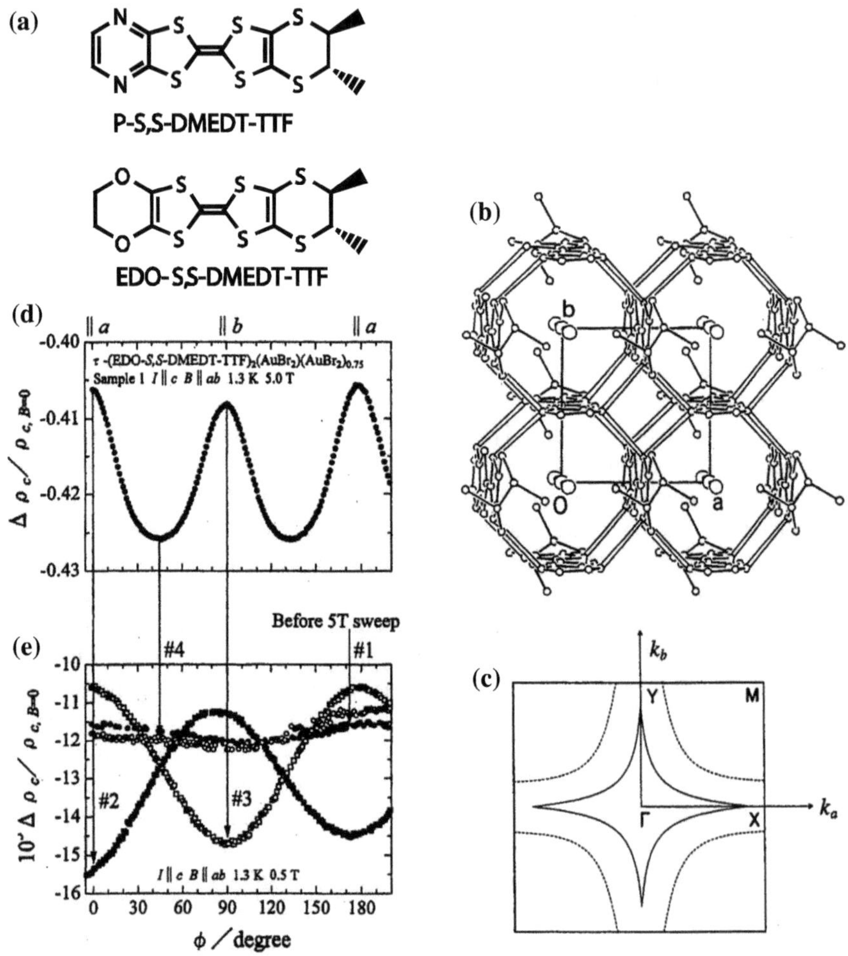


**Fig. 7.44** τ-Structure. **a** Molecular structures of P-*S*,*S*-DMEDT-TTF: pyradino-*S*,*S*-dimethylethylenedithiotetrathiafilvalene, and EDO-DMEDT-TTF: ethylenedioxy-*S*,*S*-dimethylethylenedithiotetrathiafilvalene. **b** Crystal structure, **c** Fermi surface, **d** magnetoresistance at 5 T, and **e** magnetoresistance at 0.5 T in τ-(P-*S*,*S*-DMEDT-TTF)$_2$(AuBr$_2$)$_{1+y}$ [133]

## 7.9 Conducting Metal Complexes

We have described radical cation salts based on electron donors, whereas [*M*(dmit)$_2$] is the only component of radical anion salts that show superconductivity (Fig. 7.40). The molecular structure of [*M*(dmit)$_2$] resembles TTF, where we replace the central C=C bond with a metal atom *M* = Ni or Pd [137]. (TTF)[Ni(dmit)$_2$]$_2$ was discovered first, in which TTF has 0.8+ charge, and [Ni(dmit)$_2$] is close to quarter filled (Fig. 7.45a) [138]. (TTF)[Ni(dmit)$_2$]$_2$ shows metallic conductivity at ambient pressure, but superconductivity is observed only under

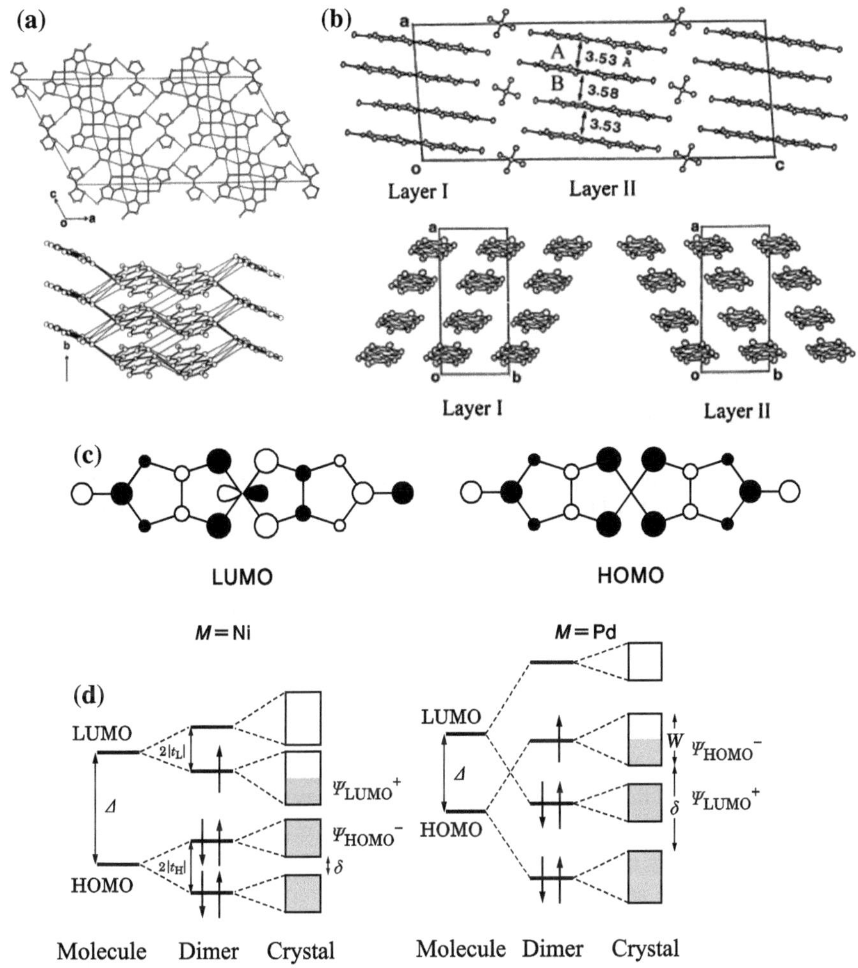


**Fig. 7.45** $[Ni(dmit)_2]$ complexes. **a** Crystal structure of $(TTF)[Ni(dmit)_2]_2$ [138], **b** Crystal structure of $(CH_3)_4N[Ni(dmit)_2]_2$ [140], **c** Molecular orbitals, and **d** energy levels in $[M(dmit)_2]_2$ [137]

pressure. (EDT-TTF)$[Ni(dmit)_2]$ is the only exceptional ambient-pressure superconductor, but superconductivity in other $[M(dmit)_2]$ salts is observed only under pressure after suppressing the metal–insulator transition [139]. $[M(dmit)_2]$ salts have stacking structures in general, though the stacking direction is slightly deviated from the ideal $\phi = 90°$ direction. The band structure is one dimensional. The HOMO and LUMO are composed of symmetrical and antisymmetrical combinations of the ligand orbitals (Fig. 7.45c), and energetically close. Thus, the weakly dimerized Ni complexes have a quarter-filled LUMO band, whereas the HOMO of the strongly dimerized Pd complexes is pushed above the lower LUMO

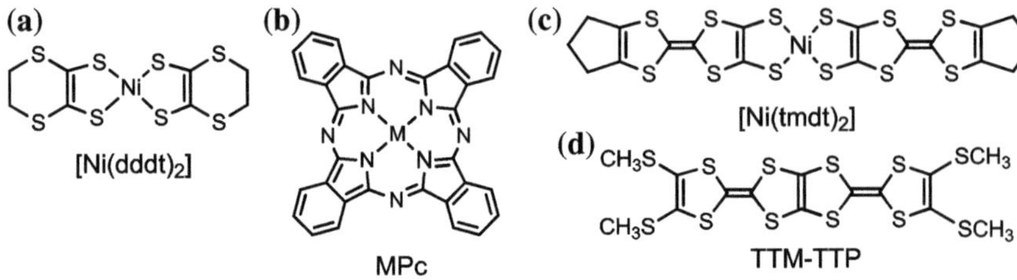


**Fig. 7.46** Molecular structures of conducting metal complexes

band and works as the conduction band (Fig. 7.45d). $(Et_2Me_2Sb)[Pd(dmit)_2]_2$ is a Mott insulator and does not show antiferromagnetism down to low temperatures because a spin liquid state is realized in a triangular lattice [111].

Redox potentials of metal dithiolate complexes largely depend on the metals and ligands [141]. Many of the nickel complexes are acceptors, but $[Ni(dddt)_2]$, in which C=C of BEDT-TTF is replaced by Ni (Fig. 7.46a), is a weak electron donor and oxidized around 0.9 V. Similarly to BEDT-TTF, this molecule forms such cation radical salts as β-$[Ni(dddt)_2]_2IBr_2$ and β″-$[Ni(dddt)_2]_3(AuBr_2)_2$. The latter complex shows metallic conductivity down to low temperatures [142].

Phthalocyanine (Pc, Fig. 7.46b) is a well-known material which has been utilized in electrophotography and CD-R. This molecule is also used in organic transistors and photovoltaics. The conductivity of pure phthalocyanine is less than $10^{-12}$ S/cm, but the charge-transfer complexes such as NiPcI and $CoPc(AsF_6)_{0.5}$ exhibit metallic conductivity [143]. (Tetraphenylphosphonium)$[FePc(CN)_2]_2$ is a semiconductor with the room-temperature conductivity of 10 S/cm, but shows giant negative magnetoresistance around 50 K, where the resistivity drops by two orders at 18 T [144]. This is ascribed to the πd-interaction between the Pc conduction electron and the Fe spin. $[Ni(mnt)_2]$ (Fig. 7.5) is a representative nickel dithiolate [141, 145, 146], and some of the complexes are metallic conductors, though the electronic structure is essentially one dimensional [137]. (Perylene)$_2[M(mnt)_2]_2$ ($M$ = Ni, Pt, Pd, and Au) shows as high room-temperature conductivity as 300–600 S/cm, but undergoes a Peierls-like metal–insulator transition at 6.5–60 K [147].

$[Ni(tmdt)_2]$ is a single-component conductor that shows metallic conductivity down to low temperatures [148]. The three-dimensional Fermi surface has been proved by the observation of the Shubnikov–de Haas oscillations. The energy band is approximately represented by assuming each ligand has an independent molecular orbital. Tetrathiapentalene donors in which two TTF parts are fused have been prepared, among which (TTM-TTP)$I_3$ (Fig. 7.46d) shows metallic conductivity in spite of the 1:1 composition. This material undergoes a metal–insulator transition at 100 K, where the 1+ charge localizes to a TTF part with leaving another TTF neutral. This "intramolecular charge order" suggests the two TTF parts behave like two independent molecules [86, 149].

When the quinoido carbon of TCNQ is replaced by nitrogen with one CN instead of two CN, we obtain dicyanoquinonediimine (DCNQI, Fig. 7.5)

[150, 151]. Although the number of CN decreases, the nitrogen is more electronegative than carbon, so DCNQI is an acceptor as strong as TCNQ. We have described metal complexes of TCNQ in Sect. 7.3, where CN coordinates to the metal (Fig. 7.9). The metal is coordinated by four CN because a TCNQ molecule has four CN. The resulting composition is 1:1, and Cu(TCNQ) is a Mott insulator. By contrast, a DCNQI molecule has two CN, and the composition is 1:2, designated as $M(DCNQI)_2$. Then, DCNQI salts are partially filled and highly conducting. The dimethyl derivative DMDCNQI is most popular, but the halogen and methoxy DCNQI's have been known. The unsubstituted DCNQI has been known, but the substituents seem to be necessary to fix the CN groups in the *trans* position. DCNQI forms a one-dimensional column, and the CN groups coordinate to copper (Fig. 7.47a). Owing to the 1:2 composition, metal DCNQI salts show conductivity higher than 100 S/cm. The salts of Li, Ag, Na, and K undergo the Peierls transitions around 100–200 K (Fig. 7.47b). However, $Cu(DMCNQI)_2$ maintains the metallic conductivity down to low temperatures (Fig. 7.47c). The copper atom has 4/3+ charge, and the mix valency as well as the direct coordination of CN mediate the intercolumnar interaction to suppress the Peierls transition. The copper mediates intercolumnar interaction and gives rise to a three-dimensional Fermi surface (Fig. 4.47d), that has been verified by the observation of the de Haas–van Alphene oscillation. Halogen substituted DCNQI enhances the Jahn–Teller distortion around copper to facilitate the Peierls transition (Fig. 4.47e). After the metal–insulator transition, the fractional copper electron has local spins, and undergoes an antiferromagnetic transition around 10 K [152, 153]. The distortion is also enhanced under pressure, and in $Cu(DMCNQI)_2$ a pressure-induced insulator phase appears at a very small pressure (Fig. 4.47f). Since the d-orbital of the mixed valent copper is related, the copper salts are regarded as a kind of πd-system, where the copper d-orbital participates in the band formation. Owing to the large $J_{\pi d}$, the Kondo temperature is much higher than room temperature (Eq. 5.54), and the copper electrons are itinerant below room temperature. It is noteworthy that the copper electrons are susceptible to halogen substitution and pressure, and in the insulating phase the copper has local spin.

*Example 7.1* In $Cu(DMDCNQI)_2$, the acceptor molecule has 2/3− charge, so the open Fermi surface derived from DMDCNQI is located at $k_c = \pi/3c$ (Fig. 7.47d). The spherical Fermi surface coming from copper is partly hybridized with the DCNQI band. Evaluate how much fraction of the Brillouin zone is occupied by the spherical Fermi surface of copper by neglecting the hybridization.

The copper charge is 4/3+, but if copper has 1+ charge, the resulting $d^{10}$ state leads to the fully occupied d levels. For 4/3+, the extra 1/3+ charge enters the d level. Thus, the volume of the hole pocket is 1/6 of the Brillouin zone. Since the open Fermi surface is located at the 1/3 position, the volume of the hole pocket is half of it [150, 151, 154, 155].

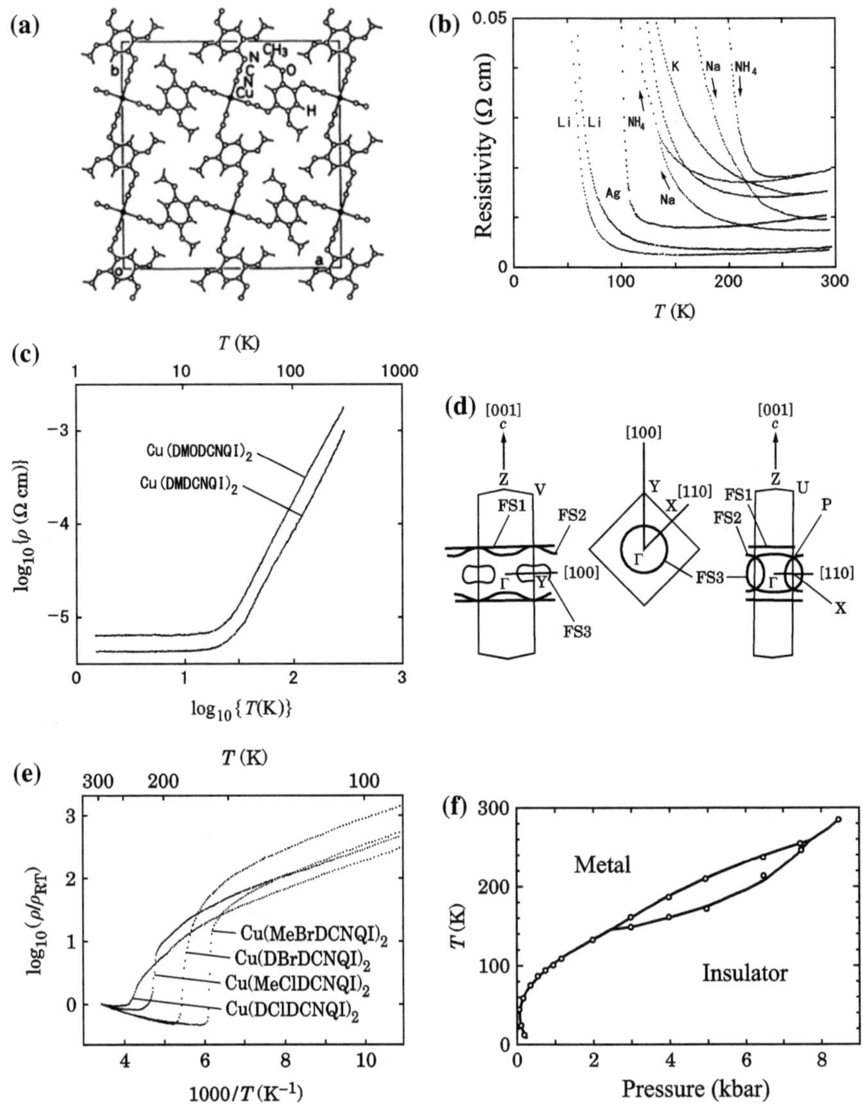


**Fig. 7.47** Conductivity of DMDCNQI complexes. **a** Crystal structure of $Cu(DMDCNQI)_2$ [152]. **b** Resistivity of the Li and Ag complexes [152]. **c** Resistivity of the Cu complexes of DMDCNQI and DMODCNQI [152]. **d** Fermi surface of $Cu(DMDCNQI)_2$. The transfer integrals are $t_{\pi\pi}$: 200, $t_{\pi d}$: 54 meV, and $\Delta_{\pi d}$: 300 meV [154, 155]. **e** Resistivity of the Cu complexes of halogen substituted DCNQI [152]. **f** Phase diagram of $Cu(DMDCNQI)_2$ [150]

## 7.10 Synthesis of Charge-Transfer Complexes

Finally, we see how to prepare charge-transfer complexes [156, 157]. There are reviews of the preparation of TTF derivatives [158, 159]. As a typical example, TMTTF is prepared as follows [160].

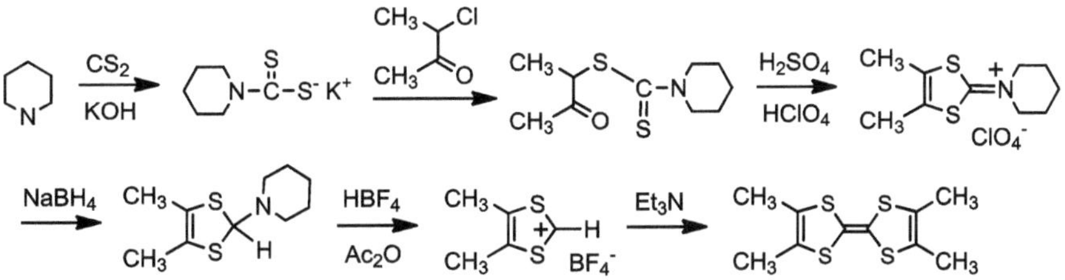


In spite of the many steps, this preparation is carried out in a short time. The $ClO_4^-$ salt is potentially explosive, so care must be taken. In particular, the dry staff in a large scale should not be handled by a metal spatula. In the final step, a proton is removed by triethylamine to generate carbene C: which couples to each other to form TTF.

BEDT-TTF is easily prepared in a large scale [161, 162].

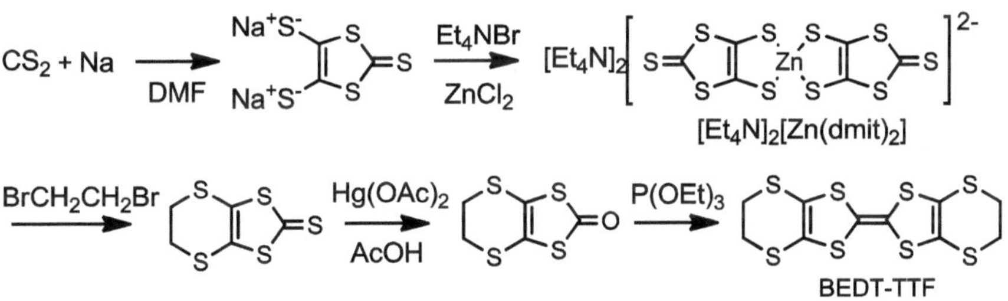


Sodium should be handled with care in the first step. The final phosphite-mediate coupling is a versatile reaction; we can obtain unsymmetrical donors using the cross-coupling of two starting components. Electron-withdrawing groups such as $-COOCH_3$ increase the reactivity, but unsubstituted 1,3-dithiole and electron-donating groups like alkyl moieties decrease the reactivity. Methoxy and ethylenedioxy groups are electron donating, while methylthio and ethylenedithio groups work as electron withdrawing. [Ni(dmit)$_2$] is also prepared from this route.

Charge-transfer complexes between an organic donor and acceptor are obtained as precipitates by mixing the corresponding solutions. To obtain the crystals, such complexes as (TTF)(chloranil) are obtained by coevaporation. Crystals of (TTF)(TCNQ) are prepared by the diffusion method (Fig. 7.48). TTF and TCNQ are placed in the arms of an H-cell, and acetonitrile is poured slowly. During several days, TTF and TCNQ dissolve gradually, and meet each other around the center to

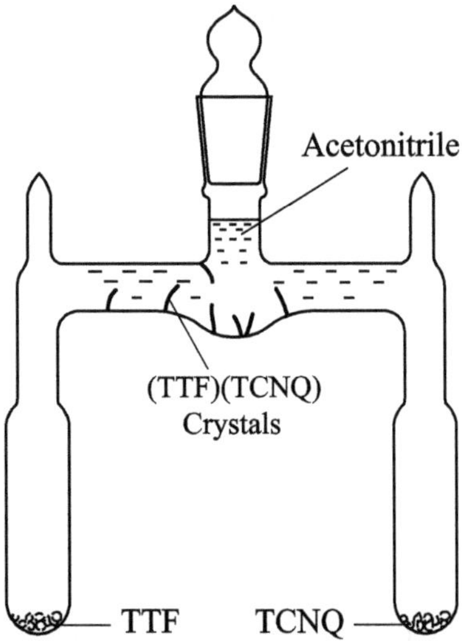


**Fig. 7.48** Crystal growth of (TTF)(TCNQ) by the diffusion method

form the charge-transfer complex. TTF and TCNQ dissolve in acetonitrile to some extent, but the complex does not dissolve because it is an ionic crystal. Then, black crystals of (TTF)(TCNQ) are obtained.

The easiest way to obtain a radical cation salt is exposure of the donor to $Br_2$ or $I_2$ vapor. Even a donor crystal is doped in this way. When a solution of donor is reacted with an oxidant such as $I_2$, $NOBF_4$, and $Fe(ClO_4)_3$, the precipitates of the corresponding salt is obtained. The most powerful method to prepare a radical cation salt is **electrochemical crystal growth** (Fig. 7.49). In an electrochemical cell equipped with 1 mm ϕ platinum wires, the donor is placed in the anode cell, and an electrolyte like $Bu_4NPF_6$ is placed in both cells. After the cell is filled with an inert gas, an organic solvent is added. A current of 0.5–2 μA is applied for one to several weeks, and the donor is oxidized on the anode. The radical cation salt between the oxidized donor and the electrolyte anion forms a kind of ionic crystal insoluble in

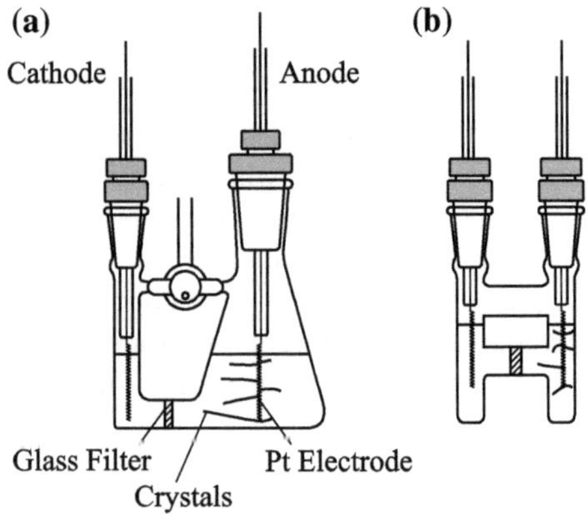


**Fig. 7.49** Cells for electrochemical crystal growth

the organic solvent, and the crystals appear on the electrode. Since the once formed crystals are conducting, the oxidation occurs on the crystals rather than the electrode, and the crystals grow. To make large crystals, it is desirable to decrease the number of seed crystals, for which the current density and electrode defects are important. However, a voltage of at least about 1 V is necessary to oxidize the donor at the electrode. When there are many polymorphisms like BEDT-TTF salts, the obtained crystal depends on the current and the solvent. There is a tendency that a polar solvent gives a phase with large charge-transfer degree.

To make a radical anion salt, $C_{60}$ is reacted with Na and K vapor in a shield tube. Cu(TCNQ) is obtained by the diffusion of CuI and TCNQ. $I^-$ reduces TCNQ and becomes $I_3^-$. $I^-$ has a reducing power, so electrocrystallization using $Bu_4NI$ as the electrolyte does not give a charge-transfer salt because the oxidized cation is reduced by $I^-$. $Bu_4NI_3$ is used as an ordinary electrolyte. Iodine has many anionic forms such as $I^-$,$I_3^-$,and $I_5^-$, and the actual form of the iodide anion has to be determined by X-ray crystallography or Raman spectroscopy [163]. Metal copper directly reacts with a solution of TCNQ and DCNQ to form Cu(TCNQ) and Cu $(DCNQI)_2$. The best way to prepare a radical anion salt is electrochemical crystal growth. $AgBF_4$, $Cu(CH_3CN)_4BF_4$, $LiClO_4$, $NaClO_4$, and $NH_4PF_6$ have been used as the electrolyte to react with TCNQ and DCNQI. To prepare $(CH_3)_4N[Ni(dmit)_2]_2$, first $(CH_3)_4N[Ni(dmit)_2]$ is chemically prepared, and is electrolyzed in the presence of $(CH_3)_4NClO_4$. (EDT-TTF)$[Ni(dmit)_2]$ is obtained from the electrocrystallization of EDT-TTF and $Bu_4N[Ni(dmit)_2]$. (TTF)$[Ni(dmit)_2]_2$ has been obtained by the diffusion of $(TTF)_3(BF_4)_2$ and $Bu_4N$ $[Ni(dmit)_2]$.

To prepare large high-quality crystals, purity of the solvent is very important, and the solvent is distilled under an inert atmosphere [164]. Acetonitrile is distilled from $P_2O_5$ and distilled again from $CaH_2$. Chlorinated solvents like chloroform, 1,1,2-trichloroethane, and chlorobenzene are stirred with several ten mL of concentrated sulfuric acid overnight. The darkened sulfuric acid is removed by decantation, and in a separatory funnel the solvent is treated with water and $NaHCO_3$ solutions several times. After dried over $CaCl_2$ at least overnight, the solvent is distilled. Such a distilled solvent decomposes in several months because the stabilizer is removed. Chlorinated solvents generate HCl by reacting with water, and HCl oxidizes TTF to $TTF^+$. We can wash TTF derivatives deposited on glassware by concentrated sulfuric acid, where the generated $TTF^+$ is easily washed out by water. THF (tetrahydrofurane) is distilled from sodium and benzophenone, which generate blue ketyl radical. However, this process is recently avoided because the presence of sodium is hazardous.

## References

1. M.S. Dresselhaus, G. Dresselhaus, Adv. Phys. **51**, 1 (2002)
2. D.D. Eley, Nature **162**, 819 (1948)
3. A.T. Vartanyan, Zh Fiz, Khim. **22**, 769 (1948)

4. H. Akamatsu, H. Inokuchi, J. Chem. Phys. **18**, 810 (1950)
5. H. Inokuchi, Bull. Chem. Soc. Jpn. **24**, 222 (1951)
6. D.D. Eley, G.D. Parfitt, M.J. Perry, D.H. Taysum, Trans. Faraday Soc. **49**, 79 (1953)
7. D.D. Eley, G.D. Parfitt, Trans. Faraday Soc. **51**, 1529 (1955)
8. H. Akamatsu, H. Inokuchi, Y. Matsunaga, Nature **173**, 168 (1954)
9. H. Inokuchi, Solid State Phys. **12**, 93 (1961)
10. K. Seki, Mol. Cryst. Liq. Cryst. **171**, 255 (1989)
11. D. Fichou (ed.), *Handbook of Oligo- and Polythiophenes* (Wiley, New York, 1999), p. 114
12. J.B. Torrance, Mol. Cryst. Liq. Cryst. **126**, 55 (1985)
13. C. Lefrou, P. Fabry, J.-C. Poignet, *Electrochemistry the Basics, with Examples* (Springer, Berlin, 2012)
14. M. Manriquez, M.D. Ward, W.M. Reiff, J.C. Calabrese, N.L. Jones, P.J. Carroll, E.E. Bunel, J.S. Miller, J. Am. Chem. Soc. **117**, 6182 (1995)
15. H.S.O. Chan, S.C. Ng, S.H. Seow, Synth. Met. **66**, 177 (1994)
16. M.L. Tang, A.D. Reicharrdt, N. Miyaki, R.M. Stoltenberg, Z. Bao, J. Am. Chem. Soc. **130**, 6064 (2008)
17. H. Meng, L. Zheng, A.J. Lovinger, B.-C. Wand, P.G.V. Patten, Z. Bao, Chem. Mater. **15**, 1778 (2003)
18. D.M. De Leew, M.M.J. Simenon, A.R. Brown, R.E.F. Einerhand, Synth. Met. **87**, 53 (1997)
19. M.L. Tang, A.D. Reicharrdt, P. Wei, Z. Bao, J. Am. Chem. Soc. **131**, 5264 (2009)
20. N. Sato, G. Saito, H. Inokuchi, Chem. Phys. **76**, 79 (1983)
21. N. Sato, K. Seki, H. Inokuchi, J. Chem. Soc. Faraday Trans. **2**(77), 1621 (1981)
22. J.-L. Brédas, R. Silbey, D.S. Boudreaux, R.R. Chance, J. Am. Chem. Soc. **105**, 6555 (1983)
23. J. McMurry, *Organic Chemistry* (Brooks/Cole, Belmont, 1996). (Electrophilic aromatic substitution)
24. R.J. Haber, W.R. Hertler, W. Mahler, L.R. Melby, R.E. Benson, W.E. Mochel, J. Am. Chem. Soc. **82**, 6408 (1960)
25. M. Konno, Y. Saito, Acta Cryst. B **1975**, 31 (2007)
26. R.A. Heintz, H. Zhao, X. Ouyang, G. Grandietti, J. Cowan, K.R. Dunber, Inorg. Chem. **38**, 144 (1999)
27. R.S. Potember, T.O. Poeler, D.O. Cowan, Appl. Phys. Lett. **34**, 405 (1979)
28. R. Foster, *Organic Charge-Transfer Complexes* (Wiley, New York, 1969)
29. F.H. Herbstein, Persp. Struc. Chem. **4**, 166 (1971)
30. A.W. Hanson, Acta Cryst. **19**, 610 (1965)
31. I. Ikemoto, H. Kuroda, Acta Cryst. **24**, 383 (1968)
32. G. Saito, P. Ferraris, Bull. Chem. Soc. Jpn **53**, 2141 (1980)
33. C.J. Fritchie, Acta Cryst. **20**, 892 (1966)
34. A.J. Epstein, E.M. Conwell, D.J. Sandman, J.S. Miller, Solid State Commun. **23**, 355 (1977)
35. F. Wudl, G.M. Smith, E.J. Hufnagwel, J. C. S. Chem. Commun. **1970**, 1453
36. R.B. Somoano, A. Gupta, V. Hadek, T. Datta, M. Jones, R. Deck, A.M. Hermann, J. Chem. Phys. **63**, 4970 (1975)
37. H. Kobayashi, K. Kobayashi, Bull. Chem. Soc. Jpn **50**, 3127 (1977)
38. R.B. Somoano, A. Gupta, V. Hadek, M. Novotny, M. Jones, T. Datta, R. Deck, A.M. Hermann, Phys. Rev. B **15**, 595 (1977)
39. L.B. Coleman, M.J. Cohen, D.J. Sandman, F.G. Yamagishi, A.F. Garito, A.J. Heeger, Solid State Commun. **12**, 1125 (1973)
40. T.J. Kistenmacher, T.E. Phillips, D.O. Cowan, Acta Crystallogr. B **30**, 763 (1974)
41. S. Kagoshima, H. Nagasawa, *One-Dimensional Conductors* (Springer, Berlin, 2012)
42. A. J. Heeger, in *Highly Conducting One-Dimensional Solids*, ed. by J.T. Devreese, R. P. Evard, V.E. van Doren, (Plenum, New York, 1979), p. 69
43. D. Allender, J.W. Bray, J. Bardeen, Phys. Rev. B **9**, 119 (1974)
44. J.C. Scott, A.F. Garito, A.J. Heeger, Phys. Rev. B **10**, 3131 (1974)
45. W.A. Little, Phys. Rev. **134**, A1416 (1964)
46. S. Kagoshima, T. Ishiguro, H. Anzai, J. Phys. Soc. Jpn. **41**, 2061 (1976)

47. A. Andriex, H.J. Schulz, D. Jérome, K. Bechgaard, Phys. Rev. Lett. **43**, 227 (1979)
48. S. Megtert, R. Comes, C. Vettier, R. Pynn, A.F. Garito, Solid State Commun. **31**, 977 (1979)
49. G. Grüner, *Density Waves in Solids* (Addison-Wesley, Boston, 1994)
50. A.N. Bloch, T.F. Carruthers, T.O. Poehler, D.O. Cowan, in *Chemistry and Physics of One-Dimensional Metals*, ed. by H.J. Keller, (Plenum, New York, 1977), p. 47
51. D. Jérome, K. Bechgaard, J. Phys. Lett. **41**, L95 (1980)
52. T. Ishiguro, K. Yamaji, G. Saito, *Organic Superconductors*, 2nd edn. (Springer, Berlin, 1998)
53. N. Thorup, G. Rindorf, H. Soling, K. Bechgaard, Acta Crystallogr. B **37**, 1236 (1981)
54. K. Bechgaard, C.S. Jacobsen, K. Mortensen, J.H. Perdersen, N. Thorup, Solid State Commun. **33**, 1119 (1980)
55. K. Mortensen, Y. Tomkiewicz, K. Bechgaard, Phys. Rev. B **25**, 3319 (1982)
56. R. Brusetti, M. Ribault, D. Jérome, K. Bechgaard, J. Phys. **43**, 801 (1982)
57. K. Bechgaard, K. Carneiro, M. Olsen, F.B. Rasmussen, C.S. Jacobsen, Phys. Rev. Lett. **46**, 852 (1981)
58. D. Jérome, Science **252**, 1509 (1991)
59. M. Takigawa, H. Yasuoka, G. Saito, J. Phys. Soc. Jpn. **56**, 873 (1987)
60. I.J. Lee, M.J. Naughton, G.M. Danner, P.M. Chaikin, Phys. Rev. Lett. **78**, 3555 (1997)
61. J.F. Kwak, J.E. Schirber, R.L. Greene, E.M. Engler, Phys. Rev. Lett. **46**, 1296 (1981)
62. T. Mori, Bull. Chem. Soc. Jpn. **71**, 2509 (1998)
63. T. Mori, H. Mori, S. Tanaka, Bull. Chem. Soc. Jpn. **72**, 179 (1999)
64. T. Mori, Bull. Chem. Soc. Jpn. **72**, 2011 (1999)
65. G. Saito, T. Enoki, K. Toriumi, H. Inokuchi, Solid State Commun. **42**, 557 (1982)
66. H. Kobayashi, A. Kobayashi, Y. Sasaki, G. Saito, T. Enoki, H. Inokuchi, J. Am. Chem. Soc. **105**, 297 (1983)
67. T. Mori, A. Kobayashi, Y. Sasaki, H. Kobayashi, G. Saito, H. Inokuchi, Bull. Chem. Soc. Jpn **54**, 627 (1984)
68. T. Mori, Bull. Chem. Soc. Jpn. **89** (2016) doi:10.1246/bcsj.20160110
69. T. Mori, A. Kobayashi, Y. Sasaki, H. Kobayashi, G. Saito, T. Enoki, H. Inokuchi, Chem. Lett. **13**, 957 (1984)
70. K. Murata, M. Tokumoto, H. Anzai, H. Bando, G. Saito, K. Kajimura, T. Ishiguro, J. Phys. Soc. Jpn. **54**, 2084 (1985)
71. V.N. Laukhin, E.E. Kostyychenko, Y.V. Suchko, I.F. Shchegolev, E.B. Yagubskii, JETP Lett. **41**, 81 (1985)
72. J.E. Schirber, L.J. Azevedo, J.F. Kwak, E.L. Venturini, P.C.W. Leung, M.A. Beno, H.H. Wang, J.M. Williams, Phys. Rev. B **33**, 1987 (1986)
73. M. Tokumoto, H. Bando, H. Anzai, G. Sito, K. Murata, K. Kajimura, T. Ishiguro, J. Phys. Soc. Jpn. **54**, 869 (1985)
74. T. Mori, H. Inokuchi, Solid State Commun. **62**, 525 (1987)
75. H. Taniguchi, M. Miyashita, K. Ushiyama, K. Satoh, N. Môri, H. Okamoto, K. Miyagawa, K. Kanoda, M. Hedo, Y. Uwatoko, J. Phys. Soc. Jpn. **72**, 468 (2003)
76. T. Mori, H. Inokuchi, Bull. Chem. Soc. Jpn. **60**, 402 (1987)
77. T. Mori, H. Inokuch, Solid State Commun. **59**, 355 (1986)
78. H.M. Yamamoto, M. Hagiwara, R. Kato, Synth. Met. **133–134**, 449 (2003)
79. T. Mori, F. Sakai, G. Saito, H. Inokuchi, Chem. Lett. **15**, 1037 (1986)
80. H. Kobayashi, R. Kato, T. Mori, A. Kobayashi, Y. Sasaki, G. Saito, T. Enoki, H. Inokuchi, Chem. Lett. **13**, 179 (1984)
81. T. Yamamoto, M. Uruichi, K. Yakushi, A. Kawamoto, Phys. Rev. B **73**, 125116 (2006)
82. H. Mori, J. Phys. Soc. Jpn. **75**, 051003 (2006)
83. M. Kurmoo, A.W. Graham, P. Day, S.J. Coles, M.B. Hursthouse, J.L. Caulfield, J. Singleton, F.L. Pratt, W. Hayes, L. Ducasse, P. Guionneau, J. Am. Chem. Soc. **117**, 12209 (1995)
84. T. Mori, H. Inokuchi, Solid State Commun. **64**, 335 (1987)

85. H. Mori, I. Hirabayashi, S. Tanaka, T. Mori, Y. Maruyama, H. Inokuchi, Solid State Commun. **80**, 411 (1991)
86. T. Mori, Chem. Rev. **104**, 4947 (2004)
87. A. Ardavan, S. Brown, S. Kagoshima, K. Kanoda, K. Kuroki, H. Mori, M. Ogata, S. Uji, J. Wosnitza, J. Phys. Soc. Jpn. **81**, 011004 (2012)
88. H. Mori, S. Tanaka, T. Mori, Phys. Rev. B **57**, 12023 (1998)
89. R. Kondo, S. Kagoshima, M. Chusho, H. Hoshino, T. Mori, H. Mori, S. Tanaka, Curr. Appl. Phys. **2**, 483 (2002)
90. T. Mori, J. Phys. Soc. Jpn. **72**, 1469 (2003)
91. T. Mori, H. Mori, I. Terasaki, J. Mater. Chem. **17**, 4343 (2007)
92. T. Mori, in *Nanomaterials*, ed. by H. Hosono, Y. Mishima, H. Takezoe, K.J.D. MacKenzie, (Elsevier, Philadelphia, 2006), p. 224
93. N. Tajima, S. Sugawara, M. Tamura, Y. Nishio, K. Kajita, J. Phys. Soc. Jpn. **75**, 051010 (2006)
94. N. Tajima, K. Kajita, Sci. Tech. Adv. Mater. **10**, 024308 (2009)
95. B. Rothaemel, L. Ferro, J.R. Cooper, J.S. Silling, Phy. Rev. B **34**, 704 (1986)
96. K. Yamamoto, A.A. Kowalska, K. Yakushi, Appl. Phys. Lett. **96**, 122901 (2010)
97. T. Kakiuchi, Y. Wakabayashi, H. Sawa, T. Takahashi, T. Nakamura, J. Phys. Soc. Jpn. **76**, 113702 (2007)
98. T. Mori, J. Phys. Soc. Jpn. **82**, 034712 (2013)
99. A. Kobayashi, S. Katayama, Y. Suzumura, Sci. Tech. Adv. Mater. **10**, 024309 (2009)
100. N. Tajima, S. Sugawara, R. Kato, Y. Nishio, K. Kajita, Phys. Rev. Lett. **102**, 176403 (2009)
101. M. Osada, J. Phys. Soc. Jpn. **77**, 084711 (2008)
102. H. Mori, S. Tanaka, M. Oshima, G. Saito, T. Mori, Y. Maruyama, H. Inokuchi, Bull. Chem. Soc. Jpn **63**, 2183 (1990)
103. J.S. Brooks, X. Chen, S.J. Klepper, S. Valfells, G.J. Athas, Y. Tanaka, T. Kinoshita, N. Kinoshita, M. Tokumoto, H. Anzai, C.C. Agosta, Phys. Rev. B **52**, 14457 (1995)
104. T. Osada, R. Yagi, A. Kawasumi, S. Kinoshita, N. Miura, M. Osada, G. Saito, Phys. Rev. B **41**, 5428 (1990)
105. R. Kato, H. Kobayashi, A. Kobayashi, S. Moriyama, Y. Nisho, K. Kajita, Y. Sasaki, Chem. Lett. **16**, 507 (1987)
106. H. Urayama, H. Yamochi, G. Saito, K. Nozawa, T. Sugano, M. Kinoshita, S. Sato, K. Oshima, A. Kawamoto, J. Tanaka, Chem. Lett. **17**, 55 (1988)
107. A.M. Kini, U. Geiser, H.H. Wang, K.D. Carlson, J.M. Williams, W.K. Kwok, K.G. Vandervoort, J.E. Thompson, D.L. Stupka, Inorg. Chem. **29**, 2555 (1990)
108. J.M. Williams, A.M. Kini, H.H. Wang, K.D. Carlson, U. Geiser, L.K. Montgomery, G. J. Pyrka, D.M. Watkins, J.M. Kommers, S.J. Boryshuk, A.V.S. Crouch, W.K. Kwok, J.E. Schirber, D.L. Overmyer, D. Jung, M.-H. Whangbo, Inorg. Chem. **29**, 3272 (1990)
109. J.M. Williams, J.R. Ferraro, R.J. Thorn, K.D. Carlson, U. Geiser, H.H. Wang, A.M. Kini, M.-H. Whangbo, *Organic Superconductors* (Prentice Hall, Upper Saddle River, 1992)
110. K. Kanoda, Hyperfine Interact. **104**, 235 (1997)
111. K. Kanoda, R. Kato, Ann. Rev. Cond. Matter. Phys. **2**, 168 (2011)
112. M. Adbel-Jawad, I. Terasaki, T. Sasaki, N. Yoneyama, N. Kobayashi, Y. Uesu, C. Hotta, Phys. Rev. B. **82**, 125119 (2010)
113. C. Hotta, Phys. Rev. B **82**, 241104R (2010)
114. T. Mori, F. Sakai, G. Saito, H. Inokuchi, Chem. Lett. **15**, 1589 (1986)
115. H. Kobayashi, T. Mori, R. Kato, A. Kobayashi, Y. Sasaki, G. Saito, H. Inokuchi, Chem. Lett. **12**, 581 (1983)
116. Y. Nogami, T. Mori, J. Phys. IV **12**, 233 (2002)
117. K. Kikuchi, M. Kikuchi, T. Namiki, K. Saito, I. Ikemoto, K. Murata, T. Ishiguro, K. Kobayashi, Chem. Lett. **16**, 931 (1987)

118. K. Kikuchi, Y. Honda, Y. Ishikawa, K. Saito, I. Ikemoto, K. Murata, H. Anzai, T. Ishiguro, Solid State Commun. **66**, 405 (1988)
119. S. Kahlich, D. Schweitzer, I. Heinen, Song En Lan, B. Nuber, H.J. Keller, K. Winzer, H.W. Helberg, Solid State Commun. **80**, 191 (1991)
120. H. Kobayashi, A. Kobayashi, P. Cassoux, Chem. Soc. Rev. **29**, 325 (2000)
121. S. Uji, H. Shinagawa, T. Terashima, T. Yakabe, Y. Terai, M. Tokumoto, A. Kobayashi, H. Tanaka, H. Kobayashi, Nature **410**, 908 (2001)
122. T. Mori, M. Katsuhara, J. Phys. Soc. Jpn. **71**, 826 (2002)
123. S. Uji, J.S. Brooks, J. Phys. Soc. Jpn. **75**, 051014 (2006)
124. V. Jaccarino, M. Peter, Phys. Rev. Lett. **9**, 290 (1962)
125. H. Fujiwara, H. Kobayashi, Bull. Chem. Soc. Jpn. **78**, 1181 (2005)
126. T. Konoike, S. Uji, T. Terashima, M. Nishimura, S. Yasuzuka, K. Enomoto, H. Fujiwara, B. Zhang, H. Kobayashi, Phys. Rev. B **70**, 094514 (2004)
127. E. Coronado, P. Day, Chem. Rev. **104**, 5419 (2004)
128. T. Kawamoto, K. Takimiya, Sci. Tech. Adv. Mater. **10**, 024303 (2009)
129. J. Yamada, H. Akutsu, H. Nishikawa, K. Kikuchi, Chem. Rev. **104**, 5057 (2004)
130. R. Mitsuhashi, Y. Suzuki, Y. Yamanari, H. Mitamura, T. Kambe, N. Ikeda, H. Okamoto, A. Fujiwara, M. Yamaji, N. Kawasaki, Y. Maniwa, Y. Kubosono, Nature **464**, 76 (2010)
131. S. Heguri, M. Kobayashi, K. Tanigaki, Phys. Rev. B **92**, 014502 (2015)
132. T. Kosugi, T. Miyake, S. Ishibashi, R. Arita, H. Aoki, J. Phys. Soc. Jpn. **78**, 113704 (2009)
133. H. Yoshino, K. Imamura, T. Sasaki, A. Oda, G.C. Papavassiliou, K. Murata, J. Phys. Soc. Jpn. **68**, 68 (1999)
134. T. Konoike, K. Ishikawa, H. Yoshino, K. Murata, T. Sasaki, G.C. Papavassiliou, Phys. Rev. B **61**, 245308 (2002)
135. H. Aizawa, K. Kuroki, H. Yoshino, K. Murata, Phys. B **405**, S27 (2010)
136. H. Yoshino, G.C. Papavassiliou, K. Murata, J. Therm. Ana. Calorim. **92**, 457 (2008)
137. R. Kato, Chem. Rev. **104**, 5319 (2004)
138. L. Brossard, M. Ribault, M. Bousseau, L. Valade, P. Cassoux, C. R. Acad. Sci. Paris, II **302**, 205 (1986)
139. H. Tajima, M. Inokuchi, A. Kobayashi, T. Ohta, R. Kato, H. Kobayashi, H. Kuroda, Chem. Lett. **22**, 1235 (1993)
140. A. Kobayashi, H. Kim, Y. Sasaki, R. Kato, H. Kobayashi, S. Moriyama, Y. Nishio, K. Kajita, W. Sasaki, Chem. Lett. **16**, 1819 (1987)
141. J.A. McCleverty, Prog. Inorg. Chem. **10**, 49 (1968)
142. E.B. Yagubskii, J. Solid State Chem. **168**, 464 (2002)
143. T. Inabe, H. Tajima, Chem. Rev. **104**, 5503 (2004)
144. N. Hanasaki, H. Tajima, M. Masuda, T. Naito, T. Inabe, Phys. Rev. B **62**, 5839 (2000)
145. H. Holm, M.J. O'Connor, Prog. Inorg. Chem. **14**, 241 (1971)
146. J.A. Cleverty, Prog. Inorg. Chem. **10**, 49 (1968)
147. R.T. Henriques, L. Alcacer, J.P. Pouget, D. Jérome, J. Phys. C **17**, 5197 (1984)
148. A. Kobayashi, E. Fujiwara, H. Kobayashi, Chem. Rev. **104**, 5243 (2004)
149. Y. Misaki, Sci. Tech. Adv. Mater. **10**, 024301 (2009)
150. R. Kato, Bull. Chem. Soc. Jpn. **73**, 515 (2000)
151. S. Hünig, E. Herberth, Chem. Rev. **104**, 5535 (2004)
152. T. Mori, H. Inokuchi, A. Kobayashi, R. Kato, H. Kobayashi, Phys. Rev. B **38**, 5913 (1988)
153. S. Tomić, D. Jérome, A. Aumüller, O.P. Erk, O.S. Hünig, J.U. von Schützll, J Phys. C **21**, L203 (1988)
154. S. Uji, T. Terashima, H. Aoki, J.S. Brooks, R. Kato, H. Sawa, S. Aonuma, M. Tamura, M. Kinoshita, Phys, Rev. B **50**, 15597 (1994)
155. H. Kojima, Z. Zhang, K.R. Dunbar, T. Mori, J. Mater. Chem. C **1**, 1781 (2013)
156. J.M. Williams, Inorg. Synth. **24**, 130 (1987)
157. J.M. Williams, Inorg. Synth. **26**, 386 (1989)

158. G. Schukat, A.M. Richter, E. Fanghänel, Sulfur Rep. **7**, 155 (1987)
159. J. Yamada, T. Sugimoto, *TTF Chemistry* (Kodansha, Tokyo, 2004)
160. A. Mas, J.-M. Fabre, E. Torreilles, K. Giral, G. Brun, Tetrahedron Lett. **18**, 2579 (1977)
161. G. Steimecke, H.-J. Sieler, R. Kirmse, E. Hoyer, Phosph. Sulf. **7**, 49 (1979)
162. J.M. Williams, Inorg. Synth. **26**, 386 (1989)
163. A.F. Wells, *Structural Inorganic Chemistry*, 5th edn. (Clarendon, Oxford, 1984)
164. W.L.F. Armarego, C. Chai, *Purification of Laboratory Chemicals*, 6th edn. (Butterworth-Heinemann, Oxford, 2009)

# Chapter 8
# Organic Semiconductors

**Abstract** Organic semiconductors are used in organic electronic devices. After discussing basic physics of organic semiconductors, organic devices such as light-emitting diodes, transistors, and solar cells are described.

**Keywords** Time-of-flight (TOF) method · Mobility · Space-charge-limited current (SCLC) · Poole–Frenkel effect · Organic light-emitting diodes (OLED) · Organic transistor · Schottky barrier · Fluorescence · Phosphorescence · Gradual channel approximation · Output characteristics · Transfer characteristics · Marcus theory · Contact resistance · Ambipolar transistors · Organic solar cell (photovoltaics) · Dye-sensitized solar cell

## 8.1 Photoconductivity: Electrophotography and Time-of-Flight Mobility

Organic semiconductors have been investigated for a long time in parallel with charge-transfer salts. However, organic semiconductor devices have developed remarkably after the 1980s. In this chapter, basic concepts of conducting phenomena in organic semiconductors are described. Semiconducting transport mediated by thermally excited carriers has been discussed in Sect. 3.3. Charge-transfer salts in which the carriers are generated by chemical doping have been described in Sect. 7.2. Here, we first investigate transport phenomena of carriers generated by photoexcitation.

When a semiconductor is irradiated by light, the photoexcited electron and hole mediate the conduction. Such conduction is called **photoconductivity** in contrast to **dark conductivity** owing to the thermally excited carriers. There is a long history of photoconductivity [1, 2]. For instance, photoconductivity of anthracene has been investigated since the beginning of the twentieth century.

If an isolated molecule is excited by light, the adjacent molecule is potentially excited by the light with the same energy. Then, the excited electron-hole pair moves in the solid as an "exciton." A relatively localized exciton is called **Frenkel**



T. Mori, *Electronic Properties of Organic Conductors*,
DOI 10.1007/978-4-431-55264-2_8

**exciton**, while largely delocalized exciton is called **Wannier exciton**. Exciton in organic materials is the Frenkel exciton, because practically a single molecule is excited. If molecules make dimers in a solid, the bonding and antibonding combinations generate two levels, and the exciton level splits to two by the **Davydov splitting**. In a semiconductor, the excitation from the valence band to the conduction band makes an exciton, but the Coulomb attraction between the electron and the hole gives rise to the binding energy, by which the exciton energy is smaller than the energy gap. In the absorption spectrum, a sharp exciton peak appears at a lower energy than the continuous interband transition.

Threshold wavelength of photoconduction is related to the edge of the absorption spectrum but influenced by the exciton formation and dissociation. Photoconduction in organic materials is important associated with the application to electrophotography and organic solar cells.

Electrophotography was invented by C. F. Carson in 1938, and is widely used in photocopy machines and laser printers. As a photoconductor for the photoreceptor, inorganic materials such as amorphous selenium and selenium compounds have been utilized. However, presently **organic photoconductors** (OPC) are mainly used.

First, the photoreceptor on a drum is charged by a voltage of 300 $\sim$ 1000 V ((1) in Fig. 7.1). Next, (2) laser or LED light makes the exposed part photoconductive and the irradiated part is discharged. In principle, the exposure is carried out directly in photocopy, but recently images are formed from the scanned digital data. Then, (3) charged toner particles are attracted on the drum. Toner particles with the same polarity as the drum charge are coated onto the exposed part, but the inverse image is obtained using the opposite charge. Finally (4) the toner is transferred to the paper, and (5) thermally fixed.

The photoreceptor drum is constructed either of a single layer photoconductor (Fig. 8.1b) or of the charge generation layer and the charge transporting layer (Fig. 8.1c). Phthalocyanin and azo dyes are used in the charge generation layer, and

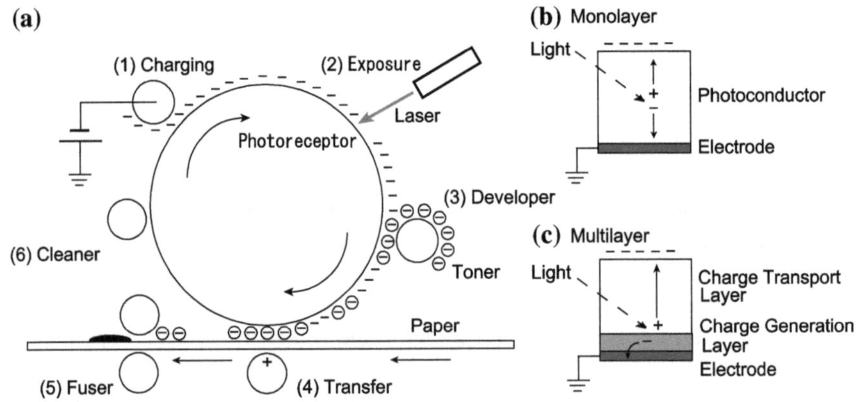


**Fig. 8.1** **a** Principle of electrophotography. **b** Monolayer type. **c** Multilayer type

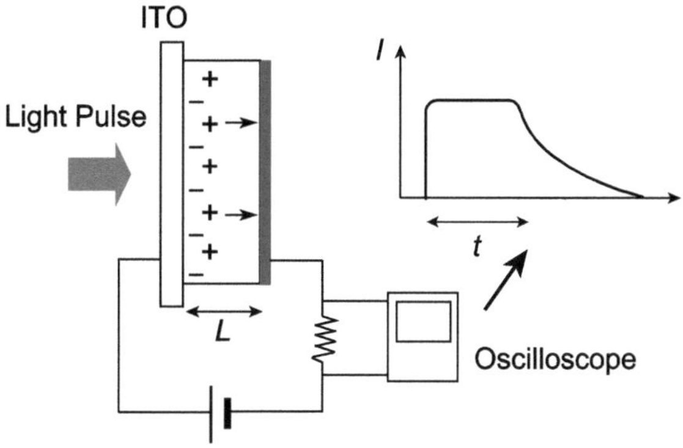


**Fig. 8.2** Mobility measurement based on the time-of-flight method

aromatic amines are used as the hole-transporting layer. Such a structure is close to organic light-emitting diodes, so the technology of organic light-emitting diodes is closely related to the electrophotography.

Mobility of an organic semiconductor is measured by the **time-of-flight** (TOF) **method** based on the transient photoconduction. An organic crystal or a thin film with the thickness of 1 $\sim$ 10 μm is sandwiched by electrodes (Fig. 8.2), and a light pulse is irradiated from the ITO side. If the absorbance of the sample is sufficiently large, the carriers are generated in the thin layer of the vicinity of the ITO electrode. Owing to the voltage applied between the electrodes, the carriers with the opposite charge move to the counter electrode, and the transient current is recorded on an oscilloscope. If we can assume the carrier layer moves to the counter electrode at a constant velocity, the drift velocity is represented by $v = \mu E = \mu V/L$. Since the transporting time is $t = L/v = L/\mu E = L^2/\mu V$, the **mobility** is obtained.

$$\mu = L^2/tV. \tag{8.1}$$

The estimated mobility of high-purity anthracene and naphthalene crystals is about 1 cm$^2$/Vs at room temperature (Table 8.1) [2, 3].

Temperature dependence of mobility in high-purity naphthalene crystals has been measured by the TOF method as shown in Fig. 8.3 [4]. The mobility increases with lowering the temperature, and reaches to several hundred cm$^2$/Vs. Such an increasing mobility indicates **band transport**. Mobility of other organic crystals is

**Table 8.1** Mobility (cm$^2$/Vs) of organic crystals in different crystal directions measured by the TOF method [3]

| | Hole *a*/*b*/*c* | Electron *a*/*b*/*c* |
|---|---|---|
| Naphthalene | 0.94/1.48/0.32 | 0.62/0.64/0.44 |
| Anthracene | 1.13/2.07/0.73 | 1.73/1.05/0.39 |
| Perylene | activated | 2.37/5.53/0.78 |
| Terphenyl | 0.6/–/0.80 | 0.34/1.2/0.25 |

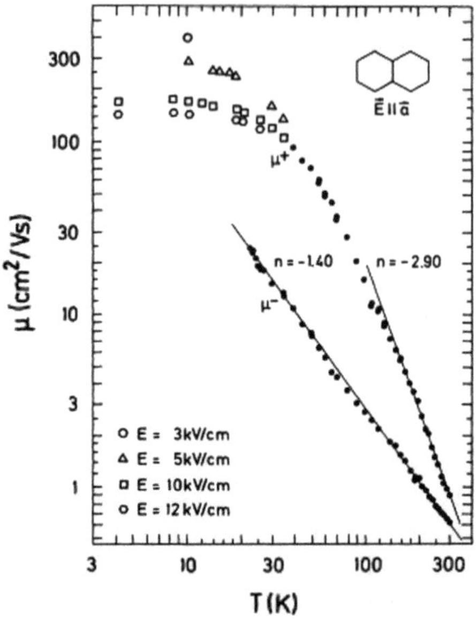


**Fig. 8.3** Temperature dependence of mobility in naphthalene measured by the TOF method [4]

typically 1 $cm^2/Vs$ at room temperature, and this value is recognized as the criterion of band transport [3, 5]. However, mobility of thin films and comparatively impure organic crystals usually decrease in an activated manner with lowering the temperature. Such **hopping conduction** is usually observed in organic materials.

## 8.2 Space Charge Limited Current

Next, we shall investigate carriers injected from the electrode. Suppose an insulating material in which thermally excited carriers are negligible. Carriers are injected from the electrode, and the current is determined by the balance of the potential generated by the carriers. Two ohmic contacts are made on a thin organic sample, and carriers are injected by applying a fairly high voltage. The direction between the electrodes is defined as $x$, and the current density $J = en\mu E$ is put in the Maxwell equation of the electric field $E$ and the charge density $n$.

$$\frac{\partial E}{\partial x} = \frac{en}{\varepsilon} = \frac{J}{\varepsilon \mu E}$$

Integration of the equation affords

$$E = \left(\frac{2J}{\varepsilon \mu}\right)^{1/2} x^{1/2}$$

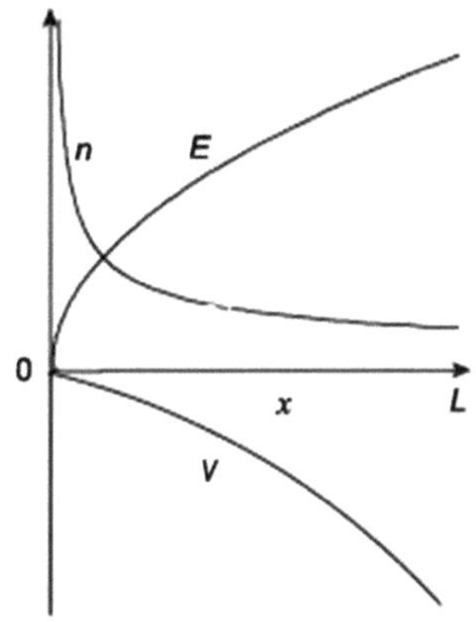


**Fig. 8.4** Distribution of $n$, $E$, and $V$ in space-charge-limited current

which represents the spacial ($x$) dependence of $E$ (Fig. 8.4). Here, we used the boundary condition; $E$ is continuous and $E = 0$ at $x = 0$. Since the carriers do not flow smoothly, the carriers are accumulated near the electrode at $x = 0$, and $E$ has a large slope around $x = 0$. Integration of this equation with respect to $x$ in the range of the thickness $L$ gives the voltage.

$$V = \int_0^L E dx = \left( \frac{8J}{9\varepsilon\mu} \right)^{1/2} L^{3/2}$$

This is solved to $J$.

$$J = \frac{9}{8} \varepsilon\mu \frac{V^2}{L^3} \tag{8.2}$$

This relation is known as **space-charge-limited current** (SCLC) [1]. The current increases in proportion to $V$ square, so Ohm's law does not apply. From this relation, we can estimate the mobility μ.

The above discussion applies to trap-free insulators, but actual insulators have traps. Owing to the trapped carriers, only a part of the injected carriers move freely. We have to multiply the fraction of the free carriers on Eq. 8.2, and $J$ decreases by this ratio. The resulting voltage–current characteristics is shown in Fig. 8.5 [6]. Ohmic current is observed generally at low voltages. With increasing $V$, the SCLC region starts, and $J$ is proportional to $V^2$. Due to the traps, $J$ is smaller than the ideal $V^2$ dependence. At higher voltages, the traps are entirely filled to enter the SCLC region with the $V^2$ dependence. At further high voltages, the current does not increase any more in the saturated region. The power law $V$ dependence in the trapped region depends on the trap distribution. In general, $J$-$V$ characteristics of an insulator are steeper than Ohm's law, and tend to bend to smaller resistance direction at high voltages [6].

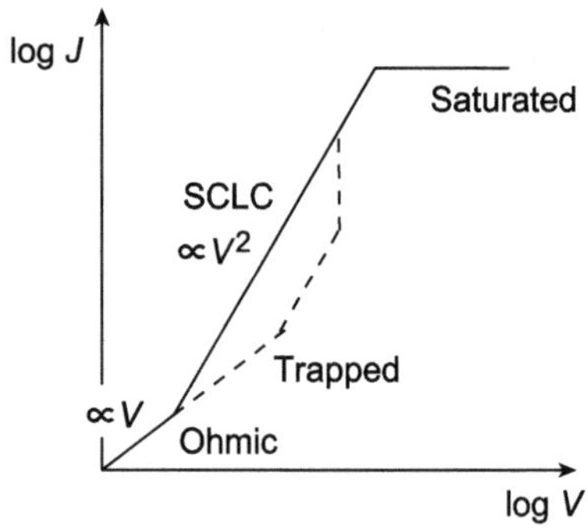


**Fig. 8.5** *V-I* characteristics in an insulator

## 8.3 Relaxation Time of Carriers

When a conductor is sandwiched by two electrodes, Ohmic current flows. However, when an insulator is sandwiched by two electrodes, a capacitor is formed, and displacement current flows to accumulate charges. We have to consider both for insulating organic materials.

$$J = \sigma E + \frac{\mathrm{d}D}{\mathrm{d}t} = \sigma E + \varepsilon_o \frac{\mathrm{d}E}{\mathrm{d}t} \quad \text{or} \quad I = \frac{V}{R} + C\frac{\mathrm{d}V}{\mathrm{d}t}$$

These relations represent a parallel circuit of resistance and capacitor. This circuit has a time constant of $\tau = \varepsilon_0/\sigma = RC$. This means, when charges are applied to an organic semiconductor, the carrier diffusion occurs during the time of $\tau$.

*Example 8.1* Calculate the time constant of a conductor with $\varepsilon \sim 1$ and $\sigma \sim 100$ S/cm, and a dielectric material with $\varepsilon \sim 1$ and $\sigma \sim 10^{-8}$ S/cm.

Dielectric constant of vacuum is $\varepsilon_0 = 8.854 \times 10^{-12}$ F/m = $8.854 \times 10^{-14}$ F/cm, and the unit of $\varepsilon_0/\sigma$ is

$$\left[\frac{\text{F/m}}{\text{C/Vsm}}\right] = \left[\frac{\text{C/Vm}}{\text{C/Vsm}}\right] = [\text{s}].$$

The time constant of the conductor is

$$\frac{(8.854 \times 10^{-14}\,\text{F/cm})}{(100\,\text{S/cm})} \approx 10^{-15}\,\text{s}.$$

The time constant of the dielectric material is

$$\frac{10 \times (8.854 \times 10^{-14}\,\text{F/cm})}{(10^{-8}\,\text{S/cm})} \approx 10^{-4}\,\text{s}.$$

Carriers in a conductor relax rapidly, but those in a dielectric material relax slowly.

When current $I$ flows across a boundary of materials with different time constants, charge

$$Q = C_2V_2 - C_1V_1 = C_2(R_2I) - C_1(R_1I) = I(\tau_2 - \tau_1)$$

is accumulated (**Maxwell–Wagner effect**). The boundary of an organic semiconductor with a metal electrode is a special case. If the organic material is completely insulating and $\sigma$ is zero, the charges are entirely accumulated at the boundary, and the system is regarded as a capacitor. If $\sigma$ is large, the system is a resistor. In the intermediate case, when the injected carriers do not exceed the thermally excited carriers, Ohmic conduction occurs, but when the former exceed the latter, charge accumulation occurs. In the case of effective charge injection, charge accumulation occurs over the organic semiconductor according to the electric field gradient; this is SCLC. The boundary between Ohm's law and SCLC in Fig. 8.5 is defined by the point at which the injected carriers exceed the thermally excited carriers. In SCLC, transport in the whole semiconductor restricts the conductivity, so SCLC is **transport limited**. By contrast, the Schottky current in the following is **injection limited**, where the charge accumulation occurs mainly at the boundary between the semiconductor and the electrode.

## 8.4 Ionization of Traps

When carriers of an organic semiconductor are mainly supplied from the trapped states, the carrier number increases with increasing the electric field, by which the electron starts to move to show nonlinearity in the conductivity at large $V$. Suppose an electric field which is applied to an electron bound to a trap by the Coulomb potential (Fig. 8.6).

$$V(r) = -eEr - \frac{e^2}{4\pi\varepsilon r} \tag{8.3}$$

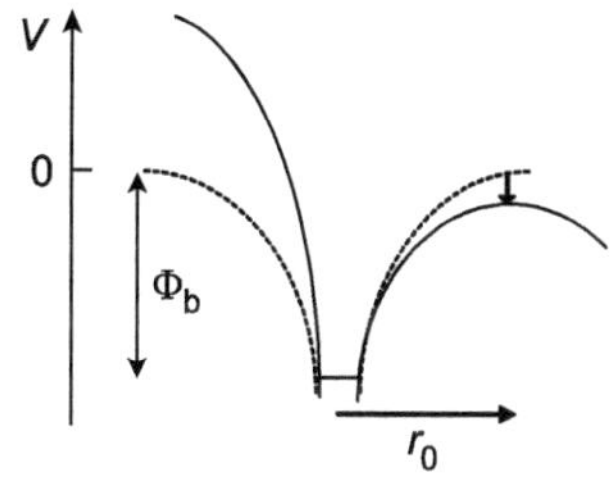


**Fig. 8.6** Trap potential. The dashed curve at zero field drops to the solid curve under an electric field

Under the electric field, the peak potential at $r_0$ is determined from $\partial V/\partial r = 0$.

$$-eE + \frac{e^2}{4\pi\varepsilon r^2} = 0 \quad \text{leads to } r_0 = \sqrt{\frac{e}{4\pi\varepsilon E}}$$

Putting $r_0$ in $V(r)$ to afford the potential barrier

$$V(r_0) = -e\sqrt{\frac{eE}{\pi\varepsilon}}$$

which is lowered from zero. Assume the conductivity is given by the standard semiconducting form.

$$J = \sigma_0 \exp\left(-\frac{e\Delta}{k_B T}\right)E$$

The drop of Δ is involved like

$$J = \sigma_0 \exp\left(-\frac{e\left(\Delta - \sqrt{eE/\pi\varepsilon}\right)}{k_B T}\right)E. \tag{8.4}$$

Then, the conductivity increases at large $E$. Such a mechanism is known as **Poole–Frenkel effect**, which explains why the conductivity of an insulator increases at high voltages [7].

Similar nonlinear conductivity has been observed in charge-ordered insulating phases in two-dimensional BEDT-TTF charge-transfer salts [8]. When the dielectric constant of the organic donor layer is much larger than the dielectric constant of the anion layer, an electron-hole pair, bound by the Coulomb interaction, is confined in the two-dimensional layer. Then, the Maxwell equation div $\boldsymbol{E} = \rho/\varepsilon_0$ is reduced to $2\pi r\varepsilon_0 E = \rho$, which is integrated to $V(r) \propto -\ln r$. Accordingly, instead of Eq. 8.3, the potential is assumed to be

$$V(r) = -eEr + U_0 \ln(r/a)$$

where $a$ is the minimal spacing represented by the lattice constant. Similar calculation to the above gives the maximum $V(r)$ at $r_0 = U_0/eE$. This $V(r_0)$ is put in Δ, and by using $r \gg a$, Eq. 8.4 is substituted by

$$J = \sigma_0 (ea/U_0)^{U_o/k_B T} E^{U_0/k_B T + 1}.$$

Here, $J$ follows a power law of $E$ like $J \propto E^n$. The index $n$ is $U_0/k_B T + 1$, which increases inversely proportional to $T$ at low temperatures. In θ-$(BEDT\text{-}TTF)_2CsZn(SCN)_4$ with $T_{MI} \sim 20$ K (Sect. 7.7.4), $n$ is as large as eight at 0.3 K, at which remarkable nonlinear conductivity following $J \propto V^8$ has been observed. Such giant nonlinear conductivity is widely observed in charge-ordered two-dimensional conductors.

## 8.5 Diode

Next, we shall investigate a PN junction. [9] A diode is made by a PN junction as shown in Fig. 8.7a. The Fermi levels of the P-type and N-type parts are different, but in the solid the Fermi levels have to be the same, so the band bending occurs as shown in Fig. 8.7b. Since $E_F$ of the N-type side is high, the N side has positive charges to lower $E_F$. The positive charges are generated in the **depletion layer**, where the band is pushed up, and the electrons in the donor levels are evacuated. The electrons enter the acceptor levels of the P-side. Accordingly, the interface polarizes into the opposite polarity to the **majority carrier**. The depletion layer is highly resistive because of the absence of the majority carriers. Thus, most resistance of a PN junction comes from the depletion layer. The junction has a potential barrier of $V_D$.

When the N side is negatively and the P-side is positively biased (Fig. 8.8), the positive charges in the depletion layer are attracted to the negative bias, and the negative charges are attracted to the positive bias. Then, current flows. The P-side levels drop due to the positive bias, and the N side goes up due to the negative bias. The barrier at the interface decreases from $V_D$ to $V_D - V$. The opposite bias increases the barrier and current does not flow. The PN junction conducts current only when a positive bias is applied to the P-side. Then, the diode works as a rectifier.

A little more quantitative treatment follows. In the N region, electron is the majority carrier, and the density is

$$N_e = N_e^0 \exp\left(-\frac{E_C - E_F}{k_B T}\right).$$

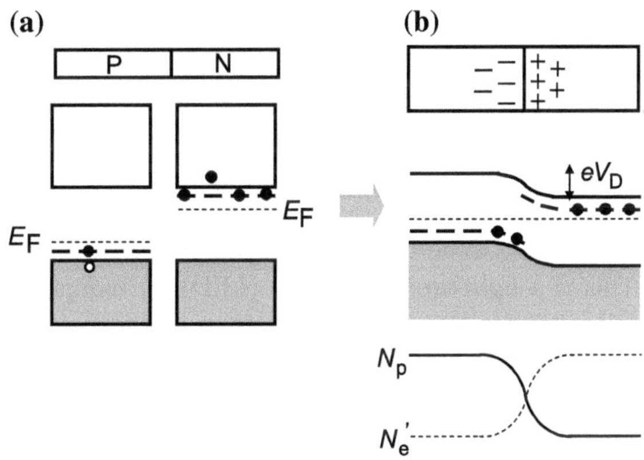


**Fig. 8.7** PN junction. **a** Energy levels before making a junction. **b** Energy levels of the PN junction

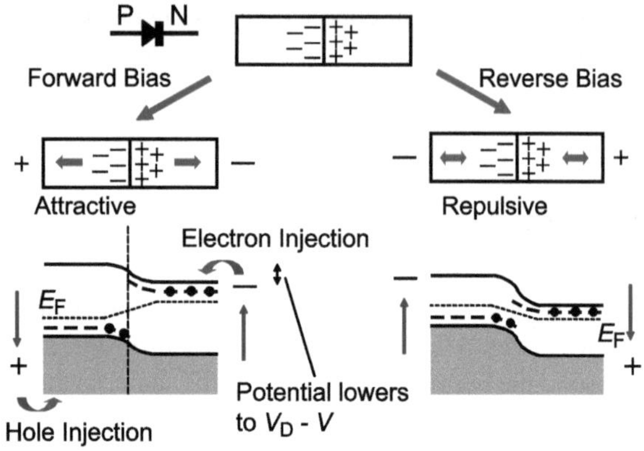


**Fig. 8.8** PN junction under the applied bias

In the P region, electron is the **minority carrier**, and the density is

$$N'_e = N_e^0 \exp\left(-\frac{E_C - E_F + eV_D}{k_B T}\right) \text{ leads to } N'_e = N_e \exp\left(-\frac{eV_D}{k_B T}\right). \tag{8.5}$$

Under the bias of $V$, this is replaced by

$$N'_e = N_e \exp\left(-\frac{e(V_D - V)}{k_B T}\right). \tag{8.6}$$

The transport of the majority carriers from the N region to the P region is the diffusion current of the electrons, which is proportional to $N_e$. The opposite electron transport from the P region to the N region is drift current of the minority carriers and in proportion to $N'_e$. Thus, the overall current is

$$I \propto N'_e - N_e = I_0(\exp(eV/k_B T) - 1) \tag{8.7}$$

where $I_0$ does not depend on $V$. This represents the characteristics of a diode as depicted in Fig. 8.9. In the inverse bias, a small current of $I_0$ flows independent of $V$. At $V = 0$, the current is zero. The same equation applies to holes.

A rectifier is not the only function of a PN junction. When a forward bias is applied to a diode, electrons flow in the N region as the majority carriers, and holes flow in the P region. At the junction region, electrons from the N region go down to the hole states coming from the P region (Fig. 8.10a). In the ordinary PN junction, the recombination energy is lost as heat, but the recombination brings about light emission in some materials. Thus, the electric energy is converted to light energy. This is a **light-emitting diode** (LED). In inorganic semiconductors, the emission efficiency is high when the bottom of the conduction band and the top of the valence band are located at the same $k$ point, which is called **direct transition type**. The emission does not occur in **indirect transition** semiconductors such as silicon and germanium. In organic semiconductors, the emission occurs when the

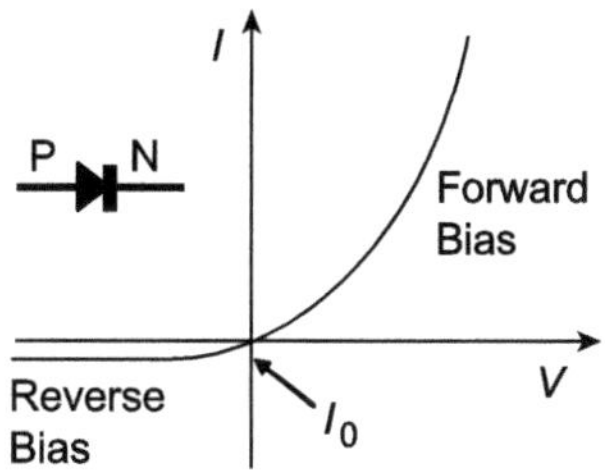


**Fig. 8.9** Voltage–current characteristics of a diode

semiconductor shows light-induced fluorescence or phosphorescence. The wavelength of the emitted light is determined by the energy gap. Widely used red LED is made of AlInGaP, but green LED from GaP. Blue light is obtained by using GaN with a little wider band gap. Accordingly, the threshold voltage of LED depends on the colors; red and green LEDs work at 2 V, while a blue LED works above 3 V. White LED is made from the blue LED combined with fluorescent dyes. LED has been used in many electric instruments for a long time, but recently the application is spreading to large screen, electric scoreboard, traffic signal, and LED illumination.

When increasing the inverse bias applied to a PN junction, and the bias is as large as the band gap, the bottom of the conduction band in the N region is located at the same level as the top of the valence band in the P region (Fig. 8.10b). If the carriers tunnel across the junction, current flow occurs abruptly above a certain inverse bias (Fig. 8.10b). Such a phenomenon is called **Zener breakdown**, and the **Zener diode** is widely used in electric instruments as a voltage standard. An LED is used by applying a forward bias, whereas a Zener diode is used applying an inverse bias.

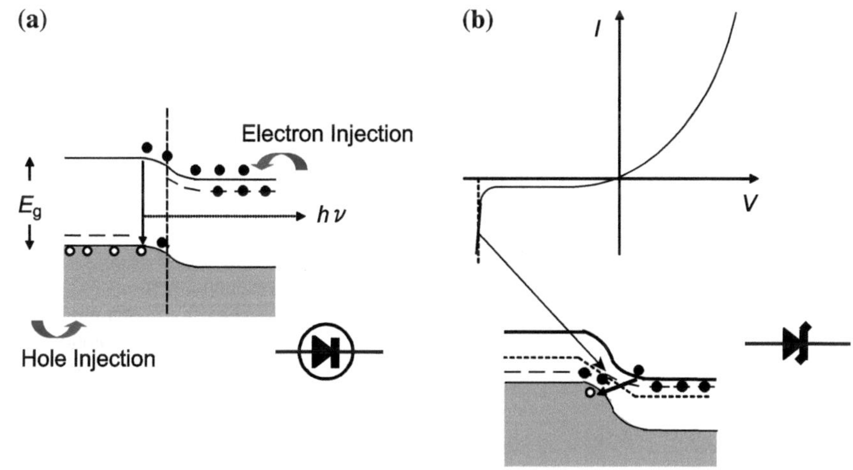


**Fig. 8.10** (**a**) Light emitting diode, and (**b**) Zener diode

## 8.6 Schottky Junction

A rectifying property occurs not only at a PN junction but also at a metal–semiconductor interface. Historically, the first semiconductor device is a point-contact diode based on a metal–semiconductor contact. Suppose a contact between a metal and an N-type semiconductor (Fig. 8.11), where the independent energy levels are represented in Fig. 8.11a. When the metal $E_F$ is lower than $E_F$ of the N-type semiconductor, a positively charged depletion region is generated by lowering the semiconductor energy levels so as to align $E_F$ at the same level as the metal $E_F$ (Fig. 8.11b). Here, the semiconductor band bends upwards, and the majority carriers, electrons, are depleted. The metal $E_F$ measured from the vacuum level is represented by the work function $\Phi_m$. On the other hand, $E_F$ of the N-type semiconductor is close the bottom of the conduction band, which is approximated by the electron affinity $\Phi_s$. Accordingly, this junction has the **Schottky barrier**, $\Phi_b = \Phi_m - \Phi_s$. When a negative bias is applied to the semiconductor side, it is attractive to the positive charges in the depletion layer, and current flows. So this is the forward bias. The negative bias lifts up the semiconductor levels, and the depletion layer disappears at the bias larger than $\Phi_b$. Then, the majority carriers, electrons, flow from the semiconductor to the metal without the Schottky barrier. In contrast, a positive bias applied to the semiconductor enhances the Schottky barrier, and bars the current flow. Thus, such a **Schottky diode** works as a rectifier.

If the original semiconductor $E_F$ is located lower than the metal $E_F$, the energy band bends downward (Fig. 8.11c). Thus, an **accumulation layer** of the majority carriers is formed. Then, the current flows for any bias. Such a junction is called **Ohmic contact**. In the forward bias of the Schottky junction, the N-type semiconductor is negatively biased. This is the same direction as a PN junction.

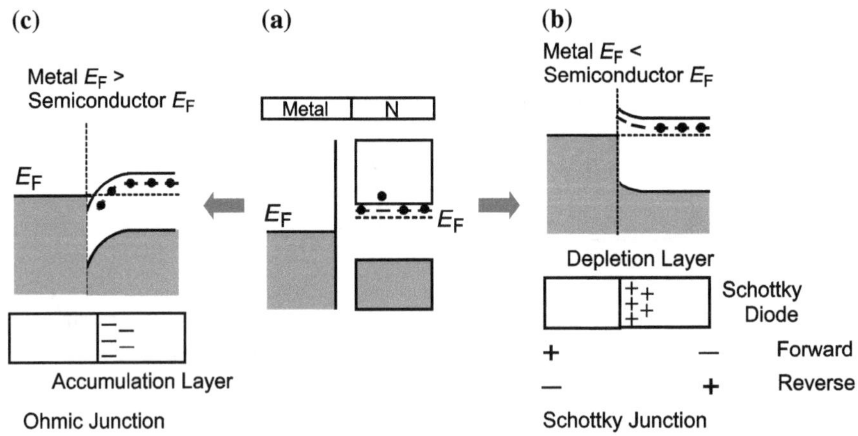


**Fig. 8.11** Schottky junction formed at a metal–semiconductor interface

A combination of a metal and a P-type semiconductor makes a similar Schottky junction, in which the positive bias to the P-type semiconductor is the forward bias.

For a little more quantitative treatment of the Schottky junction, we suppose metal electrons with the energy exceeding $\Phi_b$ are emitted in the semiconductor. This is analogous to emission of thermal electrons from a metal, where metal electrons with the energy exceeding the work function $\Phi_m$ are emitted to the vacuum. Then, we shall replace $\Phi_m$ to $\Phi_b$. Assume $x$ to be perpendicular to the metal surface, and the electrons with momenta $p_x > p$ are emitted. The resulting current is

$$J = e\int_{p_0}^{\infty} n\frac{p_x}{m}\mathrm{d}p_x.$$

Suppose $n$ to be given by the Boltzmann distribution, and the metal electron is represented by the free electron. Then, $n$ under an applied $V$ is

$$\exp\left(-\frac{E - E_\mathrm{F} - eV}{k_\mathrm{B}T}\right) = \exp\left(-\frac{p_x^2 + p_y^2 + p_z^2 - 2mE_\mathrm{F}}{2mk_\mathrm{B}T}\right)\exp\left(\frac{eV}{k_\mathrm{B}T}\right).$$

The integration along $p_y$ and $p_z$ affords $n$ and $J$

$$J = \frac{2e}{mh^3}\exp\left(\frac{eV}{k_\mathrm{B}T}\right)\int_{p_0}^{\infty} p_x\exp\left(-\frac{p_x^2 - 2mE_\mathrm{F}}{2mk_\mathrm{B}T}\right)\mathrm{d}p_x \int_{-\infty}^{\infty}\exp\left(-\frac{p_y^2}{2mk_\mathrm{B}T}\right)\mathrm{d}p_y \int_{-\infty}^{\infty}\exp\left(-\frac{p_z^2}{2mk_\mathrm{B}T}\right)\mathrm{d}p_z$$

where the factor $2/h^3$ comes from the state number. The last two integrals are calculated using

$$\int_{-\infty}^{\infty}\exp(-ax^2)\mathrm{d}x = \sqrt{\frac{\pi}{a}}$$

to give $\sqrt{2\pi mk_\mathrm{B}T}$. The integral for $p_x$ has the lower limit

$$p_0^2/2m - E_\mathrm{F} = e\Phi_b$$

to give an exponential function

$$mk_\mathrm{B}T\exp\left(-\frac{e\Phi_b}{k_\mathrm{B}T}\right).$$

The overall result is

$$J = \frac{4\pi emk_\mathrm{B}^2}{h^3}T^2\exp\left(\frac{eV}{k_\mathrm{B}T}\right)\exp\left(-\frac{e\Phi_b}{k_\mathrm{B}T}\right) = AT^2\exp\left(\frac{eV}{k_\mathrm{B}T}\right)\exp\left(-\frac{e\Phi_b}{k_\mathrm{B}T}\right)$$

where $A$ is called **Richardson constant**. Similarly to a PN junction, at $V = 0$ the inverse diffusion current flows from the semiconductor to the metal to make $J = 0$. Thus the voltage–current characteristics of the Schottky junction is

$$J = AT^2 \exp\left(-\frac{e\Phi_b}{k_B T}\right)\left(\exp\left(\frac{eV}{k_B T}\right) - 1\right). \tag{8.8}$$

When a bias is applied to the Schottky junction, the Schottky barrier is lowered as shown in Fig. 8.11d. Similarly to the Poole–Frenkel effect (Fig. 8.6), the potential maximum is lowered due to the electric field $E$.

$$\exp\left(-\frac{e\Phi_b}{k_B T}\right) \to \exp\left(-\frac{e\left(\Phi_b - \sqrt{eE/\pi\varepsilon}\right)}{k_B T}\right)$$

The SCLC, the Poole–Frenkel effect, and the Schottky barrier show similar nonlinear resistance, in which the resistance decreases at high voltages.

## 8.7 Photoelectron Spectroscopy

Energy levels of organic semiconductors are estimated from the redox potentials (Sect. 7.2), but they are more directly measured by the **ultraviolet photoelectron spectroscopy** (UPS). Photoelectron spectrum consists of peaks related to the density of states together with continuously distributed secondary electron scattering (Fig. 8.12). From the Einstein relation (Sect. 2.2), the maximum kinetic energy $E_k$ affords the work function of the metal according to $\Phi_m = E_{vac} - E_F = h\nu - E_k$. In a laboratory scale, He I radiation (21.2 eV) is used, but a wide wavelength range of light is obtained from synchrotron radiation. The cutoff appears when the electron kinetic energy just out of the material is zero ($E_k = 0$). The cutoff gives the vacuum level $E_{vac}$, which somewhat depends on the materials. For instance, a positively charged material has a large $E_{vac}$ because it needs finite energy to remove the rest electron just out of the material to the common reference energy far from the material.

Work functions of metals are shown in Fig. 8.13. In a Schottky junction, the majority carrier of the semiconductor always flows from the semiconductor to the metal, but it does not flow oppositely. The Schottky junction is formed for electrons when the metal $E_F$ is lower than the semiconductor $E_F$ because the electron flows only downward. Work function is related to the ionization energy of the metal. Metals with strong ionization tendency, such as sodium and calcium, have a very small work function (<3 eV, Fig. 8.13), and show a strong tendency to form the cation. These metals have great ability to inject electrons. In contrast, gold with a comparatively large work function has a great ability to inject holes, and is used as an anode. Many metals have work functions between 4 and 5 eV.

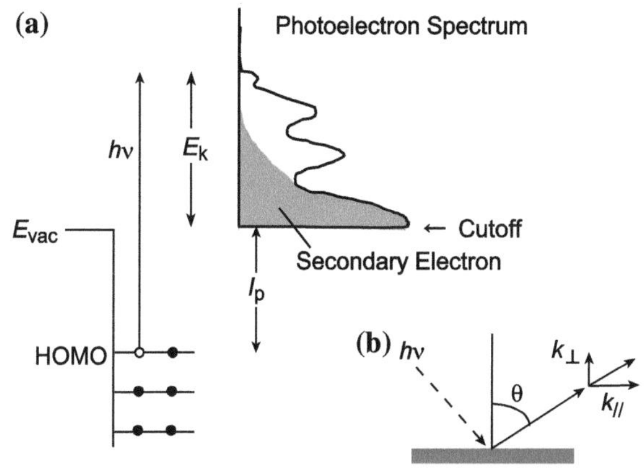


**Fig. 8.12** Ultraviolet photoelectron spectroscopy

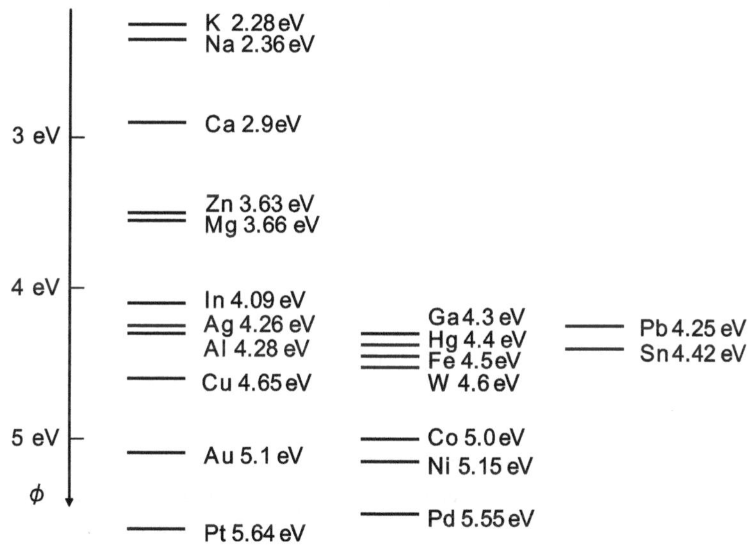


**Fig. 8.13** Work functions of metals

Among such metals, silver and aluminium with relatively small work functions are used as a cathode to inject electrons. Platinum and palladium are exceptional metals having larger work functions than gold. Work function of indium tin oxide (ITO) is about 5.0 eV, which is suitable for hole injection, and that of graphite is also located around 4.8 eV. Work function, however, changes depending on the atmosphere and the treatment. In general, oxygen in air increases the work function and deepens the $E_F$ location.

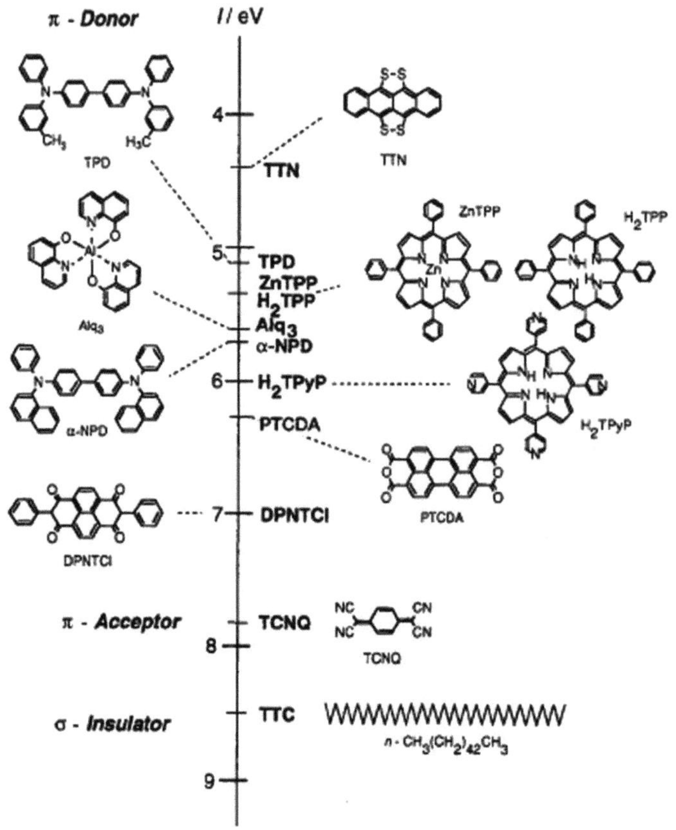


**Fig. 8.14** Ionization energies of organic semiconductors estimated by photoelectron spectroscopy [13]. This figure includes the HOMO levels of several acceptors

Photoelectron spectrum of a molecule affords the ionization energy from the maximum kinetic energy $E_k$ according to $I_p = h\nu - E_k$ (Fig. 8.12). The ionization energy corresponds to the energy of HOMO as discussed in Sects. 7.1 and 7.2. Ionization energies of several representative organic semiconductors are plotted in Fig. 8.14.

**Angle-resolved photoelectron spectroscopy** measures not only the energy but also the momentum from the angle distribution of the emitted electrons. From conservation of the electron momentum $k_{//}$ parallel to the surface (Fig. 8.12b)

$$k_{//} = \sqrt{2mE_k}\sin\theta/\hbar = \left(0.51\,\text{Å}^{-1}\right)\sqrt{E_k/\text{eV}}\sin\theta$$

$k_{//}$ is estimated from θ. Then, the θ dependence of the photoelectron spectrum, observed by changing the angle of the electron detector, affords the $k_{//}$ dependence of the state density. Accordingly, the energy band $E(k_{//})$ is observed. The vertical $k_{\perp}$

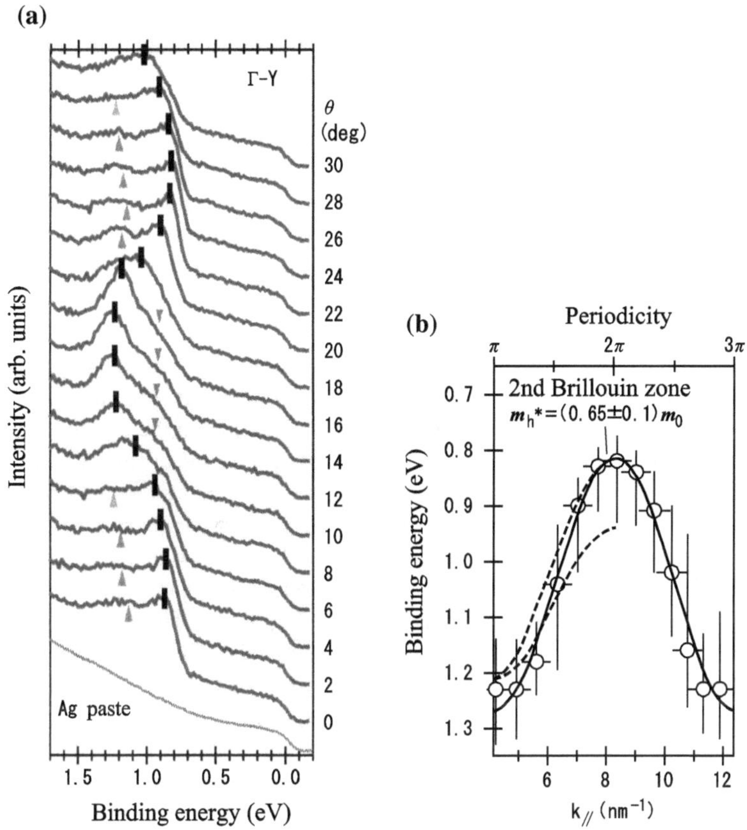


**Fig. 8.15** **a** Angle-resolved photoelectron spectrum of rubrene, and **b** the resulting energy band [12]

is also estimated if we can use variable *hν* from synchrotron radiation. Energy bands of such organic materials as pentacene, rubrene, and PTCDI (Fig. 8.26) have been obtained in this method [10–12]. Figure 8.15a shows angle-resolved photoelectron spectrum of rubrene, form which the energy band in Fig. 8.15b is extracted.

Photoelectron spectroscopy reveals occupied levels. Unoccupied levels such as LUMO are investigated by the inverse process called **inverse photoelectron spectroscopy**, where electrons are irradiated to the material, and the emitted light is analyzed. Electron affinity is estimated from this measurement.

The ordinary vacuum in the order of $10^{-5}$ Pa, generated by a turbo molecular pump, is widely used in vacuum evaporation and sublimation. This is, however, not sufficient for photoelectron spectroscopy, and instead, a high vacuum of the order of $10^{-8}$ Pa generated by an ion pump is used. In inorganic materials, clean surface is produced by cleavage, but this method is not usually applied to organic materials.

Some small-molecule organic materials sometimes show clean surface due to the relatively high vapor pressure, but such material tend to contaminate the vacuum chamber.

Photoelectron spectroscopy reveals electronic states of organic compounds deposited on a metal surface. The hole injection barrier at a metal/organic interface (Fig. 8.16a) is naively represented by $\Phi_b = I_p - \Phi_m$ using the metal work function $\Phi_m$ and the ionization energy of the organic material $I_p$. Metal work functions are listed in Fig. 8.13, and ionization energies of organic materials are shown in Fig. 8.14. From such information, we can roughly estimate the injection barrier at the metal/organic interface. $\Phi_b > 0$ results in the Schottky junction, whereas $\Phi_b < 0$ leads to an Ohmic contact.

When we measure photoelectron spectrum of an organic thin film deposited on a metal surface, $I_p$ is different depending on the molecular orientation. $I_p$ is not entirely inherent in the materials, but somewhat depends on the surface structures. Mean free path of an electron with an energy of several eV to several ten eV is as small as several nm in a solid, so photoelectron spectroscopy mainly reflects the surface state. Then, an organic film thicker than several nm on a metal surface affords photoelectron spectrum mostly coming from the organic material. After measuring the photoelectron spectrum of the metal, an organic material is deposited, and the spectrum is measured as shown in Fig. 8.16c. Thus, the cutoff shift of $\Delta$ is observed. The shift of the cutoff comes from the shift of the vacuum level (Fig. 8.16b). The difference of the right ends of the spectra (Fig. 8.16c), namely, the difference of the maximum kinetic energy $E_k$, gives the injection barrier $\Phi_b$. Positive charges at the organic interface lower the vacuum level of the organic material, and consequently lower the whole energy levels. When the organic material is an electron donor, the first monolayer is charged positively. Such **interfacial polarization** or the generation of the **interfacial potential** explains the energy level shift. In the terminology of semiconductor physics, an accumulation layer is generated at the interface. Such polarization lowers the HOMO level, and increases the hole injection barrier. When an acceptor film is deposited on a metal surface, the organic material charges negatively, and the LUMO level goes up. In most cases, the interfacial polarization occurs in such a direction to increase the carrier injection barrier. As an exception, acceptors such as $Alq_3$ on gold polarize in

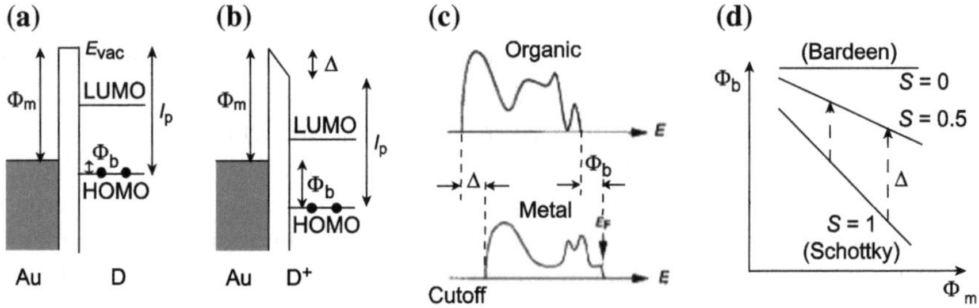


**Fig. 8.16** Energy level of an organic compound on a metal surface

the direction to lower the energy levels, and the electron injection barrier is reduced. This is the reason that gold is sometimes used for electron injection, though the work function is comparatively large.

The variation of $\Phi_b$ for metals with different $\Phi_m$ is plotted in Fig. 8.16d. Without the interfacial polarization, $\Phi_b$ and $\Phi_m$ make a line with a slope of $S = 1$ (**Schottky limit**). However, the interfacial polarization reduces $S$ to less than one. The limit at $S = 0$ is called **Bardeen limit** [13–16]. Another model assumes $S$ becomes zero when $\Phi_m$ exceeds $I_p$, where the donor is sufficiently strong to be charged positively to chancel the negative $\Phi_b$ [17]. In the actual organic materials, $S = 0$ happens not exactly at $I_p$ but at a polaron level a little smaller than $I_p$. In the case of Fig. 8.16a with obviously positive $\Phi_b$, this $\Phi_b$ works as the injection barrier without the interfacial polarization. When the metal $E_F$ is lowered below the polaron level, the organic semiconductor charges positively and the energy levels go down. Then, $\Phi_b$ never becomes negative. Such interfacial polarization is crucial in organic devices. For instance, bottom-contact organic transistors, where the organic film is deposited on the metal electrodes, are susceptible to the interfacial polarization and show much lower performance than the top-contact transistors, where the electrodes are fabricated on the organic film.

## 8.8 Organic Light-Emitting Diode

Historically, organic light-emitting devices, in which current is injected to anthracene crystals or thin films, have been investigated since the 1960s. The current organic light-emitting diodes (LED) stem from Tang's proposal in 1987 (Fig. 8.17a) [18]. This device structure is analogous to inorganic LED (Fig. 8.10a), which is, however, constructed from P- and N-doped inorganic semiconductors [19,20]. In contrast, organic LED consists of undoped organic semiconductors. The two types of organic semiconductors are called **hole-transporting layer** and **electron transporting layer**. Diamines such as NPD and TPD (Fig. 8.14) with electron donor ability are used as the hole-transporting layer, and acceptors such as $Alq_3$ are used as the electron transporting layer. A transparent conductor, ITO, is deposited on a glass substrate by sputtering or electron beam method. The work function of ITO is about 5.0 eV, which is large enough to effectively inject holes to the hole-transporting layer (Fig. 8.18). On the hole-transporting layer, an $Alq_3$ layer

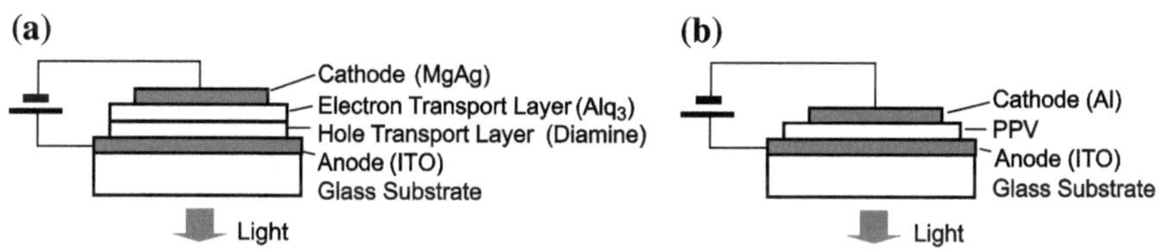


**Fig. 8.17** Structures of (**a**) small-molecule-based, and (**b**) polymer-based organic LED

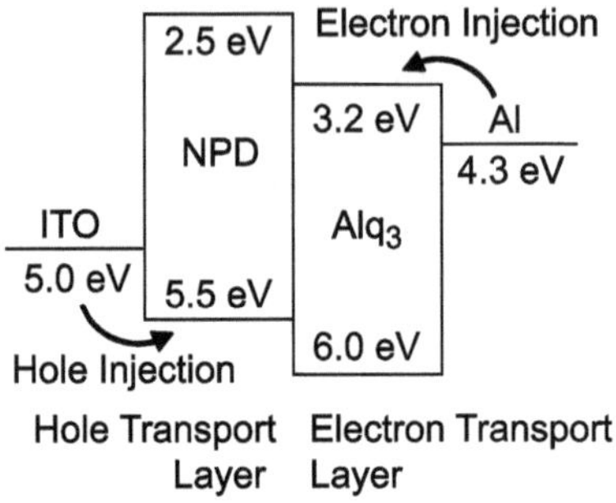


**Fig. 8.18** Energy levels of an organic LED device consisting of ITO/NPD/$Alq_3$/Al

is deposited, and finally the cathode such as MgAg and Al is fabricated. Owing to the small work functions, electrons are efficiently injected from these metals to the electron transporting layer. Around the boundary of the hole and electron transporting layers, the hole and the electron encounter. The hole-electron pair on the same molecule recombines to emit fluorescence.

Organic LED devices based on poly(p-phenylenevinylene) (PPV, Fig. 8.19a) has a monolayer structure (Fig. 8.17b) [21]. Small-molecule-based organic LEDs consisting of multilayers are successively deposited by vacuum evaporation. In contrast, polymer-based organic LEDs are formed by **spin coating**, where a solution of polymer is dropped on a high-speed spinning substrate. This method is not very suitable for a multilayer structure because the successively dropped solution tends to dissolve the former film. PPV does not dissolve in organic solvents, but MEH-PPV (Fig. 8.19b) is soluble in organic solvents. Polyfluorene (Fig. 8.19c) is also well known, and the copolymer with acene, thiophene, and benzothiazole affords various fluorescence colors (Fig. 8.19d).

$Alq_3$ has fluorescence, which emits light from the photoexcited state, but also emits light by the current injection. In contrast to the former **photoluminescence**, the

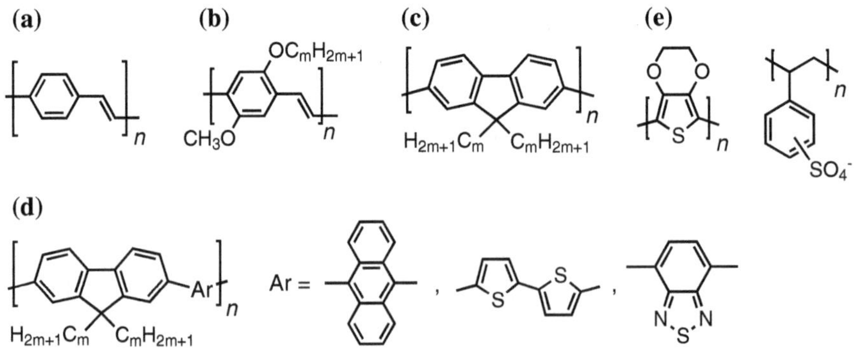


**Fig. 8.19** (**a**) PPV, (**b**) MEH-PPV, (**c**) polyfluorene, (**d**) copolymers of fluorene, and (**e**) PEDOT: PSS

latter is called **electroluminescence**. The cathode (Fig. 8.17) is composed of a reflecting metal, and the emitted light effectively goes out through the ITO. An organic LED is as thin as 100 nm, and the typical operation voltage is 10 V. From the typical current density around $10^3$ mA/cm$^2$, the resistivity is estimated to be as large as $10^8$ Ω cm. An organic LED uses a practically insulating organic film in a very thin form. Current–voltage characteristics of an organic LED are similar to that of the ordinary diode (Eq. 8.7), and the brightness is basically proportional to the current.

ITO has the resistivity of $10^{-4}$ Ω cm and the film of 150 nm thickness shows the sheet resistance of 10 Ω/sq and the transparency more than 90 %. ITO is industrially produced as a transparent conducting material, but indium is a comparatively rare element. Currently, the replacement is investigated. Anode materials other than ITO are doped zinc oxide and indium zinc oxide (IZO).

To achieve the efficient hole injection from the anode to the hole-transporting layer, sometimes a hole injection layer is inserted between the anode and the hole-transporting layer. Such a hole injection layer decreases the operation voltage. The hole injection layer is formed by diamine or PEDOT: PSS. A doped polythiophene, PEDOT: PSS (polyethylenedioxythiophene: polystylenesulfonate, Fig. 8.19e) is a highly conducting polymer, and used as a transparent electrode or a hole injection layer. PEDOT shows good solubility, and PEDOT: PSS is soluble in water. The film shows as high conductivity as several 100 S/cm. After an aqueous solution of PEDOT: PSS is spin coated, a polymer layer is easily fabricated by spin coating a solution of an organic solvent.

Diamine in the hole-transporting layer is used as an amorphous form. Condensation due to the Joule heating reduces the efficiency. Thus, a high glass transition temperature $T_g$ is desirable. $T_g$ of TPD (Fig. 8.14) is 60 °C, and $T_g$ of NPD is 95 °C. Materials having more than three triphenylamines and sterically bulky spiro groups are developed. These materials achieve $T_g$ higher than 100 °C and a long life time of the organic LED device.

Metals with small work functions like MgAg and Al are used as cathode. It has been known that a thin LiF layer between the cathode and the electron transporting layer improves the electron injection efficiency. The role of LiF is mysterious but the function is attributed to the liberated lithium. Calcium is sometimes used as the cathode in polymer LED. Electron injection layer is sometimes inserted in between the cathode and the electron transporting layer.

When about 1 % of a fluorescence dye is doped in an organic LED, the original fluorescence from the host disappears and only the fluorescence from the doped guest is observed. This is due to the excitation transfer from the host to the guest. For instance, we can control the emitted light by using a dye in a dye laser as a guest material. In general, fluorescence spectrum is a mirror image of the absorption spectrum (Fig. 8.20). For the efficient excitation transfer from the host to the guest, overlap between the host fluorescence spectrum and the guest absorption spectrum is necessary. Rubrene shows 100 % quantum yield of the fluorescence, so rubrene is used as a guest to attain 100 % quantum yield. Quantum yields of several organic materials are listed in Table 8.2. The extension of the π-skeleton tends to increase the quantum yield [22, 23].

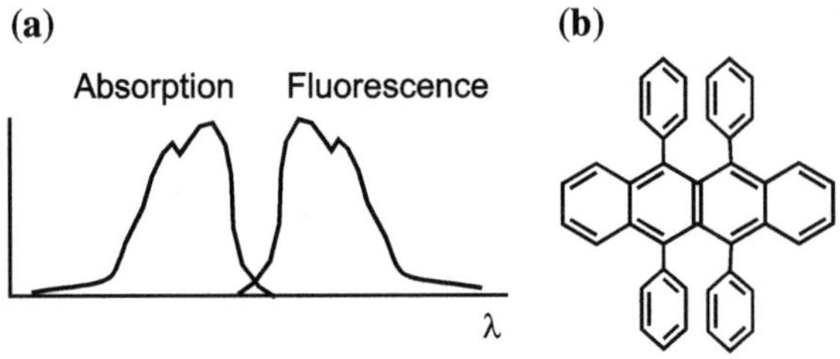


**Fig. 8.20** **a** Absorption and fluorescence spectra, and **b** rubrene

**Table 8.2** Fluorescence quantum yields [22, 23]

| | | | | | |
|---|---|---|---|---|---|
| Benzene | 0.07 | Naphthalene | 0.23 | Anthracene | 0.36 |
| Tetracene | 0.21 | Pylene | 0.65 | Perylene | 0.94 |
| Rubrene | ~1.0 | Fluorene | 0.80 | | |
| Biphenyl | 0.18 | Terphenyl | 0.93 | PPV | 0.08 |
| TPD | 0.35 | $Alq_3$ | 0.25 | Sexithiophene | 0.40 |

Light emission from a singlet excited state is called **fluorescence** (Fig. 8.21a), whereas emission from a triplet excited state is **phosphorescence**. Spins of electrically injected carriers are entirely random, to generate 25 % singlet and 75 % triplet. Thus, the light energy emitted as the fluorescence never exceeds 25 % of the whole energy. So phosphorescence materials are explored. A representative phosphorescence material is an iridium complex (Fig. 8.21b), where various emission colors are obtained by changing the substituents. Owing to the intersystem crossing from the singlet to the triplet, phosphorescence material utilizes the energy of not only the triplet state but also the singlet state, to achieve a nearly 100 % quantum yield [24]. The probability of the intersystem crossing depends on the magnitude of the spin-orbit coupling. This is the reason that phosphorescence materials contain heavy elements like iridium. Usually, a phosphorescence material is contained in a bipolar transporting layer between the hole and electron transporting layers (Fig. 8.21c). Representative bipolar transporting materials are CBP and BCP (Fig. 8.21d, e).

Light extracting efficiency $\eta_p$ is usually around 20 %. The external quantum efficiency $\eta_{ext}$ is given by the product of the internal quantum yield $\eta_{int}$ and the light extracting efficiency $\eta_p$, which becomes as small as $\eta_{ext} = \eta_{int} \times \eta_p = 25\ \% \times 20\ \% = 5\ \%$ in a fluorescence LED. A phosphorescence material with $\eta_{int} = 100\ \%$ affords $\eta_{ext} = 20\ \%$. The light extracting efficiency largely depends on the difference of refractive indexes in the organic film, glass substrate, and the outside. When the inside and outside refractive indexes are $n_i$ and $n_e$, according to Snell's law, incident light with an angle larger than the critical angle $\sin\theta = n_e/n_i$ is totally reflected and does not go out. The light from a molecule whose dipole is perpendicular to the substrate is emitted parallel to the substrate and does not go out. In order to increase $\eta_p$, microlenses and cavities are used to improve the optical structures.

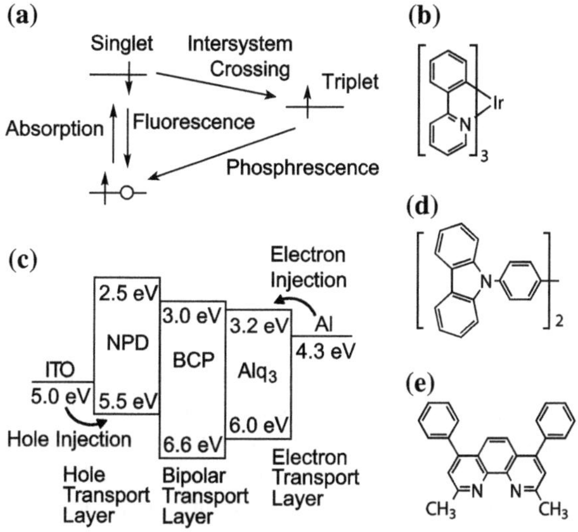


**Fig. 8.21** **a** Fluorescence and phosphorescence. **b** Phosphorescent iridium complex. **c** Organic LED with a bipolar layer. **d** Bipolar transporting material, CBP (4,4′-dicarbazoyl-1,1′-biphenyl), and **e** BCP (bathocuproine)

An organic LED display is thinner than liquid crystal displays, and the contrast is improved due to the self emitting properties. An organic LED displays are widely used in mobile phones. The merit of the easy large-area fabrication is utilized in the application to organic LED lightening.

## 8.9 Thin-Film Field-Effect Transistors

Structure of a thin-film field-effect transistor is depicted in Fig. 8.22. There are **source** and **drain** electrodes on the semiconductor thin film, and a **gate** electrode is located below the channel region between the two electrodes, separated by a thin gate insulating layer. A metal-oxide-semiconductor field-effect transistor (MOS-FET) based on crystal silicon has PN junctions at the source and drain electrodes, and uses the inverted layer of the minority carriers with the polarity opposite to the originally doped carriers of the channel region. In contrast, thin-film

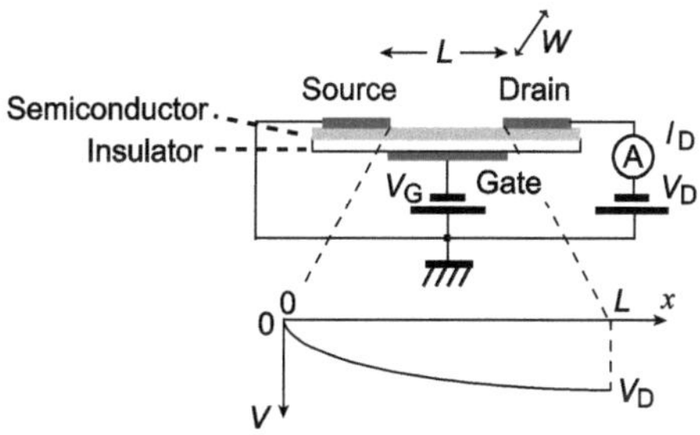


**Fig. 8.22** Structure of a field-effect transistor

transistors based on amorphous silicon and organic semiconductors have a much simpler structure than the single-crystal MOS-FET described in the usual textbooks of semiconductor devices. The distance between the source and drain electrodes is the channel length $L$, and the width is the channel width $W$. The channel semiconductor together with the source and drain electrodes are regarded as a set of conductors, which make a capacitor with the gate electrode. When a gate voltage $V_G$ is applied, charges $Q=C(V_G - V_T)$ appear at the interface of the semiconductor and the gate insulator, where $V_T$ is the **threshold voltage**. If $V_G$ is negative, holes are accumulated at the semiconductor interface. Then, drain current $I_D$ flows when the drain voltage $V_D$ is applied. Usually, the source voltage is taken as zero, and $V_G$ and $V_D$ have the same polarity. So the carriers, that are holes in Fig. 8.22, are injected from the source and flow to the drain. When $V_G$ is turned off, the field-induced carriers disappear, and $I_D$ becomes zero. Then, we can switch $I_D$ by changing $V_G$. Since the gate electrode is separated by the gate insulator from the other parts, the input resistance of a MOS-FET is very large. Therefore, a MOS-FET is low-power consuming, and widely used in digital circuits such as microprocessors and memory devices as well as active matrices in displays.

For more quantitative analysis of the FET characteristics, the voltage at a channel position $x$ is assumed to be $V(x)$ (gradual channel approximation), where $x$ is defined from the source ($x = 0$) to the drain ($x = L$). Since the voltage difference to the gate is $V_G - V(x)$, the field-induced charge is

$$Q(x) = C(V_G - V_T - V(x)) \tag{8.9}$$

where $C$ is the capacitance per area and $Q(x)$ is the induced charge per area. Conductivity of a semiconductor is represented by $\sigma = ne\mu$ (Eq. 3.17), but $ne$ corresponds to $Q(x)$, and the electric field is $E_x = -\mathrm{d}V(x)/\mathrm{d}x$. Thus, $I_D$ is given by

$$I_D = WQ(x)\mu(-\mathrm{d}V(x)/\mathrm{d}x). \tag{8.10}$$

This equation is integrated after putting Eq. 8.9

$$\int_0^L I_D \mathrm{d}x = \int_0^{V_D} WC\mu(V_G - V_T - V(x))\mathrm{d}V. \tag{8.11}$$

When integrated from the source ($x = 0$) to the drain ($x = L$), $V(x)$ changes from zero to $V_D$. Since $I_D$ does not depend on $x$, the left-hand side affords $I_D L$. Then, the integration gives

$$I_D = C\mu\frac{W}{L}\left[(V_G - V_T)V_D - \frac{1}{2}V_D^2\right]. \tag{8.12}$$

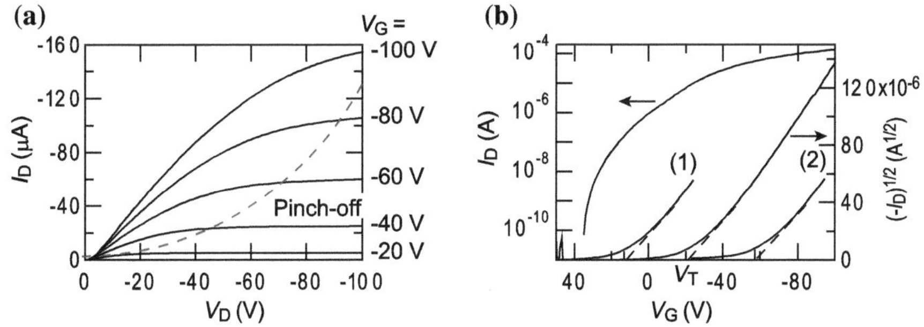


**Fig. 8.23** **a** Output, and **b** transfer characteristics in a pentacene transistor

The plot of $I_D$ versus $V_D$ for a given $V_G$ is the **output characteristics** (Fig. 8.23a). For small $V_D$, $I_D$ is a parabola of $V_D$. The slope of this parabola becomes zero at $V_D$=$V_G$ − $V_T$, above which $I_D$ is

$$I_D = \frac{1}{2} C\mu \frac{W}{L} (V_G - V_T)^2 \tag{8.13}$$

and does not increase any more with increasing $V_D$. Without the correction by $V_T$, this happens at $V_G = V_G$, where the gate-drain voltage difference is zero. Accordingly, charge is not induced by $V_G$; this is called **pinch-off**. The operation above the pinch-off ($V_D > V_G$) is called **saturation region**, in contrast to the **linear region** below the pinch-off ($V_D < V_G$). In the saturation region, $I_D$ is proportional to $V_G$ square, where the plot of $I_D$ versus $V_G$ for a given $V_D$ is called **transfer characteristics** (Fig. 8.23b). Usually, $I_D^{1/2}$ is plotted to $V_G$, and the slope gives the mobility μ according to Eq. 8.13. The intercept to the horizontal axis affords $V_T$.

$V_T$ is non zero because the carrier number is non zero even at $V_G = 0$. A transistor with finite $I_D$ even at $V_G = 0$ ((1) in Fig. 8.23b) is called normally-on type or depletion type, whereas a transistor where $I_D$ starts to increase on the right of $V_G = 0$ is called normally-off type or enhancement type ((2) in Fig. 8.23b). The case (1) happens when a lot of carriers exist without the gate voltage. The case (2) comes from a lot of deep traps, where the transistor is not turned on until all traps are filled.

## 8.10 Organic Transistors

The first organic transistors were fabricated in the 1980s [25, 26]. The performance was much improved in the 1990s mainly using oligothiophenes (Fig. 8.24a). Sexithiophene composed of $n = 6$ thiophenes is as a strong electron donor as pentacene (Fig. 7.3), but the solubility is fairly poor. So oligothiophenes with $n = 4$ to 6 are usually used. Thin films of oligothiphenes are fabricated by vacuum evaporation.

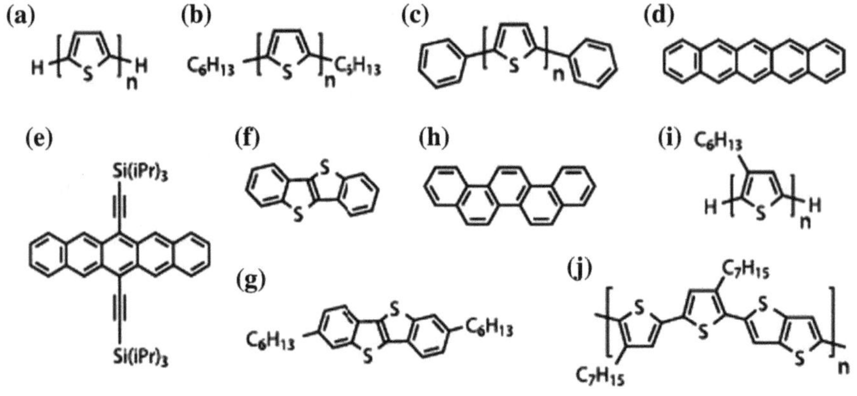


**Fig. 8.24** Organic semiconductors used in organic transistors. **a** oligothiphenes with $n = 4$: quarterthiophene, and $n = 6$: sexithiophene. **e** TIPS-pentacene: 6,13-bis(triisopropylsilyletynyl) pentacene. **f** BTBT: benzothienobenzothiophene. **i** poly(3-hexylthiophene). **j** pBTTT: poly((alkyl thieophen-2-yl) thienothiophene)

Thiophenes with terminal alkyl groups such as hexyl thiophenes show improved solubility, and the quality of the evaporated film is also improved (Fig. 8.24b). Phenyl containing oligothiophenes have increased $I_p$, and the slightly weak donor ability improves the device stability (Fig. 8.24c). There are many combinations of oligomers containing phenyl and thiophene moieties [27]. Terminal alkyl groups again improve the solubility and the thin-film properties.

Pentacene (Fig. 8.24d) is a representative transistor material showing a mobility larger than 1 $cm^2/Vs$. Pentacene HOMO is located around 4.85 eV (Fig. 7.3). Compounds with more than six phenyl rings are somewhat unstable. Pentacene is not very soluble in ordinary organic solvents, but TIPS-pentacene (Fig. 8.24e) is solution processable. Pentacene transistors are not quite stable for a long time without sealing, but BTBT (Fig. 8.24f) achieves excellent stability together with the high mobility larger than 1 $cm^2/Vs$. Alkyl BTBT transistors fabricated by the solution process attain as high mobility as 10 $cm^2/Vs$.

Phthalocyanine is a well known transistor material, and TTF derivatives represented by HMTTF (Fig. 7.16) realize high-performance transistors. A record high mobility of 40 $cm^2/Vs$ has been reported for single-crystal rubrene (Fig. 8.20b). Mobility of crystal silicon is 1500 and 450 $cm^2/Vs$ for electron and hole transport, respectively, but amorphous silicon (a-Si) achieves 1 $cm^2/Vs$, and polycrystalline silicon, made from annealed a-Si, realizes around 100 $cm^2/Vs$. Mobility of organic transistors is comparable or a little higher than that of a-Si. Since the operating speed of a transistor is determined by the mobility, high mobility is required for high-frequency circuits.

Organic LED materials are used in the amorphous form, whereas highly crystalline nature is important in high-performance organic transistors [28]. Crystalline thin films and the resulting improved performance have been achieved by such methods as elevated substrate temperatures and thin-film annealing, sometimes up

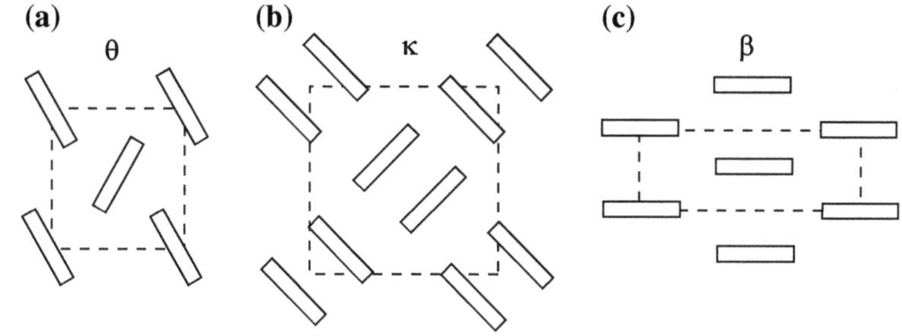


**Fig. 8.25** Representative crystal structures of organic semiconductors. **a** Herringbone. **b** Dimer. **c** Stack

to the liquid crystal phases. Oligothiphenes and pentacene have a herringbone structure (Fig. 8.25a). This is a representative structure observed in acenes, phenes like picene (Fig. 8.24f), terphenyl, and many other elongated molecules [29]. Crystalline part of polyethylene has a similar packing, so this is the most popular structure of organic compounds. This structure is analogous to the θ-phase of the BEDT-TTF salts (Sect. 7.7.4), where a uniform two-dimensional energy band is realized (Example 2.11). Since the long axes of molecules are usually perpendicular to the substrate, the conducting two-dimensional layer is parallel to the substrate. In contrast, perylene and pyrene with a little larger molecular plane have a dimer structure analogous to the κ-phase (Fig. 8.25b). Molecules like coronene with larger planes realize a stacking structure (Fig. 8.25c) similar to the β-phase of the BEDT-TTF salt. Sulfur atoms work like a large plane, so such molecules as TTF incorporating many sulfur atoms tend to have the stacking structure or the κ-like structure.

Poly(3-hexylthiophene) (P3HT, Fig. 8.24i) is a classical polymer transistor material. It is important that the hexyl groups, which improve the solubility, are attached in the same side (regioregular). P3HT is usually spin coated from the chloroform solution. pBTTT (Fig. 8.24j) attains an improved performance.

In contrast to these hole transporting-semiconductors, electron transporting organic materials are comparatively limited. Organic semiconductors are usually not used in the doped form, but the single-component material maintains either holes or electrons. Therefore, strictly speaking, organic transistor materials are classified to p-channel and n-channel, instead of p-type and n-type. This is in analogy with hole and electron transporting LED materials. Electron donating materials with relatively high HOMO are p-channel, and electron accepting materials with comparatively low LUMO are n-channel. However, it has been known that the ordinary p-channel materials such as pentacene and PPV exhibit electron transport with calcium electrodes [30, 31].

A representative n-channel material is $C_{60}$, though it is not very air stable. A thin film of $C_{60}$ is usually fabricated by vacuum evaporation, but PCBM (Fig. 8.35) is fabricated by the solution process. Fluoroalkyl othigothiophenes (Fig. 8.26a) and cyano-substituted compounds (Fig. 8.26b) show electron transporting properties.

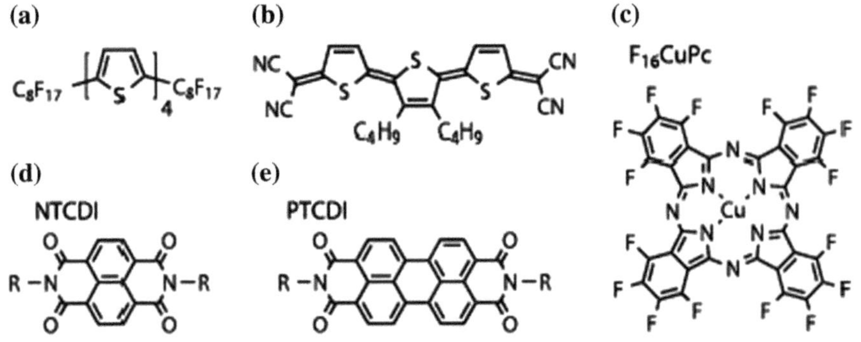


**Fig. 8.26** N-channel organic semiconductors. **d** NTCDI: naphthalene tetracarboxylic diimide. **e** PTCDI: perylene tetracarboxylic diimide

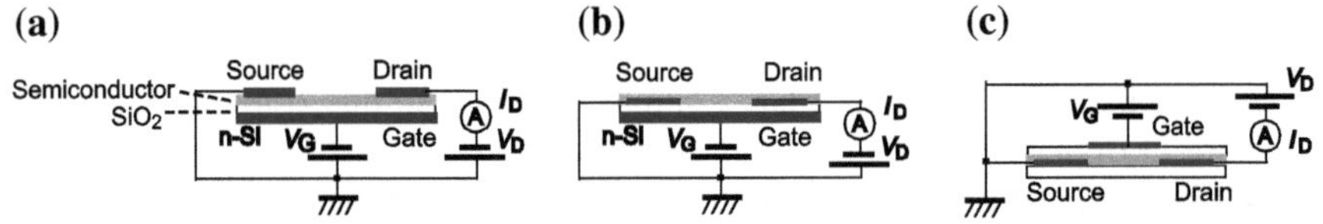


**Fig. 8.27** Structures of organic transistors. **a** Top-Contact Bootom-Gate. **b** Bootom-Contact Bottom-Gate. **c** Bottom-Contact Top-Gate

Perfluorophtharocyanine is another classical material. NTCDI and NTCDI (Fig. 8.26d, e) with various substituents are an important group n-channel materials.

Organic transistors are most easily fabricated on a doped silicon wafer with several hundred nm of an $SiO_2$ insulating layer (Fig. 8.27). Here, the high conducting silicon is used as the gate electrode. On the substrate, about 100 nm of organic semiconductor film is fabricated by vacuum evaporation or spin coating. A too thick film reduces the performance. Finally, source and drain electrodes are constructed by evaporating gold. Such a device structure is called top-contact bottom-gate type (Fig. 8.27a).

Bottom-contact transistors (Fig. 8.27b), in which the organic semiconductors are fabricated on the source and drains electrodes, usually exhibit much reduced performance in comparison with the top-contact transistors. This is because the interfacial potential $\Delta$ increases the injection barrier (Fig. 8.16). Organic semiconductor molecules are usually standing perpendicular to the hydrophobic substrate in the edge-on manner, and the conducting plane is parallel to the substrate. However, $\pi$-molecules attach on a hydrophilic surface like gold in the face-on manner. Then, the most conducting direction is perpendicular to the substrate, to reduce the lateral carrier transport. This is the second reason that the bottom-contact transistors show poor performance. Top-contact transistors maintain the desirable molecular orientation below the contact part, and do not give rise to the interfacial potential. Performance of bottom-contact transistors is improved by thiol treatment

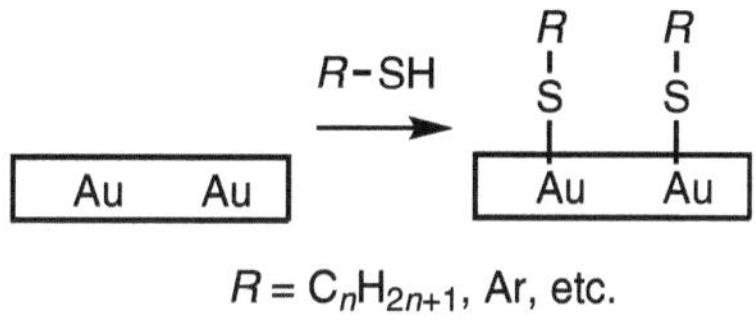


**Fig. 8.28** Thiol treatment of gold electrodes

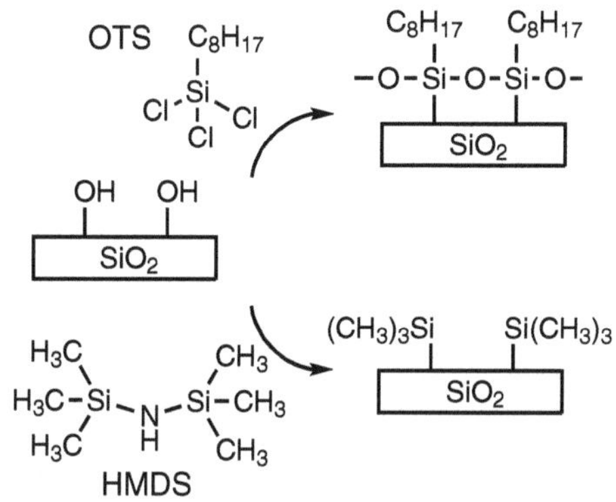


**Fig. 8.29** Self-assembled monolayer (SAM). OTS: octyltrichlorosilane, HMDS: hexamethyldisilazane

of the gold electrodes (Fig. 8.28). Other remedies are replacement of the gold electrodes by organic conducting polymers like PEDOT: PSS, organic conductors like (TTF)(TCNQ), or carbon.

Performance of organic transistors is susceptible to the gate interface. $SiO_2$ surface has a lot of OH groups, which are origin of traps, and largely reduce the transistor performance. To avoid this, the $SiO_2$ surface is treated by a **self-assembled monolayer** (SAM, Fig. 8.29). The SAM treatment not only reduces the traps but also improves the molecular alignment. Thin films of organic semiconductors are composed of μm order of crystalline domains. The domain size and alignment is improved by the SAM treatment or the annealing. The transistor performance, and particularly the threshold voltage are largely influenced by the SAM. If the SAM contains electron donating groups such as $-NH_2$, charge transfer occurs, and $V_T$ of a p-channel transistor shifts to the off side. In contrast, an electron withdrawing group like fluorine in the SAM moves $V_T$ to the on side [32].

In polycrystalline thin-film transistors, most carriers are trapped, and the transport is achieved by the thermal excitation according to the multiple trapping and release mechanism. The resulting mobility follows the thermally activated type.

$$\mu_{eff} = \mu_0 \exp\left(-\frac{E_a}{k_B T}\right)$$

Thus, an inverse temperature plot of logarithmical mobility affords approximately a straight line. $E_a$ depends on the trap depth, and changes depending on the thin film quality and the SAM treatment. $\mu_0$ is the band mobility.

Single-crystal transistors are made by putting a crystal of less than 1 μm thickness on the substrate [33]. In some single-crystal transistors, the mobility increases with lowering the temperature, and the band transport is realized. However, the band transport is observed only down to 200 K, and hopping transport occurs at lower temperatures, in contrast with the entirely band-like transport observed in TOF measurement based on high-quality crystals (Fig. 8.3). Recently, the temperature range of the band transport is extended below 100 K [34–36]. Band transport potentially shows anisotropy depending on the crystal directions, but anisotropy is not observed even in single-crystal transistors when the hopping transport is realized.

Transistor mobility is theoretically calculated on the basis of the **Marcus theory** [37–40].

$$\mu = \frac{ea^2}{k_B T} k_{ET} \quad \text{where } k_{ET} = \frac{2\pi}{\hbar} t^2 \sqrt{\frac{1}{4\pi\lambda k_B T}} \exp\left[-\frac{\lambda}{4k_B T}\right]$$

Here, $k_{ET}$ is the rate of electron transfer, and $a$ is the lattice spacing. $t$ is the transfer integral and λ is the reorganization energy (Fig. 8.30a). (1) is the energy of $D^+$ keeping the structure of the neutral D. The energy difference from the geometry optimized $D^+$ (2) is $\lambda^{(1)}$. Similarly, the energy of the optimized D (4) is subtracted from the energy of D keeping the structure of $D^+$ (3), to afford $\lambda^{(2)}$. When charge transfers from $D^+ - D$ to $D - D^+$ occurs, the "directly" excited state ((1) + (3)) has a higher energy than the structurally relaxed state ((2) + (4)). The difference is the reorganization energy $\lambda = \lambda^{(1)} + \lambda^{(2)}$. The charge transfer is regarded as a chemical reaction with an activation energy $G^*$ (Fig. 8.30b). If the potential energy of the charge transfer is approximately parabolic, $G^*$ is one quarter of λ. Then, the rate constant $k_{ET}$ is given by the activation energy λ/4, and the mobility is obtained from the above formula. λ is typically 0.1 ~ 0.3 eV.

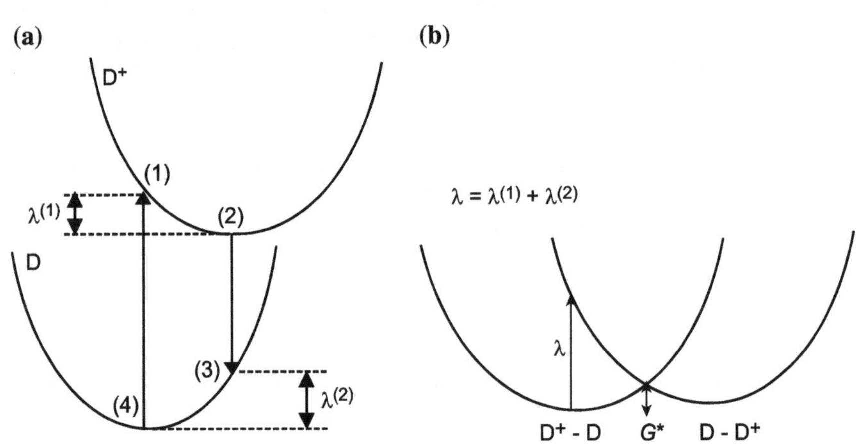


**Fig. 8.30** Reorganization energy λ in the Marcus theory

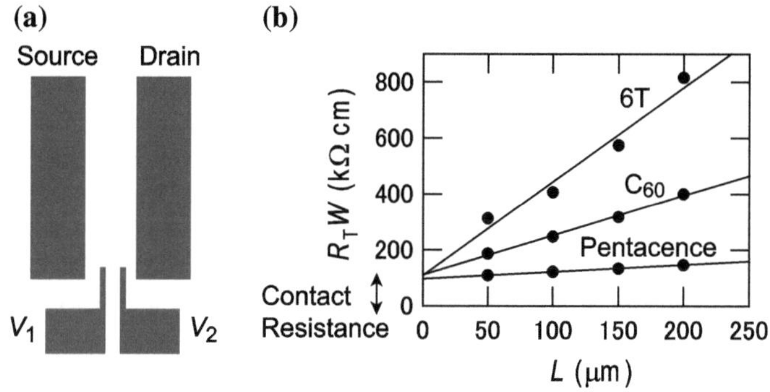


**Fig. 8.31** **a** Four-probe pattern. **b** Estimation of the contact resistance by the transfer-line method [41]

Since $\sigma = ne\mu = Q\mu$, the mobility represents the conductivity per charge. Site-dependent potential in the transistor channel has been investigated by the Kelvin probe method, which reveals that a considerable voltage drop occurs not in the channel region but at the contact interface. The ordinary transistor corresponds to a two-probe measurement, but the **contact resistance** is eliminated by the four-probe measurement (Fig. 8.31a). Another simple method to estimate the contact resistance is the transfer-line method (Fig. 8.31b); resistance of transistors with various $L$ fabricated on the same substrate is plotted, and the extrapolation to $L = 0$ affords the contact resistance. Here, the slope gives the channel resistance free from the contact resistance. The contact resistance is usually represented by the unit of Ω cm, by multiplying the channel width $W$. For instance, the contact resistance of an ordinary pentacene transistor is around 1 kΩ cm. Figure 8.31b indicates that for $L < 50$ mm, the contact resistance is larger than the channel resistance, and the apparent device performance decreases for short channel transistors. From this, the short channel limit is several μm. Using special fabrication techniques, organic transistors with a several hundred nm channel length have been reported. In contrast, the present limit of silicon transistors is as small as several ten nm.

Field-effect transistors are basically unipolar devices with a single kind of carriers. The carrier polarity is unambiguously defined by the polarity of the gate voltage. However, all organic molecules have HOMO and LUMO, and even pentacene shows electron transport with calcium electrodes [30, 31]. If the reduction potential of a reducing reagent is negative (*vs*. SHE), this reagent potentially reduces water to $H_2$ (Fig. 8.32a) as evident from the definition of SHE. Similarly, the oxidation potential of water to $O_2$ is 1.23 V (*vs*. SHE), and an oxidizing reagent with more positive potential oxidizes water [42]. Then, water is stable between 0 and 1.23 V. This is true at pH = 0, but from the Nernst equation

$$E = E^0 + \frac{RT}{F}\ln[\mathrm{H}^+] = E^0 - \frac{RT}{F}\mathrm{pH}$$

the **stability field of water** shifts to −0.41 ~ 0.82 V at pH = 7. The actual electrode reaction does not occur until the electrode potential exceeds these values considerably. The typical overpotential is 0.6 V, and the stability field is extended to −1.05 ~ 1.4 V. Based on the conversion of the redox potentials to the energy levels (Sect. 7.2), this corresponds to the energy levels, 3.15 ~ 5.6 eV. This is in agreement with the observation on various organic transistor materials; electron transport is realized when the LUMO is located below 3.15 eV, whereas, hole transport is observed when the HOMO is located above 5.6 eV [43]. This stability region is also indicated in Fig. 7.5. When the LUMO is located below 4.0 eV, the n-channel transistor is operated even in air [44]. A strong donor whose HOMO energy is above 5.0 eV tends to be oxidized automatically, and the transistor performance tends to be degraded in air. Pentacene and TTF derivatives belong to this category. Then, the HOMO level below 5.0 eV is desirable to achieve a stable p-channel transistor.

In this connection, we shall mention the energy levels of photocatalyst (Fig. 8.32b). In $TiO_2$, the photoexcited electron in the conduction band reduces $H^+$ to $H_2$, and the hole in the valence band oxidizes water to $O_2$. Then, water is "electrolyzed" by the light energy. For this reaction, the conduction and valence bands have to be located out of the stability region of water. In contrast, the energy levels of organic semiconductors have to be within the stability region of water for the stable operation.

Transistors in which both electron and hole transports are capable, are called **ambipolar transistors**, because a bipolar transistor means a transistor consisting of PNP or NPN junctions. In an ambipolar transistor, electron transport occurs for

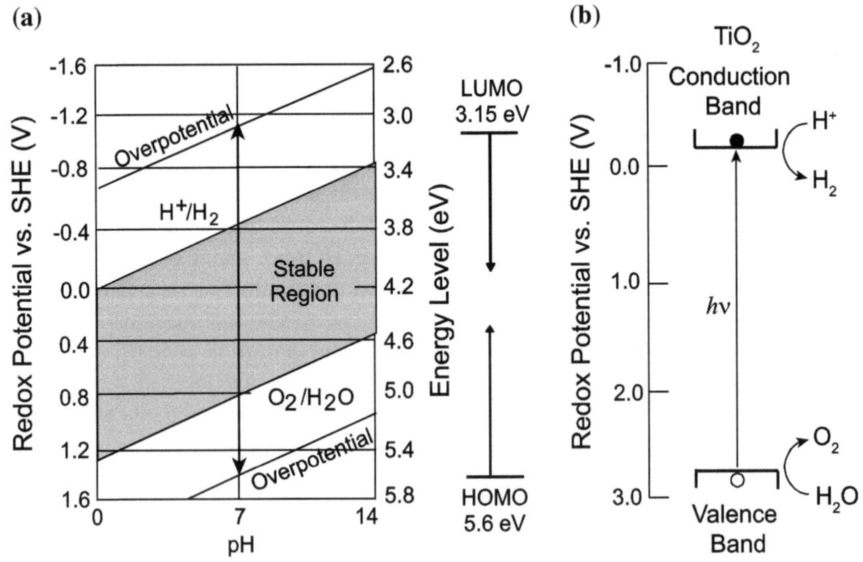


**Fig. 8.32** **a** Stability field of water. **b** Energy levels of a photocatalyst, $TiO_2$

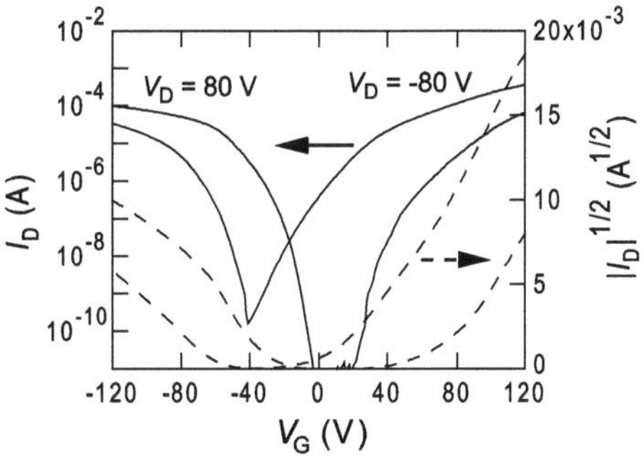


**Fig. 8.33** Transfer characteristics of an ambipolar transistor based on diphenylindigo fabricated on a thin tetratetraconetane layer [48]

positive $V_G$, and hole transport occurs for negative $V_G$ (Fig. 8.33). In principle, ambipolar transport is expected when the HOMO is located above 5.6 eV, and the LUMO is located below 3.15 eV. However, ambipolar characteristics have been usually realized only after such special treatments as annealing. In this way, ambipolar organic transistors have been reported on the basis of dicyanomethylene terthiophene (Fig. 8.26b), pentacene, $C_{60}$, and PCBM. Recently, excellent ambiploar transistors have been attained using donor–acceptor polymers, which are copolymers of a donor moiety like oligothiophenes and an acceptor unit like diketopyrrolopyrrole (Fig. 8.34) [45, 46]. Here, the HOMO-LUMO gap is determined by the donor HOMO and the acceptor LUMO [46]. The resulting HOMO-LUMO gap is as small as 1.3 eV, and the absorption edge is 980 nm. Recently, even small-molecule ambipolar transistors have been realized relatively easily by evaporating a thin layer of inert tetratetraconetane $C_{44}H_{90}$ on $SiO_2$ as a passivation layer (Fig. 8.33) [48].

In an ambipolar transistor, electrons are injected from the source, and holes are injected from the drain, and the electrons encounter the holes somewhere in the channel [43]. In a **light-emitting transistor**, light is emitted when the hole-electron pair recombines. In this true ambipolar region, the current is given by a sum of the electron current and the hole current.

$$I_D = \frac{W\mu_e C}{2L}(V_G - V_T)^2 + \frac{W\mu_h C}{2L}\left(V_D - V_G + V_T'\right)^2$$

This true ambipolar region appears only when $V_D > V_T - V_T'$, where $V_T$ is the electron threshold voltage and $V_T'$ is the hole threshold voltage [48]. It is more usual that an off region, in which either the electron transport or the hole transport does not occur, appears between $V_T$ and $V_T'$ [48].

300 nm of $SiO_2$ realizes a capacitance of 13 nF/cm$^2$. From $Q = CV$, a gate voltage of $V_G$ = 100 V affords $Q = 10^{-6}$ C/cm$^2$, and the carrier density is around $10^{13}$ /cm$^2$. Then, one carrier exists per 1000 Å$^2$. Using the typical area of an organic semiconductor, 50 Å$^2$, one carrier is located in twenty molecules. This is contrasting to charge-transfer salts, where the typical carrier density is $2 \times 10^{14}$ /cm$^2$ (Example 3.4). For high-density field-effect doping, large $C$ is necessary, which is

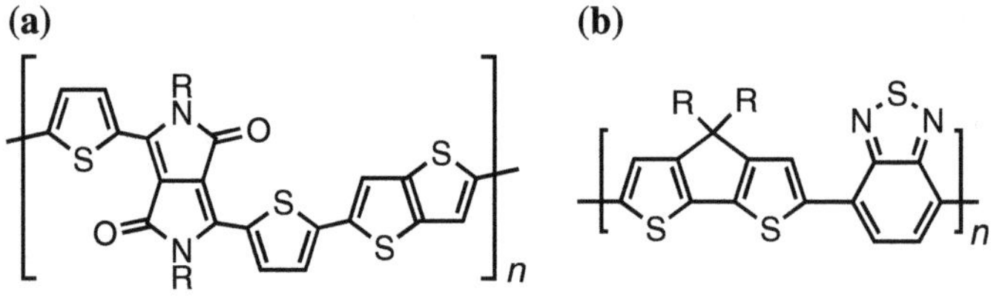


**Fig. 8.34** Donor–acceptor polymers consisting of (**a**) thiophen and diketopyrrolopyrrole, and (**b**) thiophene and benzothiazole [45, 46]

achieved by using a gate material with large ε. For this purpose, electrical double layer transistors with gate insulators consisting of polymer ionic conductors or ionic liquid have been investigated.

In a display, the signal voltage is applied successively between the $x$ and $y$ scanning lines. The device at the crossing point emits light, but in this method the device turns on in a very short period. Then, the scanned signal is memorized by a transistor-capacitor circuit to keep the device turned on until the next signal. Transistors are used in these active-matrix displays. Usual liquid crystal displays use a-Si or poly-Si transistors, but organic transistors are necessary for flexible displays fabricated on a plastic substrate. Organic semiconductors are suitable for applications to large area, flexible, and solution-process electronic devices. Organic transistors are constructed using a printing method such as ink-jet printing. Integrated circuits are printed in such a printed electronics.

## 8.11 Organic Solar Cells

In an organic LED, electric energy is converted to light. In an **organic solar cell** (**photovoltaics**), light is converted to electronic energy through the opposite process. Thus, an organic solar cell has a similar structure to an organic LED (Fig. 8.35a). Light generates an excited molecule, called exciton (Fig. 8.35b). Then, a charge-transfer state composed of $D^+$ and $A^-$ is formed (Fig. 8.35c). The positive charge is transferred between the donors, and the negative charge is transferred between the acceptors. Finally, the separated charges reach the cathode and the anode, respectively, and electric energy is taken out [49].

The charge separation occurs after the photoexcitation, but diffusion length of an organic exciton is as small as 10 nm. If the device has a layered structure similar to an organic LED, the charge separation occurs in a very narrow interfacial region between the hole and electron transporting layers. Here, the light is not efficiently absorbed in the very thin layer, and the effective charge separation is also difficult. Then, the donors and the acceptors are scrambled into nano structures with the size of about 10 nm, which is called **bulk heterojunction** (Fig. 8.35d). The whole

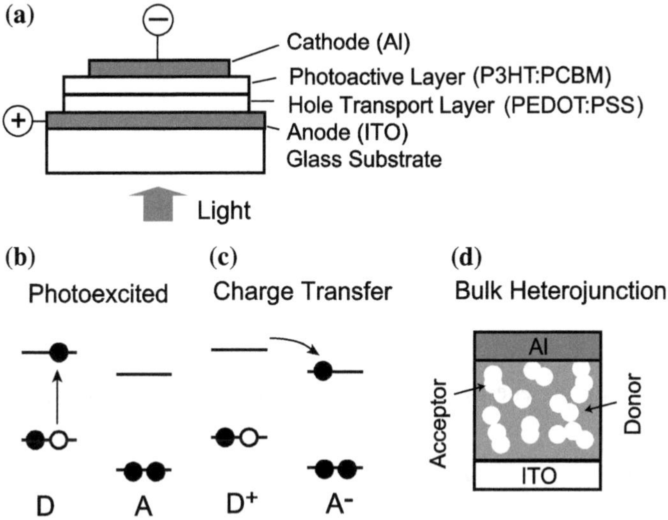


**Fig. 8.35** Organic solar cell

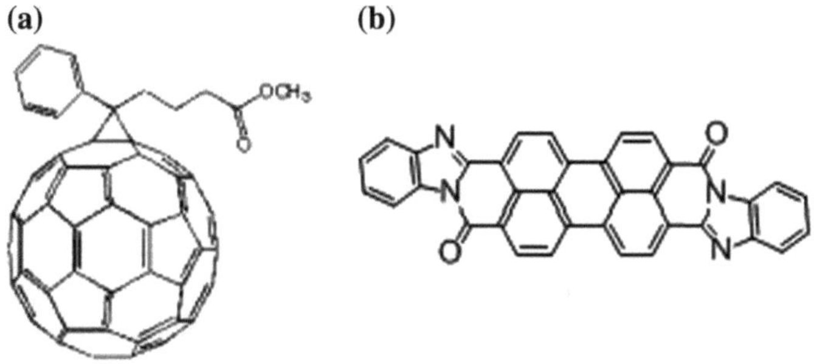


**Fig. 8.36** **a** PCBM, and **b** PTCBI *PCBM* Phenyl-$C_{61}$-butylic acid methyl ester, *PTCBI* Perylene tetracalboxylic bis(benzimidazole)

photoelectric conversion layer is close to the junctions, and the separated carriers are transported in the donor and acceptor parts to reach the electrodes.

A typical solar cell depicted in Fig. 8.35a consists of a donor, P3HT and an acceptor, PCBM. PCBM (Fig. 8.36a) is a soluble material derived from $C_{60}$ by chemical modification. Other frequently used donors are phthalocyanines, diamines, and MEH-PPV, and acceptors are $C_{60}$, PTCDI, and PTCBI (Fig. 8.36b).

Figure 8.37 is a characteristics of an organic solar cell. The generated voltage for an open output is called **open circuit voltage**, $V_{oc}$, and output current for a connected output is called **short circuit current**, $V_{oc}$. A solar cell shown in Fig. 8.35a affords $V_{oc} = 0.6$ V. The voltage–current characteristics is depicted in Fig. 8.37 for varing the load resistance. The maximum power is attained at an intermediate of $J_{max}$ and $V_{max}$, where

Fig. 8.37 Characteristics of an organic solar cell

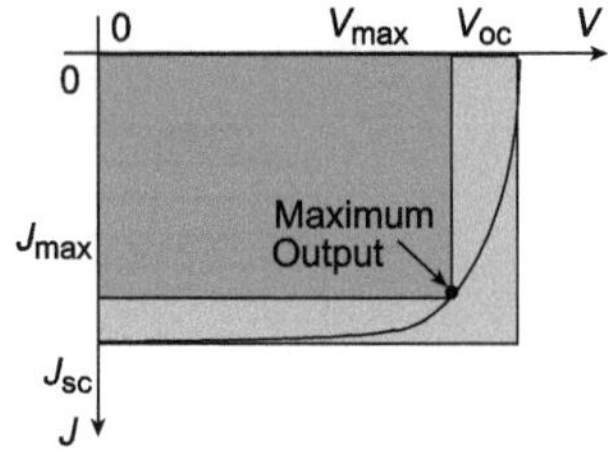


$$\mathrm{FF} = J_{\max} \times V_{\max}/(J_{\mathrm{sc}} \times V_{\mathrm{oc}})$$

is called **fill factor**. Fill factor does not exceed one, and the typical values are 0.4 ~ 0.6. Using the energy of the irradiated light, $P_{\mathrm{ino}}$, **power conversion efficiency** (PCE)

$$\mathrm{PCE} = J_{\max} \times V_{\max}/P_{\mathrm{inc}}(\%)$$

is defined. This is a ratio of the energy, but the generated electron number is divided by the input or absorbed photon number to give the **external** and **internal quantum efficiency** (EQE and IQE). EQE is obtained by multiplying the absorption efficiency to IQE.

EQE for a monochromatic light is called incident photon to current efficiency (IPCE), which is represented by

$$\frac{J_{\mathrm{SC}}}{P}\frac{1240}{\lambda}$$

where $P$ is the incident light intensity (W/cm$^2$) and λ is the wavelength (nm). 1240/λ affords the light energy in the eV unit (Sect. 1.2), by which $P$ is divided to give the incident photon number in the A/cm$^2$ unit. Since $J_{\mathrm{SC}}$ (A/cm$^2$) is the electron number, the ratio affords IPCE.

The prototypical organic solar cell, which was composed by copper phthalocyanine and PTCBI, was reported almost at the same time as the prototypical organic LED [50]. In the 1990s, bulk heterojunction was proposed, and various materials such as PCBM were developed. To improve the energy conversion efficiency, it is important to absorb long wavelength light. For instance, the HOMO-LUMO gap of P3HT is 1.9 eV, and the corresponding absorption edge is 650 nm, so the longer wavelength light is not absorbed. Since the exciton binding energy of organic semiconductors is as large as 0.3 ~ 1.0 eV, ordinarily free carriers are not efficiently produced from the photoexcitation. Then, a light with energy considerably higher than the absorption edge is required for the charge separation. Figure 8.34 shows examples of donor-acceptor polymers used in organic solar cells. A narrow HOMO-LUMO gap and high energy conversion efficiency have been achieved using such donor–acceptor polymers.

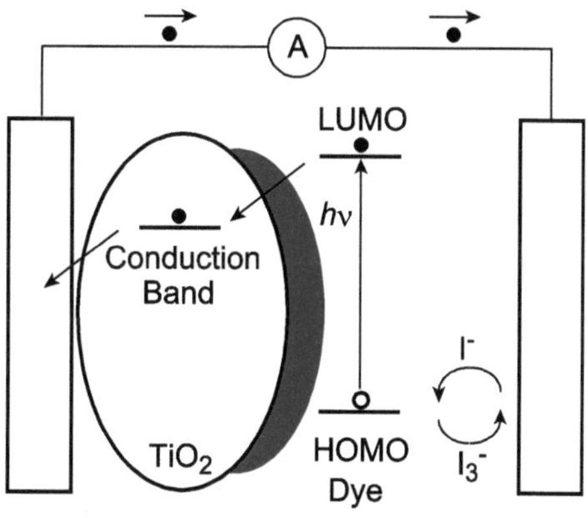


**Fig. 8.38** Dye sensitized solar cell

There is another kind of solar cell called **dye-sensitized solar cell** (Grätzel cell, Fig. 8.38) [51, 52]. In a photocatalyst (Fig. 8.32b), $TiO_2$ is photoexcited only by ultraviolet light because the energy gap of $TiO_2$ is large. This is also obvious because $TiO_2$ is a colorless material used as a white pigment. Then, an organic dye is attached on $TiO_2$ particles. The dye absorbs visible light, and the photoexcited electron is transferred to $TiO_2$, and then to an electrode. The dye molecule is recovered by the reduction of $I^-$, which accepts an electron from another electrode. Accordingly, the photoexcited energy is taken out as electric energy.

Efficiency of a silicon solar cell is 15 % for thin films and 25 % for crystals. More than 30 % has been achieved in compound semiconductors. Efficiency is 5 ∼ 10 % in organic solar cells, and more than 10 % in dye-sensitized solar cells. Recently, about 20 % has been achieved in dye-sensitized photovoltaics using perovskite, $CH_3NH_3$ $PI_3$ as a dye [53]. The industrially produced solar cells are mainly based on silicon crystals.

## 8.12 Synthesis of Conducting Polymers

**Polyacetylene** is a prototypical conducting polymer (Fig. 8.39). Polyacetylene is produced as a film with black luster by introducing acetylene gas into a Ziegler–Natta catalyst.

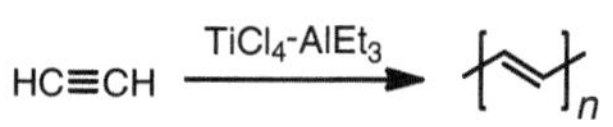


Polyacetylene is doped either in p-type (with $AsF_5$ or $Br_2$) or n-type (with alkali metals), and the doped polyacetylene shows high conductivity. In particular, the extended sample shows as high conductivity as $10^4$ S/cm together with the almost flat temperature dependence down to liquid helium temperatures. However,

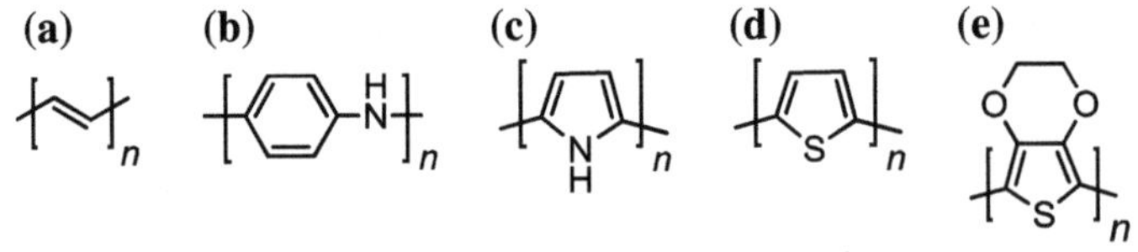


**Fig. 8.39** Conducting polymers. **a** Polyacetylene. **b** Polyaniline. **c** Polypyrrole. **d** Polythiophene. **e** PEDOT

polyacetylene is not used in practical applications because it is not very stable in ambient conditions.

**Polyaniline**, **polypyrrole**, and **polythiophene** (Fig. 8.39) are obtained by the electrochemical oxidation of the monomer solutions.

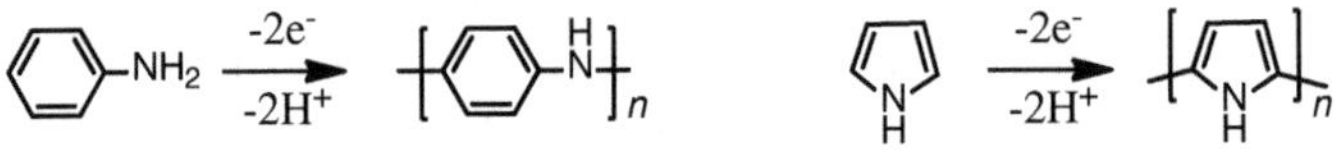


Polyaniline and polypyrrole are widely used as antistatic coating materials. A large amount of polyaniline is used in electrolytic capacitors. PEDOT: PSS is a kind of polythiophene, which is investigated as the replacement of ITO and transparent conducting films on a flexible substrate.

**Experiment 8.1** Electrochemical polymerization of conducting polymers

1. To 30 mL of 0.2 M nitric acid in a 100 mL beaker, 0.2 mL pyrrole is dissolved.
2. ITO and stainless electrodes with approximately $4 \times 1$ cm$^2$ size are connected by clips to, respectively, positive and negative terminals of a 3 V battery. The electrodes are dipped in the above solution (Fig. 8.40).
3. Electrochemically polymerized polypyrrole deposits immediately on the ITO. This polypyrrole is $NO_3^-$ doped.
4. The electrodes are moved into another beaker containing 0.2 M sodium nitrate solution after reversing the battery polarity. Then, the polypyrrole film is decolorized by dedoping (**electrochromism**).
5. Electrodes in (2) are connected to a tester instead of a battery, and the voltage is measured. The polypyrrole film works as a plastic battery.
6. The same experiments are repeated for aniline instead of pyrrole.

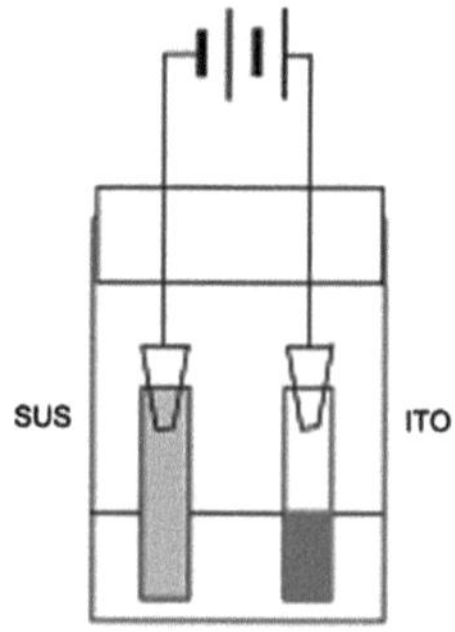


**Fig. 8.40** Electrochemical polymerization [58]

These polymers are chemically prepared by oxidizing the monomer by $FeCl_3$.

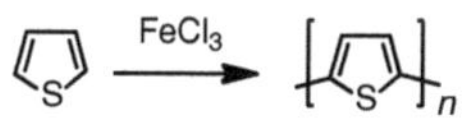


Polymerization is also achieved by the coupling reaction [54].

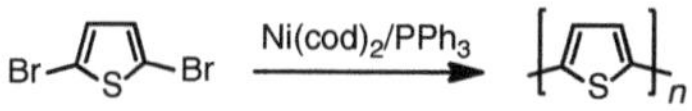


cod = cyclooctadiene

Regioregular P3HT is synthesized as follows [55, 56]:

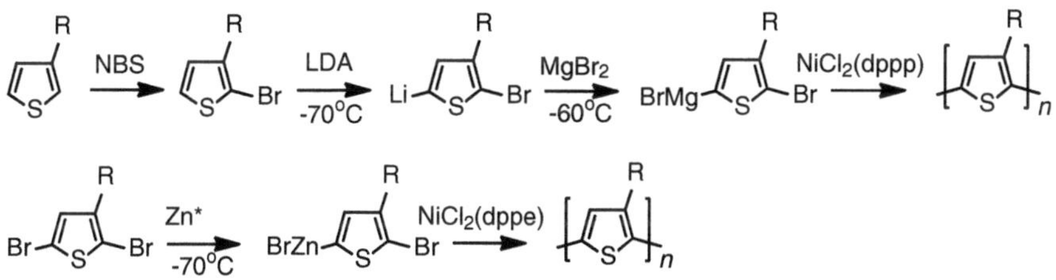


here dppp is diphenylphosphinopropane, dppe is diphenylphosphinoethane, and Zn* is Reike zinc.

Oligothiophens are prepared by repeating the reactions in Fig. 8.41. Oligomers including phenyl rings are similarly prepared. Here, the key is the coupling reaction. Ni(0) means the use of a catalyst including zerovalent nickel. Figure 8.41 shows the Tamao coupling using the Grignard reagent. Other frequently used coupling reactions are the Stille coupling.

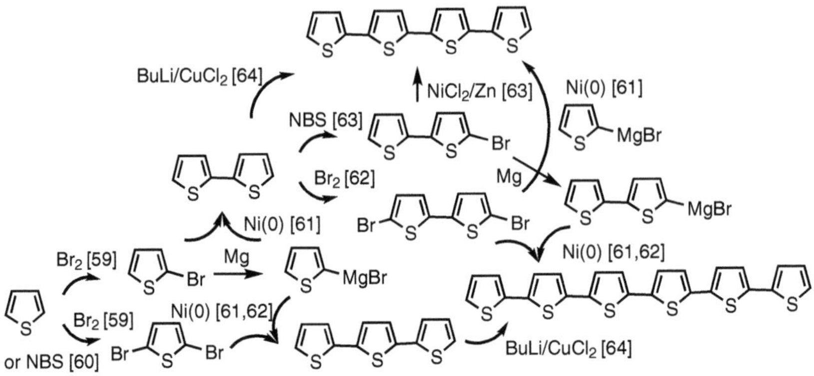


**Fig. 8.41** Synthesis of oligothiophenes

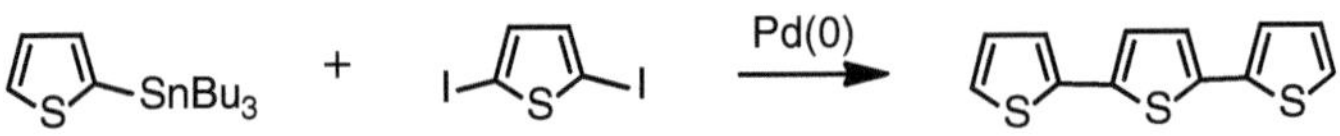


and the Suzuki coupling

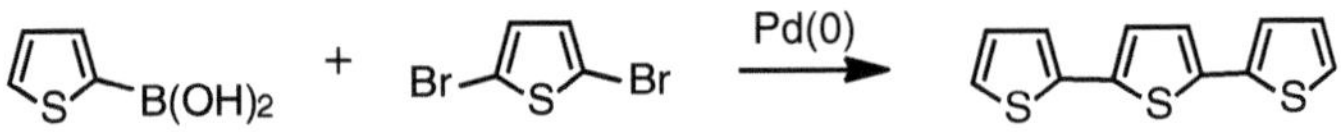


which proceed in a mild condition [57].

PCBM is prepared as follows:

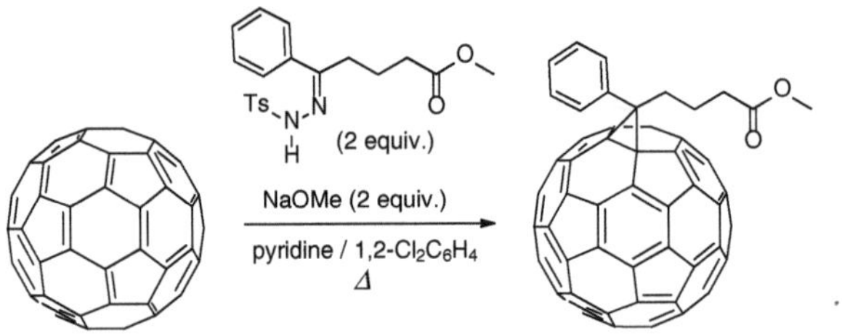


## References

1. H. Meier, *Organic Semiconductors* (Verlag Chemie, Weinheim, 1974)
2. M. Pope, C.E. Swenberg, *Electronic Processes in Organic Crystals* (Clarendon Press, Oxford, 1982)
3. D. Bimberg, I. Eisele, W. Fuchs, H. Kahlert, N. Karl, *Landort-Börnstein Numerical Data and Functional Relationships in Science and Technology*, New Series Group III 17a-i (1985)
4. W. Warta, N. Karl, Phys. Rev. B **32**, 1172 (1985)

5. Y. Maruyama, Mol. Cryst. Liq. Cryst. **171**, 287 (1989)
6. M.A. Lampert, P. Mark, *Current Injection in Solids* (Academic Press, New York, 1970), p. p44
7. J. Frenkel, Phys. Rev. **54**, 647 (1938)
8. Y. Takahide, T. Konoike, K. Enomoto, M. Nishimura, T. Terashima, S. Uji, H.M. Yamamoto, Phys. Rev. Lett. **96**, 136602 (2006)
9. S.M. Sze, *Physics of Semiconductor Devices* (Wiley, Hoboken, 1981)
10. N. Ueno, S. Kera, Prog. Surf. Sci. **83**, 490 (2008)
11. K. Seki, N. Ueno, U.O. Karlsson, R. Engelhardt, E.-E. Koch, Chem. Phys. **105**, 247 (1986)
12. S. Machida, Y. Nakayama, S. Duhm, Q. Xin, A. Funakoshi, N. Ogawa, S. Kera, N. Ueno, H. Ishii, Phys. Rev. Lett. **104**, 156401 (2010)
13. H. Ishii, K. Sugiyama, E. Ito, K. Seki, Adv. Mater. **11**, 605 (1999)
14. H. Ishii, K. Sugiyama, D. Yoshimura, E. Ito, Y. Ouchi, K. Seki, IEEE J. Selected Topics in Quant. Electron. **4**, 24 (1998)
15. H. Ishii, K. Seki, IEEE Trans. Electr. Device **44**, 1295 (1997)
16. I.G. Hill, A. Kahn, SPIE **3476**, 168 (1998)
17. S. Braun, W.R. Salaneck, M. Halman, Adv. Mater. **21**, 1450 (2009)
18. C.W. Tang, S.A. VanSlyke, Appl. Phys. Lett. **51**, 913 (1987)
19. J. Shinar (ed.), *Organic Light-Emitting Devices A Survey* (Springer, Heidelberg, 2010)
20. A. Buckley (ed.), *Organic Light-Emitting Diodes (OLED) Materials, Devices and Applications*. (Woodhead, 2013)
21. J.H. Burroughes, D.D.C. Bradley, A.R. Brown, R.N. Marks, K. Mackay, R.H. Friend, P.L. Burns, A.B. Holmes, Nature **347**, 539 (1990)
22. I.B. Berlman, *Handbook of Fluorescence Spectra of Aromatic Molecules* (Academic Press, Cambridge, 1971)
23. H. Mattoussi, H. Murata, C.D. Merritt, Y. Iizumu, J. Kido, Z.H. Kafafi, J. Appl. Phys. **86**, 2642 (1999)
24. C. Adachi, M.A. Baldo, S.R. Forrest, M.E. Thompson, Appl. Phys. Lett. **77**, 904 (2000)
25. F. Ebisawa, T. Kurokawa, S. Nara, J. Appl. Phys. **54**, 3255 (1983)
26. K. Kudo, M. Yamashita, T. Moriizumi, Jpn. J. Appl. Phys. **23**, 130 (1984)
27. I.F. Perepichka, D.F. Perepichka (ed.), *Handbook of Thiophene-Based Materials* (Wiley, Hoboken, 2009)
28. Z. Bao, J. Locklin (ed.), *Organic Field-Effect Transistors* (CRC, 2007)
29. G.R. Desiraju, A. Gavezzotti, Acta Cryst. B **45**, 473 (1989)
30. L.-L. Chua, J. Zaumseil, J.-F. Chang, E.C.-W. Ou, P.K.-H. Ho, H. Sirringhaus, R.H. Friend, Nature **434**, 194 (2005)
31. T. Yasuda, T. Goto, K. Fujita, T. Tsutsui, Mol. Cryst. Liq. Cryst. **444**, 219 (2006)
32. S. Kobayashi, T. Nishikawa, T. Takenobu, S. Mori, T. Shimada, T. Mitani, H. Shimotani, N. Yoshimoto, S. Ogawa, Y. Iwasa, Nature Mater. **3**, 317 (2004)
33. T. Hasegawa, J. Takeya, Sci. Tech. Adv. Mater. **10**, 024314 (2009)
34. C. Liu, T. Minari, X. Lu, A. Kumatani, K. Takimiya, K. Tsukagoshi, Adv. Mater. **23**, 523 (2011)
35. A. Kumatani, C. Liu, Y. Li, P. Darmawan, K. Takimiya, T. Minari, K. Tsukagoshi, Sci. Rep. **2**, 393 (2012)
36. J. Cho, T. Higashino, T. Mori, Appl. Phys. Lett. **106**, 193303 (2015)
37. S. Kera, H. Yamane, N. Ueno, Prog. Surf. Sci. **84**, 135 (2009)
38. J.-L. Brédas, D. Beljonne, V. Coropceanu, J. Cornil, Chem. Rev. **104**, 4971 (2004)
39. V. Coropceanu, J. Cornil, D.A.S. Filho, Y. Olivier, R. Silbey, J.-L. Brédas, Chem. Rev. **107**, 926 (2007)
40. E.-G. Kim, V. Coropceanu, N.E. Gruhn, R.S. Sánchez-Carrera, R. Snoeberger, A.J. Matzger, J.-L. Brédas, J. Am. Chem. Soc. **129**, 13072 (2007)
41. T. Kadoya, D. de Caro, K. Jacob, C. Faulmann, L. Valade, T. Mori, J. Mater. Chem. **21**, 18421 (2011)

42. P. Atkins, T. Overton, J. Rourke, M. Weller, F. Armstrong, *Inorganic Chemistry*, Oxford (2006)
43. M.L. Tang, A.D. Reicharrdt, P. Wei, Z. Bao, J. Am. Chem. Soc. **131**, 5264 (2009)
44. J. Zaumseil, H. Sirringhaus, Chem. Rev. **107**, 1296 (2007)
45. J.D. Yuen, F. Wudl, Energy Environ. Sci. **6**, 392 (2013)
46. Y. Zhao, Y. Guo, Y. Liu, Adv. Mater. **25**, 5372 (2013)
47. S.C. Rasmussen, M. Pomerants, in *Conjugated Polymers* ed. by T.A. Skotheim, J.R. Reynolds (CRC, 2007), p. 12
48. O. Pitayatanakul, T. Higashino, T. Kadoya, M. Tanaka, H. Kojima, M. Ashizawa, T. Kawamoto, H. Matsumoto, K. Ishikawa, T. Mori, J. Mater. Chem. C **2**, 9311 (2014)
49. C. Brabec, V. Dyakonov, U. Scherf, (ed.), *Organic Photovoltaics* (Wiley, Hoboken, 2009)
50. C.W. Tang, Appl. Phys. Lett. **48**, 183 (1986)
51. M. Grätzel, J. Photochem. Photobio. C **4**, 145 (2003)
52. A. Hagfeldt, G. Boschloo, L. Sun, L. Kloo, H. Pettersson, Chem. Rev. **110**, 6595 (2010)
53. T. Miyasaka, Chem. Lett. **44**, 720 (2015)
54. T. Yamamoto, Bull. Chem. Soc. Jpn **72**, 621 (1999)
55. R.D. McCullough, Adv. Mater. **10**, 93 (1998)
56. P. C. Ewbank, M. C. Stefan, G. Sauvé, R. D. McCullough, in *Handbook of Thiophene-Based Materials*, ed. by I. F. Perepichka and D. F. Perepichka Wiley (2009) p. 157
57. P. Bäuerle, *Handbook of Oligo- and Polythiophenes* (Wiley, Hoboken, 1999), p. 89
58. H. Goto, H. Yoneyama, F. Togashi, R. Ohta, A. Tsujimoto, E. Kita, K. Ohshima, J. Chem. Educ. **85**, 1067 (2008)
59. M.A. Keegstra, L. Brandsma, Synthesis 890 (1988)
60. R.M. Kellogg, A.P. Schaap, E.T. Harper, H. Wynbert, J. Org. Chem. **33**, 2902 (1968)
61. K. Tamao, S. Kodama, I. Nakajima, M. Kumada, A. Minato, K. Suzuki, Tetrahedron **38**, 3347 (1982)
62. C. van Phan et al., Phosph. Sulf. Silicon **46**, 153 (1989)
63. J. Nakayama, Heterocycles **26**, 1793 (1987)
64. J. Kagan, S.K. Arora, Heterocycles **20**, 1937 (1983)

# Appendix

See Tables 1 and 2.



T. Mori, *Electronic Properties of Organic Conductors*,
DOI 10.1007/978-4-431-55264-2

**Table 1** Atomic parameters $I_p$ (eV)/$\zeta$.

| | | | | | | | |
|---|---|---|---|---|---|---|---|
| H<br>1s -13.6/1.3 | | | | | | | He |
| Li<br>2s -5.4/0.65<br>2p -3.5/0.65 | Be<br>2s -10.0/0.975<br>2p -6.0/0.975 | B<br>2s -15.2/1.3<br>2p -8.2/1.3 | C<br>2s -21.4/1.625<br>2p -11.4/1.625 | N<br>2s -26.0/1.95<br>2p -13.4/1.95 | O<br>2s -32.3/2.275<br>2p -14.8/2.275 | F<br>2s -40.0/2.425<br>2p -18.1/2.425 | Ne<br>2s -40.96/2.879<br>2p -21.56/2.879 |
| Na<br>3s -5.1/0.733<br>3p -3.0/0.733 | Mg<br>3s -9.0/0.950<br>3p -4.5/0.950 | Al<br>3s -12.3/1.17<br>3p -6.5/1.17 | Si<br>3s -17.3/1.383<br>3p -9.2/1.383 | P<br>3s -18.6/1.60<br>3p -14.0/1.60 | S<br>3s -20.0/1.817<br>3p -13.3/1.817 | Cl<br>3s -30.0/2.033<br>3p -15.0/2.033 | Ar<br>3s -27.63/2.585<br>3p -15.76/2.255 |
| K<br>4s -4.34/0.874<br>4p -2.73/0.874 | Ca<br>4s 7.0/1.2<br>4p -4.01.2 | Ga | Ge<br>4s -16.0/2.16<br>4p -9.0/1.85 | As<br>4s -16.22/2.23<br>4p -12.16/1.89 | Se<br>4s -20.5/2.44<br>4p -14.4/2.07 | Br<br>4s -22.07/2.588<br>4p -13.1/2.131 | Kr<br>4s -24.36/2.829<br>4p -13.99/2.442 |
| Rb<br>5s -4.18/0.997<br>5p -2.6/0.997 | Sr<br>5s -4.93/1.21<br>5p -3.28/1.00 | In<br>5s -12.6/1.903<br>5p -6.19/1.677 | Sn | Sb<br>5s -18.8/2.323<br>5p -11.7/1.999 | Te<br>5s -20.78/2.51<br>5p -14.8/2.16 | I<br>5s -18.0/2.679<br>5p -12.7/2.322 | Xe<br>5s -21.21/2.844<br>5p -12.13/2.485 |
| Cs | Ba | Tl<br>6s -11.6/2.3<br>6p -5.0/1.6 | Pb<br>6s -15.7/2.35<br>6p -8.0/2.06 | Bi<br>6s-18.67/2.56<br>6p -7.81/2.07 | | | |

| | | | | | | | | | |
|---|---|---|---|---|---|---|---|---|---|
| Sc | Ti<br>4s -8.97/1.075<br>4p -5.44/0.6755<br>3d -10.81/<br>4.55(0.4206)<br>1.4(0.7839) | V<br>4s -8.81/1.3<br>4p -5.52/0.875<br>3d -11.0/<br>4.7(0.4755)<br>1.7(0.7052) | Cr<br>4s -8.66/1.7<br>4p -5.24/1.7<br>3d -11.22/<br>4.95(0.4876)<br>1.6(0.7205) | Mn<br>4s -8.63/1.8<br>4p-5.06/1.8<br>3d-11.59/<br>5.15(0.532)<br>1.9(0.649) | Fe<br>4s -9.17/1.9<br>4p -5.37/1.9<br>3d -12.7/<br>5.35(0.5366)<br>1.8(0.6678) | Co<br>4s -9.21/2.0<br>4p -5.29/2.0<br>3d -13.18/<br>5.55(0.5679)<br>2.1(0.6059) | Ni<br>4s -8.86/2.1<br>4p -4.9/2.1<br>3d-12.99/<br>5.75(0.5683)<br>2(0.6292) | Cu<br>4s -11.4/2.2<br>4p -6.06/2.2<br>3d -14.0/<br>5.95(0.5933)<br>2.3(0.5744) | Zn<br>4s -12.41/2.01<br>4p -6.53/1.7 |

(continued)

**Table 1** (continued)

| | | | | | | | | | |
|---|---|---|---|---|---|---|---|---|---|
| Y<br>5s -5.48/1.31<br>5p -3.51/1.08 | Zr<br>5s -9.87/1.817<br>5p -6.76/1.776<br>4d -11.18/<br>3.835(0.6211)<br>1.505(0.5796) | Nb<br>5s -10.1/1.89<br>5p -6.86/1.85<br>4d -12.1/<br>4.08(0.6401)<br>1.64(0.5516) | Mo<br>5s -8.77/1.96<br>5p -5.6/1.9<br>4d -11.06/<br>4.54(0.58988)<br>1.9(0.58988) | Tc<br>5s -10.07/2.018<br>5p -5.04/1.984<br>4d -12.82/<br>4.9(0.5715)<br>2.094(0.6012) | Ru<br>5s -8.6/2.078<br>5p-3.28/2.043<br>4d -11.12/<br>5.378(0.534)<br>2.303(0.6365) | Rh<br>5s -8.09/2.135<br>5p -4.57/2.1<br>4d -12.5/<br>4.29(0.5807)<br>1.97(0.5685) | Pd<br>5s -7.32/2.19<br>5p-3.45/2.152<br>4d -12.02/<br>5.983(0.5535)<br>2.613(0.6701) | Ag | Cd |
| La – Lu | Hf | Ta<br>6s -10.1/2.28<br>6p-6.86/2.241<br>5d -12.1/<br>4.762(0.6815)<br>1.938(0.5774) | W<br>6s -8.26/2.341<br>6p -5.17/2.309<br>5d -10.37/<br>4.982(0.694)<br>2.068(0.5631) | Re<br>6s -9.36/2.398<br>6p -5.96/2.372<br>5d -12.66/<br>5.343(0.6359)<br>2.277(0.5677) | Os<br>6s -8.17/2.452<br>6p -4.81/2.429<br>5d -11.84/<br>5.571(0.63717)<br>2.416(0.55982) | Ir<br>6s -11.36/2.5<br>6p -4.5/2.2<br>5d -12.17/<br>5.796(0.6698)<br>2.557(0.586) | Pt<br>6s -9.077/2.554<br>6p -5.475/2.554<br>5d -12.59/<br>6.013(0.6334)<br>2.696(0.5513) | Au<br>6s -10.92/2.602<br>6p -5.55/2.584<br>5d -15.07/<br>6.163(0.6851)<br>2.794(0.5696) | Hg<br>6s -13.68/2.649<br>6p -8.47/2.631<br>5d -17.5/<br>6.436(0.6438)<br>3.032(0.5215) |

$I_p$, $\zeta_1$ ($c_1$), and $\zeta_2$ ($c_2$) are listed for double $\zeta$ d orbitals, $\chi = c_1 \exp(-\zeta_1 r) + c_2 \exp(-\zeta_2 r)$

**Table 2** Physical constants

| Quantity | Symbol | | cgs | MKS | |
|---|---|---|---|---|---|
| Velocity of light | $c$ | 2.997925× | $10^{10}$ cm/s | $10^{8}$ m/s | |
| Planck constant | $h$ | 6.62607× | $10^{-27}$ erg s | $10^{-34}$ Js | |
| | $\hbar$ | 1.05457× | $10^{-27}$ erg s | $10^{-34}$ Js | |
| Charge of electron | $e$ | 1.60218 | | $10^{-19}$ C | |
| | | 4.80325× | $10^{-10}$ esu | | |
| Mass of electron | $m$ | 9.10938× | $10^{-28}$ g | $10^{-31}$ kg | |
| Mass of proton | $M$ | 1.67262× | $10^{-24}$ g | $10^{-27}$ kg | |
| Boltzmann constant | $k_B$ | 1.38065× | $10^{-16}$ erg/K | $10^{-23}$ J/K | |
| Avogadro constant | $N$ | 6.02214× | $10^{23}$ | | |
| Bohr radius | $r_0$ | 0.529177× | $10^{-8}$ cm | $10^{-10}$ m | $4\pi\varepsilon_0\hbar^2/me^2$ |
| Rydberg constant | $R_0$ | 13.6058 eV | | | $me^4/8\varepsilon_0^2h^2$ |
| Bohr magneton | $\mu_B$ | 9.27401× | $10^{-21}$ erg/G | $10^{-24}$ J/T | $e\hbar/2m$ |
| Permittivity of vacuum | $\varepsilon_0$ | 8.85419× | | $10^{-12}$ F/m | $1/\mu_0c^2$ |
| Permeability of vacuum | $\mu_o$ | 1.25664× | $10^{-6}$ Js$^2$/C$^2$m | $4\pi \times 10^{-7}$ | |
| 1 eV= | | 1.60218× | $10^{-12}$ erg | $10^{-19}$ J | $e$ |
| | | 9.6485× | $10^{11}$ erg/mol | $10^{4}$ J/mol | $eN$ |
| | Wave number | 8.0655× | $10^{3}$ cm$^{-1}$ | | $k = E/hc$ |
| | Wave length | 1.2398 μm (inverse) | | | $\lambda = hc/E$ |
| | Temperature | 1.1605× | $10^{4}$ K | | $T = E/k_B$ |
| | Frequency | 2.4180× | $10^{14}$ Hz | | $\nu = E/h$ |

MIX
Papier aus verantwortungsvollen Quellen
Paper from responsible sources
FSC® C105338

If you have any concerns about our products,
you can contact us on
**ProductSafety@springernature.com**

In case Publisher is established outside the EU,
the EU authorized representative is:
**Springer Nature Customer Service Center GmbH**
**Europaplatz 3, 69115 Heidelberg, Germany**

Printed by Libri Plureos GmbH
in Hamburg, Germany